Foundations of the Unity of Science

Volume I

Committee of Organization

Rudolf Carnap
Philipp Frank
Joergen Joergensen

Charles Morris
Otto Neurath
Louis Rougier

Advisory Committee

Niels Bohr
Egon Brunswik
J. Clay
John Dewey
Federigo Enriques
Herbert Feigl
Clark L. Hull
Waldemar Kaempffert
Victor F. Lenzen
Jan Lukasiewicz
William M. Malisoff

R. von Mises
G. Mannoury
Ernest Nagel
Arne Naess
Hans Reichenbach
Abel Rey
Bertrand Russell
L. Susan Stebbing
Alfred Tarski
Edward C. Tolman
Joseph H. Woodger

Foundations of the Unity of Science

.

Toward an
International Encyclopedia
of Unified Science

Volume I, Nos. 1-10

Edited by

Otto Neurath   Rudolf Carnap

Charles Morris

The University of Chicago Press
Chicago and London

This edition combines in two clothbound volumes the ten
monographs of volume I and nine monographs of volume II of
the *Foundations of the Unity of Science: Toward an International
Encyclopedia of Unified Science*. This work, combined and in its
several parts, was formerly titled *International Encyclopedia of
Unified Science*.

International Standard Book Number: 0-226-57586-1
Library of Congress Catalog Card Number: 56-553

The University of Chicago Press, Chicago 60637
The University of Chicago Press, Ltd., London

CONTENTS
Volume I

Preface   vii

On the History of the International Encyclopedia
of Unified Science   ix

Encyclopedia and Unified Science   1
Neurath, Bohr, Dewey, Russell, Carnap,
Morris

Foundations of the Theory of Signs   77
Charles Morris

Foundations of Logic and Mathematics   139
Rudolph Carnap

Linguistic Aspects of Science   215
Leonard Bloomfield

Procedures of Empirical Science   279
Victor F. Lenzen

Principles of the Theory of Probability   341
Ernest Nagel

Foundations of Physics   423
Philipp Frank

Cosmology   505
E. Finlay-Freundlich

Foundations of Biology   567
Felix Mainx

The Conceptual Framework of Psychology   655
Egon Brunswik

Preface

The two volumes (nineteen monographs) which comprise *Foundations of the Unity of Science* were originally conceived as the first and introductory part of the projected greater work of the *International Encyclopedia of Unified Science*. The *Encyclopedia* was in origin the idea of Otto Neurath. It was meant as a manifestation of the unity of science movement, in addition to the six International Congresses for the Unity of Science, the *Journal of Unified Science* (formerly *Erkenntnis*), and the Library of Unified Science. These last three enterprises came to a halt largely as the result of the Second World War. Neurath's death in 1945 deprived the unity of science movement of its central organizational impetus.

Original plans for the *Encyclopedia* were ambitious. In addition to the two introductory volumes, there was to be a section on the methodology of the sciences, one on the existing state of the unification of the sciences, and possibly a section on the applications of unified science to the professions. It was planned that the work in its entirety would comprise about twenty-six volumes (260 monographs). Part of an article by C. Morris, "On the History of the International Encyclopedia of Unified Science" (*Synthese*, 12, 1960, 517-21), follows this preface as an indication of the original plans for this work.

Otto Neurath's death and the aftermath of the war prevented the realization of these ambitious plans. It had been hoped that the two introductory volumes would be finished early in the 1940's. However, conditions were such that a number of changes in the original list of authors had to be made, and many of the authors who did continue were seriously delayed in finishing their monographs. So only now is the series of nineteen monographs, plus bibliography and index, making up *Foundations of the Unity of Science* brought to completion. There are no plans at this time to proceed further with the *International Encyclopedia of Unified Science*, to which these monographs were intended to be the introduction.

Rudolf Carnap
Charles Morris
1969

On the History of the International Encyclopedia of Unified Science

The original idea of the *Encyclopedia* was Otto Neurath's. In a letter of 1935 he wrote that he was at work on the project at least as early as 1920. He wrote that he first talked it over with Einstein and Hans Hahn and had early discussions about it with Carnap and Philipp Frank. In the 1930's Neurath was with the Mundaneum Institute of the Hague. He had set up the Unity of Science Institute in 1936 as a department of the Mundaneum Institute, and in 1937 this department was renamed the International Institute for the Unity of Science with Neurath, Frank, and Morris forming the executive committee. There was also set up an Organization Committee of the International Encyclopedia of Unified Science, composed of Neurath, Carnap, Frank, Joergen Joergensen, Morris, and Louis Rougier. (Also formed was an Organization Committee of the International Congresses for the Unity of Science, composed of the same persons plus L. Susan Stebbing.) The general project of the *Encyclopedia* was discussed at length at the First International Congress for the Unity of Science, Paris, September 1935, and the Congress voted approval of the project. As the above details make clear, the idea of the *Encyclopedia* long preceded this Congress.

My correspondence contains three extensive statements of Neurath's ideas about the *Encyclopedia:* a five-page outline in May 1936, a four-page statement in February 1937, and another four-page discussion in June 1938. There were many letters exchanged concerning these proposals. As will be seen, Neurath's ideas ranged very far.

In addition to the two introductory volumes (each to contain ten monographs) with the section title of *Foundations of the Unity of Science,* Neurath thought of two (and at times of three) other, larger sections. The three of us came to no agreement on proper titles for these sections, but the general plan was clear. Section 2 was to deal with *Methodological* problems involved in the special sciences and in the systematization of science, with particular stress to be laid upon the confrontation and discussion of divergent points of view. Section 3 was to concern itself with the *actual state of systematization* within the special sciences and the connections which obtained between them, with the hope that this might help toward further systematization. Neurath in 1938 was thinking of six volumes (sixty monographs) for section 2 and eight volumes (eighty monographs) for section 3. But there was even more in his bag of ideas than this.

Neurath had long planned a comprehensive *Visual Thesaurus* (sometimes he called it the Pictorial and sometimes the Isotype Thesaurus) which would be "eine Weltübersicht in Bildern." At times he thought that this might be an adjunct of the *Encyclopedia*, in which case he proposed an additional section 4 for the *Encyclopedia* which would exemplify and apply the methods and results of the preceding three sections to such fields as education, engineering, law, and medicine. Neurath proposed ten volumes for this section, each of which would have a pictorial companion in the *Visual Thesaurus*. So at the most elaborate range of his proposals Neurath was thinking of a twenty-six-volume (260-monograph) *Encyclopedia* supplemented by a ten-volume *Visual Thesaurus*.

At one time Neurath had contemplated English, French, and German editions of the work. He conceived of it as genuinely international in scope, with writers from Asiatic countries as well as from the West. It is clear from a number of his letters that he had the great French *Encyclopédie* often in mind, both with respect to the historical importance which he envisaged for the work he planned, and with respect to the difficulties the two enterprises encountered. A letter of 18 November 1944 gives a long nine-page discussion of Neurath's post-War plans for the various agencies of the unity of science movement (the *Encyclopedia*, the *Journal*, the Library, and the Congresses). My last letter from him was dated 7 November 1945. He died in London on 22 December 1945 at the age of 63.

Carnap and I did not correspond at any length with Neurath concerning his proposed fourth section for the *Encyclopedia*, although both of us had doubts about it. Section 3 was discussed only in very general terms. Nor was any agreement reached on the details of section 2, the methodology section. However, a general statement concerning this section was agreed upon in 1939; it was to be used to obtain advance subscriptions, but because of the war it was never so used. It may be of historical interest to quote some paragraphs from this statement.

Volumes III-VIII of the *International Encyclopedia of Unified Science:* The second unit will be composed of six volumes of ten monographs each. This unit will stress the problems and procedures involved in the progressive systematization of science.

Differences of opinion exist among those interested in the analysis and integration of the language of science, and it is part of the purpose of the *Encyclopedia* to present such differences. It becomes necessary at the present stage of development to stress the problems encountered and the techniques relevant to their solution—to take stock, as it were, of the contemporary situation in the analysis and the unification of scientific knowledge. This is the task of the second unit of the *Encyclopedia*. It thus exhibits in a new perspective the problems and procedures somewhat vaguely indicated in such phrases as "the logic of science" and "the methodology of science." Its six volumes of ten monographs each offer the opportunity for treatment of the subject with a comprehensiveness not found elsewhere.

Many persons will be interested in knowing the various results and opinions which exist in this field. Hence Volumes III-VIII will especially stress the controversial differences in regard to special sciences (physics, psychology, etc.), in regard to the possibilities and limitations of scientific unification, and in regard to the methods involved in scientific progress and systematization. . . . Representatives of controversial opinions will be given a chance to present their views. A special Editorial Committee will plan each of the volumes.

The content of the separate volumes will be roughly as follows:

Volume III: the most general problems and procedures which an ever-expanding unification of science encounters (construction and confirmation of theories, induction, historical attempts at scientific integration, probability, etc.).

Volume IV: the nature of logic and mathematics, and their role and place in the structure of science.

Volume V: physics.

Volume VI: biology and psychology.

Volume VII: the social and humanistic sciences.

Volume VIII: history of the scientific attitude.

We had hoped in 1939 to complete section 2 by 1944. All of the authors of Volume I as originally announced in the first monograph completed their monographs. But the second volume ran into many difficulties, and for one reason or another the studies proposed by Federigo Enriques, Jan Lukasiewicz, Arne Naess, Louis Rougier, and Louis Wirth never appeared.

Charles Morris

Encyclopedia and Unified Science

Otto Neurath, Niels Bohr, John Dewey,
Bertrand Russell, Rudolf Carnap,
Charles W. Morris

Encyclopedia and Unified Science

Contents:

OTTO NEURATH: Unified Science as Encyclopedic   PAGE
Integration 1
  I. Unity of Science Movement 1
  II. Mosaic of Empirical Science 3
  III. From Metaphysical Comprehensiveness to Empiristic Synthesis 5
  IV. Scientific Attitude and Systematization of Empirical Procedure 8
  V. Logical Analysis of Scientific Statements . . 10
  VI. Logico-empirical Integration 15
  VII. Unified Science and Encyclopedism 20
  VIII. Structure of the Encyclopedia 23

NIELS BOHR: Analysis and Synthesis in Science 28

JOHN DEWEY: Unity of Science as a Social Problem 29
  I. The Scientific Attitude 29
  II. The Social Unity of Science 32
  III. Education and the Unity of Science 35

BERTRAND RUSSELL: On the Importance of Logical Form 39

PAGE

RUDOLF CARNAP: Logical Foundations of the Uni-

ty of Science 42

    I. What Is Logical Analysis of Science? . . . 42

    II. The Main Branches of Science 45

    III. Reducibility 49

    IV. The Unity of the Language of Science . . . 52

    V. The Problem of the Unity of Laws 60

CHARLES W. MORRIS: Scientific Empiricism . . 63

    I. Method in Science 63

    II. Generalization of Scientific Method 65

    III. The Viewpoint of Scientific Empiricism . . . 68

    IV. Science and Practice 72

    V. Scientific Empiricism and Encyclopedism . . 74

Unified Science as Encyclopedic Integration

Otto Neurath

I. Unity of Science Movement

Unified science became historically the subject of this *Encyclopedia* as a result of the efforts of the unity of science movement, which includes scientists and persons interested in science who are conscious of the importance of a universal scientific attitude.

The new version of the idea of unified science is created by the confluence of divergent intellectual currents. Empirical work of scientists was often antagonistic to the logical constructions of a priori rationalism bred by philosophico-religious systems; therefore, "empiricalization" and "logicalization" were considered mostly to be in opposition—the two have now become synthesized for the first time in history. Certain empiricists began to appreciate logical analysis and construction as universal scientific aids; other thinkers, especially interested in the historical importance of scientific imagination—many call it rationalistic fancy—stress their opposition to all kinds of a priori reasoning and demand empirical tests for all their theories. The term 'logical empiricism' expresses very well this new type of synthesis; it may be used synonymously with the term 'empirical rationalism.' To further all kinds of scientific synthesis is one of the most important purposes of the unity of science movement, which is bringing together scientists in different fields and in different countries, as well as persons who have some interest in science or hope that science will help to ameliorate personal and social life. The terms 'unified science' and 'unity of science,' used before this movement came into being, are becoming more and more widespread in usage. This movement has found in the International Congresses for the Unity of Science its organized contemporary expression. These congresses, held yearly, show a new field for co-operation.[1] Physicists are familiar with co-operation in the field of physics, biologists in the field of biology; the same type of scientific co-operation is shown by these congresses, because the members of these congresses and all scientists in the movement co-operate in the field of unified science—physicists with biologists, biologists with sociologists, and other specialists with logicians and mathematicians. One important work within the wider unity of science movement will be this *Encyclopedia*.

The *International Encyclopedia of Unified Science* aims to show how various scientific activities such as observation, experimentation, and reasoning can be synthesized, and how all these together help to evolve unified science. These efforts to synthesize and systematize wherever possible are not directed at creating *the* system of science; this *Encyclopedia* continues the work of the famous French *Encyclopédie* in this and other respects.

About one hundred and ninety years ago D'Alembert wrote a *Discours préliminaire* for the French *Encyclopédie*, a gigantic work achieved by the co-operation of a great many specialists. Although a lover of the systematizing scientific mind, D'Alembert objected to the making of a universal system, just as Condillac opposed such attempts in his *Traité des systèmes*, in which he criticized the great rationalistic systems of his day. D'Alembert's idea of the procedure of empirical science was mostly based on Francis Bacon, and the idea of science in general on Newton; no comprehensive idea of "logicalization" stimulated the Encyclopedists. One must carefully look at their work as an important example of organized co-operation. Perhaps the same kind of scientific tolerance will appear in this *Encyclopedia* which appeared in the French *Encyclopédie* when D'Alembert, in his Introduction, opposed Rousseau's aggressions against science and yet expressed his pleasure that Rousseau had become a collaborator in the work. As a modern scientific man, D'Alembert stressed the degree to which all scientific activities depend upon social institutions. Today this idea is so familiar that this *Encyclopedia* will give it special attention. Succeeding generations may better be able to assess how far this present *Encyclopedia* expresses living activities, old traditions, and a rising future.

II. Mosaic of Empirical Science

Continual scientific activity throughout the centuries gives rise step by step to a distinctive intellectual environment. The history of this evolution of empirical science and scientific empiricism can be regarded as the history of a "mosaic," the pattern of which has been formed by combining new observations and new logical constructions of diverse character and origin. The generations of the "mosaicists" are not only inlaying the stones but also changing certain stones for others and varying the whole pattern. Scientific thinkers who were combating one another in social and intellectual conflicts must nevertheless often be regarded as the contributors of little stones to the same part of the whole pattern.

The history of philosophy, on the contrary, can be written as a history of philosophical systems formed by certain persons who concentrated upon and focused attention upon particular groups of ideas. Such a focalization can be more easily correlated to certain social situations than can the mosaic of empirical science to its historical environment. One can speak about a "republic of scientists" which makes a scientific pattern but not about a "republic of philosophers."

Science as a whole can be regarded as a combination of an enormous number of elements, collected little by little. One example may show this: The principle of Huyghens says that each point on an expanding sphere of light can be regarded as a center of a new expanding sphere of light. While the followers of the emission theory could use the idea of a certain periodicity, the followers of Huyghens at first lacked it; Euler used the idea of the periodicity of undulation but objected to the principle of Huyghens. It happened subsequently that Huyghens' principle, together with periodicity and other factors, formed the theory of light of the nineteenth century. In the same way, in the modern theory of economics one finds a connecting of elements of diverse origin. Smith and Ricardo did show certain correlations of the market system but did not concentrate their interest in certain decreases of wealth as did Sismondi; in that part of modern theory which deals with business cycles one finds more elements of Marxism than of Smithian ideas. Only a complicated comparative scheme could show the amalgamation of various elements, the common and different features of various theories. In a similar way one might analyze this *Encyclopedia* and show the elements which form the idea of a unified science.

Modern empiricism has grown up in the scientific tradition, in the activities of daily life, and in philosophico-religious speculations. The more or less common thought of the Middle Ages, which was mainly based on a literature written in Latin, Arabic, Hebrew, or Greek, was also influenced by other peoples outside the Mediterranean. Indian and Chinese influences were not very strong, but increased in later times. The social ideas of Quesnay, Montesquieu, Voltaire, and other thinkers were directly influenced by Chinese ideas. One also finds such an influence in philosophers, for instance, in Leibniz. Scholasticism, roughly speaking, was the native soil of all European thinking. No great attention was devoted to empirical studies within the structure of scholasticism, but one sees scientific research and reasoning arising within this structure. Men who wrote about physical problems also gave metaphysical explanations or combined empirically tested statements with purely traditional statements. Certain elements of optics in the works of Roger Bacon (about 1250) or in the works of other Scholastics, such as Dietrich of Freiberg, are of the same type as the corresponding explanations of Snellius or Descartes (about 1600); nevertheless, Roger Bacon discussed the question of the distance between the western coast of Spain and the eastern coast of Asia without arguing empirically or making particular proposals for research. Macrobius had previously analyzed Cicero's *Somnium Scipionis* and his explanations dealing with the five zones of the globe, with the antipodes and the continent "Australis";

Roger Bacon did not discuss these matters systematically and maintained only that the distance was relatively short, using quotations from Aristotle, from an apocryphal book of the Bible, and from a well-known saint, thus preparing for the discovery of America by Columbus. The fourteenth-century Oresme, a mathematician, physicist, and economist of importance, discussed a great many questions in a modern manner, but he holds a far from universal scientific attitude, as appears when his work as a whole is analyzed.

The increasing number of technical inventions and geographical discoveries, together with the increasing secularization of politics, produced a new attitude and the necessity to analyze and combine a great deal of empirical data. Machiavelli, for instance, discussed problems of social organization in an attitude far from the Scholastic tradition. A long series of historians and economists, physicists and biologists, did the same in the following centuries and worked out a modern idea of the world. Alexander von Humboldt has shown in his *Kosmos* how such empirical studies can be correlated without the help of philosophico-religious construction. Empiristic interest spread in the public not only in the field of technical activities but also in daily life: during the eighteenth century, for instance, microscopy was a hobby as photography is now. That physiognomic and phrenologic studies became popular can be regarded as a sign of a growing—though only uncritical—empirical interest.

The evolution of a comprehensive empiricism in opposition to the traditional systems has started with work in decentralized camps. Scholasticism was based on a well-organized movement, but not so the young empirical science. In the *Encyclopedia of Unified Science* this historical situation and its consequences will be demonstrated by showing the formation of the mosaic of scientific activities.

III. From Metaphysical Comprehensiveness to Empiristic Synthesis

All-embracing vision and thought is an old desire of humanity. Gnosticism was not only characterized by realistic ideas pertaining to astrology and magic but was also full of ideas dealing with angel-like "emanations" and personified concepts and qualities. Elements of these "cosmic poems"—if one may use this term—combined with elements of Platonism, Aristotelianism, and other philosophical and religious tendencies, formed the background of the medieval desire for a comprehensive viewpoint—a desire which can also be found in medieval mysticism and not only in Scholastic intellectual systems. A bewildering multiplicity of logical processes were performed by the deductions from and combinations of dogmatic texts of the Catholic church, statements from the Bible, the Church Fathers, Aristotle, and others. One finds mainly theological explanations (moral theology and other theological disciplines included) and medieval cosmology (references to heaven, purgatory, hell, and other places for the soul) in the Scholastic systems. Both *Summae* of Thomas Aquinas are representative of these structures. The logical instrument of the Scholastics was sharpened for nonempiristic purposes. Their interest in logical combination, in discussion, and in argument can be illustrated by the subjects of the trivium and the quadrivium: grammar, logic, rhetoric; arithmetic, music, geometry, and astronomy.

This interest in combining concepts and statements without empirical testing prepared a certain attitude which appeared in the following ages as metaphysical construction. The neglect of testing facts and using observation statements in connection with all systematized ideas is especially found in the different idealistic systems. One may use Hegel's *Encyclopedia* as an example of this type. His comprehensive work can be regarded as substituting a priori philosophico-religious statements for traditional theological statements, and as joining these philosophico-religious statements to metaphysically transformed empirical statements. Hegel's vigor and all-embracing enthusiasm stimulated such empirical thinkers as Feuerbach, Marx, and Engels; they received more thoughts that breathe and words that burn from Hegel than from the books in which Helvetius, Holbach, and others wrote about the world in an empirical sense. Neither Hegel nor Schelling encouraged a scientific attitude and produced logical analyses or particular theses which could be used directly in the sciences, as did, for instance, certain ideas of Descartes and Leibniz. Schelling's *Naturphilosophie* influenced chemists and other scientists; the results were such that Liebig and others had to fight Schellingism in their own scientific camp.

The large and panoramic systems of idealistic philosophy are, as it were, late branches of a deformed scholasticism. Hegelianism is a typical metaphysical system of our age; Thomism is a typical Scholastic system which is still living in the Catholic church and also attracting some persons outside the Catholic church. Both these systems have not so far shown any disposition to logicalize empirical science, to form a quasi-addition to their philosophico-religious structures. Synthesis of empirical science was not directly supported by metaphysical and theological systems, though certain stimulations came from them.

Collections of interesting subjects, biographical, geographical, philological, and other data were already known in the Middle Ages, but in the seventeenth and eighteenth centuries some enormous encyclopedias of a new type were published (not so large as the Chinese, however). The French philosophical opposition organized the great French *Encyclopédie*. It was not a *"faute de mieux* encyclopedia" in place of a comprehensive system, but an alternative to systems. Since the Encyclopedists stressed the view that empirical statements were to be used not only for science (in spite of the fact that one can find not a few metaphysical and theological explanations in the work itself) but also for engineering and other technical purposes, collections of pictures were added to the text. The French *Encyclopédie* and its empirical attitude were combated by the church and government. Persons writing papers which seemed able to stimulate social perturbation had to fear persecution. Diderot (a polyhistor such as Voltaire but, unlike him, not doing experimental research) had to permit of textual variations by the publisher who feared the powerful enemies of the *Encyclopédie*. This encyclopedia had no comprehensive unity despite the expression of a certain empirical attitude; it was organized by means of a classification of sciences, references, and other devices.

More constructive ideas formed the basis of Comte's *Philosophie positive*. Comte, familiar with mathematics and physics, on the one hand, and very interested in social problems, on the other (the term 'sociology' was coined by him), tried to combine his ordering and classification of science with historical interests. Spencer's gigantic work, the system of *Synthetic Philosophy*, is also an example of this synthetic tendency which avoids metaphysical construction (this is not the place to discuss some metaphysical arguments in Comte's and Spencer's work) but endeavors to substitute an empirical scientific whole for philosophical speculations. One must stress the fact that Spencer's interest was concentrated in biology and sociology, because it is a usual prejudice that the idea of unified science and the unity of science movement is especially based on an interest in physics. Neither the Encyclopedists nor Comte and Spencer nor similar thinkers made an attempt to organize a logical synthesis of science.

IV. Scientific Attitude and Systematization of Empirical Procedure

The parallel increase of the scientific attitude and the systematic analysis of scientific procedures prepared the way for logical empiricism. During the Renaissance modern thinkers began to be interested in the procedure of empirical science and in the scientific attitude. Leonardo da Vinci, a universal genius influenced by the Scholastics, became a man with a comprehensive scientific attitude. He worked in different fields of engineering, was interested in scientific problems, and of course in all matters connected with painting. He began to feel the common root of all empirical science and stressed, for instance, the importance of "generalization" and other scientific aids. He understood what we call the empirical procedure. The universality and versatility of Leonardo da Vinci and other thinkers, scientists, and amateurs in science were stimuli of great importance for the origin of a comprehensive empiricism and a scientific attitude.

History shows us a great many scientists whose scientific attitude is not equally maintained in the various fields of thought. Such a scientist may be very critical in his own domain, for instance in physics, but of a totally different behavior when he speaks about "free will," "social privileges," or similar traditional problems. On such occasions some scientists sharply change their criticism and exactness of arguments and their style of language. Newton, for instance, was a scientist in whom theological speculations and scientific empiricism existed partly side by side, partly in actual connection—he speaks, for instance, about space as *sensorium dei*. A comprehensive scientific attitude has come into being, within scientific activity, since the Renaissance, but neither Leonardo da Vinci, nor Galileo a hundred years after him, analyzed the rules of empirical procedure or the scientific attitude. Francis Bacon was not very active in science or engineering, but he promoted certain ideas of empiricism successfully, especially the idea of "induction" as a scientific aid, in spite of the fact that he gave very poor examples in his work and did not recognize the scientific importance of Galileo and other empirical thinkers of his time.

Modern scientific empiricism attained very late in its development a comprehensive work which analyzes empirical procedure in all scientific fields: John Stuart Mill's *A System of Logic, Ratiocinative and Inductive, Being a Connected View of the Principles of Evidence and the Methods of Scientific Investigation.* Mill does not question the fact that astronomy and social science, physics and biology, are sciences of the same type. Mill, who was familiar with the problem of utilitarianism, with political economy and governmental practice as a pupil of his father, and with Bentham (who is not even now sufficiently appreciated), could effectively use Whewell's famous *History of the Inductive Sciences from the Earliest to the Present Times.* Mill's work influenced modern empiricism despite the fact that many of his particular statements were criticized. This type of comprehensiveness is also represented by the *Principles of Science* of Stanley Jevons; he was an economist like Mill and one of the promoters of modern symbolic logic. Karl Pearson's *The Grammar of Science* details the measures and procedures of modern science. Karl Pearson was very interested in biological and sociological questions, as was his teacher, Galton, but he was also familiar with physics.

But neither Mill nor other thinkers of similar type applied logical analysis consistently to the various sciences, thus attempting to make science a whole on a "logicalized" basis. They only achieved a comprehensive understanding of all the arguments which a scientist needs if he makes generalizations and tests scientific hypotheses. This *Encyclopedia* will show modern attempts to reform generalization, classification, testing, and other scientific activities, and to develop them by means of modern logic.

V. Logical Analysis of Scientific Statements

One science after the other separated from the mother-philosophy; scientists became more capable of solving difficult scientific problems than were philosophers occupied by a great many unscientific speculations and the particular problems of their own systems. One example may demonstrate this situation. During the eighteenth century people of different interests discussed the problem of "inertia." Some were influenced by Descartes, who used the term 'motion' when a body which is in the neighborhood of a second body is later found in the neighborhood of a third body. 'Space' and 'groups of bodies' were for these people more or less the same terms and 'empty absolute space' seemed to be a meaningless term. Others used the idea on which actual physical calculation was based in these times. The question arose whether to describe "inertia" in terms of relative correlations between bodies (a certain body together with the fixed stars), in terms of absolute space, or by means of other conceptions. Euler (about 1750) discussed the kinds of argumentation and was inclined to oppose the opinion that inertia could depend upon the totality of the fixed stars—

an opinion which is similar to modern ideas and which has been developed step by step by Mach and others. Kant, discussing this group of problems and analyzing Euler's paper, did not feel that this idea about correlation between inertia and the fixed stars was important enough to be criticized and did not even mention this, for him, strange construction, being more interested in his own a priori philosophy and in supporting Newton's ideas. The "essayistic" criticism by Hume and similar thinkers loosened the firmness and coherence of compact traditional opinions, but Kant, who stimulated some scientists, formed new barriers of peculiar rigidness by focusing and regulating criticism and skepticism. The essayist-philosopher Nietzsche showed how much of an antiscientific attitude can be found in Kant's system, which reduces the power of science and thus opens the doors to metaphysical and philosophico-religious speculations.

The evolution of non-Euclidean geometry, for instance, which prepared modern theories of measuring time and space, was hardly supported by modern philosophers—another example of the inadequacy of philosophers. One can rather assume that the ideas of Gauss, Bolyai, Lobachevski, and others were impeded by Kantianism: they had to start by opposing all kinds of apriorism. Not a few philosophers opposed the theory of relativity. The new intellectual environment was prepared more by specialists in physics and mathematics, or by certain imaginative amateurs in the field of science, and by poets, than by systematic philosophers. How many people may have been educated in the field of scientific imagination by Jonathan Swift!

A long series of imaginative analyses started with the animated statue imagined by Condillac, who had been influenced in his thought by Locke, Lamettrie, and others. Imagined statues which received one sense after another are relatives of Caspar Hauser and all the many children who have been found in the woods or other isolated places. These imagined and real beings are the subject matter of an old and rich literature which helped to prepare a logico-empirical attitude by means of imaginative analysis. Helmholtz and others, opposing Kant's ideology, imagined (partly for pedagogical purposes) two-dimensional beings on a sphere discussing geometrical problems. About the middle of the nineteenth century Fechner and others fancied dreamlands of different kinds: three-dimensional beings of different ages were produced by cutting off slices from a four-dimensional sausage. Some scientists fear such imaginative analogies as unreliable guides and demand the use of more systematic analysis. Actually, the history of all these imaginations has to be regarded as a part of the history of the empiristic mosaic. One must add to these significant imaginings that of two-dimensional beings who, traversing a hill, observe a retardation region, an indifference region, and an acceleration region of a geometrically homogeneous world. One could imagine a country ("Aeonia") in which beings could be repaired as are machines. There could be beings with connected nerves, or powerful beings consisting of a brain, one muscle, and one sense organ, but using complicated mechanical devices. Poincaré's problem of similar worlds (our world, reduced or enlarged in size) was also analyzed imaginatively about the middle of the nineteenth century by Eberty (*Die Gestirne und die Weltgeschichte*, newly edited by Gregorius Itelson with an introduction by Albert Einstein), who fancied also a trip throughout the universe quicker than light, in concordance with certain thoughts of Humboldt and Babbage. Renouvier (*Uchronie, l'utopie dans l'histoire*) used imagination in the field of history: how history really happened and how it might have happened. Lichtenberg, together with Chodowiecki (well known by his graphic work) and Sonya Kovalevska, the mathematician, elaborated similar fantasies for single individuals.

From these imaginations one can enter into the problem of behavioristics, logico-empirical analysis, and poetry. One can ask whether a blind man can make a complete physical description by means of certain devices or how the sensorium of Siamese twins is formed in the common part of their body, and similar concrete questions; one can write stimulating imaginative novels as did H. G. Wells, or a book of logico-empirical analysis like Carnap's *Logischer Aufbau der Welt*. To what extent imaginative constructions will be useful in the future may remain an open question.

All these imaginative analyses and constructions formed a part of the "essayistic" analysis of scientific argumentation such as that made by Poincaré, Duhem, and others. Their modern logical analysis of scientific statements, hypotheses, and theories was prepared for by many thinkers: the physicist Brewster, for instance, said that it is of the greatest importance that the same value d, characterizing the periodicity of light, could be found in the undulation theory and in the emission theory. John F. W. Herschel was anticipating Poincaré and Duhem when he wrote in his *Preliminary Discourse on the Study of Natural Philosophy* that one might imagine such a development of Newton's theory that it could solve certain problems which seemed reserved for the undulation theory. This sort of analysis of scientific statements was strongly supported by historico-critical studies. Mach's leading books dealing with mechanics, optics, and theory of heat characterize this tendency. He, by his paradigmatic analysis of concepts such as space and inertia, furthered the evolution of Einstein's theory of relativity. He did not make new experiments for this particular purpose but used, of course, the physical knowledge of his time. The fact that Duhem, Enriques, Mach, and others were active in their special sciences in logical analysis of scientific statements, and in the historical analysis of science, suggests the idea that the "history of the history of science" should be very instructive. In centuries in which such an analysis of scientific statements was not evolved as a special activity, historians of science were often busy with the analysis of scientific statements as a preparation for their historical analysis. A history of the history of optics, for instance, from Joseph Priestley (*The History and the Present State of Discoveries Relating to Vision, Light, and Colors* [1772], written as a part of a universal history of physics) up to Ernst Mach (*Die Prinzipien der physikalischen Optik* [1921]) shows clearly the increase of logicalization.

Corresponding historical sequences may be found in all scientific fields. The systematic analysis of "planned economy," for instance, has to be based on various fundamental problems; one group of problems deals with describing various possibilities: Thomas More's *Utopia* and Francis Bacon's *Nova Atlantis* are of another type of social analysis than are the plans of Ballod-Atlanticus or Popper-Lynkeus—plans which are forerunners of the science of socio-economic planning. Such kinds of social imagination can be combined with historico-social analyses like those made by Montesquieu, Stein, Marx, and others. The structure of economics is logically not so developed as the structure of physics, and so the history of the history of economics is not so abundant in logico-empirical analysis as is the history of the history of optics. Whewell's description and discussion of Linnaeus' problems dealing with classification and systematization give a picture of the logico-empirical analysis of sciences in the time of Whewell. A corresponding description and discussion of biological ideas from Linnaeus' classifications up to Woodger's formalization of biology should put a scientist in a position to compare directly the logico-empirical analysis of our time with the logico-empirical analysis made by a historian of science a hundred years ago.

A comparison of the argumentation in cosmology, geology, physics, biology, behavioristics ("psychology"), history, and social sciences in different ages will be furthered by the unity of science movement. An increasing number of scientists are busy with such problems, and one can hope for great success if scientists analyzing various sciences co-operate with men concerned with the history and logic of science. The inconsistency of the historically given universal pattern appears immediately. Successive editions of the *Encyclopedia of Unified Science* would show progress in logical analysis of scientific statements. Perhaps a special technique will be evolved which is able to describe systematically all these changes in the different sciences, thus continuing the work of Ernst Mach, whose centennial will be celebrated this year.

VI. Logico-empirical Integration

The mosaic pattern of empirical science progressively shows more marked interconnections than in the times in which empirical studies were relatively isolated. Scientific analysis of the sciences led to the observation that an increase of logical intercorrelation between statements of the same science and between statements of different sciences is a historical fact; one finds rationalism (as a quality of our experience), as it were, empirically, and may use the term 'empirical rationalism' with this meaning in which Gregorius Itelson proposed it, not merely as "rationalism based on experience." Comprehensiveness arises thus as a scientific need and is no longer a desire for vision only. The evolving of all such logical connections and the integration of science is a new aim of science.

Logical empiricism or empirical rationalism can also be regarded as a regeneration of certain elements of a priori rationalism. The *Scientia generalis* of Leibniz was the background of his "panlogism"—if this term is permitted in this sense and can be regarded as a "secularization of logic." The driving power of panlogism in the framework of a priori rationalism depends partly upon the idea that one can anticipate by means of logical combinations the progress of empirical science and not simply fructify its results or give certain suggestions. Leibniz was the first and last of the great philosophers who planned seriously to work out a comprehensive calculus adequate for all scientific progress. He promoted a universal logicalization of the whole of human thinking by means of a general calculus and a general terminology. He worked as a scientist and also began to organize scientific co-operation by means of scientific academies, but he was far from attempting or executing a universal scientific empiricism—too busy writing his *Théodicée*, elaborating the *Monadology*, and moreover getting entangled in theological disputes and church diplomacy.

His career was closely connected with Scholastic influences. As a boy he played, as it were, with logical elements such as "notions" and "subnotions." Young Leibniz at twenty years of age published his *Ars combinatoria*, which is influenced by Raimundus Lullus and other Scholastic authors. He was later influenced by the rationalistic Descartes and other modern thinkers, but he made clear that one could successfully use certain ideas of such Scholastic thinkers as Thomas Aquinas. Leibniz, the grandfather of modern logistic, transformed the often vague logical ideas of Scholasticism, and took the first steps toward modern exactness in logic, preparing the way for a great many modern ideas in the field of mathematics and logic.

He planned to organize a large encyclopedia, together with an *Atlas universalis*, in close connection with his *Characteristica universalis*. The plan embraced not only scientific disciplines, including rational grammar, moral science, geo-politics, but also natural theology. This gigantic plan was intended to form a logically organized whole. One may say that the *Pansophia* of Comenius (who came in touch with Leibniz) together with his *Orbis pictus* can be regarded as the parallel to the *Encyclopedia* and the *Atlas* of Leibniz. Both these pairs of works are based on a philosophico-religious rationalism and correspond in a certain sense to the medieval pair: Scholastic system and the overwhelming visual presentation in a medieval Catholic church.

Since Leibniz, like other a priori rationalists, was seeking *the* system of science and *the* logical key for it, one can understand that such an ideal was strange to empiricists. Most of the logical studies of Leibniz were not published, and only a few persons, like Lambert, were interested in special logistic problems. Public opinion was against formal logic. Kant and his followers discredited formal logic and thus petrified the aversion of Galileo and others against logic as an instrument of the traditional scholasticism. The growing new logic of Boole, De Morgan, and Grassman was not supported by philosophical thinkers of this period. Bolzano, for instance, influenced Austrian scientists and pedagogues about the middle of the nineteenth century by certain of his ideas, but his important investigations in the logical field (one example: *Paradoxien des Unendlichen*) were not studied and esteemed for a long period.

A universal application of logical analysis and construction to science in general was prepared not only by the systematization of empirical procedure and the systematization of logico-empirical analysis of scientific statements, but also by the analysis of language from different points of view. A direct route leads from Scholastic analysis of language, made especially by nominalists, up to Condillac (*Essai sur l'origine des connaissances humaines*), who influenced French and English thinkers, to Bentham, whose multifarious work still lives today, and to other thinkers interested in language as an aid for our daily life and for science. The connection between modern logic and empiricism did not arise instantly. The importance of logicians of the second half of the nineteenth century, such as Venn, Schroeder, Peano, Frege, and others, will be expressly discussed in the *Encyclopedia*. A few of the modern logicians, such as Peirce and, later on, Bertrand Russell, combined the interest in logic with an interest in empiricism. Traditional idealistic philosophers did not discuss carefully or look with favor upon this new combination of logicalization and empiricalization. The fact that Peirce was a logician and simultaneously interested in empiricism was in turn important for the preparation of modern scientific empiricism in the United States. In Europe the Vienna Circle, the Berlin group, related thinkers in England and Scandinavia, the Centre de synthèse in France, and the Scientia group in Italy are evidence of the interest in this evolution of logic and empiricism. Some thinkers are mainly busy with logical calculi, such as the members of the Polish School or the Münster group. One cannot judge at the moment what elements of these and other circles of thinkers may become most essential for the future of unified science. The importance of Riemann's geometry for modern physics did not appear at once. What part will metalogic, semantics, and other disciplines play in the unification of the language of the empirical sciences?

The important opinion arose (the influence of Wittgenstein, a metaphysician in many respects, has to be mentioned in this connection) that all statements can be expressed as "scientific statements" and that one cannot speak of special "philosophic statements." Some persons proposed to use the term 'philosophize' for an activity which makes concepts and statements clear; others proposed to use the term 'philosophy' for 'logic of science.' If one takes the thesis seriously that in the field of knowledge one only has to deal with scientific statements, the most comprehensive field of statements must be that of unified science. If one does not care to avoid the term 'philosopher,' one may use it for persons engaged in unified science. Such "philosophers" may be specialists in one discipline and amateurs in others or comprehensive scientific amateurs like Voltaire, but not speculative thinkers.

It is common to all these persons that they do not join scientific statements with a second type of specific "philosophical" formulation; this attitude gives one the feeling that one is acting within the collective scientific atmosphere and not in the sphere of individual philosophemes. Voltaire mentioned that opinions which become common do not bear the names of their creators (Du Bois-Reymond added that Voltaire's name and activity are insufficiently known because "Voltairianism" is a quality of the age after Voltaire). Nietzsche stressed esteem for "the unpretentious truths," objecting to the fascinating errors of metaphysical ages. An evolved civilization likes, according to Nietzsche, the modest results found by means of exact methods which are fruitful for the whole future; and such manliness, simplicity, and temperance will characterize not only an increasing number but also the whole of humanity in the future. Moritz Schlick explained in a similar sense that the evolution of modern critical thinking is founded on an anonymous mass of thinkers, especially scientists, and that progress does not arise from the sensational philosophical systems which form an endless row, each contradicting the others.

Scientists may now build up systematical bridges from science to science, analyzing concepts which are used in different sciences, considering all questions dealing with classification, order, etc. Axiomatization of science seems to give an opportunity to make the use of fundamental terms more precise and to prepare the combination of different sciences; preliminary axiomatization has to be founded on a long evolution of science. We cannot anticipate a "final axiomatization."

Some difficulties in science, even within a special discipline, arise from the fact that one cannot always decide whether two scientists (for instance, psychologists) speak about the same or different problems, or whether they explain the same or different opinions, by means of different scientific languages. Unification of scientific language is one of the purposes of the unity of science movement. It is a question to what extent such unification can be furthered. One can perhaps reduce all scientific terms to one kind of term by means of a special logical technique. The thesis of physicalism which will be discussed in this *Encyclopedia* (see the following article by Carnap) emphasizes that it is possible to reduce all terms to well-known terms of our language of daily life. Another question is to what extent one can reduce the statements or laws of biology, behavioristics, or sociology to physicalistic statements or laws. All studies dealing with languages and scientific terminology are regarded seriously not only in connection with what one usually calls logical questions but also in connection with questions of sociology and behavioristics; one may ask, for instance, how problems discussed by the Dutch group of thinkers interested in "significs" are connected with problems of semantics and other new disciplines. A great many scientists, working in different fields, are pushing these analyses forward.

Since more and more scientists stress the fact that in the end one must test all theories by means of the language of daily life, the correlations between the calculus of theories and the language of daily life will be systematically analyzed.

Many people think that logic (or logistic) is, as it were, an antidote to metaphysical speculations; that is wrong: one can elaborate a speculative metaphysical system *more logico demonstrata*. There is no automatically acting antidote against statements which, though formulated in an empirical language, yet need scientific criticism. Of which temper one's mind is, one can show by presenting one's work.

VII. Unified Science and Encyclopedism

Science itself is supplying its own integrating glue instead of aiming at a synthesis on the basis of a "super science" which is to legislate for the special scientific activities. The historical tendency of the unity of science movement is toward a unified science departmentalized into special sciences, and not toward a speculative juxtaposition of an autonomous philosophy and a group of scientific disciplines. If one rejects the idea of such a super science as well as the idea of a pseudo-rationalistic anticipation of *the* system of science, what is the maximum of scientific co-ordination which remains possible? The answer given by the unity of science movement is: an encyclopedia of unified science. An encyclopedia (in contradistinction to an anticipated system or a system constructed a priori) can be regarded as the model of man's knowledge. For, since one cannot compare the historically given science with "the real science," the most one can achieve in integration of scientific work seems to be an encyclopedia, constructed by scientists in co-operation. It may happen that one must use in one hypothesis, destined for a particular purpose, a supposition which contradicts another supposition used in another hypothesis, destined for another particular purpose. One may try to eliminate such contradictions, but in the historically given science, and so in a real encyclopedia, these and other difficulties always appear.

Encyclopedism may be regarded as a special attitude; one may also speak of encyclopedism as a program. Encyclopedism starts with the analysis of certain groups of scientific statements; it may happen that these can be axiomatized and that this axiomatized group of statements can be combined with others expressed in a similar form. But such a system of statements must not be regarded as a model of the scientific knowledge of a given age. An encyclopedia and not a system is the genuine model of science as a whole. An encyclopedic integration of scientific statements, with all the discrepancies and difficulties which appear, is the maximum of integration which we can achieve. It is against the principle of encyclopedism to imagine that one "could" eliminate all such difficulties. To believe this is to entertain a variation of Laplace's famous demon who was supposed to have a complete knowledge of present facts sufficient for making complete predictions of the future. Such is the idea of *the system* in contrast to the idea of *an encyclopedia;* the anticipated completeness of *the* system is opposed to the stressed incompleteness of an encyclopedia.

Such encyclopedism is the expression of a certain skepticism which objects not only to metaphysical speculations but also to overstatements within the field of empirical sentences. But how many theories and predictions, especially those which are used for stimulating people to practical work, often use terms such as 'certain,' 'always,' instead of terms such as 'perhaps,' 'sometimes'! An empiricist must permit himself, if necessary, a certain vagueness. Scientism—if one may use this term, introduced by French positivists a hundred years ago—does not depend upon "exactness" but only upon the permanence of scientific criticism. New ideas of scientific importance start mostly with vague and sometimes queer explanations; they become clearer and clearer, but the theories which follow will stand in time before the door with all their new vagueness and queerness. Niels Bohr expressed this historically and pedagogically essential fact in his paradoxical manner: the law of complementarity is valid also for fruitfulness and clearness of scientific theories. Must one fear by this to encourage vague speculations? No! Persons who are interested in unscientific speculation will undertake it under all circumstances. But it is useful to avoid dogmatism and bumptiousness in scientism and empirical panlogism. One can love exactness and nevertheless consciously tolerate a certain amount of vagueness.

How can one combine such a critical and skeptical attitude with the unparalyzed driving power which is needed to attain success in social and private life? The wish to eliminate all these limitations to a comprehensive scientific attitude often leads men to hypocrisy and cant; one cannot deny that a certain practical antagonism may arise between the scientific attitude and human activity, be it in a particular or in all social orders.

Encyclopedia and Unified Science

Science forms an essential part of the rich pattern character-izing modern life, which is becoming more and more uniform. If one looks at the whole of humanity, one sees a constant in-crease of the scientific attitude in daily life during the last cen-turies, in spite of the fact that books, speeches, and propaganda dealing with metaphysical speculation show us a "to and fro." There is no direct correlation between empirical activity in life and business, on the one hand, and systematically expressed empiricism, on the other. How much modern engineering and technical activity, together with all the helpful special sciences, were evolved, for instance, in Germany during the nineteenth century and how little comprehensive scientific empiricism!

The empiricalization of daily life is increasing in all countries: cities in the United States and in Japan, highways in Mexico and Germany, armies in China and France, universities in Turkey and Italy—all show us certain common features. A meteorolo-gist trained in Denmark may become a useful collaborator to a Canadian polar expedition; English economists can discuss a Russian analysis of American business cycles; and Russian economists may object to or accept the opinions of English econ-omists about the effect of rural collectivization in the Soviet Union. If fundamental difficulties arise in discussions between engineers, generals, and scientists of different countries, they are not based on the fact that the debaters are of different na-tions; one can show that analogous fundamental differences also arise within the same nation—generally when one of the de-baters, or both of them, intends to use "superscientific" or "isolated" sentences. The fact that the scientific language is common to all these people is almost concealed by this type of discussion.

The unified-science attitude based on the simplicity and straightforwardness of scientific empiricism is concentrating its attention on generalizations and predictions made by the de-baters. What people in all the countries expect scientists to do is always to predict successfully by means of so-called scientific procedure; one hopes that a surgeon who knows about bones and veins will make a diagnosis and then perform a satisfactory operation, that a historian knows much about human history and can foretell the main results of a newly undertaken excavation, that an economist judging from the first symptoms can warn the public of an impending slump, that a political leader can systematically predict social changes which are arising.

One can state all these scientific prognostications in terms of everyday language—the language which is common to all men in the world irrespective of the fact that the scientist himself uses expressions and symbols in preparatory work which are mostly of an international character. Unified science is therefore supported, in general, by the scientific attitude which is based on the internationality of the use of the language of everyday life and on the internationality of the use of scientific language.

It may happen that people create and prefer certain terms and formulations not for universal understanding but for stimulating certain emotions, and may decide that in certain cases an emotional activity is more important than a scientific attitude. It is not the subject of a scientific explanation to support or oppose such a decision. If one prefers a comprehensive scientific attitude, this *Encyclopedia* tries to show him the spectrum of scientific thinking. Each scientifically oriented man knows very well that the elaboration of such an encyclopedia, like other activities, is influenced by wishing and fearing, but there is a difference between men who intend to discover such influences and others who do not. That leads one to a great many unsolved problems. Incompleteness and open questions arise in all parts of this work, but encyclopedism maintains, nevertheless, that the integration of science is an inevitable part of man's scientific activities.

VIII. Structure of the Encyclopedia

One may ask: "What program is common to all the collaborators of the *Encyclopedia?*" A program formed of statements accepted by all the collaborators would be narrow and would be a source of divergences in the near future. This *Encyclopedia* will show that scientists, though working in different scientific fields and in different countries, may nevertheless co-operate as successfully within unified science as when scientists co-operate within physics or biology. The *Encyclopedia* will perhaps be a mainstay of scientific empiricism as well as of the unity of science movement in the widest sense. The maximum of co-operation—that is the program! This co-operation strives to elaborate the framework of unified science. Encyclopedism based on logical empiricism was the general historical background which underlay the proposal of an international encyclopedia of unified science.[2]

The general purpose of the *International Encyclopedia of Unified Science* is to bring together material pertaining to the scientific enterprise as a whole. The work will not be a series of alphabetically arranged articles; rather will it be a series of monographs with a highly analytical index, which will make it possible to find the bit of information sought if the *Encyclopedia* is to be used as a reference work. Each monograph, sometimes written by more than one collaborator, is devoted to a particular group of problems. The collaborators and organizers of this work are concerned with the analysis and interrelation of central scientific ideas, with all problems dealing with the analysis of sciences, and with the sense in which science forms a unified encyclopedical whole. The new *Encyclopedia* so aims to integrate the scientific disciplines, so to unify them, so to dovetail them together, that advances in one will bring about advances in the others.

The *Encyclopedia* is to be constructed like an onion. The heart of this onion is formed by twenty pamphlets which constitute two introductory volumes. These volumes, entitled *Foundations of the Unity of Science*, are complete in themselves but also serve as the introduction to what will follow.

The first "layer" of the onion which will inclose this "heart," consisting of the first two volumes, is planned as a series of volumes which will deal with the problems of systematization in special sciences and in unified science—including logic, mathematics, theory of signs, linguistics, history and sociology of science, classification of sciences, and educational implications of the scientific attitude. In these volumes scientists with different opinions will be given an opportunity to explain their individual ideas in their own formulation, since it is a special aim of this work to stress the gaps in our present knowledge and the difficulties and discrepancies which are found at present in various fields of science. "Heart" and "first layer" together will be a completely self-contained unit. The following "layers" may deal with more specialized problems; the interests of the reader and the collaborators in the particular problems will lead the members of the Committee of Organization and the Advisory Committee to consider various possible lines of development. It is hoped that an *Atlas* can be worked out as an *Isotype Thesaurus* showing important facts by means of unified visual aids.[3] The plan of this *Encyclopedia* could not be based on a generally accepted classification of the sciences—indeed, the collaborators may perhaps find a new way to assemble systematically all the special sciences. The organizers and collaborators know very well that certain frontiers of sciences are unsatisfactory and that certain terms are not sufficiently defined. The *Encyclopedia* will eliminate these defects where possible.

This Introduction has aimed to show the historical position of the *Encyclopedia;* it is amplified by special articles. Rough outlines are augmented by the articles of Charles W. Morris and Rudolf Carnap, the one explaining how scientific empiricism is an even more comprehensive movement than logical empiricism, the other stressing the importance of the logical analysis of sciences. The other articles in this introductory monograph amplify some aspects of the problems connected with unified science. Niels Bohr and Bertrand Russell are concerned with the importance for the sciences of certain phases of the unity of science movement, while John Dewey stresses the wider social implications involved in the unification of the forces of science.

Without pursuing utopian ideals, an effort will be made to have the scientific language of the *Encyclopedia* as homogeneous as it is possible to make it at the present. The *Encyclopedia* will express the situation of a living being and not of a phantom; those who read the *Encyclopedia* should feel that scientists are speaking about science as a being of flesh and blood. The collaborators will certainly learn from their encyclopedical work. Suggestions from different sources will stimulate this activity, so that this *Encyclopedia* will become a platform for the discussion of all aspects of the scientific enterprise. In this way the *International Encyclopedia of Unified Science* hopes to avoid becoming a mausoleum or a herbarium, and to remain a living intellectual force growing out of a living need of men, and so in turn serving humanity.

NOTES

1. Plans were laid at the congress at Charles University in Prague (1934) for a series of annual congresses devoted to the unity of science. The proceedings of this preliminary congress were published in the 1935 *Erkenntnis* (Leipzig: F. Meiner) and in separate volume form as *Einheit der Wissenschaft* (F. Meiner). The First International Congress for the Unity of Science was held at the Sorbonne, Paris, in 1935, and the proceedings were published under the title *Actes du congrès international de philosophie scientifique* (Paris: Hermann & Cie, 1936). The proceedings of the second congress, held at Copenhagen in 1936 and devoted to the problem of causality, appeared in the 1937 *Erkenntnis* and also as an independent volume (*Das Kausalproblem* [Leipzig: F. Meiner; Copenhagen: Levin & Munksgaard]). The third congress (Paris, 1937) took the form of a conference devoted to the project of the *International Encyclopedia of Unified Science*. Annual congresses are being planned, and preparations are now being made for the fourth congress, to be held at Girton College, Cambridge, England, July 14–19, 1938, and for the fifth congress, to be held at Harvard University, September 5–10, 1939. The general theme of this congress will be the "Logic of Science," and the publication of the *Foundations of the Unity of Science* is so arranged as to provide a background for the congress.

The International Congresses for the Unity of Science are being administered by an International Committee composed of the following members:

N. Bohr (Copenhagen)	F. Gonseth (Zurich)	G. Mannoury (Amsterdam)
M. Boll (Paris)	J. Hadamard (Paris)	
H. Bonnet (Paris)	P. Janet (Paris)	R. von Mises (Istanbul)
P. W. Bridgman (Cambridge, Mass.)	H. S. Jennings (Baltimore)	C. W. Morris (Chicago)
		O. Neurath (The Hague)
E. Brunswik (Vienna, Berkeley)	J. Joergensen (Copenhagen)	C. K. Ogden (London)
		J. Perrin (Paris)
R. Carnap (Chicago)	E. Kaila (Helsingfors)	H. Reichenbach (Istanbul)
E. Cartan (Paris)	T. Kotarbinski (Warsaw)	
J. Clay (Amsterdam)	A. Lalande (Paris)	A. Rey (Paris)
M. R. Cohen (Chicago)	P. Langevin (Paris)	C. Rist (Paris)
J. Dewey (New York City)	K. S. Lashley (Cambridge, Mass.)	L. Rougier (Besançon, Cairo)
F. Enriques (Rome)	C. I. Lewis (Cambridge, Mass.)	B. Russell (Petersfield)
P. Frank (Prague)		L. S. Stebbing (London)
M. Fréchet (Paris)	J. Lukasiewicz (Warsaw)	J. H. Woodger (London)

2. See Otto Neurath, "An International Encyclopedia of Unified Science"—a paper read at the First International Congress for the Unity of Science (Paris, 1935), published in *Actes du congrès international de philosophie scientifique* (Paris: Hermann & Cie, 1936), Part II. This idea was supported and accompanied by important explanations by the members of the Encyclopedia Committee of Organization (Rudolf Carnap, Philipp Frank, Joergen Joergensen, Charles W. Morris, Otto Neurath, Louis Rougier), who spoke about the problems, the importance, and the logical basis of this project (see papers read by Charles W. Morris, Rudolf Carnap, and Philipp Frank at the same congress). The First International Congress for the Unity of Science approved the plan and expressed willingness to help in its fulfilment. See also Otto Neurath, "L'Encyclopédie comme 'modèle,'" *Revue de synthèse*, October, 1936.

3. See Otto Neurath, *International Picture Language: The First Rules of ISOTYPE* (London: Kegan Paul, 1936).

Analysis and Synthesis in Science

Niels Bohr

Notwithstanding the admittedly practical necessity for most scientists to concentrate their efforts in special fields of research, science is, according to its aim of enlarging human understanding, essentially a unity. Although periods of fruitful exploration of new domains of experience may often naturally be accompanied by a temporary renunciation of the comprehension of our situation, history of science teaches us again and again how the extension of our knowledge may lead to the recognition of relations between formerly unconnected groups of phenomena, the harmonious synthesis of which demands a renewed revision of the presuppositions for the unambiguous application of even our most elementary concepts. This circumstance reminds us not only of the unity of all sciences aiming at a description of the external world but, above all, of the inseparability of epistemological and psychological analysis. It is just in the emphasis on this last point, which recent development in the most different fields of science has brought to the foreground, that the program of the present great undertaking distinguishes itself from that of previous encyclopedic enterprises, in which stress was essentially laid on the completeness of the account of the actual state of knowledge rather than on the elucidation of scientific methodology. It is therefore to be hoped that the forthcoming *Encyclopedia* will have a deep influence on the whole attitude of our generation which, in spite of the ever increasing specialization in science as well as in technology, has a growing feeling of the mutual dependency of all human activities. Above all, it may help us to realize that even in science any arbitrary restriction implies the danger of prejudices and that our only way of avoiding the extremes of materialism and mysticism is the never ending endeavor to balance analysis and synthesis.

Unity of Science as a Social Problem

John Dewey

I. The Scientific Attitude

Anyone who attempts to promote the unity of science must ask himself at least two basic questions: "What is meant by that whose unity is to be promoted, namely, science?" and "What sort of unity is feasible or desirable?" The following pages represent the conclusions the present writer has reached in reflecting upon these two themes.

With respect to the question as to the meaning of science, a distinction needs to be made between science as attitude and method and science as a body of subject matter. I do not mean that the two can be separated, for a method is a way of dealing with subject matter and science as a body of knowledge is a product of a method. Each exists only in connection with the other. An attitude becomes psychopathic when it is not directed to objects beyond itself. What is meant is, first, that attitude and method come before the material which is found in books, journals, and the proceedings of scientific organizations; and, second, that the attitude is manifested primarily toward the objects and events of the ordinary world and only secondarily toward that which is already scientific subject matter.

Stated in other words, the scientific method is not confined to those who are called scientists. The body of knowledge and ideas which is the product of the work of the latter is the fruit of a method which is followed by the wider body of persons who deal intelligently and openly with the objects and energies of the common environment. In its specialized sense, science is an elaboration, often a highly technical one, of everyday operations. In spite of the technicality of its language and procedures, its genuine meaning can be understood only if its con-

29

nection with attitudes and procedures which are capable of being used by all persons who act intelligently is borne in mind.

On the level of common sense there are attitudes which are like those of science in its more specialized sense, while there are attitudes which are thoroughly unscientific. There are those who work by routine, by casual cut-and-try methods, those who are enslaved to dogma and directed by prejudice, just as there are those who use their hands, eyes, and ears to gain knowledge of whatever comes their way and use whatever brains they have to extract meaning from what they observe. Few would rule engineers from out the scientific domain, and those few would rest their case upon a highly dubious distinction between something called "pure" science and something else called "applied" science.

As Dr. Karl Darrow has said in his *Renaissance of Science:*

Many of the things which modern science has to tell us are fantastic and inconceivable indeed, but they have been attested by the same sort of man with the same sort of training and using the same sort of reasoning as those who have made it possible to speak over a wire with San Francisco and over the ether of space to London, to cross the Atlantic in four days by steamer and in twenty-four hours by aeroplane, to operate a railroad with power transmitted invisibly through rails, and to photograph the bones inside the body with a light no eye can see and no fire can send forth.

When the achievements of the engineer are disparaged under the name "applied" science, it is forgotten that the inquiries and the calculations required to produce these achievements are as exacting as those which generate the science called "pure." Pure science does not apply itself automatically; application takes place through use of methods which it is arbitrary to distinguish from those employed in the laboratory or observatory. And if the engineer is mentioned, it is because, once he is admitted, we cannot exclude the farmer, the mechanic, and the chauffeur, as far as these men do what they have to do with intelligent choice of means and with intelligent adaptation of means to ends, instead of in dependence upon routine and guesswork. On the other hand, it is quite possible for the scientist to be quite unscientific in forming his beliefs outside his special subject, as he does whenever he permits such beliefs to be dictated by unexamined premises accepted traditionally or caught up out of the surrounding social atmosphere.

In short, the scientific attitude as here conceived is a quality that is manifested in any walk of life. What, then, is it? On its negative side, it is freedom from control by routine, prejudice, dogma, unexamined tradition, sheer self-interest. Positively, it is the will to inquire, to examine, to discriminate, to draw conclusions only on the basis of evidence after taking pains to gather all available evidence. It is the intention to reach beliefs, and to test those that are entertained, on the basis of observed fact, recognizing also that facts are without meaning save as they point to ideas. It is, in turn, the experimental attitude which recognizes that while ideas are necessary to deal with facts, yet they are working hypotheses to be tested by the consequences they produce.

Above all, it is the attitude which is rooted in the problems that are set and questions that are raised by the conditions of actuality. The unscientific attitude is that which shuns such problems, which runs away from them, or covers them up instead of facing them. And experience shows that this evasion is the counterpart of concern with artificial problems and alleged ready-made solutions. For all problems are artificial which do not grow, even if indirectly, out of the conditions under which life, including associated living, is carried on. Life is a process which goes on in connection with an environment which is complex, physically and culturally. There is no form of interaction with the physical environment and the human environment that does not generate problems that can be coped with only by an objective attitude and an intelligent method. The home, the school, the shop, the bedside and hospital, present such problems as truly as does the laboratory. They usually present the problems in a more direct and urgent fashion. This fact is so obvious that it would be trite to mention it were it not that it shows the potential universality of the scientific attitude.

The existence of artificial problems is also an undeniable fact in human history. The existence of such problems and the expenditure of energy upon the solution of them are the chief reasons why the potentiality of scientific method is so often unrealized and frustrated. The word 'metaphysics' has many meanings, all of which are generally supposed to be so highly technical as to be of no interest to the man in the street. But in the sense that 'metaphysical' means that which is outside of experience, over and beyond it, all human beings are metaphysical when they occupy themselves with problems which do not rise out of experience and for which solutions are sought outside experience. Men are metaphysical not only in technical philosophy but in many of their beliefs and habits of thought in religion, morals, and politics. The waste of energy that results is serious enough. But this is slight compared with that which is wrought by artificial problems and solutions in preventing, deflecting, and distorting the development of the scientific attitude which is the proper career of intelligence.

II. The Social Unity of Science

When we turn from the question of what is meant by science to the question of what is meant by its unity, we seem, at first sight, to have shifted ground and to be in another field. The unity of science is usually referred to in connection with unification of the attained results of science. In this field the problem of attaining the unity of science is that of co-ordinating the scattered and immense body of specialized findings into a systematic whole. This problem is a real one and cannot be neglected. But there is also a human, a cultural, meaning of the unity of science. There is, for instance, the question of unifying the efforts of all those who exercise in their own affairs the scientific method so that these efforts may gain the force which comes from united effort. Even when an individual is or tries to be intelligent in the conduct of his own life-affairs, his efforts are hampered, often times defeated, by obstructions due not merely to ignorance but to active opposition to the scientific attitude on the part of those influenced by prejudice, dogma, class interest, external authority, nationalistic and racial sentiment, and similar powerful agencies. Viewed in this light, the problem of the unity of science constitutes a fundamentally important social problem.

At the present time the enemies of the scientific attitude are numerous and organized—much more so than appears at superficial glance. The prestige of science is indeed great, especially in the field of its external application to industry and war. In the abstract, few would come out openly and say that they were opposed to science. But this small number is no measure of the influence of those who borrow the results of science to advance by thoroughly unscientific and antiscientific methods private, class, and national interests. Men may admire science, for example, because it gives them the radio to use, and then employ the radio to create conditions that prevent the development of the scientific attitude in the most important fields of human activity—fields which suffer terribly because of failure to use scientific method. In particular, science is not welcomed but rather opposed when it "invades" (a word often used) the field now pre-empted by religion, morals, and political and economic institutions.

To bring about unity of the scientific attitude is, then, to bring those who accept it and who act upon it into active cooperation with one another. This problem transcends in importance the more technical problem of unification of the results of the special sciences. It takes precedence over the latter issue. For it is not too much to say that science, even in its more specialized sense, now stands at a critical juncture. It must move forward in order to maintain its achievements. If it stands still, it will be confined to the field in which it has already won victories and will see the fruits of its victories appropriated by those who will use them by antiscientific methods for nonhumane ends.

Accordingly, the great need is for those who are actuated by the scientific spirit to take counsel regarding the place and function of science in the total scene of life. It follows that a movement in behalf of the unity of science need not and should not lay down in advance a platform to be accepted. It is essentially a co-operative movement, so that detailed specific common standpoints and ideas must emerge out of the very processes of co-operation. To try to formulate them in advance and insist upon their acceptance by all is both to obstruct co-operation and to be false to the scientific spirit. The only thing necessary in the form of agreement is faith in the scientific attitude and faith in the human and social importance of its maintenance and expansion.

What has been said does not minimize the difficulties that arise from the great degree of isolated specialization that now characterizes science or the importance of overcoming these difficulties. To a great extent those who now pursue the different branches of science speak different languages and are not readily understood by one another. Translation from one branch to another is not easy. In consequence, workers tend to be deprived of the useful intellectual instruments that would be available in their own special work if there were a freer give and take.

But the needed work of co-ordination cannot be done mechanically or from without. It, too, can be the fruit only of co-operation among those animated by the scientific spirit. Convergence to a common center will be effected most readily and most vitally through the reciprocal exchange which attends genuine co-operative effort. The attempt to secure unity by defining the terms of all the sciences in terms of some one science is doomed in advance to defeat. In the house which science might build there are many mansions. The first task, to change the metaphor, is to build bridges from one science to another. There are many gaps to be spanned. It seems to me, however, that the great need is the linkage of the physico-chemical sciences with psychological and social fields of science through the intermediary of biology. I should probably be expressing my own view or that of a particular and perhaps small group if I said that convergence can best be attained by considering how various sciences may be brought together in common attack upon practical social problems. But it is wholly within the scope of the present theme to say that the co-operative endeavor held in view by the present movement for the unity of science is bound gradually to disclose the causes of present gaps and to indicate where and how bridges may be built across the gulfs that still separate workers in different fields.

A very short history has been enjoyed by free scientific method in comparison with the long history enjoyed by forces which have never felt the influence of science. Ideas that descend from the prescientific epoch are still with us and are crystallized in institutions. They are not to be exorcised by reiteration of the word 'science.' Every scientific worker is still subject to their influence, certainly outside his special field and sometimes even within it. Only constant critical care, exercised in the spirit of the scientific attitude, can bring about their gradual elimination. Ultimately, this criticism must be self-criticism. But the agencies and instrumentalities of self-criticism can be had only by means of as full and free co-operation with others as it is possible to secure.

The advance of scientific method has brought with it, where the influence of the method has been felt, a great increase in toleration. We are now in a world where there is an accelerated development of intolerance. Part of the cause for this growth can be found, I think, in the fact that tolerance so far has been largely a passive thing. We need a shift from acceptance of responsibility for passive toleration to active responsibility for promoting the extension of scientific method. The first step is to recognize the responsibility for furthering mutual understanding and free communication.

III. Education and the Unity of Science

It is perhaps within the scope of my theme to say something about the connection of the movement for the unity of science with education. I have already mentioned the fact that scientific method has reached a crisis in its history, due, in final analysis, to the fact that the ultra-reactionary and the ultra-radical combine, even while acclaiming the prestige of science in certain fields, to use the techniques of science to destroy the scientific attitude. The short history of science in comparison with the history of institutions that resist its application by the mere fact of their inertia has also been mentioned. These two influences combine to render the agencies of education the crucial point in any movement to bring about a greater and more progressive unity of the scientific spirit.

After a struggle, the various sciences have found a place for themselves in the educational institutions. But to a large extent they exist merely side by side with other subjects which have hardly felt the touch of science. This, however, is far from being the most depressing feature of the educational situation as respects the place of science. For it is also true that the spirit in which the sciences are often taught, and the methods of instruction employed in teaching them, have been in large measure taken over from traditional nonscientific subjects.

I mention certain things which confirm this statement. In the first place, science has barely affected elementary education. With a very few exceptions it has not touched the early years of the elementary school. Yet this is the time when curiosity is most awake, the interest in observation the least dulled, and desire for new experiences most active. It is also the period in which the fundamental attitudes are formed which control, subconsciously if not consciously, later attitudes and methods.

In the second place, scientific subjects are taught very largely as bodies of subject matter rather than as a method of universal attack and approach. There may be laboratories and laboratory exercises and yet this statement remain true. For they may be employed primarily in order that pupils acquire a certain body of information. The resulting body of information about facts and laws has a different content from that provided in other studies. But as long as the ideal is information, the sciences taught are still under the dominion of ideas and practices that have a prescientific origin and history. Laboratory exercises and class demonstrations may be a part of a regular routine of instruction, and yet accomplish little in developing the scientific habit of mind. Indeed, except in a chosen few the mere weight of information may be a load carried in the memory, not a resource for further observation and thought.

In the third place, apart from some institutions of research and graduate departments of universities which attract relatively a small number, most money and energy go into institutions in which persons are prepared for special professional pursuits. This fact is not itself objectionable, as I have already indicated in speaking of "applied" and "pure" science. But this technical education, as it is at present conducted, is directed to narrow ends rather that to the wide and liberal end of developing interest and ability to use the scientific method in all fields of human betterment. It is quite possible, unfortunately, for a person to have the advantage of this special training and yet remain indifferent to the application of the scientific attitude in fields that lie outside his own specialized calling.

The final point is a corollary. Something called by the name of "science" gets shut off in a segregated territory of its own. There are powerful special interests which strive in any case to keep science isolated so that the common life may be immune from its influence. Those who have these special interests fear the impact of scientific method upon social issues.

They fear this impact even if they have not formulated the nature and ground of their fear. But there are influences within the status of science itself in the educational system which promote its isolation. If the schools are used for the purpose of instilling belief in certain dogmas—a use in which something called "education" becomes simply an organ of propaganda—and this use continues to grow, it will be in some measure because science has not been conceived and practiced as the sole universal method of dealing intellectually with all problems. The movement to unify workers in different fields of science is itself an educative movement for those who take part in it. It is also a precondition of effort to give the scientific attitude that place in educational institutions which will create an ever increasing number of persons who habitually adopt the scientific attitude in meeting the problems that confront them.

I said that I thought that reference to education belonged within the scope of the present theme. On the one hand, the future of the scientific attitude as a socially unified force depends more upon the education of children and youth than upon any other single force. On the other hand, the teaching of science can hardly take the place which belongs to it, as an attitude of universal application, unless those who are already animated by the scientific attitude and concerned for its expansion actively co-operate. The first condition to be satisfied is that such persons bestir themselves to become aware of what the scientific attitude is and what it is about so as to become diligently militant in demonstrating its rightful claims.

The import of what has been said is that the scientific attitude and method are at bottom but the method of free and effective intelligence. The special sciences reveal what this method is and means, and what it is capable of. It is neither feasible nor desirable that all human beings should become practitioners of a special science. But is intensely desirable and under certain conditions practicable that all human beings become scientific in their attitudes: genuinely intelligent in their ways of thinking and acting. It is practicable because all normal persons have the potential germs which make this result possible. It is desirable because this attitude forms the sole ultimate alternative to prejudice, dogma, authority, and coercive force exercised in behalf of some special interest. Those who are concerned with science in its more technical meaning are obviously those who should take the lead by co-operation with one another in bringing home to all the inherent universality of scientific method.

On the Importance of Logical Form

Bertrand Russell

The instrument of mathematical logic, which has begun to be appreciated during the present century, possesses two rather different kinds of utility—one in pure mathematics, the other in the various empirical sciences. Of the former I shall say nothing, since the ground is familiar; but on the latter there are some things to be said that bear on the importance of a modern encyclopedia.

In the empirical sciences it is not so much in relation to inference that mathematical logic is useful as in relation to analysis and the apprehension of identity and difference of form. Where identity of form is of the traditional mathematical kind, its importance has long been realized. The kinetic theory of gases has been applied to the stellar universe, which, to the non-mathematical mind, appears very different from a gas. A British mathematical professor at Tokyo was led by his location to study earthquakes, and made useful applications of his results to the vibrations of the footplates of locomotives. But, where identity of form is not of the sort that can be expressed without logical symbols, men of science have been less quick to recognize it; while the general public, through logical incompetence, has been led into grave practical errors. During the Black Death the inhabitants of Siena attributed the calamity to their presumption in planning a much enlarged cathedral, oblivious of the fact that the mortality was just as great elsewhere. Similarly, in 1931, the population of every country attributed the depression to the sins of its own government; this caused a movement to the Left where there was a Right government, and to the Right where there was a Left government. Only a few impotent intellectuals observed that the phenomenon to be explained was world-wide, not local.

The distinction between macroscopic and microscopic physics, which has become important since the rise of quantum theory, suggests possibilities as regards scientific method in other fields. Although, as a mathematical ideal, macroscopic physics may be supposed deducible from the behavior of the individual atoms, it was in fact discovered first, and its laws remain valid, for most practical purposes, in spite of the discoveries of the quantum physicists. This suggests the possibility of a social science not deduced from the laws of individual behavior but based upon laws which are only valid for large numbers. The theory of evolution, in biology, is the most striking example. Economics, in so far as it is a science, is another. Vital statistics afford another field for the observation of statistical behavior; it might be thought, for instance, that there is an inverse correlation between increase and density of the population, but Australia, though confirming this as regards rabbits, negatives it as regards human beings.

Logical method has important applications to psychology. Suppose, for example, that, in order to deal with dual and multiple personality, we desire a definition of 'person' not derived from bodily continuity. We may observe that dual personality is connected with amnesia. We may define a relation M between two experiences, consisting in the fact that one is, in whole or part, a recollection of the other, or the other of the one. If N is the ancestral relation of M, all the experiences which have to a given experience the relation N may be defined as the person to whom the given experience belongs; for the student of dual and multiple personality this is probably the most convenient definition.

I said that mathematical logic has less importance in relation to scientific inference than in relation to analysis, but this statement needs qualification. Outside mathematics, the important inferences are not deductive, i.e., they are not such as mathematical logic makes. But logic can state their character with a precision which was formerly impossible. Much has been done, for example, by Carnap, in analyzing the kind of inference upon which scientific laws are based. Since all inferences of this kind are probable, not demonstrative, the study of probability, as Reichenbach insists, is of fundamental importance in scientific method.

The importance of logical form may be illustrated by what may be called the principle of the dictionary: Given two sets of propositions such that, by a suitable dictionary, any proposition of either set can be translated into a proposition of the other set, there is no effective difference between the two sets. Suppose—to take a hypothesis that I neither affirm nor deny—that all scientific propositions can be tested in terms of physics, and can also be stated on Berkeleian principles, in terms of psychology; then the question as to which of these forms of statement is the more correct has no meaning, since both or neither must be correct. Such dictionaries, which can, as a rule, only be constructed by the help of modern logic, suffice to dispose of large numbers of metaphysical questions, and thus facilitate concentration upon genuine scientific problems.

Let us take another example of the principle of the dictionary. The general principle of relativity showed that, in expressing the laws of macroscopic physics, we can transform our co-ordinates in any way we choose, so long as topological relations in space-time are preserved as topological relations among co-ordinates. It follows that the laws of macroscopic physics are topological laws, and that the introduction of number through co-ordinates is only a practical convenience, the laws being such as can, in theory, be expressed without the use of number. The old view that measurement is of the essence of science would therefore seem to be erroneous.

The unity of science, which is sometimes lost to view through immersion in specialist problems, is essentially a unity of method, and the method is one upon which modern logic throws much new light. It may be hoped that the *Encyclopedia* will do much to bring about an awareness of this unity.

Logical Foundations of the Unity of Science

Rudolf Carnap

I. What Is Logical Analysis of Science?

The task of analyzing science may be approached from various angles. The analysis of the subject matter of the sciences is carried out by science itself. Biology, for example, analyzes organisms and processes in organisms, and in a similar way every branch of science analyzes its subject matter. Mostly, however, by 'analysis of science' or 'theory of science' is meant an investigation which differs from the branch of science to which it is applied. We may, for instance, think of an investigation of scientific *activity*. We may study the historical development of this activity. Or we may try to find out in which way scientific work depends upon the individual conditions of the men working in science, and upon the status of the society surrounding them. Or we may describe procedures and appliances used in scientific work. These investigations of scientific activity may be called history, psychology, sociology, and methodology of science. The subject matter of such studies is science as a body of actions carried out by certain persons under certain circumstances. Theory of science in this sense will be dealt with at various other places in this *Encyclopedia;* it is certainly an essential part of the foundation of science.

We come to a theory of science in another sense if we study not the actions of scientists but their results, namely, science as a body of ordered knowledge. Here, by 'results' we do not mean beliefs, images, etc., and the behavior influenced by them. That would lead us again to psychology of science. We mean by 'results' certain linguistic expressions, viz., the statements asserted by scientists. The task of the theory of science in this sense will be to analyze such statements, study their kinds and relations, and analyze terms as components of those statements and theories as ordered systems of those statements. A statement is a kind of sequence of spoken sounds, written marks, or the like, produced by human beings for specific purposes. But it is possible to abstract in an analysis of the statements of science from the persons asserting the statements and from the psychological and sociological conditions of such assertions. The analysis of the linguistic expressions of science under such an abstraction is *logic of science.*

Within the logic of science we may distinguish between two chief parts. The investigation may be restricted to the forms of the linguistic expressions involved, i.e., to the way in which they are constructed out of elementary parts (e.g., words) without referring to anything outside of language. Or the investigation goes beyond this boundary and studies linguistic expressions in their relation to objects outside of language. A study restricted in the first-mentioned way is called *formal;* the field of such formal studies is called formal logic or *logical syntax.* Such a formal or syntactical analysis of the language of science as a whole or in its various branches will lead to results of the following kinds. A certain term (e.g., a word) is defined within a certain theory on the basis of certain other terms, or it is definable in such a way. A certain term, although not definable by certain other terms, is reducible to them (in a sense to be explained later). A certain statement is a logical consequence of (or logically deducible from) certain other statements; and a deduction of it, given within a certain theory, is, or is not, logically correct. A certain statement is incompatible with certain other statements, i.e., its negation is a logical consequence of them. A certain statement is independent of certain other statements, i.e., neither a logical consequence of them nor incompatible with them. A certain theory is inconsistent, i.e., some of its statements are incompatible with the other ones. The last sections of this essay will deal with the question of the unity of science from the logical point of view, studying the logical relations between the terms of the chief branches of science and between the laws stated in these branches; thus it will give an example of a syntactical analysis of the language of science.

In the second part of the logic of science, a given language and the expressions in it are analyzed in another way. Here also, as in logical syntax, abstraction is made from the psychological and sociological side of the language. This investigation, however, is not restricted to formal analysis but takes into consideration one important relation between linguistic expressions and other objects—that of designation. An investigation of this kind is called *semantics*. Results of a semantical analysis of the language of science may, for instance, have the following forms. A certain term designates a certain particular object (e.g., the sun), or a certain property of things (e.g., iron), or a certain relation between things (e.g., fathership), or a certain physical function (e.g., temperature); two terms in different branches of science (e.g., 'homo sapiens' in biology and 'person' in economics, or, in another way, 'man' in both cases) designate (or: do not designate) the same. What is designated by a certain expression may be called its *designatum*. Two expressions designating the same are called *synonymous*. The term 'true,' as it is used in science and in everyday life, can also be defined within semantics. We see that the chief subject matter of a semantical analysis of the language of science are such properties and relations of expressions, and especially of statements, as are based on the relation of designation. (Where we say 'the designatum of an expression,' the customary phrase is 'the meaning of an expression.' It seems, however, preferable to avoid the word 'meaning' wherever possible because of its ambiguity, i.e., the multiplicity of its designata. Above all, it is important to distinguish between the semantical and the psychological use of the word 'meaning.')

It is a question of terminological convention whether to use the term 'logic' in the wider sense, including the semantical analysis of the designata of expressions, or in the narrower sense of logical syntax, restricted to formal analysis, abstracting from designation. And accordingly we may distinguish between logic of science in the narrower sense, as the syntax of the language of science, and logic of science in the wider sense, comprehending both syntax and semantics.

II. The Main Branches of Science

We use the word 'science' here in its widest sense, including all theoretical knowledge, no matter whether in the field of natural sciences or in the field of the social sciences and the so-called humanities, and no matter whether it is knowledge found by the application of special scientific procedures, or knowledge based on common sense in everyday life. In the same way the term 'language of science' is meant here to refer to the language which contains all statements (i.e., theoretical sentences as distinguished from emotional expressions, commands, lyrics, etc.) used for scientific purposes or in everyday life. What usually is called science is merely a more systematic continuation of those activities which we carry out in everyday life in order to know something.

The first distinction which we have to make is that between *formal science* and *empirical science*. Formal science consists of the analytic statements established by logic and mathematics; empirical science consists of the synthetic statements established in the different fields of factual knowledge. The relation of formal to empirical science will be dealt with at another place; here we have to do with empirical science, its language, and the problem of its unity.

Let us take 'physics' as a common name for the nonbiological field of science, comprehending both systematic and historical investigations within this field, thus including chemistry, mineralogy, astronomy, geology (which is historical), meteorology, etc. How, then, are we to draw the boundary line between physics and biology? It is obvious that the distinction between these two branches has to be based on the distinction between two kinds of things which we find in nature: organisms and nonorganisms. Let us take this latter distinction as granted; it is the task of biologists to lay down a suitable definition for the term 'organism,' in other words, to tell us the features of a thing which we take as characteristic for its being an organism. How, then, are we to define 'biology' on the basis of 'organism'? We could perhaps think of trying to do it in this way: biology is the branch of science which investigates organisms and the processes occurring in organisms, and physics is the study of nonorganisms. But these definitions would not draw the distinction as it is usually intended. A law stated in physics is intended to be valid universally, without any restriction. For example, the law stating the electrostatic force as a function of electric charges and their distance, or the law determining the pressure of a gas as a function of temperature, or the law determining the angle of refraction as a function of the coefficients of refraction of the two media involved, are intended to apply to the processes in organisms no less than to those in inorganic nature. The biologist has to know these laws of physics in studying the processes in organisms. He needs them for the explanation of these processes. But since they do not suffice, he adds some other laws, not known by the physicist, viz., the specifically biological laws. Biology presupposes physics, but not vice versa.

These reflections lead us to the following definitions. Let us call those terms which we need—in addition to logico-mathematical terms—for the description of processes in inorganic nature *physical terms*, no matter whether, in a given instance, they are applied to such processes or to processes in organisms. That sublanguage of the language of science, which contains —besides logico-mathematical terms—all and only physical terms, may be called *physical language*. The system of those statements which are formulated in the physical language and are acknowledged by a certain group at a certain time is called the physics of that group at that time. Such of these statements as have a specific universal form are called *physical laws*. The physical laws are needed for the explanation of processes in inorganic nature; but, as mentioned before, they apply to processes in organisms also.

The whole of the rest of science may be called *biology (in the wider sense)*. It seems desirable, at least for practical purposes, e.g., for the division of labor in research work, to sub-divide this wide field. But it seems questionable whether any distinctions can be found here which, although not of a fundamental nature, are at least clear to about the same degree as the distinction between physics and biology. At present, it is scarcely possible to predict which subdivisions will be made in the future. The traditional distinction between bodily (or material) and mental (or psychical) processes had its origin in the old magical and later metaphysical mind-body dualism. The distinction as a practical device for the classification of branches of science still plays an important role, even for those scientists who reject that metaphysical dualism; and it will probably continue to do so for some time in the future. But when the aftereffect of such prescientific issues upon science becomes weaker and weaker, it may be that new boundary lines for subdivisions will turn out to be more satisfactory.

One possibility of dividing biology in the wider sense into two fields is such that the first corresponds roughly to what is usually called biology, and the second comprehends among other parts those which usually are called psychology and social science. The second field deals with the behavior of individual organisms and groups of organisms within their environment, with the dispositions to such behavior, with such features of processes in organisms as are relevant to the behavior, and with certain features of the environment which are characteristic of and relevant to the behavior, e.g., objects observed and work done by organisms.

The first of the two fields of biology in the wider sense may be called biology in the narrower sense, or, for the following discussions, simply *biology*. This use of the term 'biology' seems justified by the fact that, in terms of the customary classification, this part contains most of what is usually called biology, namely, general biology, botany, and the greater part of zoology. The terms which are used in this field in addition to logico-mathematical and physical terms may be called biological terms in the narrower sense, or simply *biological terms*. Since many statements of biology contain physical terms besides biological ones, the *biological language* cannot be restricted to biological terms; it contains the physical language as a sublanguage and, in addition, the biological terms. Statements and laws belonging to this language but not to physical language will be called *biological statements* and *biological laws*.

The distinction between the two fields of biology in the wider sense has been indicated only in a very vague way. At the present time it is not yet clear as to how the boundary line may best be drawn. Which processes in an organism are to be assigned to the second field? Perhaps the connection of a process with the processes in the nervous system might be taken as characteristic, or, to restrict it more, the connection with speaking activities, or, more generally, with activities involving signs. Another way of characterization might come from the other direction, from outside, namely, selecting the processes in an organism from the point of view of their relevance to achievements in the environment (see Brunswik and Ness). There is no name in common use for this second field. (The term 'mental sciences' suggests too narrow a field and is connected too closely with the metaphysical dualism mentioned before.) The term 'behavioristics' has been proposed. If it is used, it must be made clear that the word 'behavior' has here a greater extension than it had with the earlier behaviorists. Here it is intended to designate not only the overt behavior which can be observed from outside but also internal behavior (i.e., processes within the organism); further, dispositions to behavior which may not be manifest in a special case; and, finally, certain effects upon the environment. Within this second field we may distinguish roughly between two parts dealing with individual organisms and with groups of organisms. But it seems doubtful whether any sharp line can be drawn between these two parts. Compared with the customary classification of science, the first part would include chiefly psychology, but also some parts of physiology and the humanities. The second part would chiefly include social science and, further, the greater part of the humanities and history, but it has not only to deal with groups of human beings but also to deal with groups of other organisms.

For the following discussion, the terms 'psychology' and 'social science' will be used as names of the two parts because of lack of better terms. It is clear that both the question of boundary lines and the question of suitable terms for the sections is still in need of much more discussion.

III. Reducibility

The question of the unity of science is meant here as a problem of the logic of science, not of ontology. We do not ask: "Is the world one?" "Are all events fundamentally of one kind?" "Are the so-called mental processes really physical processes or not?" "Are the so-called physical processes really spiritual or not?" It seems doubtful whether we can find any theoretical content in such philosophical questions as discussed by monism, dualism, and pluralism. In any case, when we ask whether there is a unity in science, we mean this as a question of logic, concerning the logical relationships between the terms and the laws of the various branches of science. Since it belongs to the logic of science, the question concerns scientists and logicians alike.

Let us first deal with the question of terms. (Instead of the word 'term' the word 'concept' could be taken, which is more frequently used by logicians. But the word 'term' is more clear, since it shows that we mean signs, e.g., words, expressions consisting of words, artificial symbols, etc., of course with the meaning they have in the language in question. We do not mean 'concept' in its psychological sense, i.e., images or thoughts somehow connected with a word; that would not belong to logic.) We know the meaning (designatum) of a term if we know under what conditions we are permitted to apply it in a concrete case and under what conditions not. Such a knowledge of the conditions of application can be of two different kinds. In some cases we may have a merely practical knowledge, i.e., we are able to use the term in question correctly without giving a theoretical account of the rules for its use. In other cases we may be able to give an explicit formulation of the conditions for the application of the term. If now a certain term x is such that the conditions for its application (as used in the language of science) can be formulated with the help of the terms y, z, etc., we call such a formulation a *reduction statement* for x in terms of y, z, etc., and we call x *reducible* to y, z, etc. There may be several sets of conditions for the application of x; hence x may be reducible to y, z, etc., and also to u, v, etc., and perhaps to other sets. There may even be cases of mutual reducibility, e.g., each term of the set x_1, x_2, etc., is reducible to y_1, y_2, etc.; and, on the other hand, each term of the set y_1, y_2, etc., is reducible to x_1, x_2, etc.

A *definition* is the simplest form of a reduction statement. For the formulation of examples, let us use '$\equiv$' (called the symbol of equivalence) as abbreviation for 'if and only if.' Example of a definition for 'ox': 'x is an ox $\equiv$ x is a quadruped and horned and cloven-footed and ruminant, etc.' This is also a reduction statement because it states the conditions for the application of the term 'ox,' saying that this term can be applied to a thing if and only if that thing is a quadruped and horned, etc. By that definition the term 'ox' is shown to be reducible to—moreover definable by—the set of terms 'quadruped,' 'horned,' etc.

A reduction statement sometimes cannot be formulated in the simple form of a definition, i.e., of an equivalence statement, '. . . . $\equiv$ . . . . ,' but only in the somewhat more complex form 'If . . . . , then: . . . . $\equiv$ . . . . .' Thus a reduction statement is either a simple (i.e., explicit) definition or, so to speak, a conditional definition. (The term 'reduction statement' is generally used in the narrower sense, referring to the second, conditional form.) For instance, the following statement is a reduction statement for the term 'electric charge' (taken here for the sake of simplicity as a nonquantitative term), i.e., for the statement form 'the body x has an electric charge at the time t': 'If a light body y is placed near x at t, then: x has an electric charge at t $\equiv$ y is attracted by x at t.' A general way of procedure which enables us to find out whether or not a certain term can be applied in concrete cases may be called a *method of determination* for the term in question. The method of determination for a quantitative term (e.g., 'temperature') is the method of measurement for that term. Whenever we know an experimental method of determination for a term, we are in a position to formulate a reduction statement for it. To know an experimental method of determination for a term, say 'Q₃,' means to know two things. First, we must know an experimental situation which we have to create, say the state Q₁, e.g., the arrangement of measuring apparatuses and of suitable conditions for their use. Second, we must know the possible experimental result, say Q₂, which, if it occurs, will confirm the presence of the property Q₃. In the simplest case—let us leave aside the more complex cases—Q₂ is also such that its nonoccurrence shows that the thing in question does not have the property Q₃. Then a reduction statement for 'Q₃,' i.e., for the statement form 'the thing (or space-time-point) x is Q₃ (i.e., has the property Q₃) at the time t,' can be formulated in this way: 'If x is Q_1 (i.e., x and the surroundings of x are in the state Q₁) at time t, then: x is Q₃ at t ≡ x is Q₂ at t.' On the basis of this reduction statement, the term 'Q₃' is reducible to 'Q₁,' 'Q₂,' and spatio-temporal terms. Whenever a term 'Q₃' expresses the disposition of a thing to behave in a certain way (Q₂) to certain conditions (Q₁), we have a reduction statement of the form given above. If there is a connection of such a kind between Q₁, Q₂, and Q₃, then in biology and psychology in certain cases the following terminology is applied: 'To the stimulus Q₁ we find the reaction Q₂ as a symptom for Q₃.' But the situation is not essentially different from the analogous one in physics, where we usually do not apply that terminology.

Sometimes we know several methods of determination for a certain term. For example, we can determine the presence of an electric current by observing either the heat produced in the conductor, or the deviation of a magnetic needle, or the quantity of a substance separated from an electrolyte, etc. Thus the term 'electric current' is reducible to each of many sets of other terms. Since not only can an electric current be measured by measuring a temperature but also, conversely, a temperature can be measured by measuring the electric current produced by a thermo-electric element, there is mutual reducibility between the terms of the theory of electricity, on the one hand, and those of the theory of heat, on the other. The same holds for the terms of the theory of electricity and those of the theory of magnetism.

Let us suppose that the persons of a certain group have a certain set of terms in common, either on account of a merely practical agreement about the conditions of their application or with an explicit stipulation of such conditions for a part of the terms. Then a reduction statement reducing a new term to the terms of that original set may be used as a way of introducing the new term into the language of the group. This way of introduction assures conformity as to the use of the new term. If a certain language (e.g., a sublanguage of the language of science, covering a certain branch of science) is such that every term of it is reducible to a certain set of terms, then this language can be constructed on the basis of that set by introducing one new term after the other by reduction statements. In this case we call the basic set of terms a *sufficient reduction basis* for that language.

IV. The Unity of the Language of Science

Now we will analyze the logical relations among the terms of different parts of the language of science with respect to reducibility. We have indicated a division of the whole language of science into some parts. Now we may make another division cutting across the first, by distinguishing in a rough way, without any claims to exactness, between those terms which we use on a prescientific level in our everyday language, and for whose application no scientific procedure is necessary, and scientific terms in the narrower sense. That sublanguage which is the common part of this prescientific language and the physical language may be called physical thing-language or briefly *thing-language*. It is this language that we use in speaking about the properties of the observable (inorganic) things surrounding us. Terms like 'hot' and 'cold' may be regarded as belonging to the thing-language, but not 'temperature' because its determination requires the application of a technical instrument; further, 'heavy' and 'light' (but not 'weight'); 'red,' 'blue,' etc.; 'large,' 'small,' 'thick,' 'thin,' etc.

The terms so far mentioned designate what we may call observable properties, i.e., such as can be determined by a direct observation. We will call them *observable thing-predicates*. Besides such terms the thing-language contains other ones, e.g., those expressing the disposition of a thing to a certain behavior under certain conditions, e.g., 'elastic,' 'soluble,' 'flexible,' 'transparent,' 'fragile,' 'plastic,' etc. These terms—they might be called disposition-predicates—are reducible to observable thing-predicates because we can describe the experimental conditions and the reactions characteristic of such disposition-predicates in terms of observable thing-predicates. Example of a reduction statement for 'elastic': 'If the body x is stretched and then released at the time t, then: x is elastic at the time $t \equiv x$ contracts at t,' where the terms 'stretched,' 'released,' and 'contracting' can be defined by observable thing-predicates. If these predicates are taken as a basis, we can moreover introduce, by iterated application of definition and (conditional) reduction, every other term of the *thing-language*, e.g., designations of substances, e.g., 'stone,' 'water,' 'sugar,' or of processes, e.g., 'rain,' 'fire,' etc. For every term of that language is such that we can apply it either on the basis of direct observation or with the help of an experiment for which we know the conditions and the possible result determining the application of the term in question.

Now we can easily see that every term of the *physical language* is reducible to those of the thing-language and hence finally to observable thing-predicates. On the scientific level, we have the quantitative coefficient of elasticity instead of the qualitative term 'elastic' of the thing-language; we have the quantitative term 'temperature' instead of the qualitative ones 'hot' and 'cold'; and we have all the terms by means of which physicists describe the temporary or permanent states of things or processes. For any such term the physicist knows at least one method of determination. Physicists would not admit into their language any term for which no method of determination by observations were given. The formulation of such a method, i.e., the description of the experimental arrangement to be carried out and of the possible result determining the application of the term in question, is a reduction statement for that term. Sometimes the term will not be directly reduced by the reduction statement to thing-predicates, but first to other scientific terms, and these by their reduction statements again to other scientific terms, etc.; but such a reduction chain must in any case finally lead to predicates of the thing-language and, moreover, to observable thing-predicates because otherwise there would be no way of determining whether or not the physical term in question can be applied in special cases, on the basis of given observation statements.

If we come to *biology* (this term now always understood in the narrower sense), we find again the same situation. For any biological term the biologist who introduces or uses it must know empirical criteria for its application. This applies, of course, only to biological terms in the sense explained before, including all terms used in scientific biology proper, but not to certain terms used sometimes in the philosophy of biology—'a whole,' 'entelechy,' etc. It may happen that for the description of the criterion, i.e., the method of determination of a term, other biological terms are needed. In this case the term in question is first reducible to them. But at least indirectly it must be reducible to terms of the thing-language and finally to observable thing-predicates, because the determination of the term in question in a concrete case must finally be based upon observations of concrete things, i.e., upon observation statements formulated in the thing-language.

Let us take as an example the term 'muscle.' Certainly biologists know the conditions for a part of an organism to be a muscle; otherwise the term could not be used in concrete cases. The problem is: Which other terms are needed for the formulation of those conditions? It will be necessary to describe the functions within the organism which are characteristic of muscles, in other words, to formulate certain laws connecting the processes in muscles with those in their environment, or, again in still other words, to describe the reactions to certain stimuli characteristic of muscles. Both the processes in the environment and those in the muscle (in the customary terminology: stimuli and reactions) must be described in such a way that we can determine them by observations. Hence the term 'muscle,' although not definable in terms of the thing-language, is reducible to them. Similar considerations easily show the reducibility of any other biological term—whether it be a designation of a kind of organism, or of a kind of part of organisms, or of a kind of process in organisms.

The result found so far may be formulated in this way: The terms of the thing-language, and even the narrower class of the observable thing-predicates, supply a sufficient basis for the languages both of physics and of biology. (There are, by the way, many reduction bases for these languages, each of which is much more restricted than the classes mentioned.) Now the question may be raised whether a basis of the kind mentioned is sufficient even for the whole language of science. The affirmative answer to this question is sometimes called *physicalism* (because it was first formulated not with respect to the thing-language but to the wider physical language as a sufficient basis). If the thesis of physicalism is applied to biology only, it scarcely meets any serious objections. The situation is somewhat changed, however, when it is applied to psychology and social science (individual and social behavioristics). Since many of the objections raised against it are based on misinterpretations, it is necessary to make clear what the thesis is intended to assert and what not.

The question of the reducibility of the terms of psychology to those of the biological language and thereby to those of the thing-language is closely connected with the problem of the various methods used in psychology. As chief examples of methods used in this field in its present state, the physiological, the behavioristic, and the introspective methods may be considered. The *physiological approach* consists in an investigation of the functions of certain organs in the organism, above all, of the nervous system. Here, the terms used are either those of biology or those so closely related to them that there will scarcely be any doubt with respect to their reducibility to the terms of the biological language and the thing-language. For the *behavioristic approach* different ways are possible. The investigation may be restricted to the external behavior of an organism, i.e., to such movements, sounds, etc., as can be observed by other organisms in the neighborhood of the first. Or processes within the organism may also be taken into account so that this approach overlaps with the physiological one. Or, finally, objects in the environment of the organism, either observed or worked on or produced by it, may also be studied. Now it is easy to see that a term for whose determination a behavioristic method—of one of the kinds mentioned or of a related kind— is known, is reducible to the terms of the biological language, including the thing-language. As we have seen before, the formulation of the method of determination for a term is a reduction statement for that term, either in the form of a simple definition or in the conditional form. By that statement the term is shown to be reducible to the terms applied in describing the method, namely, the experimental arrangement and the characteristic result. Now, conditions and results consist in the behavioristic method either of physiological processes in the organism or of observable processes in the organism and in its environment. Hence they can be described in terms of the biological language. If we have to do with a behavioristic approach in its pure form, i.e., leaving aside physiological investigations, then the description of the conditions and results characteristic for a term can in most cases be given directly in terms of the thing-language. Hence the behavioristic reduction of psychological terms is often simpler than the physiological reduction of the same term.

Let us take as an example the term 'angry.' If for anger we knew a sufficient and necessary criterion to be found by a physiological analysis of the nervous system or other organs, then we could define 'angry' in terms of the biological language. The same holds if we knew such a criterion to be determined by the observation of the overt, external behavior. But a physiological criterion is not yet known. And the peripheral symptoms known are presumably not necessary criteria because it might be that a person of strong self-control is able to suppress these symptoms. If this is the case, the term 'angry' is, at least at the present time, not definable in terms of the biological language. But, nevertheless, it is reducible to such terms. It is sufficient for the formulation of a reduction sentence to know a behavioristic procedure which enables us—if not always, at least under suitable circumstances—to determine whether the organism in question is angry or not. And we know indeed such procedures; otherwise we should never be able to apply the term 'angry' to another person on the basis of our observations of his behavior, as we constantly do in everyday life and in scientific investigation. A reduction of the term 'angry' or similar terms by the formulation of such procedures is indeed less useful than a definition would be, because a definition supplies a complete (i.e., unconditional) criterion for the term in question, while a reduction statement of the conditional form gives only an incomplete one. But a criterion, conditional or not, is all we need for ascertaining reducibility. Thus the result is the following: If for any psychological term we know either a physiological or a behavioristic method of determination, then that term is reducible to those terms of the thing-language.

In psychology, as we find it today, there is, besides the physiological and the behavioristic approach, the so-called *introspective method*. The questions as to its validity, limits, and necessity are still more unclear and in need of further discussion than the analogous questions with respect to the two other methods. Much of what has been said about it, especially by philosophers, may be looked at with some suspicion. But the facts themselves to which the term 'introspection' is meant to refer will scarcely be denied by anybody, e.g., the fact that a person sometimes knows that he is angry without applying any of those procedures which another person would have to apply, i.e., without looking with the help of a physiological instrument at his nervous system or looking at the play of his facial muscles. The problems of the practical reliability and theoretical validity of the introspective method may here be left aside. For the discussion of reducibility an answer to these problems is not needed. It will suffice to show that in every case, no matter whether the introspective method is applicable or not, the behavioristic method can be applied at any rate. But we must be careful in the interpretation of this assertion. It is not meant as saying: 'Every psychological process can be ascertained by the behavioristic method.' Here we have to do not with the single processes themselves (e.g., Peter's anger yesterday morning) but with kinds of processes (e.g., anger). If Robinson Crusoe is angry and then dies before anybody comes to his island, nobody except himself ever knows of this single occurrence of anger. But anger of the same kind, occurring with other persons, may be studied and ascertained by a behavioristic method, if circumstances are favorable. (Analogy: if an electrically charged raindrop falls into the ocean without an observer or suitable recording instrument in the neighborhood, nobody will ever know of that charge. But a charge of the same kind can be found out under suitable circumstances by certain observations.) Further, in order to come to a correct formulation of the thesis, we have to apply it not to the kinds of processes (e.g., anger) but rather to the terms designating such kinds of processes (e.g., 'anger'). The difference might seem trivial but is, in fact, essential. We do not at all enter a discussion about the question whether or not there are kinds of events which can never have any behavioristic symptoms, and hence are knowable only by introspection. We have to do with psychological terms not with kinds of events. For any such term, say, 'Q,' the psychological language contains a statement form applying that term, e.g., 'The person . . . . is at the time . . . . in the state Q.' Then the utterance by speaking or writing of the statement 'I am now (or: I was yesterday) in the state Q,' is (under suitable circumstances, e.g., as to reliability, etc.) an observable symptom for the state Q. Hence there cannot be a term in the psychological language, taken as an intersubjective language for mutual communication, which designates a kind of state or event without any behavioristic symptom. Therefore, there is a behavioristic method of determination for any term of the psychological language. Hence every such term is reducible to those of the thing-language.

The logical nature of the psychological terms becomes clear by an analogy with those physical terms which are introduced by reduction statements of the conditional form. Terms of both kinds designate a state characterized by the disposition to certain reactions. In both cases the state is not the same as those reactions. Anger is not the same as the movements by which an angry organism reacts to the conditions in his environment, just as the state of being electrically charged is not the same as the process of attracting other bodies. In both cases that state sometimes occurs without these events which are observable from outside; they are consequences of the state according to certain laws and may therefore under suitable circumstances be taken as symptoms for it; but they are not identical with it.

The last field to be dealt with is *social science* (in the wide sense indicated before; also called social behavioristics). Here we need no detailed analysis because it is easy to see that every term of this field is reducible to terms of the other fields. The result of any investigation of a group of men or other organisms can be described in terms of the members, their relations to one another and to their environment. Therefore, the conditions for the application of any term can be formulated in terms of psychology, biology, and physics, including the thing-language. Many terms can even be defined on that basis, and the rest is certainly reducible to it.

It is true that some terms which are used in psychology are such that they designate a certain behavior (or disposition to behavior) within a group of a certain kind or a certain attitude toward a group, e.g., 'desirous of ruling,' 'shy,' and others. It may be that for the definition or reduction of a term of this kind some terms of social science describing the group involved are needed. This shows that there is not a clear-cut line between psychology and social science and that in some cases it is not clear whether a term is better assigned to one or to the other field. But such terms are also certainly reducible to those of the thing-language because every term referring to a group of organisms is reducible to terms referring to individual organisms.

The result of our analysis is that the class of observable thing-predicates is a sufficient reduction basis for the whole of the language of science, including the cognitive part of the everyday language.

V. The Problem of the Unity of Laws

The relations between the terms of the various branches of science have been considered. There remains the task of analyzing the relations between the laws. According to our previous consideration, a biological law contains only terms which are reducible to physical terms. Hence there is a common language to which both the biological and the physical laws belong so that they can be logically compared and connected. We can ask whether or not a certain biological law is compatible with the system of physical laws, and whether or not it is derivable from them. But the answer to these questions cannot be inferred from the reducibility of the terms. At the present state of the development of science, it is certainly not possible to derive the biological laws from the physical ones. Some philosophers believe that such a derivation is forever impossible because of the very nature of the two fields. But the proofs attempted so far for this thesis are certainly insufficient. This question is, it seems, the scientific kernel of the problem of vitalism; some recent discussions of this problem are, however, entangled with rather questionable metaphysical issues. The question of derivability itself is, of course, a very serious scientific problem. But it will scarcely be possible to find a solution for it before many more results of experimental investigation are available than we have today. In the meantime the efforts toward derivation of more and more biological laws from physical laws—in the customary formulation: explanation of more and more processes in organisms with the help of physics and chemistry—will be, as it has been, a very fruitful tendency in biological research.

As we have seen before, the fields of psychology and social science are very closely connected with each other. A clear division of the laws of these fields is perhaps still less possible than a division of the terms. If the laws are classified in some way or other, it will be seen that sometimes a psychological law is derivable from those of social science, and sometimes a law of social science from those of psychology. (An example of the first kind is the explanation of the behavior of adults—e.g., in the theories of A. Adler and Freud—by their position within the family or a larger group during childhood; an example of the second kind is the obvious explanation of an increase of the price of a commodity by the reactions of buyers and sellers in the case of a diminished supply.) It is obvious that, at the present time, laws of psychology and social science cannot be derived from those of biology and physics. On the other hand, no scientific reason is known for the assumption that such a derivation should be in principle and forever impossible.

Thus there is at present *no unity of laws*. The construction of one homogeneous system of laws for the whole of science is an aim for the future development of science. This aim cannot be shown to be unattainable. But we do not, of course, know whether it will ever be reached.

On the other hand, there is a *unity of language* in science, viz., a common reduction basis for the terms of all branches of science, this basis consisting of a very narrow and homogeneous class of terms of the physical thing-language. This unity of terms is indeed less far-reaching and effective than the unity of laws would be, but it is a necessary preliminary condition for the unity of laws. We can endeavor to develop science more and more in the direction of a unified system of laws only because we have already at present a unified language. And, in addition, the fact that we have this unity of language is of the greatest practical importance. The practical use of laws consists in making predictions with their help. The important fact is that very often a prediction cannot be based on our knowledge of only one branch of science. For instance, the construction of automobiles will be influenced by a prediction of the presumable number of sales. This number depends upon the satisfaction of the buyers and the economic situation. Hence we have to combine knowledge about the function of the motor, the effect of gases and vibration on the human organism, the ability of persons to learn a certain technique, their willingness to spend so much money for so much service, the development of the general economic situation, etc. This knowledge concerns particular facts and general laws belonging to all the four branches, partly scientific and partly common-sense knowledge. For very many decisions, both in individual and in social life, we need such a prediction based upon a combined knowledge of concrete facts and general laws belonging to different branches of science. If now the terms of different branches had no logical connection between one another, such as is supplied by the homogeneous reduction basis, but were of fundamentally different character, as some philosophers believe, then it would not be possible to connect singular statements and laws of different fields in such a way as to derive predictions from them. Therefore, the unity of the language of science is the basis for the practical application of theoretical knowledge.

Selected Bibliography

I. LOGICAL ANALYSIS

CARNAP, R. *Philosophy and Logical Syntax*. London, 1935. (Elementary.)
———. *Logical Syntax of Language*. London, 1937. (Technical.)

II. REDUCIBILITY

CARNAP, R. "Testability and Meaning," *Philosophy of Science*, Vols. III (1936) and IV (1937).

III. THE UNITY OF THE LANGUAGE OF SCIENCE; PHYSICALISM

Papers by NEURATH and CARNAP, *Erkenntnis* Vol. II (1932); *ibid.*, Vol. III (1933). Translation of one of these papers: CARNAP, *The Unity of Science*. London, 1934. Concerning psychology: papers by SCHLICK, HEMPEL, and CARNAP, *Revue de synthèse*, Vol. X (1935).

Scientific Empiricism

Charles W. Morris

The mind of Leibniz—which was too comprehensive for any single individual of our time—seemed to have diffused itself over the various sections. This struck me particularly in connection with the project of a scientifically philosophical Encyclopedia, advocated by Dr. Otto Neurath. Leibniz, if he were alive, would no doubt write the whole of it, but in our day different sections of it will have to be undertaken by different men. It must, however, be said that in one point of great importance the modern movement surpasses anything imagined by Leibniz or his contemporaries: I mean the combination of empiricism with mathematical method. In science, this combination has existed since the time of Galileo; but in philosophy, until our time, those who were influenced by mathematical method were anti-empirical, and the empiricists had little knowledge of mathematics. Modern science arose from the marriage of mathematics and empiricism; three centuries later the same union is giving birth to a second child, scientific philosophy, which is perhaps destined to as great a career. For it alone can provide the intellectual temper in which it is possible to find a cure for the diseases of the modern world.

These words of Bertrand Russell appeared in 1936 in the first volume of the *Actes du congrès international de philosophie scientifique* and form part of the statement in which he registered his impression of the meeting recorded in these proceedings— the First International Congress for the Unity of Science (Paris, 1935). They suggest in miniature the context in which the present *Encyclopedia* originates and its possible importance. For that reason they merit commentary.

I. Method in Science

The development of the method of experimental science is possibly the most significant intellectual contribution of Western civilization. The development, like most developments, is a long one, and there is no one moment of absolute origin or culmination. Certainly, neither experimentation nor mathematics had to wait for birth until the flowering of Western science. Nevertheless, in this flowering something of undeniable importance took place: the incorporation of mathematics and experimentation within a single method. Previously they had functioned as rival methods: mathematics as one way to get knowledge of nature, and experimental observation as another. Those scientists who advocated the former were one species of "rationalists," and the advocates of the latter were one species of "empiricists"—even the philosophical opposition of rationalism and empiricism was at bottom a reflection of what were to be taken as different scientific methods for knowing nature. Gradually, and in a way that need not at this point be traced in detail, mathematics came to lose the status of an independent method for the study of nature, while at the same time it supplanted the classical logic as a tool of analysis and as the structural basis for the scientific edifice. The important result was a double shift from a metaphysical to a methodological rationalism, and from a loose-jointed empiricism to an empiricism which utilized the techniques and the form of mathematics. Rationalism and empiricism in this way ceased to be rival methods for knowing nature and became complementary components of experimental science with its one observational-hypothetical-deductive-experimental method.

Not only did this method find place for the rational and empirical factors in the knowledge process, but the emphasis upon experimentation as against mere observation meant the breakdown of the radical opposition of theory and practice, for not only is experimentation itself a kind of practice, but it is of such a kind as to open up the possibility of a novel and systematic control of many kinds of natural processes.

The direction of this double movement within science (the incorporation of the mathematical method within the empiricist temper and the breakdown through experimentation of the dichotomy between theory and practice) is discernible in the Hellenistic period and the late Middle Ages, becomes clearly evident in Galileo, and reaches a definite expression in Newton. By the late seventeenth century the great scientists, whatever their philosophical differences, had found a place within scientific method for careful and systematic observation, mathematical theory, and experimental practice. Since that time no fundamental change in the conception of scientific method has taken place, and science has reaped a rich harvest from its attitude of mathematical experimental empiricism.

II. Generalization of Scientific Method

The attainment of a similar attitude in philosophy came more slowly and has not even yet received wide agreement. As Charles S. Peirce remarked, metaphysics has always been the ape of mathematics, and much that passes for metaphysics rests upon views of mathematics now largely discarded by the mathematicians or upon conceptions of philosophic method once made plausible by the then current conceptions of mathematics. But if philosophic rationalists were slow to see the significance of the changed state which mathematics had undergone in science, the philosophical empiricists were equally blind to the significance of this change. The history of empiricism is not an appropriate theme for an introductory article, but it can safely be said that the main energy of empiricists was spent in opposing the a priori rationalists rather than in contributing to or positively assessing the developing sciences. In fact, these philosophical empiricists at many points became entangled themselves in the speculative nets of their opponents, as is evident in their acceptance of the same superficial subjectivistic, individualistic, and atomistic conception of experience which the rationalists had proudly exhibited, ostensibly on the basis of science but actually because it seemed to show the limitations of the appeal to experience and the consequent need for other foundations for knowledge—hence the protracted and exhausting struggle of past empiricisms against the phantoms of solipsism and idealism.

The philosophical empiricists, in the main, had connection with the biological sciences rather than with mathematics or the physical sciences, and the resulting fact that they were unable satisfactorily to account for mathematics, or to make plausible the rational systematization which the physical sciences were in fact attaining, gave strong advantages to the impressive speculative rationalism of their opponents. The traditional empiricism, in addition to freeing itself from inadequate views of experience, had in fact to be supplemented in two interconnected respects before it could claim to be the philosophical equivalent of the method which science had achieved: it had to be able to assimilate and utilize the logical and mathematical tools of rationalism, and it had to be able to account for the intellectual significance of practice. The first expansion was made by logical empiricism; the second by pragmatism.

The union of formal logic and empiricism is linked with the development of symbolic or mathematical logic—another theme which at the proper place is to receive its separate treatment. This modern version of formal logic developed in the hands of philosophic rationalists who were themselves mathematicians. It arose out of the cross-fertilization of the medieval approach to logic in terms of a general theory of signs and the methods of modern mathematics, a union which is first significantly made by Leibniz[1]—the great Leibniz whose ideals of a universal scientific language, a generalized mathematical science, a calculus applicable to all reasoning, and an encyclopedia showing the logical relation between the concepts of all sciences receive a contemporary form of expression in this *Encyclopedia.* The development of this logic in the period since Leibniz has made available a logic adequate to the relational structure of science and mathematics, and has made possible new and powerful techniques of analysis. The point to be noted in the present connection is that in this development logic, like the mathematics which it generalized, came to be regarded as an instrument free from any speculative accretions, and in particular from the metaphysical (or a priori) rationalism which nourished it, and thus became available for use by empiricists—another instance of the fact mentioned in connection with the development of science that the formal disciplines become compatible with empiricism when they pass from the status of rival methods for the knowledge of nature to that of being formal linguistic structures available to the natural sciences as methodological tools. Logic thus rests, as Peirce[2] maintained, on a general theory of signs, formal logic tracing the relations between signs within a language. So conceived, logic deals with the language in which statements about nature are made, and does not itself make statements about the nonlinguistic world. Hence it is not the rival of the empirical knowledge of nature and becomes assimilable to the temper and program of empiricism.[3] The clear development of this view and the utilization of the methods of the new logic within the framework of empiricism are the significant achievements of logical empiricism as represented by the Vienna Circle (Carnap, Frank, Hahn, Neurath, Schlick) and by Russell, Wittgenstein, and Reichenbach.

This union of empiricism and methodological rationalism requires completion by one further step. Languages are developed and used by living beings operating in a world of objects, and show the influence of both the users and the objects. If, as symbolic logic maintains, there are linguistic forms whose validity is not dependent upon nonlinguistic objects, then their validity must be dependent upon the rules of the language in question, and such rules represent habits actually found in operation or set up by deliberate convention. The introduction of such terms as 'convention,' 'decision,' 'procedure,' and 'rule' involves reference to the users of signs in addition to empirical and formal factors. It has been the function of pragmatism to make explicit the instrumental significance of ideas in general and of scientific results and procedure in particular. Thus Dewey interprets even logical rules as empirical generalizations embodying methods of inquiry which have proved particularly successful for the purpose of inference and which have therefore been transformed by the users into principles accepted for the time being as stipulations for the carrying-on of future inquiry. The introduction of pragmatic considerations avoids the extremes both of empiricism and of conventionalism in logical theory while yet doing justice to both. At the same time, in making explicit the instrumental significance of ideas, it was necessary to determine the scientific usage of such terms as 'idea,' 'meaning,' 'self,' and 'experience' in the light of post-Darwinian biology and sociology;[4] and in doing this, in addition to gaining results of intrinsic significance, pragmatism helped free empiricism itself from certain pitfalls which it had encountered during its former development. The emphasis upon the relational and the functional which the biological emphasis brought with it called attention in general to the previously neglected relational and functional aspects of experience, and the realization of the social context in which mind and knowledge arise and operate made manifest the artificiality of the subjectivistic and individualistic concept of experience with which English empiricists had often operated. Pragmatism accordingly not merely brought to the forefront the pragmatical factor which complements and completes the formal and the empirical factors, but it helped to enrich the empiricist tradition through its conception of radical empiricism.

III. The Viewpoint of Scientific Empiricism

The resulting comprehensive point of view, embracing at once radical empiricism, methodological rationalism, and critical pragmatism, may appropriately be called scientific empiricism.[5] It is the generalized analogue of the point of view which has been effective in science for some centuries. It is willing and able to admit into the scope of its considerations everything involved in the scientific enterprise as such, together with the implications of this enterprise for other human interests. It is an empiricism genuinely oriented around the methods and the results of science and not dependent upon some questionable psychological theory as to the "mental" nature of experience. It is an empiricism which, because of this orientation and the use of powerful tools of logical analysis, has become positive in temper and co-operative in attitude and is no longer condemned to the negative skeptical task of showing defects in the methods and results of its opponents.

Such a point of view, characteristic in the main of this *Encyclopedia* (though, of course, not binding on its contributors)

signalizes the widest possible generalization of scientific method. The field of application of this point of view is science itself. In analogy with certain other uses of the prefix 'meta' which are current today, we may introduce the term 'metascience' as a synonym for 'the science of science.'

The attempt to make the scientific enterprise as a whole an object of scientific investigation—i.e., to develop metascience—requires consideration of the three factors involved in this enterprise. Since these factors correspond to the three components of scientific empiricism, this point of view proves to be appropriate to the task. In its most tangible form science exists as a body of written characters and spoken words. It is possible to investigate this linguistic residue of the scientists' activity purely formally, without reference to the relation of these marks and sounds to other objects or to the activity of which they are the residues. There is no mystery as to how such "abstraction" is possible: from a linguistic point of view, to abstract from some of the properties or relations of an object is simply not to talk about them. This type of investigation may, in one sense of the term, be called logical analysis; because of the ambiguity of the term 'logical,' it may be preferable to call it, in the spirit of Carnap, the syntactical investigation of the language of science. Such investigation studies the structure of scientific language: the relation between the terms and the sentences of the same and different sciences. The degree of unity or disunity of science reveals itself here in the degree to which the sciences have or can have a common linguistic structure.

But the signs which constitute scientific treatises have, to some extent at least, a correlation with objects, and the investigation of all aspects of this relation constitutes a second task of metascience. Here belong all the problems as to the·nature of this correlation and the analysis of the specific situations under which scientific terms and sentences are applicable. This may be called, in the spirit of the Polish logicians, the semantical investigation of the language of science. The unity of science is here no longer a purely formal unity, for the unity or disunity of the scientific language corresponds to some extent to the semantical relation or lack of relation of the various terms of the sciences—and so to the relations of objects.

A third concern of metascience arises from the fact that the signs which constitute the language of science are parts and products of the activity of scientists. The study of the relation of signs to scientists may be called, in the spirit of pragmatism, the pragmatical investigation of the language of science. Here belong the problems as to how the scientist operates, the connection of science as a social institution with other social institutions, and the relation of scientific activity to other activities. The question as to the unity of science is now the question as to the unity of procedures, purposes, and effects of the various sciences.

Science, as a body of signs with certain specific relations to one another, to objects, and to practice, is at once a language, a knowledge of objects, and a type of activity; the interrelated study of the syntactics, semantics, and pragmatics of the language of science in turn constitutes metascience—the science of science. Discussion of the specific signs of science must be carried on in terms of some theory of signs, and so semiotic, as the science of signs, occupies an important place in the program—indeed, the study of the actual language of science is an instance of applied semiotic. Since the institution of science is a social institution, certain of its features, as Auguste Comte realized, reveal themselves especially well in historical perspective, so that the history of science is of abiding importance within the study of science.

The elaboration of the syntactics, semantics, and pragmatics of science may rightly be regarded as the natural extension and completion of the scientific enterprise itself. It is believed that this program will appeal to scientists and will prove of importance in the development and assessment of science. It is inevitable that, in seeking for its greatest unification, science will make itself an object of scientific investigation. The fulfilment of these related tasks cannot be left to chance if science is to grow to full stature: they must be taken in hand by those familiar with the results and spirit of science—that is, scientists; they must be encouraged by scientific institutions, foundations, and associations; they must be incorporated in educational programs for the training of scientists. It is true that the individual experimenter may not be directly helped in the carrying-out of a particular experiment—though even here he can be justly held as responsible for the careful use of his linguistic tools as he is held for the careful handling of balance, microscope, or telescope. But science has never been content with isolated facts. It has ever pressed on to larger systematizations. This process will continue as long as scientific progress continues. And in this striving for the widest systematization, science inevitably has to study itself and incorporate the results of such study in its systematization. The study of science is not an intellectual luxury for scientists; it is a movement within science itself.

But if the study of science which is here contemplated is science (and so not a domain over and above science), it may equally well be regarded as philosophy. For the three-faceted point of view of scientific empiricism, and of metascience which results from its application to science, can be regarded as embracing the contemporary empiricist equivalents of the traditional fields of philosophy (logic, metaphysics, and theory of value). Logic is grounded on semiotic; metaphysics is replaced by sign analysis and unified science; and axiology becomes the scientific study of values and judgments of values. Within the general orientation of scientific empiricism, science and philosophy relinquish all claims to the possession of distinct methods or subject matters and merge their efforts within a common task: the erection, the analysis, and the assessment of unified science. Such community of effort is an ancient ideal and in part an ancient practice; what is novel today is the scale upon which fruitful co-operation is possible. Science rounds itself out humanly and scientifically in this process, and the relevance and significance which philosophy sometimes had are again regained.

IV. Science and Practice

What has been called the syntactics and the semantics of the language of science will receive immediate and extended treatment in the pages of this *Encyclopedia*. But since this will not be true to the same degree in the case of the pragmatics of the language of science, it is well to pause at this point in order both to see certain implications of the standpoint of scientific empiricism and to avoid certain possible misunderstandings.

The very statement that science walks on three legs of theory, observation, and practice will call out opposition in a certain type of mind—the statement is more frequent with 'practice' omitted. The word 'practice' is admittedly equivocal. The activity which gives rise to the sentences of science is, like any other systematic activity, a practice proceeding in terms of rules or canons. Further, the confirmation of every proposition always involves some instrument, whether this be simply the scientist himself or in addition such instruments as those involved in experimentation—and methodologically there is no important distinction between the two cases. In this (theoretically the most important) sense, all empirical science involves experimentation, and experimentation is an activity, a practice. In the third place, science is part of the practice of the community in which it is an institution, ministering—however indirectly—to the needs of the community and being affected—and very directly—in its development by the community of social institutions of which it is a part.

It is clear that any adequate account of science must take account of these psychological, methodological, and sociological aspects of scientific practice. The present work recognizes that fact. But it should also be clear that 'practice' in all three senses of the word is not an unessential factor added to the theoretical and empirical aspects of the scientific enterprise but an equally essential factor, since, at the minimum, 'confirmation' is a concept which contains irreducible pragmatical features. If this is so, it would be well for scientists to become fully aware of this factor of practice and, in becoming aware of it, to assume the entailed responsibilities.

The same point may be given an alternative formulation in terms of the notion of value. It is often said that science gives only "facts" and has nothing to do with "values." There is an element of truth in such a statement, since the pragmatical factor in language cannot be reduced at a given moment to the empirical, and since life is more than knowledge. But this is hardly the usual import of this statement, which is often made against a background which involves a sharp distinction between the natural sciences and the socio-humanistic sciences. The detailed study of the actual relations must be left for other writers. Nevertheless, it seems clear that, while a program which stresses the unity of science can admit of whatever diversity is in fact found in the various sciences (for unity does not exclude differentiation), it must naturally be skeptical of any such wholesale cleavage. Later treatments of these matters will maintain in harmony with the empiricist attitude that there is no unbridgeable gap within science between the procedures and subject matters of the natural sciences and the socio-humanistic sciences. All knowledge forms in principle one unified whole, and there exists no system of knowledge (such as metaphysics, aesthetics, ethics, religion) alongside of or superior to unified science.

This statement should not be misunderstood. Science is an activity eventuating in a certain sort of product having certain effects; but science is not the only activity of that sort. Art, morality, piety, play, work, and war are also activities with characteristic products and effects, and they must in no sense be confused with the sciences of such activities (aesthetics, ethics, science of religion, etc.). As activities they are co-ordinate with science considered as an activity, but the sciences of these activities fall within the field of unified science. In the first case, they are alternatives to the scientific attitude; in the second case, they are part of science. But an alternative is not necessarily a rival. Indeed, once science is distinguished in terms of its specific goal (reliable knowledge), then science not only does not, in general, clash with other activities but may itself further any other activity in so far as it can be furthered by knowledge.

In this respect science is the most practical of human activities. It is in opposition only to such activities as claim to usurp its own cognitive goal or which wither and die when the light of scientific investigation is turned upon them; it is at once both co-ordinate with and instrumental to the realization of the purposes of most other activities.

It is because of this relation of scientific activity to other activities that the scientific habit of mind and scientific results are of such potential promise in society at large and education in particular.[6] For this habit of mind is the best guaranty of an objective consideration of the multiplicity of factors which enter into the complex problems of contemporary man.

V. Scientific Empiricism and Encyclopedism

The standpoint of scientific empiricism is thus ample enough to embrace and to integrate the various factors which must be taken into account in an *Encyclopedia* devoted to unified science —i.e., to the scientific study of the scientific enterprise in its totality. The theory of signs gives the general background for the consideration of the language of science. The investigation of this language breaks up into the distinguishable but interrelated investigations of the syntactics, semantics, and pragmatics of the language of science. In this way, and on a comprehensive scale, science is made an object of scientific investigation; meta-science appears both as a tool for, and as an element within unified science. The attitude of scientific empiricism is simultaneously congenial to the temper of the rationalist, the empiricist, and the pragmatist, and provides the corrective to the one-sidedness of these attitudes when held in isolation. It is from the standpoint of method the complement of encyclopedism, since while it accepts the encyclopedia as the necessary form of human knowledge it yet recognizes that science strives for the greatest degree of systematization compatible with its continual growth.

This *Encyclopedia*, reflecting this inclusive standpoint, rightfully sounds the roll call of those distinguished logicians, scientists, and empiricists whom the traditional history of ideas has so shamefully neglected. But basically it aims to present through extensive co-operation the existing status and the unrealized possibilities for the integration of science. Its existence signalizes the union of scientific and philosophic traditions in a common task. The *Encyclopedia* presents a contemporary version of the ancient encyclopedic ideal of Aristotle, the Scholastics, Leibniz, the Encyclopedists, and Comte. It wishes to give satisfaction to the pervasive human interest in intellectual unity, but its common point of view permits divergences and differences in emphasis and does not blur the fact that an inseparable feature of the institution of science is constant growth. It aims to provide a basis for co-operative activity and not a panacea.

NOTES

1. See the splendid work of Louis Couturat, *La Logique de Leibniz* (Paris, 1901).

2. See especially *Collected Papers* (Cambridge, Mass., 1932), Vol. II.

3. This point was clearly elaborated by H. Hahn (see *Erkenntnis*, II [1931], 135–41; *Logik, Mathematik und Naturerkennen* [Vienna, 1933]). It may be remarked that from this point of view Leibniz' plans for a universal mathematics, a calculus of reasoning, a general characteristic, and a unified science expressed in the form of an encyclopedia all remain valid when interpreted as logical rather than as metaphysical doctrines. Leibniz' rationalistic metaphysics, which came from the simple conversion of formal logic into a metaphysics through the neglect of the criterion of the empirically meaningful, is, in terms of the present conception of the relation of logic to empiricism, no longer the necessary cosmological corollary of his logical doctrines.

4. The writings of George H. Mead are of importance in this connection, especially *Mind, Self, and Society* (Chicago, 1934).

5. This term and certain of the more general features of the point of view are characterized in the author's pamphlet, *Logical Positivism, Pragmatism, and Scientific Empiricism* (Paris: Hermann & Cie, 1937).

6. John Dewey, in particular, has devoted his life to the formulation and assessment of the social, cultural, and educational implications of the scientific habit of mind. See *Philosophy and Civilization* (New York, 1931) and his forthcoming work, *Logic: The Theory of Inquiry.*

Foundations of the Theory of Signs

Charles Morris

Foundations of the Theory of Signs

Contents:

I. Introduction

                                            PAGE

  1. Semiotic and Science 79

II. Semiosis and Semiotic

  2. The Nature of a Sign 81

  3. Dimensions and Levels of Semiosis 84

  4. Language 88

III. Syntactics

  5. The Formal Conception of Language . . . 91

  6. Linguistic Structure 94

IV. Semantics

  7. The Semantical Dimension of Semiosis . . 99

  8. Linguistic and Nonlinguistic Structures . . 104

V. Pragmatics

  9. The Pragmatical Dimension of Semiosis . . 107

  10. Individual and Social Factors in Semiosis . . 112

  11. Pragmatic Use and Abuse of Signs 116

VI. The Unity of Semiotic

  12. Meaning 121

  13. Universals and Universality 126

  14. Interrelation of the Semiotical Sciences . . 130

VII. Problems and Applications

  15. Unification of the Semiotical Sciences . . . 132

  16. Semiotic as Organon of the Sciences . . . 134

  17. Humanistic Implications of Semiotic . . . 135

Selected Bibliography 137

Foundations of the Theory of Signs

Charles Morris

Nemo autem vereri debet ne characterum contemplatio nos a rebus abducat, imo contra ad intima rerum ducet.—Gottfried Leibniz

I. Introduction

1. Semiotic and Science

Men are the dominant sign-using animals. Animals other than man do, of course, respond to certain things as signs of something else, but such signs do not attain the complexity and elaboration which is found in human speech, writing, art, testing devices, medical diagnosis, and signaling instruments. Science and signs are inseparately interconnected, since science both presents men with more reliable signs and embodies its results in systems of signs. Human civilization is dependent upon signs and systems of signs, and the human mind is inseparable from the functioning of signs—if indeed mentality is not to be identified with such functioning.

It is doubtful if signs have ever before been so vigorously studied by so many persons and from so many points of view. The army of investigators includes linguists, logicians, philosophers, psychologists, biologists, anthropologists, psychopathologists, aestheticians, and sociologists. There is lacking, however, a theoretical structure simple in outline and yet comprehensive enough to embrace the results obtained from different points of view and to unite them into a unified and consistent whole. It is the purpose of the present study to suggest this unifying point of view and to sketch the contours of the science of signs. This can be done only in a fragmentary fashion, partly because of the limitations of space, partly because of the undeveloped state of the science itself, but mainly because of the purpose which such a study aims to serve by its inclusion in this *Encyclopedia*.

Semiotic has a double relation to the sciences: it is both a science among the sciences and an instrument of the sciences. The significance of semiotic as a science lies in the fact that it is a step in the unification of science, since it supplies the foundations for any special science of signs, such as linguistics, logic, mathematics, rhetoric, and (to some extent at least) aesthetics. The concept of sign may prove to be of importance in the unification of the social, psychological, and humanistic sciences in so far as these are distinguished from the physical and biological sciences. And since it will be shown that signs are simply the objects studied by the biological and physical sciences related in certain complex functional processes, any such unification of the formal sciences on the one hand, and the social, psychological, and humanistic sciences on the other, would provide relevant material for the unification of these two sets of sciences with the physical and biological sciences. Semiotic may thus be of importance in a program for the unification of science, though the exact nature and extent of this importance is yet to be determined.

But if semiotic is a science co-ordinate with the other sciences, studying things or the properties of things in their function of serving as signs, it is also the instrument of all sciences, since every science makes use of and expresses its results in terms of signs. Hence metascience (the science of science) must use semiotic as an organon. It was noticed in the essay "Scientific Empiricism" (Vol. I, No. 1) that it is possible to include without remainder the study of science under the study of the language of science, since the study of that language involves not merely the study of its formal structure but its relation to objects designated and to the persons who use it. From this point of view the entire *Encyclopedia*, as a scientific study of science, is a study of the language of science. But since nothing can be studied without signs denoting the objects in the field to be studied, a study of the language of science must make use of signs referring to signs—and semiotic must supply the relevant signs and principles for carrying on this study. Semiotic supplies a general language applicable to any special language or sign, and so applicable to the language of science and specific signs which are used in science.

The interest in presenting semiotic as a science and as part of the unification of science must here be restricted by the practical motive of carrying the analysis only so far and in such directions as to supply a tool for the work of the *Encyclopedia*, i.e., to supply a language in which to talk about, and in so doing to improve, the language of science. Other studies would be necessary to show concretely the results of sign analysis applied to special sciences and the general significance for the unification of science of this type of analysis. But even without detailed documentation it has become clear to many persons today that man—including scientific man—must free himself from the web of words which he has spun and that language— including scientific language—is greatly in need of purification, simplification, and systematization. The theory of signs is a useful instrument for such debabelization.

II. Semiosis and Semiotic

2. The Nature of a Sign

The process in which something functions as a sign may be called *semiosis*. This process, in a tradition which goes back to the Greeks, has commonly been regarded as involving three (or four) factors: that which acts as a sign, that which the sign refers to, and that effect on some interpreter in virtue of which the thing in question is a sign to that interpreter. These three components in semiosis may be called, respectively, the *sign vehicle*, the *designatum*, and the *interpretant;* the *interpreter* may be included as a fourth factor. These terms make explicit the factors left undesignated in the common statement that a sign refers to something for someone.

A dog responds by the type of behavior (I) involved in the hunting of chipmunks (D) to a certain sound (S); a traveler prepares himself to deal appropriately (I) with the geographical region (D) in virtue of the letter (S) received from a friend. In such cases S is the sign vehicle (and a sign in virtue of its functioning), D the designatum, and I the interpretant of the interpreter. The most effective characterization of a sign is the following: S is a sign of D for I to the degree that I takes account of D in virtue of the presence of S. Thus in semiosis something takes account of something else mediately, i.e., by means of a third something. Semiosis is accordingly a mediated-taking-account-of. The mediators are *sign vehicles;* the takings-account-of are *interpretants;* the agents of the process are *interpreters;* what is taken account of are *designata.* There are several comments to be made about this formulation.

It should be clear that the terms 'sign,' 'designatum,' 'interpretant,' and 'interpreter' involve one another, since they are simply ways of referring to aspects of the process of semiosis. Objects need not be referred to by signs, but there are no designata unless there is such reference; something is a sign only because it is interpreted as a sign of something by some interpreter; a taking-account-of-something is an interpretant only in so far as it is evoked by something functioning as a sign; an object is an interpreter only as it mediately takes account of something. The properties of being a sign, a designatum, an interpreter, or an interpretant are relational properties which things take on by participating in the functional process of semiosis. Semiotic, then, is not concerned with the study of a particular kind of object, but with ordinary objects in so far (and only in so far) as they participate in semiosis. The importance of this point will become progressively clearer.

Signs which refer to the same object need not have the same designata, since that which is taken account of in the object may differ for various interpreters. A sign of an object may, at one theoretical extreme, simply turn the interpreter of the sign upon the object, while at the other extreme it would allow the interpreter to take account of all the characteristics of the object in question in the absence of the object itself. There is thus a potential sign continuum in which with respect to every object or situation all degrees of semiosis may be expressed, and the question as to what the designatum of a sign is in any given situation is the question of what characteristics of the object or situation are actually taken account of in virtue of the presence of the sign vehicle alone.

A sign must have a designatum; yet obviously every sign does not, in fact, refer to an actual existent object. The difficulties which these statements may occasion are only apparent difficulties and need no introduction of a metaphysical realm of "subsistence" for their solution. Since 'designatum' is a semiotical term, there cannot be designata without semiosis—but there can be objects without there being semiosis. The designatum of a sign is the kind of object which the sign applies to, i.e., the objects with the properties which the interpreter takes account of through the presence of the sign vehicle. And the taking-account-of may occur without there actually being objects or situations with the characteristics taken account of. This is true even in the case of pointing: one can for certain purposes point without pointing to anything. No contradiction arises in saying that every sign has a designatum but not every sign refers to an actual existent. Where what is referred to actually exists as referred to the object of reference is a *denotatum*. It thus becomes clear that, while every sign has a designatum, not every sign has a denotatum. A designatum is not a thing, but a kind of object or class of objects—and a class may have many members, or one member, or no members. The denotata are the members of the class. This distinction makes explicable the fact that one may reach in the icebox for an apple that is not there and make preparations for living on an island that may never have existed or has long since disappeared beneath the sea.

As a last comment on the definition of sign, it should be noted that the general theory of signs need not commit itself to any specific theory of what is involved in taking account of something through the use of a sign. Indeed, it may be possible to take 'mediated-taking-account-of' as the single primitive term for the axiomatic development of semiotic. Nevertheless, the account which has been given lends itself to treatment from the point of view of behavioristics, and this point of view will be adopted in what follows. This interpretation of the definition of sign is not, however, necessary. It is adopted here because such a point of view has in some form or other (though not in the form of Watsonian behaviorism) become widespread among psychologists, and because many of the difficulties which the history of semiotic reveals seem to be due to the fact that through most of its history semiotic linked itself with the faculty and introspective psychologies. From the point of view of behavioristics, to take account of D by the presence of S involves responding to D in virtue of a response to S. As will be made clear later, it is not necessary to deny "private experiences" of the process of semiosis or of other processes, but it is necessary from the standpoint of behavioristics to deny that such experiences are of central importance or that the fact of their existence makes the objective study of semiosis (and hence of sign, designatum, and interpretant) impossible or even incomplete.

3. Dimensions and Levels of Semiosis

In terms of the three correlates (sign vehicle, designatum, interpreter) of the triadic relation of semiosis, a number of other dyadic relations may be abstracted for study. One may study the relations of signs to the objects to which the signs are applicable. This relation will be called the *semantical dimension of semiosis*, symbolized by the sign 'D_{sem}'; the study of this dimension will be called *semantics*. Or the subject of study may be the relation of signs to interpreters. This relation will be called the *pragmatical dimension of semiosis*, symbolized by 'D_p,' and the study of this dimension will be named *pragmatics*.

One important relation of signs has not yet been introduced: the formal relation of signs to one another. This relationship was not, in the preceding account, explicitly incorporated in the definition of 'sign,' since current usage would not seem to eliminate the possibility of applying the term 'sign' to something which was not a member of a system of signs—such possibilities are suggested by the sign aspects of perception and by various apparently isolated mnemonic and signaling devices.

Nevertheless, the interpretation of these cases is not perfectly clear, and it is very difficult to be sure that there is such a thing as an isolated sign. Certainly, potentially, if not actually, every sign has relations to other signs, for what it is that the sign prepares the interpreter to take account of can only be *stated* in terms of other signs. It is true that this statement need not be made, but it is always in principle capable of being made, and when made relates the sign in question to other signs. Since most signs are clearly related to other signs, since many apparent cases of isolated signs prove on analysis not to be such, and since all signs are potentially if not actually related to other signs, it is well to make a third dimension of semiosis co-ordinate with the other two which have been mentioned. This third dimension will be called the *syntactical dimension of semiosis*, symbolized by 'D_{syn},' and the study of this dimension will be named *syntactics*.

It will be convenient to have special terms to designate certain of the relations of signs to signs, to objects, and to interpreters. '*Implicates*' will be restricted to D_{syn}, '*designates*' and '*denotes*' to D_{sem}, and '*expresses*' to D_p. The word 'table' implicates (but does *not* designate) 'furniture with a horizontal top on which things may be placed,' designates a certain kind of object (furniture with a horizontal top on which things may be placed), denotes the objects to which it is applicable, and expresses its interpreter. In any given case certain of the dimensions may actually or practically vanish: a sign may not have syntactical relations to other signs and so its actual implication becomes null; or it may have implication and yet denote no object; or it may have implication and yet no actual interpreter and so no expression—as in the case of a word in a dead language. Even in such possible cases the terms chosen are convenient to refer to the fact that certain of the possible relations remain unrealized.

It is very important to distinguish between the relations which a given sign sustains and the signs used in talking about such relations—the full recognition of this is perhaps the most important general practical application of semiotic. The functioning of signs is, in general, a way in which certain existences take account of other existences through an intermediate class of existences. But there are levels of this process which must be carefully distinguished if the greatest confusion is not to result. Semiotic as the science of semiosis is as distinct from semiosis as is any science from its subject matter. If x so functions that y takes account of z through x, then we may say that x is a sign, and that x designates z, etc.; but here 'sign,' and 'designates' are signs in a higher order of semiosis referring to the original and lower-level process of semiosis. What is now designated is a certain relation of x and z and not z alone; x is designated, z is designated, and a relation is designated such that x becomes a sign and z a designatum. Designation may therefore occur at various levels, and correspondingly there are various levels of designata; 'designation' reveals itself to be a sign within semiotic (and specifically within semantics), since it is a sign used in referring to signs.

Semiotic as a science makes use of special signs to state facts about signs; it is a language to talk about signs. Semiotic has the three subordinate branches of syntactics, semantics, and pragmatics, dealing, respectively, with the syntactical, the semantical, and the pragmatical dimensions of semiosis. Each of these subordinate sciences will need its own special terms; as previously used 'implicates' is a term of syntactics, 'designates' and 'denotes' are terms of semantics, and 'expresses' is a term of pragmatics. And since the various dimensions are only aspects of a unitary process, there will be certain relations between the terms in the various branches, and distinctive signs will be necessary to characterize these relations and so the process of semiosis as a whole. 'Sign' itself is a strictly semiotical term, not being definable either within syntactics, semantics, or pragmatics alone; only in the wider use of 'semiotical' can it be said that all the terms in these disciplines are semiotical terms.

It is possible to attempt to systematize the entire set of terms and propositions dealing with signs. In principle, semiotic could be presented as a deductive system, with undefined terms and primitive sentences which allow the deduction of other sentences as theorems. But though this is the form of presentation to which science strives, and though the fact that semiotic deals exclusively with relations makes it peculiarly fit for treatment by the new logic of relations, yet it is neither advisable nor possible in the present monograph to attempt this type of exposition. It is true that much has been accomplished in the general analysis of sign relations by the formalists, the empiricists, and the pragmatists, but the results which have been attained seem to be but a small part of what may be expected; the preliminary systematization in the component fields has hardly begun. For such reasons, as well as because of the introductory function of this monograph, it has not seemed advisable to attempt a formalization of semiotic which goes much beyond the existing status of the subject, and which might obscure the role which semiotic is fitted to play in the erection of unified science.

Such a development remains, however, as the goal. Were it obtained it would constitute what might be called *pure semiotic*, with the component branches of pure syntactics, pure semantics, and pure pragmatics. Here would be elaborated in systematic form the metalanguage in terms of which all sign situations would be discussed. The application of this language to concrete instances of signs might then be called *descriptive semiotic* (or syntactics, semantics, or pragmatics as the case may be). In this sense the present *Encyclopedia*, in so far as it deals with the language of science, is an especially important case of descriptive semiotic, the treatment of the structure of that language falling under descriptive syntactics, the treatment of the relation of that language to existential situations falling under descriptive semantics, and the consideration of the relation of that language to its builders and users being an instance of descriptive pragmatics. The *Encyclopedia* as a whole, from the point of view expressed in this monograph, falls within the province of pure and descriptive semiotic.

4. Language

The preceding account is applicable to all signs, however simple or complex. Hence it is applicable to languages as a particular kind of sign system. The term 'language,' in common with most terms which have to do with signs, is ambiguous, since its characterization may be given in terms of the various dimensions. Thus the formalist is inclined to consider any axiomatic system as a language, regardless of whether there are any things which it denotes, or whether the system is actually used by any group of interpreters; the empiricist is inclined to stress the necessity of the relation of signs to objects which they denote and whose properties they truly state; the pragmatist is inclined to regard a language as a type of communicative activity, social in origin and nature, by which members of a social group are able to meet more satisfactorily their individual and common needs. The advantage of the three-dimensional analysis is that the validity of all these points of view can be recognized, since they refer to three aspects of one and the same phenomenon; where convenient the type of consideration (and hence of abstraction) can be indicated by 'L_{syn},' 'L_{sem},' or 'L_p.' It has already been noted that a sign may not denote any actual objects (i.e., have no denotatum) or may not have an actual interpreter. Similarly, there may be languages, as a kind of sign complex, which at a given time are applied to nothing, and which have a single interpreter or even no interpreter, just as an unoccupied building may be called a house. It is not possible, however, to have a language if the set of signs have no syntactical dimension, for it is not customary to call a single sign a language. Even this case is instructive, for in terms of the view expressed (namely, that potentially every sign has syntactical relations to those signs which would state its designatum, that is, the kind of situation to which it is applicable) even an isolated sign is potentially a linguistic sign. It could also be said that an isolated sign has certain relations to itself, and so a syntactical dimension, or that having a null syntactical dimension is only a special case of having a syntactical dimension. These possibilities are important in showing the degree of independence of the various dimensions and consequently of L_{syn}, L_{sem}, and L_p. They also show that there is no absolute cleft between single signs, sentential signs, and languages—a point which Peirce especially stressed.

A language, then, as a system of interconnected signs, has a syntactical structure of such a sort that among its permissible sign combinations some can function as statements, and sign vehicles of such a sort that they can be common to a number of interpreters. The syntactical, semantical, and pragmatical features of this characterization of language will become clearer when the respective branches of semiotic are considered. It will also become clear that just as an individual sign is completely characterized by giving its relation to other signs, objects, and its users, so a language is completely characterized by giving what will later be called the syntactical, semantical, and pragmatical rules governing the sign vehicles. For the moment it should be noted that the present characterization of language is a strictly semiotical one, involving reference to all three dimensions; much confusion will be avoided if it is recognized that the word 'language' is often used to designate some aspect of what is language in the full sense. The simple formula, $L = L_{syn} + L_{sem} + L_p$, helps to clarify the situation.

Languages may be of various degrees of richness in the complexity of their structure, the range of things they designate, and the purposes for which they are adequate. Such natural languages as English, French, German, etc., are in these respects the richest languages and have been called *universal languages*, since in them everything can be represented. This very richness may, however, be a disadvantage for the realization of certain purposes. In the universal languages it is often very difficult to know within which dimension a certain sign is predominantly functioning, and the various levels of symbolic reference are not clearly indicated. Such languages are therefore ambiguous and give rise to explicit contradictions— facts which in some connections (but not in all!) are disadvantageous. The very devices which aid scientific clarity may weaken the potentialities for the aesthetic use of signs, and vice versa. Because of such considerations it is not surprising that men have developed certain special and restricted languages for the better accomplishment of certain purposes: mathematics and formal logic for the exhibition of syntactical structure, empirical science for more accurate description and prediction of natural processes, the fine and applied arts for the indication and control of what men have cherished. The everyday language is especially weak in devices to talk about language, and it is the task of semiotic to supply a language to meet this need. For the accomplishment of their own ends these special languages may stress certain of the dimensions of sign-functioning more than others; nevertheless, the other dimensions are seldom if ever completely absent, and such languages may be regarded as special cases falling under the full semiotical characterization of language which has been suggested.

The general origin of systems of interconnected signs is not difficult to explain. Sign vehicles as natural existences share in the connectedness of extraorganic and intraorganic processes. Spoken and sung words are literally parts of organic responses, while writing, painting, music, and signals are the immediate products of behavior. In the case of signs drawn from materials other than behavior or the products of behavior—as in the sign factors in perception—the signs become interconnected because the sign vehicles are interconnected. Thunder becomes a sign of lightning and lightning a sign of danger just because thunder and lightning and danger are, in fact, interconnected in specific ways. If w expects x on the presence of y, and z on the presence of x, the interconnectedness of the two expectations makes it very natural for w to expect z on the presence of y. From the interconnectedness of events on the one hand, and the interconnectedness of actions on the other, signs become interconnected, and language as a system of signs arises. That the syntactical structure of language is, in general, a function both of objective events and of behavior, and not of either alone, is a thesis which may be called *the dual control of linguistic structure*. This thesis will receive elaboration later, but it should be already evident that it gives a way of avoiding the extremes of both conventionalism and the traditional empiricism in accounting for linguistic structure. For the reasons given, sets of signs tend to become systems of signs; this is as true in the case of perceptual signs, gestures, musical tones, and painting as it is in the case of speech and writing. In some cases the systematization is relatively loose and variable and may include subsystems of various degrees of organization and interconnectedness; in others it is relatively close and stable, as in the case of mathematical and scientific languages. Given such sign structures, it is possible to subject them to a three-dimensional analysis, investigating their structure, their relation to what they denote, and their relations to their interpreters. This will now be done in general terms, discussing in turn the syntactics, semantics, and pragmatics of language, but keeping in mind throughout the relation of each dimension, and so each field of semiotic, to the others. Later, after making use of the abstractions involved in this treatment, we will specifically stress the unity of semiotic.

III. Syntactics

5. The Formal Conception of Language

Syntactics, as the study of the syntactical relations of signs to one another in abstraction from the relations of signs to objects or to interpreters, is the best developed of all the branches of semiotic. A great deal of the work in linguistics proper has been done from this point of view, though often unconsciously and with many confusions. Logicians have from the earliest times been concerned with inference, and this, though historically overlaid with many other considerations, involves the study of the relations between certain combinations of signs within a language. Especially important has been the early presentation by the Greeks of mathematics in the form of a deductive or axiomatic system; this has kept constantly before men's attention the pattern of a closely knit system of signs such that by means of operations upon certain initial sets all the other sets of signs are obtained. Such formal systems presented the material whose considerations made inevitable the development of syntactics. It was in Leibniz the mathematician that linguistic, logical, and mathematical considerations jointly led to the conception of a general formal art (*speciosa generalis*) which included the general characteristic art (*ars characteristica*), essentially a theory and art of so forming signs that all consequences of the corresponding "ideas" could be drawn by a consideration of the signs alone, and the general combinatory art (*ars combinatoria*), a general calculus giving a universal formal method of drawing the consequences from signs. This unification and generalization of mathematical form and method has received since Leibniz' time a remarkable extension in symbolic logic, through the efforts of Boole, Frege, Peano, Peirce, Russell, Whitehead, and others, while the theory of such syntactical relations has received its most elaborate contemporary development in the logical syntax of Carnap. For present purposes only the most general aspect of this point of view need be mentioned, especially since Carnap treats this question in Volume I, Numbers 1 and 3.

Logical syntax deliberately neglects what has here been called the semantical and the pragmatical dimensions of semiosis to concentrate upon the logico-grammatical structure of language, i.e., upon the syntactical dimension of semiosis. In this type of consideration a "language" (i.e., L_{syn}) becomes any set of things related in accordance with two classes of rules: *formation rules*, which determine permissible independent combinations of members of the set (such combinations being called sentences), and *transformation rules*, which determine the sentences which can be obtained from other sentences. These may be brought together under the term '*syntactical rule*.' Syntactics is, then, the consideration of signs and sign combinations in so far as they are subject to syntactical rules. It is not interested in the individual properties of the sign vehicles or in any of their relations except syntactical ones, i.e., relations determined by syntactical rules.

Investigated from this point of view, languages have proved to be unexpectedly complex, and the point of view unexpectedly fruitful. It has been possible accurately to characterize primitive, analytic, contradictory, and synthetic sentences, as well as demonstration and derivation. Without deserting the formal point of view, it has proved possible to distinguish logical and descriptive signs, to define synonymous signs and equipollent sentences, to characterize the content of a sentence, to deal with the logical paradoxes, to classify certain types of expressions, and to clarify the modal expressions of necessity, possibility, and impossibility. These and many other results have been partially systematized in the form of a language, and most of the terms of logical syntax may be defined in terms of the notion of consequence. The result is that there is today available a more precise language for talking about the formal dimension of languages than has ever before existed. Logical syntax has given results of high intrinsic interest and furnished a powerful analytical tool; it will be used extensively in the analysis of the language of science in this *Encyclopedia*.

Our present interest, however, is solely with the relation of logical syntax to semiotic. It is evident that it falls under syntactics; it has indeed suggested this name. All the results of logical syntax are assimilable by syntactics. Further, it is without doubt the most highly developed part of syntactics, and so of semiotic. In its spirit and method it has much to contribute to semantics and pragmatics, and there is evidence that its influence is at work in these fields.

Many of its specific results have analogues in the other branches of semiotic. As an illustration let us use the term '*thing-sentence*,' to designate any sentence whose designatum does not include signs; such a sentence is about things and may be studied by semiotic. On this usage none of the sentences of the semiotical languages are thing-sentences. Now Carnap has made clear the fact that many sentences which are apparently thing-sentences, and so about objects which are not signs, turn out under analysis to be pseudo thing-sentences which must be interpreted as syntactical statements about language. But in analogy to these quasi-syntactical sentences there are corresponding quasi-semantical and quasi-pragmatical sentences which appear to be thing-sentences but which must be interpreted in terms of the relation of signs to designata or the relation of signs to interpreters.

Syntactics is in some respects easier to develop than its coordinate fields, since it is somewhat easier, especially in the case of written signs, to study the relations of signs to one another as determined by rule than it is to characterize the existential situations under which certain signs are employed or what goes on in the interpreter when a sign is functioning. For this reason the isolation of certain distinctions by syntactical investigation gives a clue for seeking their analogues in semantical and pragmatical investigations.

In spite of the importance thus ascribed to logical syntax, it cannot be equated with syntactics as a whole. For it (as the term 'sentence' shows) has limited its investigation of syntactical structure to the type of sign combinations which are dominant in science, namely, those combinations which from a semantical point of view are called statements, or those combinations used in the transformation of such combinations. Thus on Carnap's usage commands are not sentences, and many lines of verse would not be sentences. 'Sentence' is not, therefore, a term which in his usage applies to every independent sign combination permitted by the formation rules of a language—and yet clearly syntactics in the wide sense must deal with all such combinations. There are, then, syntactical problems in the fields of perceptual signs, aesthetic signs, the practical use of signs, and general linguistics which have not been treated within the framework of what today is regarded as logical syntax and yet which form part of syntactics as this is here conceived.

6. Linguistic Structure

Let us now consider more carefully linguistic structure, invoking semantics and pragmatics where they may be of help in clarifying the syntactical dimension of semiosis.

Given a plurality of signs used by the same interpreter, there is always the possibility of certain syntactical relations between the signs. If there are two signs, S_1 and S_2, so used that S_1 (say 'animal') is applied to every object to which S_2 (say 'man') is applied, but not conversely, then in virtue of this usage the semiosis involved in the functioning of S_1 is included in that of S_2; an interpreter will respond to an object denoted by 'man' with the responses he would make to an object denoted by 'animal,' but in addition there are certain responses which would not be made to any animal to which 'man' was not applicable and which would not be made to an animal to which certain other terms (such as 'amoeba') were applicable. In this way terms gain relations among themselves corresponding to the relations of the responses of which the sign vehicles are a part, and these modes of usage are the pragmatical background of the formation and transformation rules. The syntactical structure of a language is the interrelationship of signs caused by the interrelationship of the responses of which the sign vehicles are products or parts. The formalist substitutes for such responses their formulation in signs; when he begins with an arbitrary set of rules, he is stipulating the interrelationship of responses which possible interpreters must have before they can be said to be using the language under consideration.

In so far as a single sign (such as a particular act of pointing) can denote only a single object, it has the status of an index; if it can denote a plurality of things (such as the term 'man'), then it is combinable in various ways with signs which explicate or restrict the range of its application; if it can denote everything (such as the term 'something'), then it has relations with every sign, and so has universal implication, that is to say, it is implicated by every sign within the language. These three kinds of signs will be called, respectively, *indexical signs*, *characterizing signs*, and *universal signs*.

Signs may thus differ in the degree to which they determine definite expectations. To say 'something is being referred to' does not give rise to definite expectations, does not allow taking account of what is being referred to; to use 'animal' with no further specification awakens certain sets of response, but they are not particularized sufficiently to deal adequately with a specific animal; it is an improvement in the situation to use 'man,' as is evident in the contrast between knowing that an animal is coming and that a man is coming; finally, the use of 'this' in an actual situation with the supplementary help of bodily orientation directs behavior upon a specific object but gives a minimum of expectations concerning the character of what is denoted. Universal signs may have a certain importance in allowing one to talk in general of the designata of signs without having to specify the sign or designatum; the difficulty of attempting to avoid such terms as 'object,' 'entity,' and 'something' shows the value of such terms for certain purposes. More important, however, is the combination of indexical and characterizing signs (as in 'that horse runs') since such a combination gives the definiteness of reference of the indexical sign plus the determinateness of the expectation involved in the characterizing sign. It is the complex forms of such combinations that are dealt with formally in the sentences of logical and mathematical systems, and to which (considered semantically) the predicates of truth and falsity apply. This importance is reflected in the fact that all formal systems show a differentiation of two kinds of signs corresponding to indexical and characterizing signs. Further, the fact that the determinateness of expectation can be increased by the use of additional signs is reflected in the fact that linguistic structures provide a framework which permits of degrees of specification and makes clear the sign relations involved.

To use terms suggested by M. J. Andrade, it may be said that every sentence contains a *dominant sign* and certain *specifiers*, these terms being relative to each other, since what is a dominant sign with respect to certain specifiers may itself be a specifier with respect to a more general dominant sign—thus 'white' may make the reference to horses more specific, while 'horse' may itself be a specifier with respect to 'animal.' Since an adequate taking-account-of-something demands an indication of both its location and (relevant) properties, and since the relevant degree of specification is obtained by a combination of characterizing signs, a sentence capable of truth and falsity involves indexical signs, a dominant characterizing sign with possibly characterizing specifiers, and some signs to show the relation of the indexical and characterizing signs to one another and to the members of their own class. Hence the general formula of such a sentence:

Dominant characterizing sign [characterizing specifiers (indexical signs)]

In such a sentence as 'That white horse runs slowly,' spoken in an actual situation with indexical gestures, 'runs' may be taken as the dominant sign, and 'slowly' as a characterizing specifier specifies 'runs'; 'horse' similarly specifies the possible cases of 'runs slowly,' 'white' carries the specification further, and 'that' in combination with the indexical gesture serves as an indexical sign to locate the object to which the dominant sign as now specified is to be applied. The conditions of utterance might show that 'horse' or some other sign is to be taken as the dominant sign, so that pragmatical considerations determine what, in fact, is the dominant sign. The dominant sign may even be more general than any which have been mentioned: it may be a sign to show that what follows is a declaration or a belief held with a certain degree of conviction. Instead of the use of the indexical sign in an actual situation, characterizing signs might be so used as to inform the hearer how to supply the indexical sign: 'Find the horse such that . . . . ; it is that horse to which reference is being made'; or 'Take any horse; then that horse. . . . .' In case a set of objects is referred to, the reference may be to all of the set, to a portion, or to some specified member or members; terms such as 'all,' 'some,' 'three,' together with indexical signs and descriptions, perform this function of indicating which of the possible denotata of a characterizing sign are referred to. There need not be only a single indexical sign; in such a sentence as '*A* gave *B* to *C*,' there are three correlates of the triadic relation to be specified by indexical signs, either used alone or in connection with other devices.

The sign 'to' in the sentence '*A* gave *B* to *C*' serves as an occasion for stressing an important point: to have intelligible sign combinations it is necessary to have special signs within the language in question to indicate the relation of other signs, and such signs, being in the language in question, must be distinguished from those signs in the language of syntactics which designate these relations. In the English examples which have been given, the 's' in 'runs,' the 'ly' in 'slowly,' the position of 'that' and 'white' with reference to the position of 'horse,' the positions of '*A*' and '*B*' before and after the dominant sign 'gives,' the position of 'to' before '*C*' all furnish indications as to which sign specifies which other sign, or which indexical sign denotes which correlate of the relation, or which signs are indexical signs and which are characterizing signs. Pauses, speech melodies, and emphasis help to perform such functions in spoken language; punctuation marks, accents, parentheses, italics, size of letter, etc., are similar aids in written and printed languages. Such signs within the language perform primarily a pragmatical function, but the term 'parenthesis' and its implicates occur in the metalanguage. The metalanguage must not be confused with a language to which it refers, and in the language itself a distinction must be made between those signs whose designata fall outside the language and those signs which indicate the relation of other signs.

All the distinctions which have been recognized as involved in the functioning of language in the full semiotical sense are reflected in the features of language which syntactics has thus far studied. Syntactics recognizes classes of signs, such as individual constants and variables, and predicate constants and variables, which are the formal correlates of various kinds of indexical and characterizing signs; the operators correspond to class specifiers; dots, parentheses, and brackets are devices within the language for indicating certain relations between the signs; terms such as 'sentence,' 'consequence,' and 'analytic' are syntactical terms for designating certain kinds of sign combinations and relations between signs; sentential (or "propositional") functions correspond to sign combinations lacking certain indexical specifiers necessary for complete sentences ("propositions"); the formation and transformation rules correspond to the way in which signs are combined or derived from one another by actual or possible users of the language. In this way the formalized languages studied in contemporary logic and mathematics clearly reveal themselves to be the formal structure of actual and possible languages of the type used in making statements about things; at point after point they reflect the significant features of language in actual use. The deliberate neglect by the formalist of other features of language, and the ways in which language changes, is an aid in isolating a particular object of interest: linguistic structure. The formal logician differs from the grammarian only in his greater interest in the types of sentences and transformation rules operative in the language of science. The logician's interest needs to be supplemented by the grammarian's type of interest and by attention to sign combinations and transformations in fields other than science if the whole domain of syntactics is to be adequately explored.

IV. Semantics

7. The Semantical Dimension of Semiosis

Semantics deals with the relation of signs to their designata and so to the objects which they may or do denote. As in the case of the other disciplines dealing with signs, a distinction may be made between its pure and descriptive aspects, pure semantics giving the terms and the theory necessary to talk about the semantical dimension of semiosis, descriptive semantics being concerned with actual instances of this dimension. The latter type of consideration has historically taken precedence over the former; for centuries linguists have been concerned with the study of the conditions under which specific words were employed, philosophical grammarians have tried to find the correlates in nature of linguistic structures and the differentiation of parts of speech, philosophical empiricists have studied in more general terms the conditions under which a sign can be said to have a denotatum (often in order to show that the terms of their metaphysical opponents did not meet these conditions), discussions of the term 'truth' have always involved the question of the relation of signs to things—and yet, in spite of the

99

length of this history, relatively little has been done in the way of controlled experimentation or in the elaboration of a suitable language to talk about this dimension. The experimental approach made possible by behavioristics offers great promise in determining the actual conditions under which certain signs are employed; the development of the language of semantics has been furthered by recent discussions of the relation of formal linguistic structures to their "interpretations," by attempts (such as those of Carnap and Reichenbach) to formulate more sharply the doctrine of empiricism, and by the efforts of the Polish logicians (notably Tarski) to define formally in a systematic fashion certain terms of central importance within semantics. Nevertheless, semantics has not yet attained a clarity and systematization comparable to that obtained by certain portions of syntactics.

Upon consideration, this situation is not surprising, for a rigorous development of semantics presupposes a relatively highly developed syntactics. To speak of the relation of signs to the objects they designate presupposes, in order to refer both to signs and to objects, the language of syntactics and the thing-language. This reliance upon syntactics is particularly evident in discussing languages, for here a theory of formal linguistic structure is indispensable. For example, the constantly recurring question as to whether the structure of language is the structure of nature cannot properly be discussed until the terms 'structure' and 'structure of a language' are clear; the unsatisfactoriness of historical discussions of this question are certainly in part due to the lack of such preliminary clarification as syntactics has today supplied.

A sign combination such as ' 'Fido' designates A' is an instance of a sentence in the language of semantics. Here ' 'Fido' ' denotes Fido (i.e., the sign or the sign vehicle and not a non-linguistic object), while 'A' is an indexical sign of some object (it might be the word 'that' used in connection with some directive gesture). ' 'Fido' ' is thus a term in the metalanguage denoting the sign 'Fido' in the object language; 'A' is a term in the thing-language denoting a thing. 'Designates' is a semantical term, since it is a characterizing sign designating a relation between a sign and an object. Semantics presupposes syntactics but abstracts from pragmatics; whether dealing with simple signs or complex ones (such as a whole mathematical system), semantics limits itself to the semantical dimension of semiosis.

In considering this dimension, the most important addition to the preceding account lies in the term '*semantical rule.*' Unlike the formation and transformation rules, which deal with certain sign combinations and their relations, 'semantical rule' designates within semiotic a rule which determines under which conditions a sign is applicable to an object or situation; such rules correlate signs and situations denotable by the signs. A sign denotes whatever conforms to the conditions laid down in the semantical rule, while the rule itself states the conditions of designation and so determines the designatum (the class or kind of denotata). The importance of such rules has been stressed by Reichenbach as definitions of co-ordination, and by Ajdukiewicz as empirical rules of meaning; the latter insists that such rules are necessary to characterize uniquely a language, since with different semantical rules two persons might share the same formal linguistic structure and yet be unable to understand each other. Thus, in addition to the syntactical rules, the characterization of a language requires the statement of the semantical rules governing the sign vehicles singly and in combination (it will later become clear that the full semiotical characterization of a language demands in addition the statement of what will be called pragmatical rules).

Rules for the use of sign vehicles are not ordinarily formulated by the users of a language, or are only partially formulated; they exist rather as habits of behavior, so that only certain sign combinations in fact occur, only certain sign combinations are derived from others, and only certain signs are applied to certain situations. The explicit formulation of rules for a given language requires a higher order of symbolization and is a task of descriptive semiotic; it would be a very difficult task to formulate, for instance, the rules of English usage, as may be seen if one even tries to formulate the conditions under which the words 'this' and 'that' are used. It is natural, therefore, that attention has been chiefly devoted to fragments of the common languages and to languages which have been deliberately constructed.

A sign has a semantical dimension in so far as there are semantical rules (whether formulated or not is irrelevant) which determine its applicability to certain situations under certain conditions. If this usage is stated in terms of other signs, the general formula is as follows: The sign vehicle '*x*' designates the conditions *a*, *b*, *c* . . . . under which it is applicable. The statement of those conditions gives the semantical rule for '*x*.' When any object or situation fulfils the required conditions, then it is denoted by '*x*.' The sign vehicle itself is simply one object, and its denotation of other objects resides solely in the fact that there are rules of usage which correlate the two sets of objects.

The semantical rule for an indexical sign such as pointing is simple: the sign designates at any instant what is pointed at. In general, an indexical sign designates what it directs attention to. An indexical sign does not characterize what it denotes (except to indicate roughly the space-time co-ordinates) and need not be similar to what it denotes. A characterizing sign characterizes that which it can denote. Such a sign may do this by exhibiting in itself the properties an object must have to be denoted by it, and in this case the characterizing sign is an *icon;* if this is not so, the characterizing sign may be called a *symbol.* A photograph, a star chart, a model, a chemical diagram are icons, while the word 'photograph,' the names of the stars and of chemical elements are symbols. A "concept" may be regarded as a semantical rule determining the use of characterizing signs. The semantical rule for the use of icons is that they denote those objects which have the characteristics which they themselves have—or more usually a certain specified set of their characteristics. The semantical rule for the use of symbols must be stated in terms of other symbols whose rules or usages are not in question, or by pointing out specific objects which serve as models (and so as icons), the symbol in question then being employed to denote objects similar to the models.

It is the fact that the semantical rule of usage for a symbol can be stated in terms of other symbols which makes possible (to use Carnap's term) the reduction of one scientific term to others (or, better, the construction of one term upon others) and thus the systematization of the language of science. It is because indexical signs are indispensable (for symbols ultimately involve icons, and icons indices) that such a program of systematization as physicalism proposes is forced to terminate the process of reduction by the acceptance of certain signs as primitive terms whose semantical rules of usage, determining their applicability to things indicated by indices, must be taken for granted but cannot, within that particular systematization, be stated.

The semantical rule for the use of a sentence involves reference to the semantical rules of the component sign vehicles. A sentence is a complex sign to the effect that the designatum of the indexical component is also a designatum of the component which is a characterizing sign. The designatum of a sentence is thus the designatum-of-an-indexical-sign-as-the-designatum-of-a-characterizing-sign; when the situation conforms to the semantical rule of a sentence, the situation is a denotatum of that sentence (and the sentence may then be said to be true of that situation).

The difference between indices, icons, and symbols (sentences being compounds of other signs) is accounted for by different kinds of semantical rules. Things may be regarded as the designata of indexical signs, properties as the designata of one-place characterizing signs, relations as the designata of two- (or more) place characterizing signs, facts or state of affairs as designata of sentences, and entities or beings as the designata of all signs whatsoever.

It is because a sign may have a rule of usage to determine what it can denote without actually being so used that there can be signs which in fact denote nothing or have null denotation. It was previously noted that the very notion of sign involves that of designatum, but not that there be actually existing objects which are denoted. The designatum of a sign is such things which the sign *can* denote, i.e., such objects or situations which according to the semantical rule of usage could be correlated to the sign vehicle by the semantical relation of denotation. It is now clear, as formerly it could not be, that the *statement* of what would constitute a designatum of a certain sign must itself make use of terms with syntactical relations, since the semantical rule of usage states what the sign in question signifies by using the sign in relation to other signs. 'Designatum' is clearly a semiotical term, while the question as to whether there are objects of such and such a kind is a question to be answered by considerations which go beyond semiotic. The failure to keep separate the statements of semiotic from thing-sentences has led to many pseudo thing-sentences. To say that there is a "realm of subsistence" in addition to, but on a par with, the realm of existences, since "When we think, we must think about something," is a quasi-semantical statement: it seems to speak about the world in the same way that physics does, but actually the statement is an ambiguous form of a semantical sentence, namely, the sentence that for every sign that can denote something a semantical rule of usage can be formulated which will state the conditions under which the sign is applicable. This statement, analytically correct within semantics, does not in any sense imply that there are objects denoted by every such sign—objects which are "subsistential" when not existential.

8. Linguistic and Nonlinguistic Structures

One of the oldest and most persistent theories is that languages mirror (correspond with, reflect, are isomorphic with) the realm of nonlinguistic objects. In the classical tradition it was often held that this mirroring was threefold: thought reflected the properties of objects; and spoken language, composed of sounds which had been given a representative function by mind, in turn reflected the kinds and relations of mental phenomena and so the realm of nonmental objects.

It goes without saying that such a persistent tradition as lies behind the doctrine in question must have something to commend it; it is, nevertheless, significant that this tradition has progressively weakened and has even been repudiated by some of its most vigorous former champions. What light can the general semiotical point of view throw on the situation? In attempting to answer this question, it will be seen that the heart of the matter lies in the fact that the only relevant correlation which exists between signs and other objects is that established by semantical rules.

It seems plausible that the excesses and difficulties of the attempt to find a complete semantical correlation between linguistic signs and other objects lies in the neglect or oversimplification of the syntactical and pragmatical dimensions of semiosis. It has been noted that the very possibility of language requires that there be some special signs to indicate the syntactical relations of other signs in the language. Examples of such signs are pauses, intonations, order of signs, prepositions, affixes, suffixes, etc. Such signs function predominantly in the syntactical and pragmatical dimensions; in so far as they have a semantical dimension, they denote sign vehicles and not nonlinguistic objects. It need not be denied that such signs might help to establish some kind of isomorphism between the remaining signs and nonlinguistic objects, for such isomorphism might be much more complicated than the relation of a model to that of which it is a model. Spatial relations of signs might not correspond to spatial relations between things, but there might be a correlating relation such that for every spatial relation between signs there holds some other relation between the objects denoted by the signs. Such possibilities are open to investigation and should be specifically explored; if they do not hold for all signs, they may hold for certain of them, namely, for such as have semantical rules correlating them with nonlinguistic situations. Nevertheless, the defenders of isomorphism have not shown that such is the case, or that such must be the case if language is to be possible.

The unconvincingness of the general theory increases if notice is taken of such signs as 'all,' 'some,' 'the,' 'not,' 'point at infinity,' '−1.' The first three terms indicate how much of the class determined by some characterizing sign is to be taken account of. The term 'not' is primarily of practical importance, since it allows reference to something other than what is specifically referred to without specifying what the other is. So clarified semantically, the practical importance of the term is obvious, but it is not theoretically necessary in a language, and certainly no existential "negative facts" need be invoked to correspond to it. The mathematical terms mentioned are commonly regarded as signs added to the language so that certain operations, otherwise impossible in certain cases, are always possible, and certain formulas, otherwise needing qualification, can be stated in their full generality.

There are also many signs in a common language which indicate the reaction of the user of the signs to the situation being described (as the 'fortunately' in 'Fortunately, he came'), or even to the signs he is himself using in the description (as in expressing his degree of confidence in a statement). Such terms within discourse have a semantical dimension only at a higher level of semiosis, since the pragmatical dimension of a process of semiosis is not denoted in that process but only in one of a higher level. As in the case of the predominantly syntactical features of a language, the predominantly pragmatical features should not be confounded with those elements correlated by means of semantical rules with the nonlinguistic objects which are being denoted. The traditional versions of isomorphism failed to distinguish the various dimensions of semiosis and the various levels of languages and designata. To what extent some qualified version of the thesis may be held can only be determined after it is formulated. But it is clear that, when a language as a whole is considered, its syntactical structure is a function of both pragmatic and empirical considerations and is not a bare mirroring of nature considered in abstraction from the users of the language.

The main point of the discussion is not to deny that all the signs in a language may have designata and so a semantical dimension but rather to call attention to the fact that the designata of signs in a given discourse (and so the objects denoted, if there are such) do not stand at the same level: the designata of some signs must be sought at the level of semiotic rather than at the level of the thing-language itself; in the given discourse such signs simply indicate (but do not designate) relations of the other signs to one another or to the interpreter— in Scholastic terms they bring something of material and simple supposition into the functioning of terms in personal supposition. The strata of signs are as complex and as difficult to unravel as geological strata; the scientific and psychological effects of unraveling them may be as great in the former case as it has been in the latter.

So much for a bare indication of the field of semantics. The precise analysis of semantical terms, their formal systematization, and the question of the applicability of semantics to domains other than the language of science (for instance, to aesthetic signs) obviously are not possible in an introductory account. If pragmatical factors have appeared frequently in pages belonging to semantics, it is because the current recognition that syntactics must be supplemented by semantics has not been so commonly extended to the recognition that semantics must in turn be supplemented by pragmatics. It is true that syntactics and semantics, singly and jointly, are capable of a relatively high degree of autonomy. But syntactical and semantical rules are only the verbal formulations within semiotic of what in any concrete case of semiosis are habits of sign usage by actual users of signs. 'Rules of sign usage,' like 'sign' itself, is a semiotical term and cannot be stated syntactically or semantically.

V. Pragmatics

9. The Pragmatical Dimension of Semiosis

The term 'pragmatics' has obviously been coined with reference to the term 'pragmatism.' It is a plausible view that the permanent significance of pragmatism lies in the fact that it has directed attention more closely to the relation of signs to their users than had previously been done and has assessed more profoundly than ever before the relevance of this relation in understanding intellectual activities. The term 'pragmatics' helps to

107

signalize the significance of the achievements of Peirce, James, Dewey, and Mead within the field of semiotic. At the same time, 'pragmatics' as a specifically semiotical term must receive its own formulation. By 'pragmatics' is designated the science of the relation of signs to their interpreters. 'Pragmatics' must then be distinguished from 'pragmatism,' and 'pragmatical' from 'pragmatic.' Since most, if not all, signs have as their interpreters living organisms, it is a sufficiently accurate characterization of pragmatics to say that it deals with the biotic aspects of semiosis, that is, with all the psychological, biological, and sociological phenomena which occur in the functioning of signs. Pragmatics, too, has its pure and descriptive aspects; the first arises out of the attempt to develop a language in which to talk about the pragmatical dimension of semiosis; the latter is concerned with the application of this language to specific cases.

Historically, rhetoric may be regarded as an early and restricted form of pragmatics, and the pragmatical aspect of science has been a recurrent theme among the expositors and interpreters of experimental science. Reference to interpreter and interpretation is common in the classical definition of signs. Aristotle, in the *De interpretatione*, speaks of words as conventional signs of thoughts which all men have in common. His words contain the basis of the theory which became traditional: The interpreter of the sign is the mind; the interpretant is a thought or concept; these thoughts or concepts are common to all men and arise from the apprehension by mind of objects and their properties; uttered words are then given by the mind the function of directly representing these concepts and indirectly the corresponding things; the sounds chosen for this purpose are arbitrary and vary from social group to social group; the relations between the sounds are not arbitrary but correspond to the relations of concepts and so of things. In this way throughout much of its history the theory of signs was linked with a particular theory of thought and mind, so much so that logic, which has always been affected by current theories of signs, was often conceived as dealing with concepts—a view made precise in the

Scholastic doctrine of logical terms as terms of second intention. Even Leibniz' insistence upon the empirical study of the sign vehicle as determined by rule was not a repudiation of the dominant tradition but merely an insistence that in this way a new and better technique could be obtained for analyzing concepts than by the attempt to inspect thought directly.

In the course of time most of the tenets of this traditional version of pragmatics were questioned, and today they would be accepted only with serious qualifications. The change in point of view has been most rapid as a result of the implications for psychology of the Darwinian biology—implications which received an early interpretation in pragmatism. Charles S. Peirce, whose work is second to none in the history of semiotic, came to the conclusion that in the end the interpretant of a symbol must reside in a habit and not in the immediate physiological reaction which the sign vehicle evoked or in the attendant images or emotions—a doctrine which prepared the way for the contemporary emphasis on rules of usage. William James stressed the view that a concept was not an entity but a way in which certain perceptual data functioned representatively and that such "mental" functioning, instead of being a bare contemplation of the world, is a highly selective process in which the organism gets indications as to how to act with reference to the world in order to satisfy its needs or interests. George H. Mead was especially concerned with the behavior involved in the functioning of linguistic signs and with the social context in which such signs arise and function. His work is the most important study from the point of view of pragmatism of these aspects of semiosis. John Dewey's instrumentalism is the generalized version of the pragmatists' emphasis upon the instrumental functioning of signs or "ideas."

If from pragmatism is abstracted the features of particular interest to pragmatics, the result may be formulated somewhat as follows: The interpreter of a sign is an organism; the interpretant is the habit of the organism to respond, because of the sign vehicle, to absent objects which are relevant to a present problematic situation as if they were present. In virtue of semiosis an organism takes account of relevant properties of absent objects, or unobserved properties of objects which are present, and in this lies the general instrumental significance of ideas. Given the sign vehicle as an object of response, the organism expects a situation of such and such a kind and, on the basis of this expectation, can partially prepare itself in advance for what may develop. The response to things through the intermediacy of signs is thus biologically a continuation of the same process in which the distance senses have taken precedence over the contact senses in the control of conduct in higher animal forms; such animals through sight, hearing, and smell are already responding to distant parts of the environment through certain properties of objects functioning as signs of other properties. This process of taking account of a constantly more remote environment is simply continued in the complex processes of semiosis made possible by language, the object taken account of no longer needing to be perceptually present.

With this orientation, certain of the terms which have previously been used appear in a new light. The relation of a sign vehicle to its designatum is the actual taking-account in the conduct of the interpreter of a class of things in virtue of the response to the sign vehicle, and what are so taken account of are designata. The semantical rule has as its correlate in the pragmatical dimension the habit of the interpreter to use the sign vehicle under certain circumstances and, conversely, to expect such and such to be the case when the sign is used. The formation and transformation rules correspond to the actual sign combinations and transitions which the interpreter uses, or to stipulations for the use of signs which he lays down for himself in the same way in which he attempts to control deliberately other modes of behavior with reference to persons and things. Considered fron the point of view of pragmatics, a linguistic structure is a system of behavior: corresponding to analytical sentences are the relations between sign responses to the more inclusive sign responses of which they are segments; corresponding to synthetical sentences are those relations between sign responses which are not relations of part to whole. The indexical signs (or their substitutes) in a sign combination direct the attention of the interpreter to parts of the environment; the dominant characterizing sign determines some general response (expectation) to these parts; the characterizing specifiers delimit the general expectation, the degree of specification and the choice of the dominant sign being determined with respect to the problem at hand. If the indexical and characterizing functions are both performed, the interpreter is judging and the sign combination is a judgment (corresponding to the sentence of syntactics and the statement or proposition of semantics). To the degree that what is expected is found as expected the sign is confirmed; expectations are, in general, only partially confirmed; there may be, in addition, various degrees of indirect confirmation that what is indexically referred to has the properties it was expected to have. In general, from the point of view of behavior, signs are "true" in so far as they correctly determine the expectations of their users, and so release more fully the behavior which is implicitly aroused in the expectation or interpretation.

Such statements go somewhat beyond pragmatics proper into the strictly semiotical question as to the interrelation of the dimensions—a topic yet to be specifically discussed. Pragmatics itself would attempt to develop terms appropriate to the study of the relation of signs to their users and to order systematically the results which come from the study of this dimension of semiosis. Such terms as 'interpreter,' 'interpretant,' 'convention' (when applied to signs), 'taking-account-of' (when a function of signs), 'verification,' and 'understands' are terms of pragmatics, while many strictly semiotical terms such as 'sign,' 'language,' 'truth,' and 'knowledge' have important pragmatical components. In a systematic presentation of semiotic, pragmatics presupposes both syntactics and semantics, as the latter in turn presupposes the former, for to discuss adequately the relation of signs to their interpreters requires knowledge of the relation of signs to one another and to those things to which they refer their interpreters. The unique elements within pragmatics would be found in those terms which, while not strictly semiotical, cannot be defined in syntactics or semantics; in the clarification of the pragmatical aspect of various semiotical terms; and in the statement of what psychologically, biologically, and sociologically is involved in the occurrence of signs. Attention may now be turned to some aspects of this latter problem.

10. Individual and Social Factors in Semiosis

The topic in question may be approached, and a possible objection forestalled, by asking why there is any need of adding pragmatics to semantics; since semantics deals with the relation of signs to objects, and since interpreters and their responses are natural objects studied by the empirical sciences, it would seem as if the relation of signs to interpreters fell within semantics. The confusion here arises from the failure to distinguish levels of symbolization and to separate—in the use of 'object'— semiotical from nonsemiotical terms. Everything that is designatable is subject matter for a (in principle) unified science, and in this sense all the semiotical sciences are parts of unified science. When descriptive statements are made about any dimension of semiosis, the statements are in the semantical dimension of a higher level of semiosis and so are not necessarily of the same dimension that is being studied. Statements in pragmatics about the pragmatical dimension of specific signs are functioning predominantly in the semantical dimension. The fact that the pragmatical dimension becomes a designatum for a higher-level process of description does not signify that the interpretant of a sign at any given level is a designatum of that particular sign. The interpretant of a sign is the habit in virtue of which the sign vehicle can be said to designate certain kinds of objects or situations; as the method of determining the set of objects the sign in question designates, it is not itself a member of that set. Even the language of a unified science which would contain an account of the pragmatical dimension would not at the moment of use denote its own pragmatical dimension, though at a higher level of usage the account given of the pragmatical dimension may be found applicable to the pragmatical dimension of the lower level. Since the pragmatical dimension is involved in the very existence of the relation of designation, it cannot itself be put within the semantical dimension. Semantics does not deal with all the relations of signs to objects but, as a semiotical science, deals with the relation of signs to their designata; pragmatics, dealing with another relation of signs, cannot be put within semantics alone or in combination with syntactics. This conclusion is completely independent of the relation of physical and biological existences; the distinction of the semantical and pragmatical dimensions is a semiotical distinction and has nothing to do with the relation of biology and physics.

The point can perhaps be made sharper if we introduce the term *'pragmatical rule.'* Syntactical rules determine the sign relations between sign vehicles; semantical rules correlate sign vehicles with other objects; pragmatical rules state the conditions in the interpreters under which the sign vehicle is a sign. Any rule when actually in use operates as a type of behavior, and in this sense there is a pragmatical component in all rules. But in some languages there are sign vehicles governed by rules over and above any syntactical or semantical rules which may govern those sign vehicles, and such rules are pragmatical rules. Interjections such as 'Oh!,' commands such as 'Come here!,' value terms such as 'fortunately,' expressions such as 'Good morning!,' and various rhetorical and poetical devices occur only under certain definite conditions in the users of the language; they may be said to express such conditions, but they do not denote them at the level of semiosis in which they are actually employed in common discourse. The statement of the conditions under which terms are used, in so far as these cannot be formulated in terms of syntactical and semantical rules, constitutes the pragmatical rules for the terms in question.

The full characterization of a language may now be given: *A language in the full semiotical sense of the term is any intersubjective set of sign vehicles whose usage is determined by syntactical, semantical, and pragmatical rules.*

Interpretation becomes especially complex, and the individual and social results especially important, in the case of linguistic signs. In terms of pragmatics, a linguistic sign is used in combination with other signs by the members of a social group; a language is a social system of signs mediating the responses of members of a community to one another and to their environment. To understand a language is to employ only those sign combinations and transformations not prohibited by the usages of the social group in question, to denote objects and situations as do the members of this group, to have the expectations which the others have when certain sign vehicles are employed, and to express one's own states as others do—in short, to understand a language or to use it correctly is to follow the rules of usage (syntactical, semantical, and pragmatical) current in the given social community.

There is a further stipulation often made in connection with the linguistic sign: it must be capable of voluntary use for the function of communicating. Such terms as 'voluntary' and 'communication' need more extended analysis than is here possible, but Mead's account, in *Mind, Self, and Society,* of the linguistic sign (which he calls the significant symbol) seems to cover the point intended in this stipulation. According to Mead, the primary phenomenon out of which language in the full human sense emerges is the gesture, especially the vocal gesture. The gesture sign (such as a dog's snarl) differs from such a nongestural sign as thunder in the fact that the sign vehicle is an early phase of a social act and the designatum a later phase of this act (in this case the attack by the dog). Here one organism prepares itself for what another organism—the dog—is to do by responding to certain acts of the latter organism as signs; in the case in question the snarl is the sign, the attack is the designatum, the animal being attacked is the interpreter, and the preparatory response of the interpreter is the interpretant. The utility of such gesture signs is limited by the fact that the sign is not a sign to the producer as it is to the receiver: the dog which snarls does not respond to his snarl as does his opponent; the sign is not held in common and so is not a linguistic sign.

On the other hand, the important characteristic of the vocal gesture lies precisely in the fact that the emitter of the sound himself hears the sound just as others do. When such sounds become connected with social acts (such as a fight, a game, a festival), the various participants in the act have through this common sign, and in spite of their differentiated functions within the act, a common designatum. Each participant in the common activity stimulates himself by his vocal gestures as he stimulates others. Couple this with what Mead termed the temporal dimension of the nervous system (namely, an earlier but more slowly aroused activity may initiate a later and more rapid activity which in turn furthers or checks the complete arousal of the first activity), and one obtains a possible explanation of how linguistic signs serve for voluntary communication. To use one of Mead's frequent examples, we may consider the situation of a person noticing smoke in a crowded theater. Smoke is a nongestural sign of fire, and its perception calls out to some degree responses appropriate to fire. But further, the spoken word 'fire,' as a response which is connected with a whole set of responses to fire, tends to be uttered. Since this is a linguistic sign, the utterer begins to respond toward this tendency toward utterance as other members of his social group would respond—to run toward an exit, to push, and perhaps trample over, others blocking the way, etc. But the individual, in virtue of certain fundamental attitudes, will respond either favorably or unfavorably to these tendencies and will thus check or further the tendency to say 'Fire!'

In such a case it is said that the man "knew what he was about," that he "deliberately used (or did not use) a certain sign to communicate to others," that he "took account of others." Mead would generalize from such common usages: from his point of view "to have a mind" or "to be conscious of something" was equivalent to "using linguistic signs." It is through such signs that the individual is able to act in the light of consequences to himself and to others, and so to gain a certain amount of control over his own behavior; the presentation of possible consequences of action through the production of linguistic signs becomes a factor in the release or inhibition of the action which has (or seems to have) such consequences. It is in such processes that the term 'choice' gains its clarification—and also whatever distinction is to be made between senders and receivers of linguistic signs. Since the linguistic sign is socially conditioned, Mead, from the standpoint of his social behaviorism, regarded the individual mind and self-conscious self as appearing in a social process when objective gestural communication becomes internalized in the individual through the functioning of vocal gestures. Thus it is through the achievements of the community, made available to the individual by his participation in the common language, that the individual is able to gain a self and mind and to utilize those achievements in the furtherance of his interests. The community benefits at the same time in that its members are now able to control their behavior in the light of the consequences of this behavior to others and to make available to the whole community their own experiences and achievements. At these complex levels of semiosis, the sign reveals itself as the main agency in the development of individual freedom and social integration.

11. Pragmatic Use and Abuse of Signs

When a sign produced or used by an interpreter is employed as a means of gaining information about the interpreter, the point of view taken is that of a higher process of semiosis, namely, that of descriptive pragmatics. Psychoanalysis among the psychologies, pragmatism among the philosophies, and now the sociology of knowledge among the social sciences have made this way of looking at signs a common possession of educated persons. Newspaper statements, political creeds, and philosophical systems are increasingly being looked at in terms of the interests which are expressed and served by the production and use of the signs in question. The psychoanalyist is interested in dreams for the light they throw upon the dreamer; the sociologist of knowledge is interested in the social conditions under which doctrines and systems of doctrine are current. In neither case is the interest in the question whether the dreams or doctrines are true in the semantical sense of the term, i.e., whether there are situations which the dreams and the doctrines may be said to denote. Such studies, together with many others, have confirmed over a wide range the general thesis of pragmatism as to the instrumental character of ideas.

Any sign whatever may be looked at in terms of the psychological, biological, and sociological conditions of its usage. The sign expresses but does not denote its own interpretant; only at a higher level is the relation of the sign to the interpreter itself made a matter for designation. When this is done and a correlation found, the sign becomes of individual and social diagnostic value, and so a new sign at a higher level of semiosis. Signs as well as things not signs can become diagnostic signs: the fact that a patient has a fever shows certain things about his condition; equally well the fact that a certain sign is used by someone expresses that person's condition, for the interpretant of the sign is part of the conduct of the individual. In such cases the same sign vehicle may be functioning as two signs, interpreted by the patient as referring to its denotata and by the diagnostician as referring to the interpretant involved in the patient's sign.

Not only may all signs be regarded in terms of pragmatics, but it is also perfectly legitimate for certain purposes to use signs simply in order to produce certain processes of interpretation, regardless of whether there are objects denoted by the signs or even whether the sign combinations are formally possible in terms of the formation and transformation rules of the language in which the sign vehicles in question are normally used. Some logicians seem to have a generalized fear of contradictions, forgetting that, while contradictions frustrate the normal uses of deduction, they may be perfectly compatible with other interests. Even linguistic signs have many other uses than that of communicating confirmable propositions: they may be used in many ways to control the behavior of one's self or of other users of the sign by the production of certain interpretants. Commands, questions, entreaties, and exhortations are of this sort, and to a large degree the signs used in the literary, pictorial, and plastic arts. For aesthetic and practical purposes the effective use of signs may require rather extensive variations from the use of the same sign vehicles most effective for the purposes of science. Scientists and logicians may be excused if they judge signs in terms of their own purposes, but the semiotician must be interested in all the dimensions and all the uses of signs; the syntactics, semantics, and pragmatics of the signs used in literature, the arts, morality, religion, and in value judgments generally are as much his concern as studies of the signs used in science. In one case as in the other the usage of the sign vehicle varies with the purpose to be served.

If semiotic must defend the legitimacy for certain purposes of a concern for the effect of the sign on those who will interpret it, it must equally set itself the task of unmasking confusion of these various purposes which signs serve, whether the confusion be unintentional or deliberate. Just as properly syntactical or semantical statements may masquerade in a form which causes them to appear as statements about nonlinguistic objects, so may pragmatical statements thus masquerade; they then become, as quasi-pragmatical statements, one particular form of pseudo thing-sentences. In the clearly dishonest cases a purpose is accomplished by giving the signs employed the characteristics of statements with syntactical or semantical dimensions, so that they seem to be rationally demonstrated or empirically supported when in fact they are neither. An intellectual intuition, superior to scientific method, may be invoked to bolster up the validity of what is apparently affirmed. The masquerading may not be of one dimension in terms of others but within the pragmatical dimension itself; a purpose that cannot fully stand the light of scrutiny expresses itself in a form suitable to other purposes: aggressive acts of individuals and social groups often drape themselves in the mantle of morality, and the declared purpose is often not the real one. A peculiarly intellectualistic justification of dishonesty in the use of signs is to deny that truth has any other component than the pragmatical, so that any sign which furthers the interest of the user is said to be true. In terms of the preceding analysis it should be clear that 'truth' as commonly used is a semiotical term and cannot be used in terms of any one dimension unless this usage is explicitly adopted. Those who like to believe that 'truth' is a strictly pragmatical term often refer to the pragmatists in support of their view and naturally fail to note (or to state) that pragmatism as a continuation of empiricism is a generalization of scientific method for philosophical purposes and could not hold that the factors in the common usage of the term 'truth' to which attention was being drawn rendered nonexistent previously recognized factors. Certain of James's statements taken in isolation might seem to justify this perversion of pragmatism, but no one can seriously study James without seeing that his doctrine of truth was in principle semiotical: he clearly recognized the need of bringing in formal, empirical, and pragmatic factors; his main difficulty was in integrating these factors, since he lacked the base which a developed theory of signs provides. Dewey has specifically denied the imputed identification of truth and utility. Pragmatism has insisted upon the pragmatical and pragmatic aspects of truth; the perversion of this doctrine into the view that truth has only such aspects is an interesting case of how the results of a scientific analysis may be distorted to lend credibility to quasi-pragmatical statements.

Pseudo thing-sentences of the quasi-pragmatical type are not for the most part deliberate deception of others by the use of signs but cases of unconscious self-deception. Thus a philosopher with certain imperious needs may from a relatively small empirical base construct an elaborate sign system, perhaps in mathematical form, and yet the great majority of terms may be without semantical rules of usage; the impression that the system is about the world, and perhaps superior in truth to science, comes from the confusion of analytic and synthetic sentences and from the illusion that the congenial attitudes evoked by the signs constitute semantical rules. A somewhat similar manifestation is found in mythology, but without the evident influence of scientific types of expression.

A particularly interesting aberration of the semiotical processes takes place in certain phenomena studied by psychopathology. Signs normally take the place of objects they designate only to a limited extent; but if for various reasons interests cannot be satisfied in the objects themselves, the signs come more and more to take the place of the object. In the aesthetic sign this development is already evident, but the interpreter does not actually confuse the sign with the object it designates: the described or painted man is called a man, to be sure, but with more or less clear recognition of the sign status—it is only a painted or described man. In the magical use of signs the distinction is less clearly made; operations on the sign vehicle take the place of operations on the more elusive object. In certain kinds of insanity the distinction between the designatum and denotata vanishes; the troublesome world of existences is pushed aside, and the frustrated interests get what satisfaction they can in the domain of signs, oblivious in varying degrees to the restrictions of consistency and verifiability imposed by the syntactical and semantical dimensions. The field of psychopathology offers great opportunities for applications of, and contributions to, semiotic. A number of workers in this field have already recognized the key place which the concept of sign holds. If, following the lead of the pragmatist, mental phenomena be equated with sign responses, consciousness with reference by signs, and rational (or "free") behavior with the control of conduct in terms of foreseen consequences made available by signs, then psychology and the social sciences may recognize what is distinctive in their tasks and at the same time see their place within a unified science. Indeed, it does not seem fantastic to believe that the concept of sign may prove as fundamental to the sciences of man as the concept of atom has been for the physical sciences or the concept of cell for the biological sciences.

VI. The Unity of Semiotic

12. Meaning

We have been studying certain features of the phenomenon of sign-functioning by making use of the abstractions involved in distinguishing syntactics, semantics, and pragmatics—just as biologists study anatomy, ecology, and physiology. While we have recognized explicitly the abstractions involved and have constantly correlated the three subdisciplines of semiotic, we must now draw even more explicitly the unity of semiotic into the focus of attention.

In a wide sense, any term of syntactics, semantics, or pragmatics is a semiotical term; in a narrow sense, only those terms are semiotical which cannot be defined in any of the component fields alone. In the strict sense 'sign,' 'language,' 'semiotic,' 'semiosis,' 'syntactics,' 'truth,' 'knowledge,' etc., are semiotical terms. What of the term 'meaning'? In the preceding discussion the term 'meaning' has been deliberately avoided. In general it is well to avoid this term in discussions of signs; theoretically, it can be dispensed with entirely and should not be incorporated into the language of semiotic. But since the term has had such a notorious history, and since in its consideration certain important implications of the present account can be made clear, the present section is devoted to its discussion.

The confusion regarding the "meaning of 'meaning' " lies in part in the failure to distinguish with sufficient clarity the dimension of semiosis which is under consideration, a situation which also obtains in the confusions as to the terms 'truth' and 'logic.' In some cases 'meaning' refers to designata, in other cases to denotata, at times to the interpretant, in some cases to what a sign implicates, in some usages to the process of semiosis as such, and often to significance or value. Similar confusions are found in the common usages of 'designates,' 'signifies,' 'indicates,' 'expresses,' and in various attempts by linguists to define such terms as 'sentence,' 'word,' and 'part of speech.' The most charitable interpretation of such confusions is to suggest that for the major purposes which the everyday languages serve it has not been necessary to denote with precision the various factors in semiosis—the process is merely referred to in a vague way by the term 'meaning.' When, however, such vague usages are taken over into domains where an understanding of semiosis is important, then confusion results. It then becomes necessary to either abandon the term 'meaning' or to devise ways to make clear the usage in question. Semiotic does not rest upon a theory of "meaning"; the term 'meaning' is rather to be clarified in terms of semiotic.

Another factor in the confusion is a psychological-linguistic one: men in general find it difficult to think clearly about complex functional and relational processes, a situation reflected in the prevalence of certain linguistic forms. Action centers around handling things with properties, and the fact that these things and properties appear only in complex contexts is a much later and more difficult realization. Hence the naturalness of what Whitehead has called the fallacy of simple location. In the present case this takes the form of looking for meanings as one would look for marbles: a meaning is considered as one thing among other things, a definite something definitely located somewhere. This may be sought for in the designatum, which thus becomes transformed in certain varieties of·"realism" into a special kind of object—a "Platonic idea" inhabiting the "realm of subsistence," perhaps grasped by a special faculty for intuiting "essences"; or it may be sought for in the interpretant, which then becomes transformed in conceptualism into a concept or idea inhabiting a special domain of mental entities whose relation to the "psychical states" of individual interpreters becomes very difficult to state; or in desperation caused by contemplation of the previous alternatives it may be sought in the sign vehicle—though historically few if any "nominalists" have held this position. As a matter of fact, none of these positions has proved satisfactory and none of them is demanded. As semiotical terms, neither 'sign vehicle,' 'designatum,' nor 'interpretant' can be defined without reference to one another; hence they do not stand for isolated existences but for things or properties of things in certain specifiable functional relations to other things or properties. A "psychical state," or even a response, is not as such an interpretant but becomes such only in so far as it is a "taking-account-of-something" evoked by a sign vehicle. No object is as such a denotatum but becomes one in so far as it is a member of the class of objects designatable by some sign vehicle in virtue of the semantical rule for that sign vehicle. Nothing is intrinsically a sign or a sign vehicle but becomes such only in so far as it permits something to take account of something through its mediation. Meanings are not to be located as existences at any place in the process of semiosis but are to be characterized in terms of this process as a whole. 'Meaning' is a semiotical term and not a term in the thing-language; to say that there are meanings in nature is not to affirm that there is a class of entities on a par with trees, rocks, organisms, and colors, but that such objects and properties function within processes of semiosis.

This formulation also avoids another persistent stumbling block, namely, the belief that meaning is in principle personal, private, or subjective. Such a view historically owes much to the assimilation of the conceptualistic position within an associational psychology which itself uncritically accepted the current metaphysical view of the subjectivity of experience. Persons such as Ockham and Locke were well aware of the importance of habit in the functioning of signs, but as the associational psychology came more and more to reduce mental phenomena to combinations of "psychical states," and to conceive these states as within the individual's "mind" and only accessible to that mind, meaning itself came to be considered in the same terms. Meanings were inaccessible to observation from without, but individuals somehow managed to communicate these private mental states by the use of sounds, writing, and other signs.

The notion of the subjectivity of experience cannot be here analyzed with the detail the problem merits. It is believed, however, that such an analysis would show that 'experience' itself is a relational term masquerading as a thing-name. x is an experience if and only if there is some y (the experiencer) which

123

stands in the experience relation to x. If E is an abbreviation for 'experience relation,' then the class of y's such that y stands in the relation of E to something or other is the class of *experiencers*, and the x's to which something or other stands in the relation E constitute the class of *experiences*. An experience is not, then, a special class of objects on a par with other objects, but objects in a certain relation. The relation E will not here be exhaustively characterized (that is a central task for empiricism), but as a first approximation it can be said that to experience something is to take account of its properties by appropriate conduct; the experience is *direct* to the degree that this is done by direct response to the something in question, and *indirect* to the degree that it is done through the intermediacy of signs. For y_1 to experience x_1 it is sufficient that y_1Ex_1 holds; there is *conscious experience* if y_1Ex_1 is an experience (e.g., if $y_1E[y_1Ex_1]$ holds), otherwise the experience is *unconscious*. An experience x_1 is *de facto subjective* with respect to y_1 if y_1 is the only one who stands in the relation E to x_1; an experience x_1 is *intrinsically subjective* with respect to y_1, relative to a certain state of knowledge, if the known laws of nature permit the deduction that no other y can stand in this relation to x_1. An experience is *de facto intersubjective* if it is not *de facto* subjective, and it is *potentially intersubjective* if it is not intrinsically subjective. It should be noted that with such usages a person may not be able directly to experience aspects of himself that others can directly experience, so that the line between subjective and intersubjective experience in no sense coincides with the distinction between experiencers and external objects.

What bearing does this (tentative and preliminary) analysis have on the question of meaning? It may be admitted, if the facts warrant it, that there are certain experiences which are *de facto* subjective as far as direct experience is concerned and that this may even be true of the direct experience of the process of semiosis; there would be nothing surprising in the conclusion that, if I am the interpreter of a particular sign, there are then aspects of the process of interpretation which I can directly experience but which others cannot. The important point is that such a conclusion would not be in opposition to the thesis of *the potential intersubjectivity of every meaning.* The fact that y_1 and y_2 do not stand in the relation of direct experience to each other's respective direct experience of x_1 does not prevent them both from directly experiencing x_1, or from indirectly designating (and so indirectly experiencing) by the use of signs the experience relations in which the other stands—for under certain circumstances an object which cannot be directly experienced can, nevertheless, be denoted. Applying this result to the case of a particular sign, y_1 and y_2 may differ in their direct experience of the meaning situation and yet have the same meaning in common and, in general, be able to decide what the other means by a particular sign and the degree to which the two meanings are the same or different. For the determination of the meaning of S_1 (where S_1 is a sign vehicle) to y_1 it is not necessary that an investigator become y_1 or have his experiences of S_1: it is sufficient to determine how S_1 is related to other signs used by y_1, under what situations y_1 uses S_1 for purposes of designation, and what expectations y_1 has when he responds to S_1. To the degree that the same relations hold for y_2 as for y_1, then S_1 has the same meaning to y_1 and y_2; to the degree that the relations in question differ for y_1 and y_2, then S_1 has a different meaning.

In short, since the meaning of a sign is exhaustively specified by the ascertainment of its rules of usage, the meaning of any sign is in principle exhaustively determinable by objective investigation. Since it is then possible, if it seems wise, to standardize this usage, the result is that the meaning of every sign is potentially intersubjective. Even where the sign vehicle is intrinsically subjective there can be indirect confirmation that there is such a sign vehicle with such and such meaning. It is true that in practice the determination of meaning is difficult and that the differences in sign usages among persons of even the same social group may be rather great. But it is theoretically important to realize that the subjectivity of certain experiences, and even experiences of semiosis, is compatible with the possibility of an objective and exhaustive determination of any meaning whatsoever.

Having introduced the term 'meaning' only provisionally in order to bring out the implications of the position here taken, the use of the term will now be discontinued—it adds nothing to the set of semiotical terms. It may be pointed out that the preceding argument shows the agreement of what will be called *sign analysis* with the requirements of scientific investigation. Sign analysis is the determination of the syntactical, semantical, and pragmatical dimensions of specific processes of semiosis; it is the determination of the rules of usage of given sign vehicles. Logical analysis is, in the widest sense of the term 'logic,' identical with sign analysis; in narrower usages, logical analysis is some part of sign analysis, such as the study of the syntactical relations of the sign vehicle in question. Sign analysis (i.e., descriptive semiotic) can be carried on in accordance with all recognized principles of scientific procedure.

13. Universals and Universality

Certain aspects of the "universality" (or generality) of signs have long attracted attention, and their explanation has been a source of many philosophical disputes. By viewing the phenomena vaguely referred to under the overworked terms 'universals' and 'universality' through the prism of semiotical analysis, the various components of the problems may be separated and their relations seen.

The subject may be approached in terms of Peirce's distinction between a *sinsign* and a *legisign:* a sinsign is a particular something functioning as a sign, while a legisign is a "law" functioning as a sign. A particular series of marks at a specific place, such as 'house,' is a sinsign; such a specific set of marks is not, however, the English word *house,* for this word is "one," while its instances or replicas are as numerous as the various employments of the word. It is a law or habit of usage, a "universal" as over against its particular instances. Peirce was very much impressed by this situation and made the difference basic in his classification of signs; it gave an instance in the domain of signs of the phenomena of law (habit, Thirdness, mediation) upon whose objectivity Peirce was so insistent.

The account which has here been given is compatible with this general emphasis; the preceding section should have made clear that semiosis, as a functional process, is just as real and objective as are the component factors which function in the process. It must also be admitted that in a given instance of semiosis in which, say, 'house' functions as a sign vehicle, this sinsign or this particular instance of semiosis is not identical with the legisign *house*. What, then, is a legisign and where in semiosis are "universals" and "universality" to be found? In general, the answer must be that there is an element of universality or generality in all the dimensions and that confusion results here as elsewhere when these are not distinguished and when statements in the metalanguage are confounded with statements in the thing-language.

It is experimentally confirmable that in a given process of semiosis various sign vehicles may be substituted for the original sign vehicle without the occurrence of any relevant change in the remainder of the process. The metronome beat to which an animal is conditioned may move faster or slower within certain limits without the response of the animal undergoing change; the spoken word 'house' may be uttered at different times by the same or different persons, with various tonal changes, and yet will awaken the same response and be used to designate the same objects. If the word is written, the sizes may vary greatly, the letters may differ in style, the media used may be of various colors. The question of the limits of such variation and what remains constant within this range is in a given case very difficult to determine even by the use of the most careful experimental techniques, but of the fact of variability there is no doubt possible. Strictly speaking, the sign vehicle is only that aspect of the apparent sign vehicle in virtue of which semiosis takes place; the rest is semiotically irrelevant. To say that a given sign vehicle is "universal" (or general) is merely to say that it is one of a class of objects which have the property or properties necessary to arouse certain expectations, to combine in specified ways with other sign vehicles, and to denote certain objects, i.e., that it is one member of a class of objects all of which are subject to the same rules of sign usage. Thus 'house' and 'HOUSE' may be the same sign vehicles, but 'house' and 'Haus' are not; the fact that 'the house is red' conforms to the rules of English while 'the Haus is red' does not, shows that the sign vehicles are not the same, since the rules of usage are (in part) different. None of the disciplines concerned with signs is interested in the complete physical description of the sign vehicle but is concerned with the sign vehicle only in so far as it conforms to rules of usage.

In any specific case of semiosis the sign vehicle is, of course, a definite particular, a sinsign; its "universality," its being a legisign, consists only in the fact, statable in the metalanguage, that it is one member of a class of objects capable of performing the same sign function.

Another component of the problem enters in connection with the semantical dimension. The designatum of a sign is the class of objects which a sign can denote in virtue of its semantical rule. The rule may allow the sign to be applied to only one object, or to many but not to all, or to everything. Here "universality" is simply the potentiality of denoting more than one object or situation. Since such a statement is semantical, a statement can be made in terms of the converse of the relation of denotation: it can then be said that objects have the property of universality when denotable by the same sign. In so far as a number of objects or situations permit of a certain sign being applied, they conform to the conditions laid down in the semantical rule; hence there is something equally true of all of them, and in this respect or to this degree they are the same— whatever differences they may have are irrelevant to the particular case of semiosis. 'Universality (or generality) of objects' is a semantical term, and to talk as if 'universality' were a term in the thing-language, designating entities ("universals") in the world, is to utter pseudo thing-sentences of the quasi-semantical type. This fact was recognized in the Middle Ages in the doctrine that 'universality' was a term of second intention rather than of first; in contemporary terms, it is a term within semiotic and not a term in the thing-language. In the thing-language there are simply terms whose rules of usage make them applicable to a plurality of situations; expressed in terms of objects it can only be said that the world is such that often a number of objects or situations can be denoted by a given sign.

A similar situation appears in syntactics, where the relations of sign vehicles are studied in so far as these relations are determined by formation or transformation rules. A combination of sign vehicles is a particular, but it may share its form with other combinations of sign vehicles, i.e., a number of combinations of different sign vehicles may be instances of the same formation or transformation rule. In this case the particular sign combination has a formal or syntactical universality.

From the standpoint of pragmatics two considerations are relevant to the problem in hand. One is the correlative of the semantical situation which has already been described. The fact that certain sign vehicles may denote many objects corresponds to the fact that expectations vary in determinateness, so that a number of objects may satisfy an expectation. One expects a nice day tomorrow—and a number of weather conditions will meet the expectation. Hence, while a response in a particular situation is specific, it is a true statement within pragmatics that similar responses are often called out by a variety of sign vehicles and are satisfied by a variety of objects. From this point of view the interpretant (in common with any habit) has a character of "universality" which contrasts to its particularity in a specific situation. There is a second aspect of sign universality distinguishable in pragmatics, namely, the social universality which lies in the fact that a sign may be held in common by a number of interpreters.

It is accordingly necessary to distinguish in the universality appropriate to semiosis five types of universality. Since the term 'universality' has such a variety of usages, and is clearly inappropriate in some of the five cases, the term 'generality' will be used instead. There are, then, five types of sign generality: *generality of sign vehicle, generality of form, generality of denotation*, *generality of the interpretant*, and *social generality*. The central point is that each of these kinds of generality can be stated only within semiotic; generality is accordingly a relational concept, since all the branches of semiotic investigate only relations. To speak of something as a "general" or a "universal" is merely to use a pseudo thing-sentence instead of the unambiguous semiotical expression; such terms can only signify that the something in question stands to something or other in one of the relations embodied in the five kinds of sign generality which have been distinguished. In this way there is kept what is significant in the historical emphases of nominalism, realism, and conceptualism, while yet avoiding the last traces of the substantive or entitive conception of generality by recognizing the level of discourse appropriate to discussions of generality and the relational character of the terms employed at this level.

14. Interrelation of the Semiotical Sciences

Since the current tendency is in the direction of specialized research in syntactics, semantics, or pragmatics, it is well to stress emphatically the interrelations of these disciplines within semiotic. Indeed, semiotic, in so far as it is more than these disciplines, is mainly concerned with their interrelations, and so with the unitary character of semiosis which these disciplines individually ignore.

One aspect of the interrelation is indicated in the fact that while each of the component disciplines deals in one way or another with signs, none of them can define the term 'sign' and, hence, cannot define themselves. 'Syntactics' is not a term within syntactics but is a strictly semiotical term—and the same is true of 'semantics' and 'pragmatics.' Syntactics speaks of formation and transformation rules, but rules are possible modes of behavior and involve the notion of interpreter; 'rule' is, therefore, a pragmatical term. Semantics refers explicitly only to signs as designating objects or situations, but there is no such relation without semantical rules of usage, and so again the notion of interpreter is implicitly involved. Pragmatics deals directly only with signs as interpreted, but 'interpreter' and 'interpretant' cannot be defined without the use of 'sign vehicle' and 'designatum'—so that all of these terms are strictly semiotical terms. Such considerations—themselves only a few among many possible ones—show that, while the component semiotical disciplines do not as sciences refer to one another, yet they can be characterized and distinguished only in terms of the wider science of which they are components.

It is also true that a person who studies some dimension of semiosis uses terms which have all three dimensions and employs the results of the study of the other dimensions. The rules which govern the sign vehicles of the language being studied must be understood, and 'understanding' is a pragmatical term. The rules for combining and transforming possible sign vehicles cannot be composed merely of possible sign vehicles but must actually function as signs. In descriptive syntactics there must be signs to denote the sign vehicles being studied, and the aim must be to make true statements about these sign vehicles—but 'denote' and 'true' are not syntactical terms. Semantics will study the relation of a sign combination to what it denotes or can denote, but this involves the knowledge of the structure of the sign combination and the semantical rules in virtue of which the relation of denotation may obtain. Pragmatics cannot go far without taking account of the formal structures for which it should seek the pragmatical correlate, and of the relation of signs to objects which it seeks to explain through the notion of habit of usage. Finally, the languages of syntactics, semantics, and pragmatics have all three dimensions: they designate some aspect of semiosis, they have a formal structure, and they have a pragmatical aspect in so far as they are used or understood.

The intimate relation of the semiotical sciences makes semiotic as a science possible but does not blur the fact that the subsciences represent three irreducible and equally legitimate points of view corresponding to the three objective dimensions of semiosis. Any sign whatsoever may be studied from any of the three standpoints, though no one standpoint is adequate to the full nature of semiosis. Thus in one sense there is no limit to either point of view, i.e., no place at which an investigator must desert one standpoint for another. This is simply because they are studies of semiosis from different points of view; in fastening attention upon one dimension, each deliberately neglects the aspects of the process discernible in terms of the other standpoints. Syntactics, semantics, and pragmatics are components of the single science of semiotic but mutually irreducible components.

VII. Problems and Applications

15. Unification of the Semiotical Sciences

There remains the task of briefly showing the problems which remain open within semiotic and the possible fields of application. These may be roughly grouped under three headings: unification of the semiotical sciences, semiotic as organon of the sciences, and humanistic implications of semiotic. The remarks which follow aim merely to be suggestive—to indicate directions rather than solutions.

The account which has been given has been adapted to the purposes of an introduction. Large areas of the field were ignored, exactitude in statement was often sacrificed to avoid lengthy preliminary analysis, and the consideration of the examples which were introduced was carried only so far as to illuminate the point at issue. Even though the larger outlines of semiotic be correct, it is still far from the condition of an advanced science. Progress will require collaboration by many investigators. There is need both for fact-finders and for systematizers. The former must make clear the conditions under which semiosis occurs and what precisely takes place in the process; the latter must in the light of available facts develop a precise systematized theoretical structure which future fact-finders can in turn use. One theoretical problem of importance lies in the relation of the various kinds of rules. The theory of signs which has been given opens up many points of contact with the concrete work of biologists, psychologists, psychopathologists, linguists, and social scientists. Systematization can profitably make use of symbolic logic; for, since semiotic deals throughout with relations, it is peculiarly amenable to treatment in terms of the logic of relations. The work of fact-finders and systematizers is equally important and must go hand in hand; each provides material for the other.

Semioticians should find the history of semiotic useful both as a stimulus and as a field of application. Such hoary doctrines as the categories, the transcendentals, and the predicables are early sallies into semiotical domains and should be clarified by the later developments. Hellenistic controversies over the admonitive and the indicative sign, and the medieval doctrines of intention, imposition, and supposition are worth reviving and interpreting. The history of linguistics, rhetoric, logic, empiricism, and experimental science offers rich supplementary material. Semiotic has a long tradition, and in common with all sciences it should keep alive its history.

In the development of semiotic the disciplines which now are current under the names of logic, mathematics, and linguistics can be reinterpreted in semiotical terms. The logical paradoxes, the theory of types, the laws of logic, the theory of probability, the distinction of deduction, induction, and hypothesis, the logic of modality—all such topics permit of discussion within the theory of signs. In so far as mathematics is knowledge of linguistic structures, and not simply identified with some (or all) of such structures, it too may be considered as part of semiotic. Linguistics clearly falls within semiotic, dealing at present with certain aspects of the complex sign structures which constitute languages in the full semiotical sense of that term. It is possible that the admittedly unsatisfactory situation with respect to such terms as 'word,' 'sentence,' and 'part of speech' can be clarified in terms of the sign functions which various linguistic devices serve. Ancient projects of a universal grammar take on a new and defensible form when translated into the study of the way all languages perform similar sign functions by the use of different devices.

Logic, mathematics, and linguistics can be absorbed in their entirety within semiotic. In the case of certain other disciplines this may occur only in part. Problems which are often classed

133

as epistemological or methodological fall in large part under semiotic: thus empiricism and rationalism are at heart theories as to when the relation of denotation obtains or may be said to obtain; discussions of truth and knowledge are inseparably linked with semantics and pragmatics; a discussion of the procedures of scientists, when more than a chapter in logic, psychology, or sociology, must relate these procedures to the cognitive status of the statements which result from their application. In so far as aesthetics studies a certain functioning of signs (such as iconic signs whose designata are values), it is a semiotical discipline with syntactical, semantical, and pragmatical components, and the distinction of these components offers a base for aesthetic criticism. The sociology of knowledge is clearly part of pragmatics, and so is rhetoric; semiotic is the framework in which to fit the modern equivalents of the ancient trivium of logic, grammar, and rhetoric. It has already been suggested that psychology and the human social sciences may find part (if not the entire) basis of their distinction from other biological and social sciences in the fact that they deal with responses mediated by signs. The development of semiotic is itself a stage in the unification of sciences dealing in whole or in part with signs; it may also play an important role in bridging the gap between the biological sciences, on the one hand, and the psychological and human social sciences, on the other, and in throwing a new light upon the relation of the so-called "formal" and "empirical" sciences.

16. Semiotic as Organon of the Sciences

Semiotic holds a unique place among the sciences. It may be possible to say that every empirical science is engaged in finding data which can serve as reliable signs; it is certainly true that every science must embody its results in linguistic signs. Since this is so, the scientist must be as careful with his linguistic tools as he is in the designing of apparatus or in the making of observations. The sciences must look to semiotic for the concepts and general principles relevant to their own problems of sign analysis. Semiotic is not merely a science among sciences but an organon or instrument of all the sciences.

This function can be performed in two ways. One is by making training in semiotic a regular part of the equipment of the scientist. In this way a scientist would become critically conscious of his linguistic apparatus and develop careful habits in its use. The second way is by specific investigations of the languages of the special sciences. The linguistically expressed results of all the sciences is part of the subject matter of descriptive semiotic. Specific analyses of certain basic terms and problems in the various sciences will show the working scientist whatever relevance semiotic has in these fields more effectively than any amount of abstract argument. Other essays in the *Encyclopedia* may be regarded as contributing such studies. Current scientific formulations embody many pseudo problems which arise from the confusion of statements in the language of semiotic and the thing-language—recent discussions of indeterminism and complementarity in the physical sciences abound in illustrations. Empirical problems of a nonlinguistic sort are not solved by linguistic considerations, but it is important that the two kinds of problems not be confused and that nonlinguistic problems be expressed in such a form as aids their empirical solution. The classical logic thought of itself as the organon of the sciences but was, in fact, unable to play the role it set itself; contemporary semiotic, embodying in itself the newer logical developments and a wide variety of approaches to sign phenomena, may again attempt to assume the same role.

17. Humanistic Implications of Semiotic

Signs serve other purposes than the acquisition of knowledge, and descriptive semiotic is wider than the study of the language of science. Corresponding to the various purposes which signs serve, there have developed more or less specialized languages which follow to some extent the various dimensions of semiosis. Thus the mathematical form of expression is well adapted to stress the interrelation of terms in a language, letting the relation to objects and interpreters recede into the background; the language of empirical science is especially suitable for the description of nature; the languages of morality, the fine arts, and the applied arts are especially adapted to the control of behavior, the presentations of things or situations as objects of interest, and the manipulation of things to effect desired eventuations. In none of these cases are any of the dimensions of semiosis absent; certain of them are simply subordinated and partially transformed by the emphasis upon one of the dimensions. Mathematical propositions may have an empirical aspect (many indeed were discovered empirically), and mathematical problems may be set by problems in other fields, but the language of mathematics subordinates these factors in order to better accomplish the task it is developed to fulfil. Empirical science is not really concerned with simply getting all true statements possible (such as the statement of the area of each mark on this page) but in getting important true statements (i.e., statements that, on the one hand, furnish a secure base for prediction and, on the other hand, that aid in the creation of a systematic science)—but the language of empirical science is adapted to expressing the truth and not the importance of its statements. Lyric poetry has a syntax and uses terms which designate things, but the syntax and the terms are so used that what stand out for the reader are values and evaluations. The maxims of the applied arts rest on true propositions relevant to the accomplishment of certain purposes ("to accomplish x, do so and so"); moral judgments may similarly have an empirical component but, in addition, assume the desirability of reaching a certain end and aim to control conduct ("You ought to have your child vaccinated," i.e., "Taking the end of health for granted, vaccination is in the present situation the surest way of realizing that end, so have it done").

Semiotic provides a basis for understanding the main forms of human activity and their interrelationship, since all these activities and relations are reflected in the signs which mediate the activities. Such an understanding is an effective aid in avoiding confusion of the various functions performed by signs. As Goethe said, "One cannot really quarrel with any form of representation"—provided, of course, that the form of representation does not masquerade as what it is not. In giving such understanding, semiotic promises to fulfil one of the tasks which traditionally has been called philosophical. Philosophy has often sinned in confusing in its own language the various functions which signs perform. But it is an old tradition that philosophy should aim to give insight into the characteristic forms of human activity and to strive for the most general and the most systematic knowledge possible. This tradition appears in a modern form in the identification of philosophy with the theory of signs and the unification of science, that is, with the more general and systematic aspects of pure and descriptive semiotic.

Selected Bibliography

AJDUKIEWICZ, K. "Sprache und Sinn," *Erkenntnis,* Vol. IV (1934).

BENJAMIN, A. C. *The Logical Structure of Science,* chaps. vii, viii, and ix. London, 1936.

CARNAP, R. *Philosophy and Logical Syntax.* London, 1935.

———. *Logical Syntax of Language.* Vienna, 1934; London, 1937.

———. "Testability and Meaning," *Philosophy of Science.* Vol. III (1936); *ibid.,* Vol. IV (1937).

CASSIRER, E. *Die Philosophie der symbolischen Formen.* 3 vols. Berlin, 1923 ff.

EATON, R. M. *Symbolism and Truth.* Cambridge, Mass., 1925.

GÄTSCHENBERGER, R. *Zeichen.* Stuttgart, 1932.

HUSSERL, E. *Logische Untersuchungen,* Vol. II, Part I. 4th ed. Halle, 1928.

KOKOSZYNSKA, M. "Über den absoluten Wahrheitsbegriff und einige andere semantische Begriffe," *Erkenntnis,* Vol. VI (1936).

MEAD, G. H. *Mind, Self, and Society.* Chicago, 1934.

———. *The Philosophy of the Act.* Chicago, 1938.

MORRIS, C. W. *Logical Positivism, Pragmatism, and Scientific Empiricism.* Paris, 1937.

OGDEN, C. K., and RICHARDS, I. A. *The Meaning of 'Meaning.'* London, 1923.

PEIRCE, C. S. *Collected Papers,* esp. Vol. II. Cambridge, Mass., 1931 ff.

REICHENBACH, H. *Experience and Prediction,* chaps. i and ii. Chicago, 1938.

SCHLICK, M. *Gesammelte Aufsätze, 1926–1936.* Vienna, 1938.

TARSKI, A. "Grundlegung der wissenschaftlichen Semantik," *Actes du congres international de philosophe scientifique.* Paris, 1936.

———. "Der Wahrheitsbegriff in den formalisierten Sprachen," *Studia philosophica,* Vol. I (1935).

WITTGENSTEIN, L. *Tractatus logico-philosophicus.* London, 1922.

Foundations of Logic and Mathematics

Rudolf Carnap

Foundations of Logic and Mathematics

Contents:

I. LOGICAL ANALYSIS OF LANGUAGE: SEMANTICS AND
SYNTAX   PAGE
  1. Theoretical Procedures in Science   143
  2. Analysis of Language 145
  3. Pragmatics of Language B   147
  4. Semantical Systems   148
  5. Rules of the Semantical System B-S   .   .   . 150
  6. Some Terms of Semantics   153
  7. L-Semantical Terms   154
  8. Logical Syntax 158
  9. The Calculus B-C 160

II. CALCULUS AND INTERPRETATION
  10. Calculus and Semantical System 163
  11. On the Construction of a Language System   . 166
  12. Is Logic a Matter of Convention?   168

III. CALCULI AND THEIR APPLICATION IN EMPIRICAL
SCIENCE
  13. Elementary Logical Calculi   171
  14. Higher Logical Calculi   175
  15. Application of Logical Calculi   177

Contents

16. General Remarks about Nonlogical Calculi (Axiom Systems) 179
17. An Elementary Mathematical Calculus . . . 180
18. Higher Mathematical Calculi 184
19. Application of Mathematical Calculi . . . 186
20. The Controversies over "Foundations" of Mathematics 190
21. Geometrical Calculi and Their Interpretations 193
22. The Distinction between Mathematical and Physical Geometry 195
23. Physical Calculi and Their Interpretations . 198
24. Elementary and Abstract Terms 203
25. "Understanding" in Physics 209

SELECTED BIBLIOGRAPHY 211

INDEX OF TERMS 213

Foundations of Logic and Mathematics

Rudolf Carnap

I. Logical Analysis of Language: Semantics and Syntax

1. Theoretical Procedures in Science

The activities of a scientist are in part practical: he arranges experiments and makes observations. Another part of his work is theoretical: he formulates the results of his observations in sentences, compares the results with those of other observers, tries to explain them by a theory, endeavors to confirm a theory proposed by himself or somebody else, makes predictions with the help of a theory, etc. In these theoretical activities, deduction plays an important part; this includes calculation, which is a special form of deduction applied to numerical expressions. Let us consider, as an example, some theoretical activities of an astronomer. He describes his observations concerning a certain planet in a report, O_1. Further, he takes into consideration a theory T concerning the movements of planets. (Strictly speaking, T would have to include, for the application to be discussed, laws of some other branches of physics, e.g., concerning the astronomical instruments used, refraction of light in the atmosphere, etc.) From O_1 and T, the astronomer deduces a prediction, P; he calculates the apparent position of the planet for the next night. At that time he will make a new observation and formulate it in a report O_2. Then he will compare the prediction P with O_2 and thereby find it either confirmed or not. If T was a new theory and the purpose of the procedure described was to test T, then the astronomer will take the confirmation of P by O_2 as a partial confirmation for T; he will apply the same procedure again and again and thereby obtain either an increasing degree of confirmation for T or else a disconfirmation. The same deduction of P from O_1 and T is made in the case where T is already scientifically acknowledged on the

143

basis of previous evidence, and the present purpose is to obtain a prediction of what will happen tomorrow. There is a third situation in which a deduction of this kind may be made. Suppose we have made both the observations described in O_1 and in O_2; we are surprised by the results of the observation described in O_2 and therefore want an explanation for it. This explanation is given by the theory T; more precisely, by deducing P from O_1 and T and then showing that O_2 is in accordance with P ("What we have observed is exactly what we had to expect").

These simple examples show that the chief theoretical procedures in science—namely, testing a theory, giving an explanation for a known fact, and predicting an unknown fact—involve as an essential component deduction and calculation; in other words, the application of logic and mathematics. (These procedures will later be discussed more in detail, especially in §§ 15, 19, and 23.) It is one of the chief tasks of this essay to make clear the role of logic and mathematics as applied in empirical science. We shall see that they furnish instruments for deduction, that is, for the transformation of formulations of factual, contingent knowledge. However, logic and mathematics not only supply rules for transformation of factual sentences but they themselves contain sentences of a different, nonfactual kind. Therefore, we shall have to deal with the question of the nature of logical and mathematical theorems. It will become clear that they do not possess any factual content. If we call them true, then another kind of truth is meant, one not dependent upon facts. A theorem of mathematics is not tested like a theorem of physics, by deriving more and more predictions with its help and then comparing them with the results of observations. But what else is the basis of their validity? We shall try to answer these questions by examining how theorems of logic and mathematics are used in the context of empirical science.

The material on which the scientist works in his theoretical activities consists of reports of observations, scientific laws and theories, and predictions; that is, formulations in language which describe certain features of facts. Therefore, an analysis of theoretical procedures in science must concern itself with language and its applications. In the present section, in preparing for the later task, we shall outline an analysis of language and explain the chief factors involved. Three points of view will be distinguished, and accordingly three disciplines applying them, called pragmatics, semantics, and syntax. These will be illustrated by the analysis of a simple, fictitious language. In the later sections the results of these discussions will be applied in an analysis of the theoretical procedure of science, especially from the point of view of calculi, their interpretation, and their application in empirical science.

2. Analysis of Language

A language, as, e.g., English, is a system of activities or, rather, of habits, i.e., dispositions to certain activities, serving mainly for the purposes of communication and of co-ordination of activities among the members of a group. The elements of the language are signs, e.g., sounds or written marks, produced by members of the group in order to be perceived by other members and to influence their behavior. Since our final interest in this essay concerns the language of science, we shall restrict ourselves to the theoretical side of language, i.e., to the use of language for making assertions. Thus, among the different kinds of sentences, e.g., commands, questions, exclamations, declarations, etc., we shall deal with declarative sentences only. For the sake of brevity we shall call them here simply *sentences.*

This restriction to declarative sentences does not involve, in the investigation of processes accompanying the use of language, a restriction to theoretical thinking. Declarative sentences, e.g., 'This apple is sour', are connected not only with the theoretical side of behavior but also with emotional, volitional, and other factors. If we wish to investigate a language as a human activity, we must take into consideration all these factors connected with speaking activities. But the sentences, and the signs (e.g., words) occurring in them, are sometimes involved in still another relation. A sign or expression may con-

145

cern or designate or describe something, or, rather, he who uses the expression may intend to refer to something by it, e.g., to an object or a property or a state of affairs; this we call the *designatum* of the expression. (For the moment, no exact definition for 'designatum' is intended; this word is merely to serve as a convenient, common term for different cases—objects, properties, etc.—whose fundamental differences in other respects are not hereby denied.) Thus, three components have to be distinguished in a situation where language is used. We see these in the following example: (1) the action, state, and environment of a man who speaks or hears, say, the German word 'blau'; (2) the word 'blau' as an element of the German language (meant here as a specified acoustic [or visual] design which is the common property of the many sounds produced at different times, which may be called the tokens of that design); (3) a certain property of things, viz., the color blue, to which this man—and German-speaking people in general—intends to refer (one usually says, "The man means the color by the word", or "The word means the color for these people", or ". . . . within this language").

The complete theory of language has to study all these three components. We shall call *pragmatics* the field of all those investigations which take into consideration the first component, whether it be alone or in combination with the other components. Other inquiries are made in abstraction from the speaker and deal only with the expressions of the language and their relation to their designata. The field of these studies is called *semantics*. Finally, one may abstract even from the designata and restrict the investigation to formal properties—in a sense soon to be explained—of the expressions and relations among them. This field is called *logical syntax*. The distinction between the three fields will become more clear in our subsequent discussions.

That an investigation of language has to take into consideration all the three factors mentioned was in recent times made clear and emphasized especially by C. S. Peirce, by Ogden and Richards, and by Morris (see Vol. I, No. 2). Morris made it the basis for the three fields into which he divides semiotic (i.e., the general theory of signs), namely, pragmatics, semantics, and syntactics. Our division is in agreement with his in its chief features. For general questions concerning language and its use compare also Bloomfield, Volume I, No. 4.

3. Pragmatics of Language B

In order to make clear the nature of the three fields and the differences between them, we shall analyze an example of a language. We choose a fictitious language B, very poor and very simple in its structure, in order to get simple systems of semantical and syntactical rules.

Whenever an investigation is made about a language, we call this language the *object-language* of the investigation, and the language in which the results of the investigation are formulated the *metalanguage*. Sometimes object-language and metalanguage are the same, e.g., when we speak in English about English. The theory concerning the object-language which is formulated in the metalanguage is sometimes called metatheory. Its three branches are the pragmatics, the semantics, and the syntax of the language in question. In what follows, B is our object-language, English our metalanguage.

Suppose we find a group of people speaking a language B which we do not understand; nor do they understand ours. After some observation, we discover which words the people use, in which forms of sentences they use them, what these words and sentences are about, on what occasions they are used, what activities are connected with them, etc. Thus we may have obtained the following results, numbered here for later reference.

Pragm. 1.—Whenever the people utter a sentence of the form '. . . ist kalt', where '. . .' is the name of a thing, they intend to assert that the thing in question is cold.

Pragm. 2a.—A certain lake in that country, which has no name in English, is usually called 'titisee'. When using this name, the people often think of plenty of fish and good meals.

Pragm. 2b.—On certain holidays the lake is called 'rumber';

147

when using this name, the people often think—even during good weather—of the dangers of storm on the lake.

Pragm. 3.—The word 'nicht' is used in sentences of the form 'nicht . . .', where '. . .' is a sentence. If the sentence '. . .' serves to express the assertion that such and such is the case, the whole sentence 'nicht . . .' is acknowledged as a correct assertion if such and such is not the case.

In this way we slowly learn the designata and mode of use of all the words and expressions, especially the sentences; we find out both the cause and the effect of their utterance. We may study the preferences of different social groups, age groups, or geographical groups in the choice of expressions. We investigate the role of the language in various social relations, etc.

The pragmatics of language B consists of all these and similar investigations. Pragmatical observations are the basis of all linguistic research. We see that pragmatics is an empirical discipline dealing with a special kind of human behavior and making use of the results of different branches of science (principally social science, but also physics, biology, and psychology).

4. Semantical Systems

We now proceed to restrict our attention to a special aspect of the facts concerning the language B which we have found by observations of the speaking activities within the group who speak that language. We study the relations between the expressions of B and their designata. On the basis of those facts we are going to lay down a system of rules establishing those relations. We call them *semantical rules*. These rules are not unambiguously determined by the facts. Suppose we have found that the word 'mond' of B was used in 98 per cent of the cases for the moon and in 2 per cent for a certain lantern. Now it is a matter of our decision whether we construct the rules in such a way that both the moon and the lantern are designata of 'mond' or only the moon. If we choose the first, the use of 'mond' in those 2 per cent of cases was right—with respect to our rules; if we choose the second, it was wrong. The facts do not determine whether the use of a certain expression is right or wrong but only how often it occurs and how often it leads to the effect intended, and the like. A question of right or wrong must always refer to a system of rules. Strictly speaking, the rules which we shall lay down are not rules of the factually given language B; they rather constitute a language system corresponding to B which we will call the *semantical system B-S*. The language B belongs to the world of facts; it has many properties, some of which we have found, while others are unknown to us. The language system B-S, on the other hand, is something constructed by us; it has all and only those properties which we establish by the rules. Nevertheless, we construct B-S not arbitrarily but with regard to the facts about B. Then we may make the empirical statement that the language B is to a certain degree in accordance with the system B-S. The previously mentioned pragmatical facts are the basis—in the sense explained—of some of the rules to be given later (Pragm. 1 for SD 2*a* and SL 1, Pragm. 2*a,b* for SD 1*a*, Pragm. 3 for SL 2).

We call the elements of a semantical system *signs;* they may be words or special symbols like '0', '+', etc. A sequence consisting of one or several signs is called an *expression*. As signs of the system B-S we take the words which we have found by our observations to be words of B or, rather, only those words which we decide to accept as "correct." We divide the signs of B-S—and, in an analogous way, those of any other semantical system—into two classes: *descriptive* and *logical* signs. As descriptive signs we take those which designate things or properties of things (in a more comprehensive system we should classify here also the relations among things, functions of things, etc.). The other signs are taken as logical signs: they serve chiefly for connecting descriptive signs in the construction of sentences but do not themselves designate things, properties of things, etc. Logical signs are, e.g., those corresponding to English words like 'is', 'are', 'not', 'and', 'or', 'if', 'any', 'some', 'every', 'all'. These unprecise explanations will suffice here. Our later discussions will show some of the differentiae of the two classes of signs.

Semantics as an exact discipline is quite new; we owe it to the very fertile school of contemporary Polish logicians. After some of this group, especially Lesniewski and Ajdukiewicz, had discussed semantical questions, Tarski, in his treatise on truth, made the first comprehensive systematic investigation in this field, giving rise to very important results.

5. Rules of the Semantical System B-S

In order to show how semantical rules are to be formulated and how they serve to determine truth conditions and thereby give an interpretation of the sentences, we are going to construct the semantical rules for the system B-S. As preliminary steps for this construction we make a classification of the signs and lay down rules of formation. Each class is defined by an enumeration of the signs belonging to it. The signs of B-S are divided into descriptive and logical signs. The descriptive signs of B-S are divided into names and predicates. Names are the words 'titisee', 'rumber', 'mond', etc. (here a complete list of the names has to be given). Predicates are the words 'kalt', 'blau', 'rot', etc. The logical signs are divided into logical constants ('ist', 'nicht', 'wenn', 'so', 'fuer', 'jedes') and variables ('x', 'y', etc.). For the general description of forms of expressions we shall use blanks like '. . .', '- - -', etc. They are not themselves signs of B-S but have to be replaced by expressions of B-S. If nothing else is said, a blank stands for any expression of B-S. A blank with a subscript 'n', 'p', 's', or 'v' (e.g., '. . .$_n$') stands for a name, a predicate, a sentence, or a variable, respectively. If the same blank occurs several times within a rule or a statement, it stands at all places for the same expression.

The rules of formation determine how sentences may be constructed out of the various kinds of signs.

Rules of formation.—An expression of B-S is called a *sentence* (in the semantical sense) or a *proposition* of B-S, if and only if it has one of the following forms, F 1–4. F 1: '. . .$_n$ ist - - -$_p$' (e.g., 'mond ist blau'); F 2: 'nicht. . .' (e.g., 'nicht mond ist blau'); F 3: 'wenn . . .$_s$, so - - -$_s$' (e.g., 'wenn titisee ist rot, so mond ist kalt'); F 4: 'fuer jedes . .$_v$, - . . -', where '- . . -' stands for an expression which is formed out of a sentence not containing a variable by replacing one or several names by the variable

'. .ᵥ' (e.g., 'fuer jedes *x*, *x* ist blau'; 'fuer jedes *y*, wenn *y* ist blau, so *y* ist kalt'). The partial sentence in a sentence of the form F 2 and the two partial sentences in a sentence of the form F 3 (indicated above by blanks) are called *components* of the whole sentence. In order to indicate the components of a sentence in case they are themselves compound, commas and square brackets are used when necessary.

Rules B-SD. Designata of descriptive signs:

SD 1. The *names* designate things, and especially

    a) each of the thing-names 'titisee' and 'rum⸍ɔer' designates the lake at such and such a longitude and latitude.

    b) 'mond' designates the moon.

    Etc. [Here is to be given a complete list of rules for all the names of B-S.]

SD 2. The *predicates* designate properties of things, and especially

    a) 'kalt' designates the property of being cold.

    b) 'blau' designates the property of being blue.

    c) 'rot' designates the property of being red.

    Etc. [for all predicates].

Rules B-SL. Truth conditions for the sentences of B-S. These rules involve the *logical signs*. We call them the L-semantical rules of B-S.

SL 1. 'ist', form F 1. A sentence of the form '. . .ₙ ist - - -ₚ' is true if and only if the thing designated by '. . .ₙ' has the property designated by '- - -ₚ'.

SL 2. 'nicht', form F 2. A sentence of the form 'nicht . . .ₛ' is true if and only if the sentence '. . .ₛ' is not true.

SL 3. 'wenn' and 'so', form F 3. A sentence of the form 'wenn . . .ᵣ, so - - -ₛ' is true if and only if '. . .ᵣ' is not true or '- - -ₛ' is true.

SL 4. 'fuer jedes', form F 4. A sentence of the form 'fuer jedes . .ᵥ, - . . -', where '- . . -' is an expression formed out of a sentence by replacing one or several names by the variable '. .ᵥ', is true if and only if all sentences of the following kind are true: namely, those sentences constructed out of the expression '- . . -' by replacing the variable '. .,' at all places where it occurs within that expression by a name, the same for all places; here names of any things may be taken, even of those for which there is no name in the list of names in B-S. (Example: The sentence 'fuer jedes x, x ist blau' is true if and only if every sentence of the form '. . ., ist blau' is true; hence, according to SL 1, if and only if everything is blue.)

The rule SL 1, in combination with SD, provides direct truth conditions for the sentences of the simplest form; direct, since the rule does not refer to the truth of other sentences. SL 2–4 provide indirect truth conditions for the compound sentences by referring to other sentences and finally back to sentences of the simplest form. Hence the rules B-SD and SL together give a general definition of '*true* in B-S' though not in explicit form. (It would be possible, although in a rather complicated form, to formulate an explicit definition of 'true in B-S' on the basis of the rules given.) A sentence of B-S which is not true in B-S is called *false* in B-S.

If a sentence of B-S is given, one can easily construct, with the help of the given rules, a direct *truth-criterion* for it, i.e., a necessary and sufficient condition for its truth, in such a way that in the formulation of this condition no reference is made to the truth of other sentences. Since to know the truth conditions of a sentence is to know what is asserted by it, the given semantical rules determine for every sentence of B-S what it asserts—in usual terms, its "meaning"—or, in other words, how it is to be translated into English.

Examples: (1) The sentence 'mond ist blau' is true if and only if the moon is blue. (2) The sentence 'fuer jedes x, wenn x ist blau, so x ist kalt' is true if and only if every thing—not only those having a name in B-S—either is not blue or is cold; in other words, if all blue things are cold. Hence, this sentence asserts that all blue things are cold; it is to be translated into the English sentence 'all blue things are cold'.

Therefore, we shall say that we *understand* a language system, or a sign, or an expression, or a sentence in a language system, if we know the semantical rules of the system. We shall also say that the semantical rules give an *interpretation* of the language system.

We have formulated the semantical rules of the descriptive signs by stating their designata, for the logical signs by stating truth conditions for the sentences constructed with their help. We may mention here two other ways of formulating them which are often used in the practice of linguistics and logic. The first consists in giving *translations* for the signs and, if necessary, for the complex expressions and sentences, as it is done in a dictionary. The second way consists in stating *designata* throughout, not only for the descriptive signs as in SD, but also for expressions containing the logical signs, corresponding to SL. Example (corresponding to SL 1): A sentence of the form '. . .$_n$ ist - - -$_p$' designates (the state of affairs) that the thing designated by '. . .$_n$' has the property designated by '- - -$_p$'.

6. Some Terms of Semantics

We shall define some more terms which belong to the metalanguage and, moreover, to the semantical part of the metalanguage (as is seen from the fact that the definitions refer to the semantical rules). Any semantical term is relative to a semantical system and must, in strict formulation, be accompanied by a reference to that system. In practice the reference may often be omitted without ambiguity (thus we say, e.g., simply 'synonymous' instead of 'synonymous in B-S').

Two expressions are said to be semantically synonymous, or briefly, *synonymous*, with each other in a semantical system S if they have the same designatum by virtue of the rules of S. Hence, according to SD 1a, the signs 'titisee' and 'rumber' are semantically synonymous with one another in B-S. They are, however, not what we might call pragmatically synonymous in B, as is shown by Pragm. 2a,b. Since the transition from pragmatics to semantics is an abstraction, some properties drop out of consideration and hence some distinctions disappear. Because of the semantical synonymity of the names mentioned, the sentences 'titisee ist kalt' and 'rumber ist kalt' are also semantically synonymous. These two sentences have the same truth conditions, although different pragmatical conditions of application. Suppose that the lake is cold and hence the sentence 'titisee ist kalt' is true. Then the sentence 'rumber is kalt' is also true, even if sinfully spoken on a working day. If this happened by mistake, people would tell the speaker that he is right in his belief but that he ought to formulate it—i.e., the same belief—in another way.

We shall apply the semantical terms to be defined not only to sentences but also to classes of sentences. In what follows we shall use 'S_1', 'S_2', etc., for sentences; 'C_1', 'C_2', etc., for classes of sentences; 'T_1', 'T_2', etc., stand both for sentences and for classes of sentences. (These 'S' and 'C' with subscripts have nothing to do with the same letters without subscripts, which we use for semantical systems and calculi, e.g., 'B-S' and 'B-C'.) We understand the assertion of a class of sentences C_1 as a simultaneous assertion of all the sentences belonging to C_1; therefore, we make the following definition: a *class* of sentences C_1 is called *true* if all sentences of C_1 are true; false, if at least one of them is false. T_1 and T_2 (i.e., two sentences, or two classes of sentences, or one sentence and one class) are called *equivalent* with each other, if either both are true or both are false. T_2 is called an *implicate* of T_1, if T_1 is false or T_2 is true. T_1 is said to *exclude* T_2 if not both are true.

7. L-Semantical Terms

Let us compare the following two sentences: 'Australia is large' (S_1) and 'Australia is large or Australia is not large' (S_2). We see that they have a quite different character; let us try to give an exact account of their difference. We learn S_1 in geography but S_2 in logic. In order to find out for each of these sentences whether it is true or false, we must, of course, first understand the language to which it belongs. Then, for S_1 we have to know, in addition, some facts about the thing whose name occurs in it, i.e., Australia. Such is not the case for S_2. Whether Australia is large or small does not matter here; just by understanding S_2 we become aware that it must be right. If we agree to use the same term 'true' in both cases, we may express their difference by saying that S_1 is factually (or empirically) true while S_2 is logically true. These unprecise explanations can easily be transformed into precise definitions by replacing the former reference to understanding by a reference to semantical rules. We call a sentence of a semantical system S (logically true or) *L-true* if it is true in such a way that the semantical rules of S suffice for establishing its truth. We call a sentence (logically false or) *L-false* if it is false in such a way that the semantical rules suffice for finding that it is false. The two terms just defined and all other terms defined on their basis we call *L-semantical terms*. If a sentence is either L-true or L-false, it is called *L-determinate*, otherwise (L-indeterminate or) *factual*. (The terms 'L-true', 'L-false', and 'factual' correspond to the terms 'analytic', 'contradictory', and 'synthetic', as they are used in traditional terminology, usually without exact definitions.) If a factual sentence is true, it is called (factually true or) *F-true;* if it is false, (factually false or) *F-false*. Every sentence which contains only logical signs is L-determinate. This is one of the chief characteristics distinguishing logical from descriptive signs. (Example: 'For every object x and every property F, if x is an F then x is an F' is L-true. There are no sentences of this kind in the system B-S.)

Classification of sentences of a semantical system:

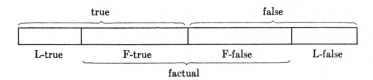

Examples of sentences in B-S: (1) We found earlier (§ 5) that the sentence 'mond ist blau' (S_1) is true in B-S if and only if the moon is blue. Hence, in order to find out whether S_1 is true or false, not only must we know the rules of B-S but we have to make observations of the moon. Hence S_1 is not L-determinate but factual. (2) Let us analyze the sentence 'wenn mond ist blau, so mond is blau' (S_2). According to rule SL 3, a 'wenn-so' sentence is true if its first component is not true or its second component is true. Now, if S_1 is true, the second component of S_2 is true, and hence S_2 is true; and if S_1 is not true, then the first component of S_2 is not true, and hence S_2 is again true. Thus S_2 is true in any case, independently of the facts concerning the moon; it is true merely in virtue of rule SL 3. Therefore S_2 is L-true. (3) The sentence 'nicht, wenn mond ist blau, so mond ist blau' (S_3) has S_2 as its com-

Foundations of Logic and Mathematics

ponent; and we found S_2 to be true on the basis of SL 3. Therefore, according to SL 2, S_3 is not true but false. And, moreover, it is false not because some fact happens to be the case but merely by virtue of the rules SL 3 and 2. Hence, S_3 is L-false.

Terminological remark.—The use of the word 'true' in everyday language and in philosophy is restricted by some to factual sentences, while some others use it in a wider sense, including analytic sentences. We adopted here the wider use; it is more customary in modern logic (e.g., 'truth function', 'truth-value-table'), and it turns out to be much more convenient. Otherwise, we should always have to say in the semantical rules and in most of the semantical theorems 'true or analytic' instead of 'true'. Semantical rules stating truth-conditions in the sense of 'F-true' would become very complicated and indeed indefinite.

The definitions given can easily be transferred to classes of sentences. C_1 is called L-true if it is possible to find out that C_1 is true with the help of the semantical rules alone, hence if all sentences of C_1 are L-true. C_1 is called L-false if it is possible to find out with the help of the semantical rules that C_1 is false, i.e., that at least one sentence of C_1 is false (in this case, however, all sentences of C_1 may be factual). If C_1 is either L-true or L-false, it is called L-determinate, otherwise factual.

If the semantical rules suffice to show that T_2 is an implicate of T_1, we call T_2 an *L-implicate* of T_1. This relation of L-implication is one of the fundamental concepts in logical analysis of language. The criterion for it can also be formulated in this way: the semantical rules exclude the possibility of T_1 being true and T_2 false; or in this way: according to the semantical rules, if T_1 is true, T_2 must be true. This last formulation of the criterion shows that L-implication, as defined here, is essentially the same as what is usually called logical consequence or deducibility or strict implication or entailment, although the form of the definitions of these terms may be different. Our definition is a semantical one as it refers to the semantical rules. Later we shall discuss the possibility of defining a corresponding syntactical term.

Examples: (1) 'mond ist rot' (S_1); 'wenn mond ist rot, so titisee ist kalt' (S_2); 'titisee ist kalt' (S_3). We shall see that S_3 is an L-implicate of the class C_1 consisting of S_1 and S_2. According to the definition of 'implicate' (§ 6), if S_3 is true, S_3 is an implicate of C_1. The same holds if S_1 is false because C_1 is then also false. The only remaining case is that S_1 is true and S_3 is false. In this case, according to rule SL 3 (§ 5), S_2 is false and, hence, C_1 is false too, and S_3 is an implicate of C_1. Thus we have found, without examining the facts described by the sentences, and merely by referring to the semantical rules, that S_3 is an implicate of C_1. Therefore, S_3 is an L-implicate of C_1. (2) 'fuer jedes x, x ist blau' (S_4); 'mond ist blau' (S_5). We shall see that S_5 is an L-implicate of S_4. If S_5 is true, S_5 is an implicate of S_4. And if S_5 is not true, then according to SL 4 (§ 5), S_4 is not true, and, hence, S_5 is again an implicate of S_4. We found this result by merely referring to a semantical rule. Therefore, S_5 is an L-implicate of S_4.

T_1 and T_2 are said to be *L-equivalent* if the semantical rules suffice to establish their equivalence, in other words, if T_1 and T_2 are L-implicates of each other. L-equivalent sentences have the same truth conditions; therefore, they say the same thing, although the formulations may be quite different.

Example: 'mond ist kalt' (S_1); 'nicht, mond ist kalt' (S_2); 'nicht, nicht, mond ist kalt' (S_3). These sentences are factual; the semantical rules do not suffice for finding out their truth or falsity. But they suffice for showing that S_1 and S_3 are equivalent. If S_1 is true, S_2 is, according to SL 2 (§ 5), false, and hence S_3 true. Therefore, in this case, S_1 and S_3 are equivalent. And, if S_1 is false, then S_2 is true and S_3 is false; hence, S_1 and S_3 are again equivalent. Thus, on the basis of the semantical rules, S_1 and S_3 cannot be other than equivalent. Therefore they are L-equivalent.

If S_1 is an L-true sentence, then the truth of S_1 can be established without any regard to the facts, e.g., to the properties of the things whose names occur in S_1. Therefore, S_1 does not convey any information about facts; this is sometimes formulated by saying that an L-true sentence has no factual content. Suppose S_2 to be an L-implicate of the class of sentences C_1. Then S_2 is an implicate of C_1, and hence, if the sentences of C_1 are true, S_2 is also true; and, moreover, this relation between C_1 and S_2 can be found to hold without taking into account any facts. Therefore, S_2 does not furnish any new information concerning facts that were not already given by C_1. This is sometimes expressed by saying that logical deduction does not increase the factual content of the premises. The two characteristics just explained of L-truth and L-implication (which have been especially emphasized by Wittgenstein) are very important for a clear understanding of the relation between logic and empirical knowledge. We shall see later that they hold also for mathematical theorems and mathematical deductions even if applied in empirical science (§ 19).

8. Logical Syntax

We distinguished three factors in the functioning of language: the activities of the speaking and listening persons, the designata, and the expressions of the language. We abstracted from the first factor and thereby came from pragmatics to semantics. Now we shall abstract from the second factor also and thus proceed from semantics to syntax. We shall take into consideration only the expressions, leaving aside the objects, properties, states of affairs, or whatever may be designated by the expressions. The relation of designation will be disregarded entirely. As this relation is the basis of the whole semantical system, it might seem as if nothing would be left. But we shall soon see that this is not the case.

A definition of a term in the metalanguage is called *formal* if it refers only to the expressions of the object-language (or, more exactly, to the kinds of signs and the order in which they occur in the expressions) but not to any extralinguistic objects and especially not to the designata of the descriptive signs of the object-language. A term defined by a formal definition is also called formal, as are questions, proofs, investigations, etc., in which only formal terms occur. We call the formal theory of an object-language, formulated in the metalanguage, the *syntax* of the object-language (or the logical syntax, whenever it seems necessary to distinguish this theory from that part of linguistics which is known as syntax but which usually is not restricted to formal terms). A formal definition, term, analysis, etc., is then also called syntactical.

The definitions of all semantical terms refer directly or indirectly to designata. But some of these terms—e.g., 'true', 'L-true', 'L-implicate'—are attributed not to designata but only to expressions; they designate properties of, or relations between, expressions. Now our question is whether it is possible to define within syntax, i.e., in a formal way, terms which correspond more or less to those semantical terms, i.e., whose extensions coincide partly or completely with theirs. The development of syntax—chiefly in modern symbolic logic—has led to an affirmative answer to that question. Especially is the possibility of defining in a formal way terms which completely correspond to 'L-true' and 'L-implicate' of fundamental importance. This shows that logical deduction can be completely formalized.

A *syntactical system* or *calculus* (sometimes also called a formal deductive system or a formal system) is a system of formal rules which determine certain formal properties and relations of sentences, especially for the purpose of formal deduction. The simplest procedure for the construction of a calculus consists in laying down some sentences as primitive sentences (sometimes called postulates or axioms) and some rules of inference. The primitive sentences and rules of inference are used for two purposes, for the construction of proofs and of derivations. We shall call the sentences to which the proofs lead *C-true* sentences (they are often called provable or proved sentences or theorems of the calculus). A derivation leads from any not necessarily C-true sentences, called the premises, to a sentence, called the conclusion. We shall call the conclusion a *C-implicate* of the class of premises (it is sometimes called derivable or derived or [formally] deducible or deduced from the premises or a [formal] consequence of the premises). A calculus may (but usually does not) also contain rules which determine certain sentences as *C-false*. If the rules of a calculus determine some sentence as both C-true and C-false, the calculus is called *inconsistent;* otherwise *consistent*. (If, as is usually done, no rules for 'C-false' are given, the calculus cannot be inconsistent.) In order to explain this procedure, we shall construct the calculus B-C as an example.

Logical syntax has chiefly grown out of two roots, one being formal logic, founded by Aristotle, the other the axiomatic method, initiated by Euclid. The general idea of operations with calculi goes back to Leibniz; since the middle of the last century it has been developed in the systems of symbolic logic into a comprehensive discipline. Among the founders of symbolic logic, or logistic, Boole (1854) is especially to be mentioned. More comprehensive systems (including the higher functional calculus [see § 14]) were created by Schroeder (1890), Frege (1893), Peano (1895), and Whitehead and Russell (1910). Frege was the first to formulate explicitly and to fulfil strictly the requirement of formality, i.e., of a formulation of rules of logic without any reference to designata. Hilbert considerably developed the axiomatic method, in its application both to geometry (see § 21) and to classical mathematics (see §§ 18 and 20).

9. The Calculus B-C

While the sentences of a semantical system are interpreted, assert something, and therefore are either true or false, within a calculus the sentences are looked at from a purely formal point of view. In order to emphasize this distinction, we sometimes call sentences as elements of a semantical system *propositions* and as elements of a calculus *formulas*.

We constructed earlier a semantical system B-S on the basis of the language B, but not, as we have seen, uniquely determined by B. Analogously, we shall now construct a calculus B-C on the basis of B. As preliminary steps for the construction of the syntactical rules proper, which we shall then call rules of transformation, we have to make a *classification* of the signs of B-C and to lay down *syntactical rules of formation* F_C 1–4. But they correspond exactly to the classification and the rules of formation F 1–4 of B-S (§ 5); these rules were already formal. Therefore we shall not write them down again.

Calculus B-C. Rules of Transformation:

PS.   A sentence of B-C is called a *primitive sentence* of B-C, if it has one of the following forms, PS 1–4:

PS 1. 'wenn . . . , so [wenn nicht . . . , so - - -]'.

PS 2. 'wenn [wenn nicht . . . , so . . .], so . . .'.

PS 3. 'wenn [wenn . . . , so - - -], so [wenn [wenn - - -, so . - . -], so [wenn . . . , so . - . -]]'.

PS 4. 'wenn [fuer jedes . . , - . . -], so - . - . -'; here '. .' is a variable, '- . - . -' is a sentence which does not contain 'fuer jedes' but contains a name '. - .' one or several times, and '- . . -' is an expression constructed out of '- . - . -' by replacing '. - .' at one or several (not necessarily all) places by the variable '. .'. (Examples: [1] 'wenn [fuer jedes x, x ist rot], so mond ist rot'; [2] see sentence (3) in the first example of a derivation, at the end of this section.)

R. Rules of Inference: The relation of direct derivability holds if and only if one of the following conditions is fulfilled.

R 1. Rule of Implication: From 'wenn . . . , so - - -' and '. . .', '- - -' is directly derivable in B-C.

R 2. Rule of Synonymity: The words 'titisee' and 'rumber' may be exchanged at any place (i.e., if S_2 is constructed out of S_1 by replacing one of those words at one place by the other one, then S_2 is directly derivable from S_1 in B-C).

A *proof* in B-C is a sequence of sentences of B-C such that each of them is either a primitive sentence or directly derivable from one or two sentences preceding it in the sequence. A sentence S_1 of B-C is called *provable* in B-C if it is the last sentence of a proof in B-C. A sentence of B-C is called *C-true* in B-C if and only if it is provable in B-C; a sentence '. . .' is called *C-false* in B-C if and only if 'nicht . . .' is provable in B-C. (For B-C, provability and C-truth coincide, and likewise derivability and C-implication; for other calculi, this is in general not the case, as we shall see.)

A *derivation* in B-C with a class C_1 of premisses is a sequence of sentences of B-C such that each of them is either a sentence of C_1 or a primitive sentence or directly derivable from one or two sentences preceding it in the sequence. The last sentence of a derivation is called its *conclusion*. S_2 is called *derivable* from C_1 and also a *C-implicate* of C_1 if it is the conclusion of a derivation with the class of premisses C_1.

Both the rules of formation and the rules of transformation of B-C do not in any way refer to designata; they are strictly formal. Nevertheless, they have been chosen with regard to B-S in such a way that the extension of the terms 'C-true', 'C-false', and 'C-implicate' in B-C coincides with that of 'L-true', 'L-false', and 'L-implicate', respectively, in B-S. There are an infinite number of other possible choices of primitive sentences and rules of inference which would lead to the same result. This result gives the practical justification for our choice of the rules of B-C. A calculus in itself needs no justification; this point will be discussed later.

The calculus B-C corresponds to a restricted form of the so-called lower functional calculus, as constructed by Hilbert and Bernays. PS 1–3 and R 1 correspond to the so-called sentential calculus. That the lower functional calculus is complete, i.e., that it exhausts the extension of L-truth and L-implication, has been shown by Gödel.

Example of a proof in B-C. If in the following sequence the blank '. . .' is always replaced by the same sentence, e.g., 'titisee ist blau', the sequence fulfils the conditions—as shown by the remarks on the left side—and therefore is a proof. Hence any sentence of the form 'wenn . . . , so . . .' is provable and C-true in B-C, e.g., 'wenn titisee ist blau, so titisee ist blau'.

PS 1   wenn . . . , so [wenn nicht . . . , so . . .]   (1)

PS 2   wenn [wenn nicht . . . , so . . .], so . . .   (2)

PS 3   wenn [wenn . . . , so [wenn nicht . . . , so . . .]],

so [wenn [wenn [wenn nicht . . . , so . . .], so . . .],

so [wenn . . . , so . . .]]   (3)

(here, 'wenn nicht . . . , so . . .' has been taken for '- - -', and '. . .' for '. - .-')

(1)(3) R 1   wenn [wenn [wenn nicht . . . , so . . .], so . . .],

so [wenn . . . , so . . .]   (4)

(2)(4) R 1   wenn . . . , so . . .   (5)

First example of a derivation in B-C:

Premises { titisee ist blau   (1)

{ fuer jedes x, [wenn x ist blau, so x ist kalt]   (2)

PS 4   wenn [fuer jedes x, [wenn x ist blau, so x ist kalt]], so [wenn titisee ist blau, so titisee ist kalt]   (3)

(2)(3) R 1   wenn titisee ist blau, so titisee ist kalt   (4)

(1)(4) R 1   *Conclusion:* titisee ist kalt   (5)

If we interpret these sentences as in B-S, (1) says that a certain object is blue, (2) says that all blue things are cold (see example [2] at the end of § 5), (5) says that that object is cold. Here, however, the conclusion is derived from the premises in a formal way, i.e., without making use of an interpretation.

Second example of a derivation in B-C:

Premises { wenn mond ist blau, so mond ist kalt   (1)

{ nicht mond ist kalt   (2)

Provable: wenn [wenn mond ist blau, so mond ist kalt], so [wenn nicht mond ist kalt, so nicht mond ist blau]   (3)

(1)(3) R 1 wenn nicht mond ist kalt, so nicht mond ist blau   (4)
(2)(4) R 1 *Conclusion:* nicht mond ist blau   (5)

(3) is a provable sentence. To save space, we do not give its proof here. Suppose that the proof of (3) has been constructed earlier, then the example shows how its result can be used in a derivation. According to the definitions previously given for 'proof' and 'derivation', any proof may also occur as a part of a derivation. If this happens, we can abbreviate the derivation; we write in the derivation not all the sentences of the proof, whose last sentence we intend to use, but only this one sentence, as we have done in the example given with sentence (3). In this way a sentence which has been proved once can be used in derivations again and again. Later, in the discussion of the application of calculi in empirical science we shall come back to this application of proved sentences in derivations (§ 19).

II. Calculus and Interpretation

10. Calculus and Semantical System

We shall investigate the relations which may hold between a calculus and a semantical system. Sometimes we shall use as examples the calculus B-C and the semantical system B-S as discussed before. Suppose a calculus is given—it may be designated by 'Z-C' or briefly 'C'—and a semantical system—designated by 'Z-S' or 'S'. We call S an *interpretation* of C if the rules of S determine truth criteria for all sentences of C; in other words, if to every formula of C there is a corresponding proposition of S; the converse is not required.

Suppose S fulfils the following condition: for any T_1, T_2, T_3, and T_4, if T_2 is a C-implicate of T_1 in C, T_2 is an implicate of T_1 in S; if T_3 is C-true in C, it is true in S; if T_4 is C-false in C, it is false in S. If an interpretation S of C fulfils the condition stated, we call it a *true interpretation* of C; otherwise a *false interpretation*. If the semantical rules suffice to show that S is a true interpretation of C, then we call S an *L-true interpretation* of C. In this case C-implication becomes L-implication; every C-true sentence becomes L-true, and every C-false sentence becomes L-false. If, on the other hand, these semantical rules suffice to show that S is a false interpretation, we call S an *L-false interpretation*. If S is an interpretation but neither an

L-true nor an L-false interpretation of C, we call S a *factual interpretation* of C. In this case, in order to find out whether the interpretation is true, we have to find out whether some factual sentences are true; for this task we have to carry out empirical investigations about facts. An interpretation S of C is called a *logical interpretation* if all sentences of C become logical sentences of S (i.e., sentences containing logical signs only), otherwise a *descriptive interpretation*. A logical interpretation is always L-determinate. Applying these definitions to the system of our former example: B-S is a true and, moreover, L-true, and descriptive interpretation of B-C.

The class of the sentences which are C-true in C is, interpreted by S, a class of assertions; we call it the *theory correlated* to C by S. If the interpretation is true, L-true or logical, respectively, the correlated theory is likewise true, L-true or logical, respectively; the converse does not hold generally.

Previously we had a semantical system B-S and then constructed a calculus B-C "in accordance with" B-S. What was meant by this can now be formulated: we intended to construct B-C in such a way that B-S is a true interpretation of B-C. It is easy to see that for any given semantical system S it is possible to construct a calculus C of that kind. All we have to do is to select partial domains, as small as we wish, of the extensions of 'implicate in S', 'true in S', and 'false in S' (usually the null class), and then lay down formal definitions of 'C-implicate', 'C-true', and possibly 'C-false', in such a way that their extensions correspond to these partial domains. On the other hand, it is an important problem whether it is possible to construct for a given system S a calculus C such that C is not only in accordance with S, in the sense explained, but that the extensions of 'C-implicate', 'C-true', and (if defined at all) 'C-false' coincide with those of 'L-implicate', 'L-true', and possibly 'L-false,' respectively. If this is the case, we call C an *L-exhaustive calculus* with respect to S. Thus B-C is L-exhaustive with respect to B-S. (We do not define a term for the case that the extensions of 'C-implicate', 'C-true', and 'C-false' coincide with those of 'implicate', 'true', and 'false' because that would be impossible for any somewhat richer language system, e.g., for any language system of a branch of science.)

In order to answer the question of the possibility of an L-exhaustive calculus, we have to distinguish two fundamentally different kinds of rules of transformation, which we call finite and transfinite rules. By *finite rules* we understand those of the customary kind: primitive sentences and rules of inference each of which refers to a finite number of premises (in most cases one or two). Almost all rules used by logicians up to the present time are finite. Finite rules are applied in the construction of proofs and derivations of the usual kind, which are finite sequences of sentences, as we have seen in the examples in B-C. A rule of transformation is called *transfinite* if it refers to an infinite number of premises. Because of this number being infinite, a transfinite rule cannot be used within a proof or derivation; a procedure of deduction of an entirely new kind is necessary. We call a calculus finite if all its rules of transformation are finite, otherwise transfinite. It may be remarked that some logicians reject transfinite rules.

We shall make the following terminological distinction: the terms 'C-implicate' and 'C-true' are applied generally with respect both to finite and to transfinite calculi. On the other hand, we shall restrict the corresponding terms 'derivable' and 'provable' to finite calculi. Thus we call T_2 a C-implicate of T_1 in C, if it is possible to obtain T_2 from the premises T_1 by a procedure of deduction of any kind in C; and we call T_3 C-true if it is possible to obtain T_3 by a procedure of deduction without premises. If C is a finite calculus— as, e.g., B-C—the deduction takes the form of a finite sequence of sentences, either a derivation or a proof. In this case T_2 is called, moreover, derivable from T_1, and T_3 is called, moreover, provable.

Now we come back to the problem whether it is possible to construct for a given semantical system S an L-exhaustive calculus C. The answer can now be formulated (but not proved here). The answer depends upon the degree of complexity of S; more precisely, it depends upon whether there are in S a sentence S_2 and an infinite class of sentences C_1 such that S_2 is an L-implicate of C_1 but not an L-implicate of any finite subclass of C_1. (Example. S contains a name for every object of an infinite domain: 'a_1', 'a_2', 'a_3', etc. 'P' is a descriptive predicate. C_1 is the [infinite] class of all sentences of the form '. . . is a P' where '. . .' is one of the object names. S_2 is the sentence 'for every x, x is a P'.) If this is not the case, then there is a finite L-exhaustive calculus C. If, however, it is the case, an L-exhaustive calculus C can be constructed if and only if transfinite rules are admitted. For, because C_1 is infinite, S_2 cannot be derivable from C_1. If we decide in a given case to admit transfinite rules, we have to accept the complications and methodological difficulties connected with them. It was first shown by Gödel that a calculus of the ordinary kind (in our terminology, a finite calculus) cannot be constructed for the whole of arithmetic.

11. On the Construction of a Language System

We found earlier that the pragmatical description of a language gives some suggestions for the construction of a corresponding semantical system without, however, determining it. Therefore, there is a certain amount of freedom for the selection and formulation of the semantical rules. Again, if a semantical system S is given and a calculus C is to be constructed in accordance with S, we are bound in some respects and free in others. The rules of formation of C are given by S. And in the construction of the rules of transformation we are restricted by the condition that C must be such that S is a true interpretation of C, as discussed before. But this still leaves some range of choice. We may, for instance, decide that the class of C-true sentences is to be only a proper subclass of the class of L-true sentences, or that it is to coincide with that class (as we did in constructing B-C), or that it is to go beyond that class and comprehend some factual sentences, e.g., some physical laws. When the extensions of 'C-true' and 'C-implicate' are decided, there is still some possibility of choice in the construction of the rules, e.g., primitive sentences and rules of inference, leading to those extensions. This choice, however, is not of essential importance, as it concerns more the form of presentation than the result.

If we are concerned with a historically given language, the pragmatical description comes first, and then we may go by abstraction to semantics and (either from semantics or immediately from pragmatics) to syntax. The situation is quite different if we wish to construct a language (or rather a language system, because we lay down rules), perhaps with the intention of practical application, as for making communications or formulating a scientific theory. Here we are not bound by a previous use of language, but are free to construct in accordance with our wishes and purposes. The construction of a language system Z may consist in laying down two kinds of rules, the semantical rules (Z-S or briefly S) and the syntactical rules (calculus Z-C or C). As a common basis for both, according to our former discussion, we have to make a classification of the signs which we intend to use and lay down rules of formation Z-F. Z-S consists of two parts, rules for the descriptive signs (Z-SD or SD) and rules for the logical signs (Z-SL or SL).

In constructing the system Z, we can proceed in two different ways—different as to the order of S and C. Here the order is not unessential, for, if we have chosen some rules arbitrarily, we are no longer free in the choice of others.

The first method consists in first constructing S and then constructing C. We start with a classification of the kinds of signs which we want, and rules F determining the forms of sentences which we intend to use. Then we lay down the rules SD; we choose objects, properties, etc., for which we wish to have direct designations, and then signs to designate these objects, properties, etc. Next we construct the rules SL; we choose signs to be used as logical signs and state for each of them the conditions of the truth of the sentences constructed with its help. (As mentioned before, we may also proceed by indicating the translations of the sentences containing logical signs, or giving their designata.) After this we proceed to syntax and construct the calculus C, e.g., by stating primitive sentences and rules of inference. It has been explained already that, if S is given or constructed, we are limited in constructing C in some essential respects, because C must be such that S is a true interpretation of C; but we are free in other respects.

The *second method* for constructing Z is first to construct C and then S. We begin again with a classification of signs and a system F of syntactical rules of formation, defining 'sentence in C' in a formal way. Then we set up the system C of syntactical rules of transformation, in other words, a formal definition of 'C-true' and 'C-implicate'. Since so far nothing has been determined concerning the single signs, we may choose these definitions, i.e., the rules of formation and of transformation, in any way we wish. With respect to a calculus to be constructed there is only a question of expedience or fitness to purposes chosen, but not of correctness. This will be discussed later.

Then we add to the uninterpreted calculus C an interpretation S. Its function is to determine truth conditions for the sentences of C and thereby to change them from formulas to propositions. We proceed in the following way. It is already determined by the rules F which expressions are formulas in C. Now we have to stipulate that each of them is also a proposition in S. By the syntactical classification of the signs it is not yet completely settled which signs are logical and which descriptive. In many cases there is still a considerable amount of freedom of choice in this respect, as we shall see later in some examples. After having stated which signs are to be logical and which descriptive, we construct the rules SL for the logical signs. Here our choice is restricted to some extent by the requirement that the interpretation must be true.

Finally we establish the rules SD for the descriptive signs. Here we have to take into account the classification of signs. We choose the designata for each kind of signs and then for each sign of that kind. We may begin with individual names. First we choose a field of objects with which we wish to deal in the language to be constructed, e.g., the persons of a certain group, the towns of a certain country, the colors, geometrical structures, or whatever else. Then we determine for each individual name, as its designatum, one object of the class chosen. Then, for each predicate, we choose a possible property of those objects, etc. In this way, a designatum for every descriptive sign is chosen. If we decide to make S an L-true interpretation of C, we have a great amount of freedom for the choice of the rules SD. Otherwise, we find some essential restrictions. If some of the C-true formulas are to become factual propositions, they must be factually true. Therefore, in this case, on the basis of our factual knowledge about the objects which we have chosen as subject matter of Z, we have to take care that the interpretations for the descriptive names, predicates, etc., i.e., their designata, are chosen in such a way that those factual C-true sentences are actually true.

12. Is Logic a Matter of Convention?

There has been much controversial discussion recently on the question whether or not logic is conventional. Are the rules on which logical deduction is based to be chosen at will and, hence, to be judged only with respect to convenience but not to correctness? Or is there a distinction between objectively right and objectively wrong systems so that in constructing a system of rules we are free only in relatively minor respects (as, e.g., the way of formulation) but bound in all essential respects? Obviously, the question discussed refers to the rules of an interpreted language, applicable for purposes of communication; nobody doubts that the rules of a pure calculus, without regard to any interpretation, can be chosen arbitrarily. On the basis of our former discussions we are in a position to answer the question. We found the possibility—which we called the second method— of constructing a language system in such a way that first a calculus C is established and then an interpretation is given by adding a semantical system S. Here we are free in choosing the rules of C. To be sure, the choice is not irrelevant; it depends upon C whether the interpretation can yield a rich language or only a poor one.

We may find that a calculus we have chosen yields a language which is too poor or which in some other respect seems unsuitable for the purpose we have in mind. But there is no question of a calculus being right or wrong, true or false. A true interpretation is possible for any given consistent calculus (and hence for any calculus of the usual kind, not containing rules for 'C-false'), however the rules may be chosen.

On the other hand, those who deny the conventional character of logic, i.e., the possibility of a free choice of the logical rules of deduction, are equally right in what they mean if not in what they say. They are right under a certain condition, which presumably is tacitly assumed. The condition is that the "meanings" of the logical signs are given before the rules of deduction are formulated. They would, for instance, insist that the rule R 1 of B-C ('from 'wenn . . . , so - - -' and '. . .', '- - -' is directly derivable' [§ 9]) is necessary; that it would be wrong to change it arbitrarily, e.g., into R 1*: 'from 'wenn . . ., so - - -' and 'nicht . . .', '- - -' is directly derivable'. What they presumably mean is that the rule R 1* is incorrect on the basis of the presupposed "meaning" of the signs 'wenn', 'so', and 'nicht'. Thus they have in mind the procedure which we called the first method (§ 11): we begin by establishing the semantical rules SL or assume them as given—obviously this is meant by saying that the "meaning" is given—and then we ask what rules of deduction, i.e., syntactical rules of transformation, would be in accordance with the presupposed semantical rules. In this order of procedure, we are, as we have seen, indeed bound in the choice of the rules in all essential respects. Thus we come to a reconciliation of the opposing views. And it seems to me that an agreement should easily be attainable in the other direction as well. The anti-conventionalists would certainly not deny that the rule R 1* can also be chosen and can lead to correct results, provided we interpret the logical signs in a different way (in the example given, we could interpret 'wenn . . . , so - - -', e.g., as '. . . or - - -').

The result of our discussion is the following: logic or the rules of deduction (in our terminology, the syntactical rules of transformation) can be chosen arbitrarily and hence are conventional if they are taken as the basis of the construction of the language system and if the interpretation of the system is later superimposed. On the other hand, a system of logic is not a matter of choice, but either right or wrong, if an interpretation of the logical signs is given in advance. But even here, conventions are of fundamental importance; for the basis on which logic is constructed, namely, the interpretation of the logical signs (e.g., by a determination of truth conditions) can be freely chosen.

It is important to be aware of the conventional components in the construction of a language system. This view leads to an unprejudiced investigation of the various forms of new logical systems which differ more or less from the customary form (e.g., the intuitionist logic constructed by Brouwer and Heyting, the systems of logic of modalities as constructed by Lewis and others, the systems of plurivalued logic as constructed by Lukasiewicz and Tarski, etc.), and it encourages the construction of further new forms. The task is not to decide which of the different systems is "the right logic" but to examine their formal properties and the possibilities for their interpretation and application in science. It might be that a system deviating from the ordinary form will turn out to be useful as a basis for the language of science.

III. Calculi and Their Application in Empirical Science

13. Elementary Logical Calculi

For any given calculus there are, in general, many different possibilities of a true interpretation. The practical situation, however, is such that for almost every calculus which is actually interpreted and applied in science, there is a certain interpretation or a certain kind of interpretation used in the great majority of cases of its practical application. This we will call the *customary interpretation* (or kind of interpretation) for the calculus. In what follows we shall discuss some calculi and their application. We classify them according to their customary interpretation in this way: logical calculi (in the narrower sense), mathematical, geometrical, and (other) physical calculi. The customary interpretation of the logical and mathematical calculi is a logical, L-determinate interpretation; that of the geometrical and physical calculi is descriptive and factual. The mathematical calculi are a special kind of logical calculi, distinguished merely by their greater complexity. The geometrical calculi are a special kind of physical calculi. This classification is rather rough and is only meant to serve a temporary, practical purpose.

To the logical calculi (in the narrower sense) belong most of the calculi of elementary structure used in symbolic logic, above all, the so-called sentential calculus and the so-called lower functional calculus. The *sentential calculus* has approximately the structure of B-C with F 4 and PS 4 omitted. The customary interpretation corresponds to the rules B-SL 2, 3. The form mostly used contains, however, only those signs which are logical in the customary interpretation, corresponding to the English words 'not', 'if', 'or', 'and', and the like, and sentential variables. The *lower functional calculus* (or predicate calculus) contains the sentential calculus and, in addition, general sentences with individual variables, namely, universal sentences (interpretation: 'for every x, ... ') and existential sentences (interpretation: 'there is an x such that ... '). Within symbolic logic, this calculus too is mostly used without descriptive signs but with three kinds of variables: sentential variables, individual variables (as in B-C), and predicate variables. The customary interpretation is a logical one, as given by B-SL. In the case of the logical calculi here explained the customary interpretation is the only one which is ever used practically. (If the calculi are supplemented in a certain way, it is even the only possible true interpretation.) Therefore, we shall call it the *normal interpretation* of the logical calculus.

If a calculus C is constructed with the intention of using it mostly or exclusively with a certain interpretation S, it may often seem convenient to use as signs of C not artificial symbols but those words of the word-language whose ordinary use is approximately in acccord with the interpretation intended (a word with exact accordance will usually not be available). Then we have in C the same sentences as in the interpreted language S, which is perhaps to be applied in science; "the same sentences" as to the wording, but in C they are formulas, while they are propositions in S. This procedure is mostly chosen in geometrical and other physical calculi (for examples see end of § 17, beginning of § 22).

In what follows we shall do the same for the logical calculus (where, for good reasons, it is usually not done). Thus, instead of symbols, we shall use the words 'not', 'if', etc. It has been shown (by H. M. Sheffer) that two primitive signs are sufficient, namely, 'excludes' (to be interpreted later) and 'for every'. It is not necessary to take as many primitive signs as we did in B-C, corresponding to 'not', 'if—then', 'for every'. The other logical signs of the logical calculus can be introduced by definitions. The primitive signs mentioned and all signs defined with their help are called logical constants. We shall use three kinds of variables: sentential variables ('p', 'q', etc.), individual variables ('x', 'y', etc., as in B-C), and predicate variables ('F',

'*G*', etc.). For a sentential variable a sentence may be substituted, for an individual variable an individual name, for a predicate variable a predicate, and for '*Fx*' an expression of sentential form containing the variable '*x*'.

A *definition* is a rule of a calculus which serves for introducing a new sign. In simpler cases the rule states that the new sign is to be taken as an abbreviation for a certain expression consisting only of old signs (i.e., primitive signs or signs defined earlier). In other cases the rule states that sentences containing the new sign and old signs are to be taken as abbreviations for certain sentences containing old signs only. Rules of the first kind are called explicit definitions (e.g., Defs. 11, 12, and 13 in § 14); those of the second kind are called definitions in use (e.g., Defs. 1–7, below); we shall use still another kind of definition, the so-called recursive definitions frequently found in arithmetic (e.g., Defs. 14 and 15 in § 14). The definitions in a calculus are, so to speak, additional rules of transformation, either primitive sentences or rules of inference, according to their formulation; they are added in order to provide shorter expressions. If a calculus C contains definitions and the interpretation S contains semantical rules for the primitive signs of C, the interpretation of the defined signs need not be given explicitly. The definitions, together with those rules of S, determine the truth conditions of the sentences containing the defined signs and thereby the interpretation of these signs.

We shall formulate the definitions here in this form: ' '. . .' for '- - -' '. This means that '. . .' is to serve as an abbreviation for '- - -', i.e., that '. . .' and '- - -', and likewise two expressions constructed out of '. . .' and '- - -' by the same substitutions, may always be replaced by each other. In this calculus, we take as simplest form of sentences in the beginning '*Fx*' (e.g., 'city Chicago' instead of 'Chicago is a city'); the usual form with 'is a' is introduced later by Definition 7.

The expressions included in parentheses serve merely to facilitate understanding; in the exact formulation they have to be omitted. The brackets and commas, however, are essential; they indicate the structure of the sentence (cf. § 5).

Def. 1. 'not p' for 'p excludes p'.
Def. 2. 'p or q' for 'not p, excludes, not q'.
Def. 3. 'p and q' for 'not [p excludes q]'.
Def. 4. 'if p then q' for 'not p, or q'.
Def. 5. 'p if and only if q' for '[if p then q] and [if q then p]'.
Def. 6. 'for some x, Fx' for 'not [for every x, not Fx]'.
Def. 7. 'x is an F' for 'Fx'.

The rules of transformation of the sentential calculus and the functional calculus will not be given here. They are not essentially different from those of B-C. It has been shown (by J. Nicod) that, if 'excludes' is taken as primitive sign, one primitive sentence is sufficient for the sentential calculus. For the lower functional calculus we have to add one more primitive sentence for 'for every', analogous to PS 4 in B-C.

The *normal interpretation* for the logical calculus is a logical one. Therefore, if interpreted, it is, so to speak, a skeleton of a language rather than a language proper, i.e., one capable of describing facts. It becomes a factual language only if supplemented by descriptive signs. These are then interpreted by SD-rules, and the logical constants by SL-rules. As SL-rules for the lower functional calculus we can state the following two rules for the two primitive signs. For the sentential calculus the first rule suffices.

1. A sentence of the form '. . . excludes - - -' is true if and only if not both '. . .' and '- - -' are true.
2. A sentence of the form 'for every . . . , - - -' is true if and only if all individuals have the property designated by '- - -' with respect to the variable '. . .'. (The individuals are the objects of the domain described, which is to be determined by an SD-rule.)

The interpretation of the defined signs 'not', etc., is determined by rule (1) and Definition 1, etc. The interpretation of 'not' and 'if—then' is easily seen to be the same as that of 'nicht', and 'wenn—so' in B-SL. (The truth conditions here given by rule [1] and Definitions 1–5 are the same as those which in symbolic logic usually are stated with the help of truth-value tables for the corresponding symbols, the so-called connectives.)

14. Higher Logical Calculi

The lower functional calculus can be enlarged to the higher functional calculus by the addition of predicates of higher levels. The predicates occurring in the lower functional calculus are now called predicates of first level; they designate properties of first level, i.e., properties of individuals. Now we introduce predicates of second level, which designate properties of second level, i.e., properties of properties of first level; predicates of third level designating properties of third level, etc. Further, new kinds of variables for these predicates of higher levels are introduced. (In the subsequent definitions we shall use as variables for predicates of second level 'm' and 'n', for predicates of third level 'K'.) Expressions of the form 'for every . . .', and analogously 'for some . . .' (Def. 6), are now admitted not only for individual variables but also for predicate variables of any level. Some new rules of transformation for these new kinds of variables have to be added. We shall not give them here. Some of them are still controversial.

The *normal interpretation* of the higher functional calculus can again be given by two semantical rules. Rule (1) is kept, as the sentential calculus remains the basis for the higher functional calculus. Rule (2) must be replaced by the subsequent rule (2*), because of the extended use of 'for every'. For individual variables, (2*) is in accordance with (2). (It may be remarked that there are some controversies and unsolved problems concerning the properties of higher levels.)

2*. A sentence of the form 'for every . . . , - - -' is true if and only if all entities belonging to the range of the variable '. . .' have the property designated by '- - -' with respect to '. . .'. (To the range of an individual variable belong all individuals, to the range of a predicate variable of level r belong all properties of level r.)

To the definitions which we stated in the lower functional calculus, new ones can now be added which make use of predicates and variables of higher levels. We shall first give some rough explanations of the new expressions and later the definitions. First, identity can be defined; '$x = y$' is to say that x is the same object as y; this is defined by 'x and y have all properties in common' (Def. 8). Then we shall define the concept of a cardinal number of a property, restricting ourselves, for the sake of simplicity, to finite cardinal numbers. 'F is an m' is to say that the property F has the cardinal number m; i.e., that there are m objects with the property F. This concept is defined by a recursive definition (for finite cardinals only). 'F is a 0' is defined as saying that no object has the property F (Def. 9a). Then 'F is an m^+', where 'm^+' designates the next cardinal number greater than m, i.e., $m + 1$, is defined in the following way in terms of 'm': there is a property G with the cardinal number m such that all objects which have the property G, and, in addition, some object x, but no other objects, have the property F (Def. 9b). A property K of numbers is called hereditary if, whenever a number m is a K, $m + 1$ is also a K. Then 'm is a finite cardinal number' can be defined (as Frege has shown) in this way: m has all hereditary properties of 0 (Def. 10). The numerals '1', '2', etc., can easily be defined by '0^+', '1^+', etc. (Def. 11, etc.). The sum ('$m + n$') and the product ('$m \times n$') can be defined by recursive definitions, as is customary in arithmetic (Defs. 14 and 15).

Def. 8. '$x = y$' for 'for every (property) F, if x is an F then y is an F'.
        Analogously for any higher level.

Def. 9a. 'F is a 0' for 'not [for some x, x is an F]'.

    b. 'F is an m^+' for 'for some G, for some x, for every y [[y is an F if and only if [y is a G or $y = x$]] and G is an m and, not x is a G].

Def. 10. 'm is a finite cardinal number' for 'for every (property of numbers) K, if [0 is a K and, for every n [if n is a K then n^+ is a K]] then m is a K'.

Def. 11. '1' for '0^+'.

Def. 12. '2' for '1^+'.

Def. 13. '3' for '2^+'.

        Analogously for any further numeral.

Def. 14a. '$m + 0$' for 'm'.

    b. '$m + n^+$' for '$[m + n]^+$'.

Def. 15a. '$m \times 0$' for '0'.

    b. '$m \times n^+$' for '$[m \times n] + m$'.

For the reasons mentioned before we have used, instead of arbitrary symbols, words whose ordinary use agrees approximately with the interpretation intended. It is, however, to be noticed that their exact interpretation in our language system is not to be derived from their ordinary use but from their definition in connection with the semantical rules (1) and (2*).

We see that it is possible to define within the logical calculus signs for numbers and arithmetical operations. It can further be shown that all theorems of ordinary arithmetic are provable in this calculus, if suitable rules of transformation are established.

The method of constructing a calculus of arithmetic within a logical calculus was first found by Frege (1884) and was then developed by Russell (1903) and Whitehead (1910). (Defs. 9–15 are, in their essential features, in accordance with Frege and Russell, but make use of some simplifications due to the recent development of symbolic logic.) We shall later outline another form of an arithmetical calculus (§ 17) and discuss the problem of mathematics more in detail (§ 20).

15. Application of Logical Calculi

The chief function of a logical calculus in its application to science is not to furnish logical theorems, i.e., L-true sentences, but to guide the deduction of factual conclusions from factual premisses. (In most presentations of logical systems the first point, the proofs, is overemphasized; the second, the derivations, neglected.)

For the following discussions we may make a rough distinction between *singular* and *universal* sentences among factual sentences. By a singular sentence of the language of science or of an interpreted calculus we mean a sentence concerning one or several things (or events or space-time-points), describing, e.g., a property of a thing or a relation between several things. By a universal sentence we mean a sentence concerning all objects of the field in question, e.g., all things or all space-time-points. A report about a certain event or a description of a certain landscape consists of singular sentences; on the other hand, the so-called laws of nature in any field (physics, biology, psychology, etc.) are universal. The simplest kind of an application of the

177

logical calculus to factual sentences is the derivation of a singular sentence from other singular sentences (see, e.g., the second example of a derivation in B-C, end of § 9). Of greater practical importance is the deduction of a singular sentence from premisses which include both singular and universal sentences. We are involved in this kind of a deduction if we explain a known fact or if we predict an unknown fact. The form of the deduction is the same for these two cases. We have had this form in the first example of a derivation in B-C (§ 9); we find it again in the following example, which contains, besides signs of the logical calculus, some descriptive signs. In an application of the logical calculus, some descriptive signs have to be introduced as primitive; others may then be defined on their basis. SD-rules must then be laid down in order to establish the interpretation intended by the scientist. Premiss (3) is the law of thermic expansion in qualitative formulation. In later examples we shall apply the same law in quantitative formulation (D_1 in § 19; D_2 in § 23).

Premisses:
1. *c* is an iron rod.
2. *c* is now heated.
3. for every *x*, if *x* is an iron rod and *x* is heated, *x* expands.

Conclusion: 4. *c* now expands.

A deduction of this form can occur in two practically quite different kinds of situations. In the first case we may have found (4) by observation and ask the physicist to explain the fact observed. He gives the *explanation* by referring to other facts (1) and (2) and a law (3). In the second case we may have found by observation the facts (1) and (2) but not (4). Here the deduction with the help of the law (3) supplies the prediction (4), which may then be tested by further observations.

The example given shows only a very short deduction, still more abbreviated by the omission of the intermediate steps between premisses and conclusion. But a less trivial deduction consisting of many steps of inference has fundamentally the same nature. In practice a deduction in science is usually made by a few jumps instead of many steps. It would, of course, be practically impossible to give each deduction which occurs the form of a complete derivation in the logical calculus, i.e., to dissolve it into single steps of such a kind that each step is the application of one of the rules of transformation of the calculus, including the definitions. An ordinary reasoning of a few seconds would then take days. But it is essential that this dissolution is theoretically possible and practically possible for any small part of the process. Any critical point can thus be put under the logical microscope and enlarged to the degree desired. In consequence of this, a scientific controversy can be split up into two fundamentally different components, a factual and a logical (including here the mathematical). With respect to the logical component the opponents can come to an agreement only by first agreeing upon the rules of the logical calculus to be applied and the L-semantical rules for its interpretation, and by then applying these rules, disregarding the interpretation of the descriptive signs. The discussion, of course, need not concern the whole calculus; it will be sufficient to expand the critical part of the controversial deduction to the degree required by the situation. The critical point will usually not be within the elementary part of the logical calculus (to which all examples of derivations discussed above belong), but to a more complex calculus, e.g., the higher, mathematical part of the logical calculus, or a specific mathematical calculus, or a physical calculus. This will be discussed later; then the advantage of the formal procedure will become more manifest.

16. General Remarks about Nonlogical Calculi (Axiom Systems)

In later sections we shall discuss certain other calculi which are applied in science. The logical calculus explained previously is distinguished from them by the fact that it serves as their basis. Each of the nonlogical calculi to be explained later consists, strictly speaking, of two parts: a logical *basic calculus* and a *specific calculus* added to it. The basic calculus could be approximately the same for all those calculi; it could consist of the sentential calculus and a smaller or greater part of the functional calculus as previously outlined. The specific partial calculus does not usually contain additional rules of inference but only additional primitive sentences, called *axioms*. As the basic calculus is essentially the same for all the different specific calculi, it is customary not to mention it at all but to describe only the specific part of the calculus. What usually is called an *axiom system* is thus the second part of a calculus whose character as a part is usually not noticed. For any of the mathematical and physical axiom systems in their ordinary form it is necessary to add a logical basic calculus. Without its help it would not be possible to prove any theorem of the system or to carry out any deduction by use of the system. Not only is a basic logical calculus tacitly presupposed in the customary formulation of an axiom system but so also is a special interpretation of the logical calculus, namely, that which we called the normal interpretation. An axiom system contains, besides the logical constants, other constants which we may call its specific or axiomatic constants. Some of them are taken as primitive; others may be defined. The definitions lead back to the primitive specific signs and logical signs. An interpretation of an axiom system is given by semantical rules for some of the specific signs, since for the logical signs the normal interpretation is presupposed. If semantical rules for the primitive specific signs are given, the interpretation of the defined specific signs is indirectly determined by these rules together with the definitions. But it is also possible—and sometimes convenient, as we shall see—to give the interpretation by laying down semantical rules for another suitable selection of specific signs, not including the primitive signs. If all specific signs are interpreted as logical signs, the interpretation is a logical and L-determinate one; otherwise it is a descriptive one. (Every logical interpretation is L-determinate; the converse does not always hold.)

17. An Elementary Mathematical Calculus

We take here as mathematical calculi those whose customary interpretation is mathematical, i.e., in terms of numbers and functions of numbers. As an example, we shall give the classical axiom system of Peano for (elementary) arithmetic. It is usually called an axiom system of arithmetic because in its customary interpretation it is interpreted as a theory of natural numbers, as we shall see. This interpretation is, however, by no means the only important one. The logical basic calculus presupposed has to include the lower functional calculus and some part of the higher, up to expressions 'for every F' for predicate variables of first level and Definition 8 for ' = ' (§ 14). The specific primitive signs are 'b', 'N', '$'$'. (The following axioms, of course, are, within the calculus, independent of any interpretation. Nevertheless, the reader who is not familiar with them will find it easier to conceive their form and function by looking at their interpretation given below.)

Axiom System of Peano:

P 1. b is an N.

P 2. For every x, if x is an N, then x' is an N.

P 3. For every x, y, if [x is an N and y is an N and $x' = y'$] then $x = y$.

P 4. For every x, if x is an N, then, not $b = x'$.

P 5. For every F, if $[b$ is an F and, for every x [if x is an F then x' is an F]] then [for every y, if y is an N then y is an F].

　　(Briefly: if F is any property of b which is hereditary [from x to x'] then all N are F.)

It is easy to see that any number of true *interpretations* of this calculus can be constructed. We have only to choose any infinite class, to select one of its elements as the beginning member of a sequence and to state a rule determining for any given member of the sequence its immediate successor. (An order of elements of this kind is called a progression.) Then we interpret in this way: 'b' designates the beginning member of the sequence; if '...' designates a member of the sequence then '...$'$' designates its immediate successor; 'N' designates the class of all members of the sequence that can be reached from the beginning member in a finite number of steps. It can easily be shown that in any interpretation of this kind the five axioms become true.

　　Example: 'b' designates August 14, 1938; if '...' designates a day, '...$'$' designates the following day; 'N' designates the class (supposed to be infinite) of all days from August 14, 1938, on. This interpretation of the Peano system is descriptive, while the customary one is logical.

The *customary* *interpretation* of the Peano system may first be formulated in this way: 'b' designates the cardinal number 0; if '. . .' designates a cardinal number n, then '. . .' ' designates the next one, i.e., $n+1$; 'N' designates the class of finite cardinal numbers. Hence in this interpretation the system concerns the progression of finite cardinal numbers, ordered according to magnitude. Against the given semantical rule ' 'b' designates the cardinal number 0' perhaps the objection will be raised that the cardinal number 0 is not an object to which we could point, as to my desk. This remark is right; but it does not follow that the rule is incorrect. We shall give the interpretation in another way, with the help of a translation.

In the investigation of calculi the procedure of *translation* of one calculus into another is of great importance. A system of rules of translation of the calculus K_2 into the calculus K_1 determines for each primitive sign of K_2 an expression of K_1 called its correlated expression, and for each kind of variable in K_2 its correlated kind of variable in K_1. The rules must be such that the result of translating any sentence in K_2 is always a sentence in K_1. The translation is called C-true if the following three conditions are fulfilled: (1) every C-true sentence in K_2 becomes, if translated, C-true in K_1; (2) every C-false sentence in K_2 becomes C-false in K_1; (3) if the relation of C-implication in K_2 holds among some sentences, then the relation of C-implication in K_1 holds among those into which they are translated. If we have an interpretation I_1 for the calculus K_1, then the translation of K_2 into K_1 determines in connection with I_1 an interpretation I_2 for K_2. I_2 may be called a *secondary interpretation*. If the translation is C-true and the (primary) interpretation I_1 is true, I_2 is also true.

We shall now state rules of translation for the Peano system into the higher functional calculus and thereby give a secondary interpretation for that system. The logical basic calculus is translated into itself; thus we have to state the correlation only for the specific primitive signs. As correlates for 'b', '' '', 'N', we take '0', '+', 'finite cardinal number'; for any variable a variable two levels higher. Accordingly, the five axioms are translated into the following sentences of the logical calculus.

P′ 1. 0 is a finite cardinal number.

P′ 2. For every m, if m is a finite cardinal number, then m^+ is a finite cardinal number.

P′ 3. For every m, n, if [m is a finite cardinal number and n is a finite cardinal number and $m^+ = n^+$] then $m = n$.

P′ 4. For every m, if m is a finite cardinal number, then, not $0 = m^+$.

P′ 5. For every K, if [0 is a K and, for every m [if m is a K then m^+ is a K]] then [for every n, if n is a finite cardinal number then n is a K].

The *customary interpretation* of the Peano system can now be formulated in another way. This interpretation consists of the given translation together with the normal interpretation of the higher functional calculus up to the third level. (P′ 5 contains a variable of this level.)

The whole interpretation is thus built up in the following way. We have two L-semantical rules for the primitive signs 'excludes' and 'for every' of the logical calculus, indicating truth conditions (rules [1] and [2*] in § 14). Then we have a chain of definitions leading to Definitions 9*a* and *b* and 11 for '0', '+', and 'finite cardinal number' (§ 14). Finally we have rules of translation which correlate these defined signs of the logical calculus to the primitive signs '*b*', '′', and '*N*' of the Peano system.

If we assume that the normal interpretation of the logical calculus is true, the given secondary interpretation for the Peano system is shown to be true by showing that the correlates of the axioms are C-true. And it can indeed be shown that the sentences P′ 1–5 are provable in the higher functional calculus, provided suitable rules of transformation are established. As the normal interpretation of the logical calculus is logical and L-true, the given interpretation of the Peano system is also logical and L-true.

We can now define signs within the Peano axiom system which correspond to the signs '0', '1', etc., '+', etc., of the logical calculus. For greater clarity we distinguish them by the subscript '*P*'. (In an arithmetical calculus, however—whether in the form of Peano's or some other—one ordinarily does not use arbitrary symbols like '*b*' or '0_P', '*b*'' or '1_P', '$+_P$', etc., but, because of the customary interpretation, the corresponding signs of the ordinary language '0', '1', '+', etc.)

Def. P 1. '0_p' for '*b*'.
Def. P 2. '1_p' for '*b*''.
Def. P 3. '2_p' for '*b*'''.
   Etc.
Def. P 4a. '$x +_p 0_p$' for '*x*'.
     b. '$x +_p y$'' for '$[x +_p y]$''.
Def. P 5a. '$x \times_p 0_p$' for '0_p'.
     b. '$x \times_p y$'' for '$[x \times_p y] +_p x$'.

Thus the natural numbers and functions of them can be defined both in the logical calculus and in a specific arithmetical calculus, e.g., that of Peano. And the theorems of ordinary arithmetic are provable in both calculi. (Strictly speaking, they are not the same theorems in the different calculi, but corresponding theorems; if, however, the same signs are used—and, as mentioned before, this is convenient and usual—then corresponding theorems consist even of the same signs.)

18. Higher Mathematical Calculi

On the basis of a calculus of the arithmetic of natural numbers the whole edifice of classical mathematics can be erected without the use of new primitive signs. Whether a specific calculus of arithmetic or the logical calculus is taken as a basis does not make an essential difference, once the translation of the first into the second is established. It is not possible to outline here the construction of mathematics; we can make only a few remarks. There are many different possibilities for the introduction of further kinds of numbers. A simple method is the following one. The integers (positive and negative) are defined as pairs of natural numbers, the fractions as pairs of integers, the real numbers as classes either of integers or of fractions, the complex numbers as pairs of real numbers. Another way of introducing any one of these kinds of numbers consists in constructing a new specific calculus in which the numbers of that kind are taken as individuals, like the natural numbers in the Peano calculus.

This has been done especially for the real numbers. A specific calculus of this kind can be translated in one way or another into a more elementary specific calculus or into the logical calculus. (Example: The individual expressions of a specific calculus of real numbers may be translated into expressions for classes of integers or of fractions either in the Peano calculus or in the logical calculus.) For each of the kinds of numbers, functions (summation, multiplication, etc.) can be defined. Further, the concept of limit can be defined, and with its help the fundamental concepts of the infinitesimal calculus, the differential coefficient, and the integral.

If a mathematical calculus is based on the Peano calculus by the use of definitions, then its customary interpretation is determined by that of the latter. If, on the other hand, a mathematical calculus is constructed as an independent specific calculus, we can give an interpretation for it by translating it either into an enlarged Peano system or into an enlarged logical calculus (as indicated above for a calculus of real numbers.) Here we can scarcely speak of "the" customary interpretation, but only of the set of customary interpretations. Their forms may differ widely from one another; but they have in common the character of logical interpretations. If the interpretation is given by a translation either into the Peano system with reference to its customary interpretation or by a translation into the logical calculus with reference to its normal interpretation, this character is obvious. In a customary interpretation of a mathematical calculus every sign in it is interpreted as a logical sign, and hence every sentence consists only of logical signs and is therefore L-determinate (see § 7).

If we choose the form of the construction of mathematics within the logical calculus, we do not even need a translation; the interpretation is simply the normal interpretation of the logical calculus. In this case every mathematical sign is defined on the basis of the two primitive signs of the logical calculus, and hence every mathematical sentence is an abbreviation for a sentence containing, besides variables, only those two signs. In most cases, though, this sentence would be so long that it would not be possible to write it down within a lifetime. Therefore, the abbreviations introduced in the construction of mathematics are not only convenient but practically indispensable.

19. Application of Mathematical Calculi

The application of mathematical calculi in empirical science is not essentially different from that of logical calculi. Since mathematical sentences are, in the customary interpretation, L-determinate, they cannot have factual content; they do not convey information about facts which would have to be taken into consideration besides those described in empirical science. The function of mathematics for empirical science consists in providing, first, forms of expression shorter and more efficient than non-mathematical linguistic forms and, second, modes of logical deduction shorter and more efficient than those of elementary logic.

Mathematical calculi with their customary interpretation are distinguished from elementary logical calculi chiefly by the occurrence of numerical expressions. There are two procedures in empirical science which lead to the application of numerical expressions: counting and measurement (cf. Lenzen, Vol. I, No. 5, §§ 4 and 5). Counting is ascertaining the cardinal number of a class of single, separate things or of events. Measuring is ascertaining the value of a magnitude for a certain thing or place at a certain time. For each physical magnitude, e.g., length, weight, temperature, electric field, etc., there are one or several methods of measurement. The result of a measurement is a fraction or a real number. (Irrational real numbers can also occur, but only if, besides direct measurement, calculation is applied.) If a deduction has to do with results of counting, we may apply, besides an elementary logical calculus, a calculus of elementary arithmetic. If it has to do with results of measurements, we may apply a calculus of analysis, i.e., of real numbers.

Let us look at a very simple example of a logico-mathematical deduction. We apply a certain part of the higher functional calculus and an arithmetical calculus. We presuppose for the following derivation that in this arithmetical calculus the sentence

'3+6 = 9' (7) has been proved earlier. Whether we take the arithmetical calculus in the form of a part of the higher functional calculus (as in § 14) or in the form of a specific calculus (as in § 17) does not make any essential difference; in both cases sentence (7) is provable. In order to keep in closer contact with ordinary language, we use the following definition: 'there are *m F*'s' for '*F* is an *m*'; further, we write 'n.i.t.r.' for 'now in this room'.

Premisses	1. There are 3 students n.i.t.r.
	2. There are 6 girls n.i.t.r.
	3. For every x [x is a person n.i.t.r. if and only if [x is a student n.i.t.r. or x is a girl n.i.t.r.]].
	4. For every x [if x is a girl n.i.t.r., then, not x is a student n.i.t.r.].
Defs. 1–9, 14	5. For every F, G, H, m, n [if [m and n are finite cardinal numbers and G is an m and H is an n and for every x [x is an F if and only if, x is a G or x is an H] and for every y [if y is a G then, not y is an H]] then F is an $m+n$].
	[This says that, if a class F is divided into two parts, G and H, the cardinal number of F is the sum of the cardinal numbers of G and H.]
(1)(2)(3)(4)(5)	6. There are 3+6 persons n.i.t.r.
Arithmet. theorem:	7. 3+6 = 9.
(6)(7) *Conclusion:*	8. There are 9 persons n.i.t.r.

The premisses of this derivation describe some facts empirically established by observation (including counting). The conclusion is also a factual sentence; but its content, the amount of factual information it conveys, does not go beyond that of the premisses. We have discussed earlier (at the end of § 9) the application of proved theorems in a derivation; here (5) and (7) are examples of this method. These sentences do not contribute to the factual content of the conclusion; they merely help in transforming the premisses into the conclusion. To say that the result (8) is "calculated" from the data (1)–(4), means just this: it is obtained by a formal procedure involving a mathematical calculus. The effect of the application of a mathematical calculus is always, as in this example, the possibility of presenting in a shorter and more easily apprehensible way facts already known.

Here an objection will perhaps be raised. That the application of mathematics consists merely in a transformation of the premises without adding anything to what they say about the facts, may be true in trivial cases like the example given. If, however, we predict, with the help of mathematics, a future event, do we not come to a new factual content? Let us discuss an example of a derivation of this kind. The derivation—called D_1—leads to the prediction of a thermic expansion as in a former example (§ 15), but now with quantitative determinations. The premises of D_1 relate the results of measurements of the temperature of an iron rod at two time-points and its length at the first; further, the law of thermic expansion is one of the premisses, but now in quantitative formulation; and, finally, there is included a statement of the coefficient of thermic expansion. The conclusion states the amount of the expansion of the rod. We shall not represent D_1 here in detail because a similar derivation D_2 will be discussed later (§ 23); the premises of D_1 are not only the sentences (1)–(5) of D_2 but also (6) and (10); the conclusion in D_1 is the same as in D_2. In D_1 a calculus of real numbers (or at least of fractions) is applied. The conclusion describes a fact which has not yet been observed but could be tested by observations. Now, the question is whether the derivation D_1 does not lead, with the help of a mathematical calculus, to a factual content beyond that of the premises. This might seem so if we look only at the singular sentences among the premisses. But two laws also belong to the premises of D_1 (the sentences [6] and [10] of D_2). They are universal; they say that certain regularities hold not only in the cases so far observed but at any place at any time. Thus, these sentences are very comprehensive. The conclusion merely restates what is already stated by the universal premisses for all cases and hence also for the present case, but now explicitly for this case. Thus, the logico-mathematical derivation merely evolves what is implicitly involved in the premises. To be sure, if we state a new law on the basis of certain observations, the law says much more than the observation sentences known; but this is not a deduction. If, on the other hand, a law is used within a derivation with the help of a logico-mathematical calculus, then the law must be among the premises, and hence the conclusion does not say more than the premises. The situation is different in the application of a physical calculus, as we shall see later (§ 23).

On the basis of the presupposed interpretation, the premises and the conclusion of the derivation D_1 are factual. But D_1 also contains sentences which are proved in a logico-mathematical calculus and hence, when interpreted, are L-true, e.g., the sentences which in D_2 occur as (7) and (13) (§ 23). As explained before, derivations are immensely simplified by the method of laying down for any future use certain partial sequences occurring in many derivations and containing only provable sentences. Each sequence of this kind is a proof of its last sentence; wherever it occurs in other proofs or derivations it may be represented by its last element, i.e., the theorem proved. Thus a logical or mathematical theorem is, regarded from the point of view of its application in empirical science, a device or tool enabling us to make a very complex and long chain of applications of the rules of the calculus by one stroke, so to speak. The theorem is itself, even when interpreted, not a factual statement but an instrument facilitating operations with factual statements, namely, the deduction of a factual conclusion from factual premises. The service which mathematics renders to empirical science consists in furnishing these instruments; the mathematician not only produces them for any particular case of application but keeps them in store, so to speak, ready for any need that may arise.

It is important to notice the distinction between 'primitive sentence' and 'premiss'. A primitive sentence of a calculus C (no matter whether it belongs to the basic calculus or is one of the specific axioms, and no matter whether, in an interpretation, it becomes L-true or factual) is stated as C-true by the rules of the calculus C. Therefore, it has to become a true proposition in any adequate (i.e., true) interpretation of C. The premises of a derivation D in C, on the other hand, need not be C-true in C or true in a true interpretation of C. It is merely shown by D that a certain other sentence (the conclusion of D) is derivable from the premises of D and must therefore, in a true interpretation, be true *if* the premises happen to be true; but whether this is the case is not determined by D.

20. The Controversies over "Foundations" of Mathematics

There have been many discussions in modern times about the nature of mathematics in general and of the various kinds of numbers, and, further, about the distinction and relationship between knowledge in mathematics and knowledge in empirical science. In the course of the last century, mathematicians found that all mathematical signs can be defined on the basis of the signs of the theory of natural numbers.

The fundamental concepts of the infinitesimal calculus (differential coefficient and integral) were defined by Cauchy and Weierstrass in terms of the calculus of real numbers, with the help of the concept 'limit (of a sequence of real numbers)'. Thereby they succeeded in entirely eliminating the dubious concept of "infinitely small magnitudes" and thus giving the infinitesimal calculus a rigorous basis in the theory of real numbers. The next step was made by Frege and Russell, who defined real numbers as classes of natural numbers or of fractions. (Fractions can easily be defined as pairs of natural numbers.)

The reduction mentioned was entirely inside of mathematics. Therefore, it left the more general and fundamental problems unanswered. These have been discussed especially during the last fifty years, usually under the heading "foundations of mathematics". Among the different doctrines developed in this field, three are outstanding and most often discussed; they are known as logicism, formalism, and intuitionism. We will indicate briefly some characteristic features of the three movements. *Logicism* was founded by Frege and developed by Russell and Whitehead. Its chief thesis is that mathematics is a branch of logic. This thesis was demonstrated by constructing a system for the whole of classical mathematics within a logical calculus (see § 14 and some remarks in § 18). Truth conditions for the primitive signs of the logical calculus were given; thereby an interpretation for the whole mathematical system was determined. In this interpretation all mathematical signs became logical signs, all mathematical theorems L-true propositions.

Formalism, founded by Hilbert and Bernays, proposed, in contradistinction to logicism, to construct the system of classical mathematics as a mere calculus without regard to interpretation. The theory developed is called metamathematics; it is, in our terminology, a syntax of the language system of mathematics, involving no semantics. Hilbert's system is a combination of a logical basic calculus with a specific mathematical calculus using as specific primitive signs '0' and '′' as did Peano's system (§ 17). The controversy between the two doctrines concerning the question whether first to construct logic and then mathematics within logic without new primitive signs, or both simultaneously, has at present lost much of its former appearance of importance. We see today that the logico-mathematical calculus can be constructed in either way and that it does not make much difference which one we choose. If the method of logicism is chosen, constructing the system of mathematics as a part of the logical calculus, then by the normal interpretation of the latter we get an interpretation, and moreover the customary one, of the former. The formalists have not concerned themselves much with the question how the mathematical calculus, if constructed according to their method, is to be interpreted and applied in empirical science. As already explained (§ 17), the interpretation can be given by rules of translation for the specific primitive signs into the logical calculus. Another way would be to lay down L-semantical rules for these signs, stating the truth conditions for the descriptive sentences in which they occur. Formalists do not give an interpretation for the mathematical calculus and even seem to regard it as impossible for the nonelementary parts of the calculus, but they emphasize very much the need for a proof of the consistency for the mathematical calculus and even regard it as the chief task of metamathematics. There is some relation between the two questions; if a proof of consistency for a calculus can be given, then a true interpretation and application of the calculus is logically possible. So far, a proof of consistency has been given only for a certain part of arithmetic; the most comprehensive one has been constructed by Gentzen (1936).

Gödel has shown (1931) that it is not possible to construct a proof for the consistency of a calculus C containing arithmetic, within a metalanguage possessing no other logical means (forms of expression and modes of deduction) than C. Hilbert's aim was to construct the proof of consistency in a "finitist" metalanguage (similar to an intuitionist system, see below). At the present, it is not yet known whether this aim can be reached in spite of Gödel's result. In any case, the concept of "finitist logic" is in need of further clarification.

The doctrine of *intuitionism* was originated by Brouwer (1912) and Weyl (1918) on the basis of earlier ideas of Kronecker and Poincaré. This doctrine rejects both the purely formal construction of mathematics as a calculus and the interpretation of mathematics as consisting of L-true sentences without factual content. Mathematics is rather regarded as a field of mental activities based upon "pure intuition". A definition, a sentence, or a deduction is only admitted if it is formulated in "constructive" terms; that is to say, a reference to a mere possibility is not allowed unless we know a method of actualizing it. Thus, for instance, the concept of provability (in the mathematical system) is rejected because there is no method which would lead, for any given sentence S, either to a proof for S or to a proof for the negation of S. It is only allowed to call a sentence proved after a proof has been constructed. For similar reasons, the principle of the excluded middle, the indirect proof of purely existential sentences, and other methods are rejected. In consequence, both elementary logic and classical mathematics are considerably curtailed and complicated. However, the boundary between the admissible and the nonadmissible is not stated clearly and varies with the different authors.

Concerning mathematics as a pure calculus there are no sharp controversies. These arise as soon as mathematics is dealt with as a system of "knowledge"; in our terminology, as an interpreted system. Now, if we regard interpreted mathematics as an instrument of deduction within the field of empirical knowledge rather than as a system of information, then many of the controversial problems are recognized as being questions not of truth but of technical expedience. The question is: Which form of the mathematical system is technically most suitable for the purpose mentioned? Which one provides the greatest safety?

If we compare, e.g., the systems of classical mathematics and of intuitionistic mathematics, we find that the first is much simpler and technically more efficient, while the second is more safe from surprising occurrences, e.g., contradictions. At the present time, any estimation of the degree of safety of the system of classical mathematics, in other words, the degree of plausibility of its principles, is rather subjective. The majority of mathematicians seem to regard this degree as sufficiently high for all practical purposes and therefore prefer the application of classical mathematics to that of intuitionistic mathematics. The latter has not, so far as I know, been seriously applied in physics by anybody.

The problems mentioned cannot here be discussed more in detail. Such discussion is planned for a later volume of this *Encyclopedia*. A more detailed discussion can be found in those of the books which deal with mathematics mentioned in the "Selected Bibliography" at the end of this monograph.

21. Geometrical Calculi and Their Interpretations

When we referred to mathematics in the previous sections, we did not mean to include geometry but only the mathematics of numbers and numerical functions. Geometry must be dealt with separately. To be sure, the geometrical calculi, aside from interpretation, are not fundamentally different in their character from the other calculi and, moreover, are closely related to the mathematical calculi. That is the reason why they too have been developed by mathematicians. But the customary interpretations of geometrical calculi are descriptive, while those of the mathematical calculi are logical.

A geometrical calculus is usually constructed as an axiom system, i.e., a specific calculus presupposing a logical calculus (with normal interpretation). Such a calculus describes a structure whose elements are left undetermined as long as we do not make an interpretation. The geometrical calculi describe many different structures. And for each structure, e.g., the Euclidean, there are many different possible forms of calculi describing it. As an example let us consider an axiom system of Euclidean geometry. We choose a form having six primitive signs; three for classes of individuals, 'P_1', 'P_2', 'P_3', and three for relations, 'I', 'B', 'K'. We write '$I(x,y)$' for 'the relation I holds between x and y', and '$B(x,y,z)$' for 'the (triadic) relation B holds for x,y,z'. We will give only a few examples out of the long series of axioms:

G 1. For every x, y [if [x is a P_1 and y is a P_1] then, for some z [z is a P_2 and $I(x,z)$ and $I(y,z)$]].

G 2. For every x [if x is a P_3 then, for some y [y is a P_1 and, not $I(y,x)$]].

G 3. For every x, y, z [if $B(x,y,z)$ then, not $B(y,x,z)$].

G 4. For every x, y, z [if [x is a P_1 and y is a P_2 and z is a P_3 and $I(x,z)$ and $I(y,z)$ and, not $I(x,y)$] then there is (exactly) 1 u such that [u is a P_2 and $I(x,u)$ and $I(u,z)$ and, for every t [if $I(t,u)$ then, not $I(t,y)$]]]]. (Euclidean parallel axiom.)

For a geometrical calculus there are many interpretations, and even many quite different and interesting interpretations, some of them logical, some descriptive. The *customary inter-pretation* is descriptive. It consists of a translation into the physical calculus (to be dealt with in the next section) together with the customary interpretation of the physical calculus. Rules of translation: (1) 'P_1' is translated into 'point', (2) 'P_2' into 'straight line', (3) 'P_3' into 'plane', (4) '$I(x,y)$' into 'x is lying on y' (incidence), (5) '$B(x,y,z)$' into 'the point x is between the points y and z on a straight line', (6) '$K(x,y,u,v)$' into 'the segment x,y is congruent with the segment u,v (i.e., the distance between x and y is equal to the distance between u and v)'. It is to be noticed that the words 'point', etc., are here signs of the physical calculus in its customary interpretation. Hence we may think of a point as a place in the space of nature; straight lines may be characterized by reference to light rays in a vacuum or to stretched threads; congruence may be characterized by referring to a method of measuring length, etc. Thus the specific signs of a geometrical calculus are interpreted as descriptive signs. (On the other hand, the specific signs of a mathematical calculus are interpreted as logical signs, even if they occur in descriptive factual sentences stating the results of counting or measuring; see, e.g., the logical sign '3', defined by Def. 13, § 14, occurring in premiss [1], § 19.) The axioms and theorems of a geometrical calculus are translated into descriptive, factual propositions of interpreted physics; they form a theory which we may call *physical geometry*, because it is a branch of physics, in contradistinction to mathematical geometry i.e., the geometrical calculus. As an example, the four axioms stated above are translated into the following sentences of the physical calculus (formulated here, for simplicity, in the forms of ordinary language).

PG 1. For any two points there is a straight line on which they lie.

PG 2. For any plane there is a point not lying on it.

PG 3. If the points x, y, and z lie on a straight line and x is between y and z, then y is not between x and z.

PG 4. If the point x and the line y lie in the plane z, but x not on y, then there is one and only one line u in the plane z such that x lies on u and no point is both on u and y (hence u is the parallel to y through x).

22. The Distinction between Mathematical and Physical Geometry

The distinction between mathematical geometry, i.e., the calculus, and physical geometry is often overlooked because both are usually called geometry and both usually employ the same terminology. Instead of artificial symbols like 'P_1', etc., the words 'point', 'line', etc., are used in mathematical geometry as well. The axioms are then not formulated like G 1–4 but like PG 1–4, and hence there is no longer any difference in formulation between mathematical and physical geometry. This procedure is very convenient in practice—like the analagous procedure in the mathematical calculus, mentioned previously— because it saves the trouble of translating, and facilitates the understanding and manipulating of the calculus. But it is essential to keep in mind the fundamental difference between mathematical and physical geometry in spite of the identity of formulation. The difference becomes clear when we take into consideration other interpretations of the geometrical calculus.

Of especial importance for the development of geometry in the past few centuries has been a certain translation of the geometrical calculus into the mathematical calculus. This leads, in combination with the customary interpretation of the mathematical calculus, to a logical interpretation of the geometrical calculus. The translation was found by Descartes and is known as analytic geometry or geometry of coordinates. 'P_1' (or, in ordinary formulation, 'point') is translated into 'ordered triple of real numbers'; 'P_3' ('plane') into 'class of ordered triples of real numbers fulfilling a linear equation', etc. The axioms, translated in this way, become C-true sentences of the mathematical calculus; hence the translation is C-true. On the basis of the customary interpretation of the mathematical calculus, the axioms and theorems of geometry become L-true propositions.

The difference between mathematical and physical geometry became clear in the historical development by the discovery of non-Euclidean geometry, i.e., of axiom systems deviating from the Euclidean form by replacing the parallel axiom (G 4) by some other axiom incompatible with it. It has been shown that each of these systems, although they are incompatible with one another, does not contain a contradiction, provided the Euclidean system is free from contradictions. This was shown by giving a translation for each of the non-Euclidean systems into the Euclidean system. Mathematicians regarded all these systems on a par, investigating any one indifferently. Physicists, on the other hand, could not accept this plurality of geometries; they asked: "Which one is true? Has the space of nature the Euclidean or one of the non-Euclidean structures?" It became clear by an analysis of the discussions that the mathematician and the physicist were talking about different things, although they themselves were not aware of this in the beginning. Mathematicians have to do with the geometrical calculus, and with respect to a calculus there is no question of truth and falsity. Physicists, however, are concerned with a theory of space, i.e., of the system of possible configurations and movements of bodies, hence with the interpretation of a geometrical calculus. When an interpretation of the specific signs is established—and, to a certain extent, this is a matter of choice—then each of the calculi yields a physical geometry as a theory with factual content. Since they are incompatible, at most one can be true (truth of a class of sentences [see § 6]). The theories are factual.

The truth conditions, determined by the interpretation, refer to facts. Therefore, it is the task of the physicist, and not of the mathematician, to find out whether a certain one among the theories is true, i.e., whether a certain geometrical structure is that of the space of nature. (Of course, the truth of a system of physical geometry, like that of any other universal factual sentence or theory, can never be known with absolute certainty but at best with a high degree of confirmation.) For this purpose, the physicist has to carry out experiments and to see whether the predictions made with the help of the theory under investigation, in connection with other theories confirmed and accepted previously, are confirmed by the observed results of the experiments. The accuracy of the answer found by the physicist is, of course, dependent upon the accuracy of the instruments available. The answer given by classical physics was that the Euclidean system of geometry is in accordance with the results of measurements, within the limits of the accuracy of observations. Modern physics has modified this answer in the general theory of relativity by stating that the Euclidean geometry describes the structure of space, though not exactly, yet with a degree of approximation sufficient for almost all practical purposes; a more exact description is given by a certain non-Euclidean system of geometry. Physical geometry is in its methods not fundamentally different from the other parts of physics. This will become still more obvious when we shall see how other parts of physics can also take the form of calculi (§ 23).

The doctrine concerning geometry acknowledged by most philosophers in the past century was that of Kant, saying that geometry consists of "synthetic judgments a priori", i.e., of sentences which have factual content but which, nevertheless, are independent of experience and necessarily true. Kant attributed the same character also to the sentences of arithmetic. Modern logical analysis of language, however, does not find any sentences at all of this character. We may assume that the doctrine is not to be understood as applying to the formulas of a calculus; there is no question of truth with respect to them

197

because they are not assertions; in any case they are not synthetic (i.e., factual). The doctrine was obviously meant to apply to arithmetic and geometry as theories, i.e., interpreted systems, with their customary interpretations. Then, however, the propositions of arithemetic are, to be sure, independent of experience, but only because they do not concern experience or facts at all; they are L-true (analytic), not factual (synthetic). For geometry there is also, as mentioned before, the possibility of a logico-mathematical interpretation; by it the sentences of geometry get the same character as those of mathematics. On the basis of the customary interpretation, however, the sentences of geometry, as propositions of physical geometry, are indeed factual (synthetic), but dependent upon experience, empirical. The Kantian doctrine is based on a failure to distinguish between mathematical and physical geometry. It is to this distinction that Einstein refers in his well-known dictum: "So far as the theorems of mathematics are about reality they are not certain; and so far as they are certain they are not about reality."

The question is frequently discussed whether arithmetic and geometry, looked at from the logical and methodological point of view, have the same nature or not. Now we see that the answer depends upon whether the calculi or the interpreted systems are meant. There is no fundamental difference between arithmetic and geometry as calculi, nor with respect to their *possible* interpretations; for either calculus there are both logical and descriptive interpretations. If, however, we take the systems with their *customary* interpretation—arithmetic as the theory of numbers and geometry as the theory of physical space—then we find an important difference: the propositions of arithmetic are logical, L-true, and without factual content; those of geometry are descriptive, factual, and empirical.

23. Physical Calculi and Their Interpretations

The method described with respect to geometry can be applied likewise to any other part of physics: we can first construct a calculus and then lay down the interpretation intended in the form of semantical rules, yielding a physical theory as an interpreted system with factual content. The customary formulation of a physical calculus is such that it presupposes a logico-mathematical calculus as its basis, e.g., a calculus of real numbers in any of the forms discussed above (§ 18). To this basic calculus are added the specific primitive signs and the axioms, i.e., specific primitive sentences, of the physical calculus in question.

Thus, for instance, a calculus of mechanics of mass points can be constructed. Some predicates and functors (i.e., signs for functions) are taken as specific primitive signs, and the fundamental laws of mechanics as axioms. Then semantical rules are laid down stating that the primitive signs designate, say, the class of material particles, the three spatial coordinates of a particle x at the time t, the mass of a particle x, the class of forces acting on a particle x or at a space point s at the time t. (As we shall see later [§ 24], the interpretation can also be given indirectly, i.e., by semantical rules, not for the primitive signs, but for certain defined signs of the calculus. This procedure must be chosen if the semantical rules are to refer only to observable properties.) By the interpretation, the theorems of the calculus of mechanics become physical laws, i.e., universal statements describing certain features of events; they constitute physical mechanics as a theory with factual content which can be tested by observations. The relation of this theory to the calculus of mechanics is entirely analogous to the relation of physical to mathematical geometry. The customary division into theoretical and experimental physics corresponds roughly to the distinction between calculus and interpreted system. The work in theoretical physics consists mainly in constructing calculi and carrying out deductions within them; this is essentially mathematical work. In experimental physics interpretations are made and theories are tested by experiments.

In order to show by an example how a deduction is carried out with the help of a physical calculus, we will discuss a calculus which can be interpreted as a theory of thermic expansion. To the primitive signs may belong the predicates 'Sol' and

199

Foundations of Logic and Mathematics

'Fe', and the functors 'lg', 'te', and 'th'. Among the axioms may be A 1 and A 2. (Here, 'x', 'β' and the letters with subscripts are real number variables; the parentheses do not contain explanations as in former examples, but are used as in algebra and for the arguments of functors.)

A 1. For every $x, t_1, t_2, l_1, l_2, T_1, T_2, \beta$ [if [x is a Sol and lg(x,t_1) = l_1 and lg(x,t_2) = l_2 and te(x,t_1) = T_1 and te(x,t_2) = T_2 and th(x) = β] then $l_2 = l_1 \times (1 + \beta \times (T_2 - T_1))$].

A 2. For every x, if [x is a Sol and x is a Fe] then th(x) = 0.000012.

The *customary interpretation*, i.e., that for whose sake the calculus is constructed, is given by the following semantical rules. 'lg(x,t)' designates the length in centimeters of the body x at the time t (defined by the statement of a method of measurement); 'te(x,t)' designates the absolute temperature in centigrades of x at the time t (likewise defined by a method of measurement); 'th(x)' designates the coefficient of thermic expansion for the body x; 'Sol' designates the class of solid bodies; 'Fe' the class of iron bodies. By this interpretation, A 1 and A 2 become physical laws. A 1 is the law of thermic expansion in quantitative form, A 2 the statement of the coefficient of thermic expansion for iron. As A 2 shows, a statement of a physical constant for a certain substance is also a universal sentence. Further, we add semantical rules for two signs occuring in the subsequent example: the name 'c' designates the thing at such and such a place in our laboratory; the numerical variable 't' as time coordinate designates the time-point t seconds after August 17, 1938, 10:00 A.M.

Now we will analyze an example of a derivation within the calculus indicated. This derivation D_2 is, when interpreted by the rules mentioned, the deduction of a prediction from premisses giving the results of observations. The construction of the derivation D_2, however, is entirely independent of any interpretation. It makes use only of the rules of the calculus, namely, the physical calculus indicated together with a calculus of real numbers as basic calculus. We have discussed, but not written down, a similar derivation D_1 (§ 19), which, however, made use only of the mathematical calculus. Therefore the physical laws used had to be taken in D_1 as premisses. But here in D_2 they belong to the axioms of the calculus (A 1 and A 2, occurring as [6] and [10]). Any axiom or theorem proved in a physical calculus may be used within any derivation in that calculus without belonging to the premisses of the derivation, in exactly the same way in which a proved theorem is used within a derivation in a logical or mathematical calculus, e.g., in the first example of a derivation in § 19 sentence (7), and in D_1 (§ 19) the sentences which in D_2 are called (7) and (13). Therefore only singular sentences (not containing variables) occur as premisses in D_2. (For the distinction between premisses and axioms see the remark at the end of § 19.)

Derivation D_2:

Premisses	1. c is a Sol.
	2. c is a Fe.
	3. $\mathrm{te}(c,0)=300$.
	4. $\mathrm{te}(c,600)=350$.
	5. $\mathrm{lg}(c,0)=1{,}000$.

Axiom A 1 6. For every x, t_1, t_2, l_1, l_2, T_1, T_2, β $\big[$if $[x$ is a Sol and $\mathrm{lg}(x,t_1)=l_1$ and $\mathrm{lg}(x,t_2)=l_2$ and $\mathrm{te}(x,t_1)=T_1$ and $\mathrm{te}(x,t_2)=T_2$ and $\mathrm{th}(x)=\beta]$ then $l_2=l_1\times(1+\beta\times(T_2-T_1))\big]$.

Proved mathem. theorem: 7. For every l_1, l_2, T_1, T_2, β $\big[l_2-l_1=l_1\times\beta\times(T_2-T_1)$ if and only if $l_2=l_1\times(1+\beta\times(T_2-T_1))\big]$.

(6)(7) 8. For every x, t_1, ... (as in [6]) ... $\big[$if $[\text{- - -}]$ then $l_2-l_1=l_1\times\beta\times(T_2-T_1)\big]$.

(1)(3)(4)(8) 9. For every l_1, l_2, β $\big[$if $[\mathrm{th}(c)=\beta$ and $\mathrm{lg}(c,0)=l_1$ and $\mathrm{lg}(c,600)=l_2]$ then $l_2-l_1=l_1\times\beta\times(350-300)\big]$.

Axiom A 2 10. For every x, if $[x$ is a Sol and x is a Fe] then $\mathrm{th}(x)=0.000012$.

(1)(2)(10) 11. $\mathrm{th}(c)=0.000012$.

(9)(11)(5) 12. For every l_1, l_2 $\big[$if $[\mathrm{lg}(c,0)=l_1$ and $\mathrm{lg}(c,600)=l_2]$ then $l_2-l_1=1{,}000\times0.000012\times(350-300)\big]$.

Proved mathem. theorem: 13. $1{,}000\times0.000012\times(350-300)=0.6$.

(12)(13) *Conclusion:* 14. $\mathrm{lg}(c,600)-\mathrm{lg}(c,0)=0.6$.

On the basis of the interpretation given before, the premisses are singular sentences concerning the body c. They say that c is a solid body made of iron, that the temperature of c was at 10:00 A.M. 300° abs., and at 10:10 A.M. 350° abs., and that the length of c at 10:00 A.M. was 1,000 cm. The conclusion says that the increase in the length of c from 10:00 to 10:10 A.M. is 0.6 cm. Let us suppose that our measurements have confirmed the premisses. Then the derivation yields the conclusion as a prediction which may be tested by another measurement.

Any physical theory, and likewise the whole of physics, can in this way be presented in the form of an interpreted system, consisting of a specific calculus (axiom system) and a system of semantical rules for its interpretation; the axiom system is, tacitly or explicitly, based upon a logico-mathematical calculus with customary interpretation. It is, of course, logically possible to apply the same method to any other branch of science as well. But practically the situation is such that most of them seem at the present time to be not yet developed to a degree which would suggest this strict form of presentation. There is an interesting and successful attempt of an axiomatization of certain parts of biology, especially genetics, by Woodger (Vol. I, No. 10). Other scientific fields which we may expect to be soon accessible to this method are perhaps chemistry, economics, and some elementary parts of psychology and social science.

Within a physical calculus the mathematical and the physical theorems, i.e., C-true formulas, are treated on a par. But there is a fundamental difference between the corresponding *mathematical* and the *physical propositions* of the physical theory, i.e., the system with customary interpretation. This difference is often overlooked. That physical theorems are sometimes mistaken to be of the same nature as mathematical theorems is perhaps due to several factors, among them the fact that they contain mathematical symbols and numerical expressions and that they are often formulated incompletely in the form of a mathematical equation (e.g., A 1 simply in the form of the last equation occurring in it). A mathematical proposition may contain only logical signs, e.g., 'for every m, n, $m + n = n + m$', or descriptive signs also, if the mathematical calculus is applied in a descriptive system. In the latter case the proposition, although it contains signs not belonging to the mathematical calculus, may still be provable in this calculus, e.g., '$\lg(c) + \lg(d) = \lg(d) + \lg(c)$' ('lg' designates length as before). A physical proposition always contains descriptive signs, because otherwise it could not have factual content; in addition, it usually contains also logical signs. Thus the difference between mathematical theorems and physical theorems in the interpreted system does not depend upon the kinds of signs occurring but rather on the kind of truth of the theorems. The truth of a mathematical theorem, even if it contains descriptive signs, is not dependent upon any facts concerning the designata of these signs. We can determine its truth if we know only the semantical rules; hence it is L-true. (In the example of the theorem just mentioned, we need not know the length of the body c.) The truth of a physical theorem, on the other hand, depends upon the properties of the designata of the descriptive signs occuring. In order to determine its truth, we have to make observations concerning these designata; the knowledge of the semantical rules is not sufficient. (In the case of A 2, e.g., we have to carry out experiments with solid iron bodies.) Therefore, a physical theorem, in contradistinction to a mathematical theorem, has factual content.

24. Elementary and Abstract Terms

We find among the concepts of physics—and likewise among those of the whole of empirical science—differences of abstractness. Some are more elementary than others, in the sense that we can apply them in concrete cases on the basis of observations in a more direct way than others. The others are more abstract; in order to find out whether they hold in a certain case, we have to carry out a more complex procedure, which, however, also finally rests on observations. Between quite elementary concepts and those of high abstraction there are many intermediate levels. We shall not try to give an exact definition for 'degree of abstractness'; what is meant will become sufficiently clear by the following series of sets of concepts, proceeding from elementary to abstract concepts: bright, dark, red, blue, warm, cold, sour, sweet, hard, soft (all concepts of this first set are meant as properties of things, not as sense-data); coincidence; length; length of time; mass, velocity, acceleration, density, pressure; temperature, quantity of heat; electric charge, electric current, electric field; electric potential, electric resistance, coefficient of induction, frequency of oscillation; wave function.

Suppose that we intend to construct an interpreted system of physics—or of the whole of science. We shall first lay down a calculus. Then we have to state semantical rules of the kind SD for the specific signs, i.e., for the physical terms. (The SL-rules are presupposed as giving the customary interpretation of the logico-mathematical basic calculus.) Since the physical terms form a system, i.e., are connected with one another, obviously we need not state a semantical rule for each of them. For which terms, then, must we give rules, for the elementary or for the abstract ones? We can, of course, state a rule for any term, no matter what its degree of abstractness, in a form like this: 'the term 'te' designates temperature', provided the meta-language used contains a corresponding expression (here the word 'temperature') to specify the designatum of the term in question. But suppose we have in mind the following purpose for our syntactical and semantical description of the system of physics: the description of the system shall teach a layman to understand it, i.e., to enable him to apply it to his observations in order to arrive at explanations and predictions. A layman is meant as one who does not know physics but has normal senses and understands a language in which observable properties of things can be described (e.g., a suitable part of everyday non-scientific English). A rule like 'the sign '*P*' designates the property of being blue' will do for the purpose indicated; but a rule like 'the sign '*Q*' designates the property of being electrically charged' will not do. In order to fulfil the purpose, we have to give semantical rules for elementary terms only, connecting them with observable properties of things. For our further discussion we suppose the system to consist of rules of this kind, as indicated in the following diagram.

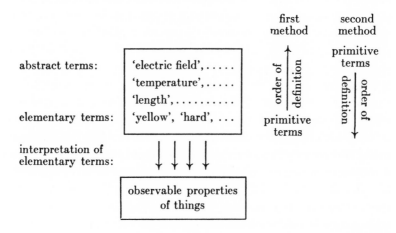

Now let us go back to the construction of the calculus. We have first to decide at which end of the series of terms to start the construction. Should we take elementary terms as primitive signs, or abstract terms? Our decision to lay down the semantical rules for the elementary terms does not decide this question. Either procedure is still possible and seems to have some reasons in its favor, depending on the point of view taken. The *first method* consists in taking elementary terms as primitive and then introducing on their basis further terms step by step, up to those of highest abstraction. In carrying out this procedure, we find that the introduction of further terms cannot always take the form of explicit definitions; conditional definitions must also be used (so-called reduction sentences [see Vol. I, No. 1, p. 50]). They describe a method of testing for a more abstract term, i.e., a procedure for finding out whether the term is applicable in particular cases, by referring to less abstract terms. The first method has the advantage of exhibiting clearly the connection between the system and observation and of making it easier to examine whether and how a given term is empirically founded. However, when we shift our attention from the terms of the

205

system and the methods of empirical confirmation to the laws, i.e., the universal theorems, of the system, we get a different perspective. Would it be possible to formulate all laws of physics in elementary terms, admitting more abstract terms only as abbreviations? If so, we would have that ideal of a science in sensationalistic form which Goethe in his polemic against Newton, as well as some positivists, seems to have had in mind. But it turns out—this is an empirical fact, not a logical necessity— that it is not possible to arrive in this way at a powerful and efficacious system of laws. To be sure, historically, science started with laws formulated in terms of a low level of abstractness. But for any law of this kind, one nearly always later found some exceptions and thus had to confine it to a narrower realm of validity. The higher the physicists went in the scale of terms, the better did they succeed in formulating laws applying to a wide range of phenomena. Hence we understand that they are inclined to choose the *second method*. This method begins at the top of the system, so to speak, and then goes down to lower and lower levels. It consists in taking a few abstract terms as primitive signs and a few fundamental laws of great generality as axioms. Then further terms, less and less abstract, and finally elementary ones, are to be introduced by definitions; and here, so it seems at present, explicit definitions will do. More special laws, containing less abstract terms, are to be proved on the basis of the axioms. At least, this is the direction in which physicists have been striving with remarkable success, especially in the past few decades. But at the present time, the method cannot yet be carried through in the pure form indicated. For many less abstract terms no definition on the basis of abstract terms alone is as yet known; hence those terms must also be taken as primitive. And many more special laws, especially in biological fields, cannot yet be proved on the basis of laws in abstract terms only; hence those laws must also be taken as axioms.

Now let us examine the result of the interpretation if the first or the second method for the construction of the calculus is chosen. In both cases the semantical rules concern the elementary signs. In the first method these signs are taken as primitive. Hence, the semantical rules give a complete interpretation for these signs and those explicitly defined on their basis. There are, however, many signs, especially on the higher levels of abstraction, which can be introduced not by an explicit definition but only by a conditional one. The interpretation which the rules give for these signs is in a certain sense incomplete. This is due not to a defect in the semantical rules but to the method by which these signs are introduced; and this method is not arbitrary but corresponds to the way in which we really obtain knowledge about physical states by our observations.

If, on the other hand, abstract terms are taken as primitive—according to the second method, the one used in scientific physics—then the semantical rules have no direct relation to the primitive terms of the system but refer to terms introduced by long chains of definitions. The calculus is first constructed floating in the air, so to speak; the construction begins at the top and then adds lower and lower levels. Finally, by the semantical rules, the lowest level is anchored at the solid ground of the observable facts. The laws, whether general or special, are not directly interpreted, but only the singular sentences. For the more abstract terms, the rules determine only an *indirect interpretation*, which is—here as well as in the first method—incomplete in a certain sense. Suppose 'B' is defined on the basis of 'A'; then, if 'A' is directly interpreted, 'B' is, although indirectly, also interpreted completely; if, however, 'B' is directly interpreted, 'A' is not necessarily also interpreted completely (but only if 'A' is also definable by 'B').

To give an example, let us imagine a calculus of physics constructed, according to the second method, on the basis of primitive specific signs like 'electromagnetic field', 'gravitational field', 'electron', 'proton', etc. The system of definitions will then lead to elementary terms, e.g., to 'Fe', defined as a class of regions in which the configuration of particles fulfils certain conditions, and 'Na-yellow' as a class of space-time regions in which the temporal distribution of the electromagnetic field fulfils certain conditions. Then semantical rules are laid down stating that 'Fe' designates iron and 'Na-yellow' designates a specified yellow color. (If 'iron' is not accepted as sufficiently elementary, the rules can be stated for more elementary terms.) In this way

207

the connection between the calculus and the realm of nature, to which it is to be applied, is made for terms of the calculus which are far remote from the primitive terms.

Let us examine, on the basis of these discussions, the example of a derivation D_2 (§ 23). The premisses and the conclusion of D_2 are singular sentences, but most of the other sentences are not. Hence the premisses and the conclusion of this as of all other derivations of the same type can be directly interpreted, understood, and confronted with the results of observations. More of an interpretation is not necessary for a practical application of a derivation. If, in confronting the interpreted premisses with our observations, we find them confirmed as true, then we accept the conclusion as a prediction and we may base a decision upon it. The sentences occurring in the derivation between premisses and conclusion are also interpreted, at least indirectly. But we need not make their interpretation explicit in order to be able to construct the derivation and to apply it. All that is necessary for its construction are the formal rules of the calculus. This is the advantage of the method of formalization, i.e., of the separation of the calculus as a formal system from the interpretation. If some persons want to come to an agreement about the formal correctness of a given derivation, they may leave aside all differences of opinion on material questions or questions of interpretation. They simply have to examine whether or not the given series of formulas fulfils the formal rules of the calculus. Here again, the function of calculi in empirical science becomes clear as instruments for transforming the expression of what we know or assume.

Against the view that for the application of a physical calculus we need an interpretation only for singular sentences, the following objection will perhaps be raised. Before we accept a derivation and believe its conclusion we must have accepted the physical calculus which furnishes the derivation; and how can we decide whether or not to accept a physical calculus for application without interpreting and understanding its axioms? To be sure, in order to pass judgment about the applicability of a given physical calculus we have to confront it in some way or other with observation, and for this purpose an interpretation is necessary. But we need no explicit interpretation of the axioms, nor even of any theorems. The empirical examination of a physical theory given in the form of a calculus with rules of interpretation is not made by interpreting and understanding the axioms and then considering whether they are true on the basis of our factual knowledge. Rather, the examination is carried out by the same procedure as that explained before for obtaining a prediction. We construct derivations in the calculus with premisses which are singular sentences describing the results of our observations, and with singular sentences which we can test by observations as conclusions. The physical theory is indirectly confirmed to a higher and higher degree if more and more of these predictions are confirmed and none of them is disconfirmed by observations. Only singular sentences with elementary terms can be directly tested; therefore, we need an explicit interpretation only for these sentences.

25. "Understanding" in Physics

The development of physics in recent centuries, and especially in the past few decades, has more and more led to that method in the construction, testing, and application of physical theories which we call *formalization,* i.e., the construction of a calculus supplemented by an interpretation. It was the progress of knowledge and the particular structure of the subject matter that suggested and made practically possible this increasing formalization. In consequence it became more and more possible to forego an "intuitive understanding" of the abstract terms and axioms and theorems formulated with their help. The possibility and even necessity of abandoning the search for an understanding of that kind was not realized for a long time. When abstract, nonintuitive formulas, as, e.g., Maxwell's equations of electromagnetism, were proposed as new axioms, physicists endeavored to make them "intuitive" by constructing a "model", i.e., a way of representing electromagnetic microprocesses by an analogy to known macro-processes, e.g., movements of visible things. Many attempts have been made in this direction, but without satisfactory results. It is important to realize that the discovery of a model has no more than an aesthetic or didactic or at best a heuristic value, but is not at all essential for a successful application of the physical theory. The demand for an intuitive understanding of the axioms was less and less fulfilled when the development led to the general theory of relativity and then to quantum mechanics, involving the wave function. Many people, including physicists, have a feeling of regret and disappointment about this. Some, especially philosophers, go so far as even to contend that these modern theories, since they are not intuitively understandable, are not at all theories about nature but "mere formalistic constructions", "mere calculi". But this is a fundamental misunderstanding of the function of a physical theory. It is true a theory must not be a "mere calculus" but possess an interpretation, on the basis of which it can be applied to facts of nature. But it is sufficient, as we have seen, to make this interpretation explicit for elementary terms; the interpretation of the other terms is then indirectly determined by the formulas of the calculus, either definitions or laws, connecting them with the elementary terms. If we demand from the modern physicist an answer to the question what he means by the symbol 'ψ' of his calculus, and are astonished that he cannot give an answer, we ought to realize that the situation was already the same in classical physics. There the physicist could not tell us what he meant by the symbol 'E' in Maxwell's equations. Perhaps, in order not to refuse an answer, he would tell us that 'E' designates the electric field vector. To be sure, this statement has the form of a semantical rule, but it would not help us a bit to understand the theory. It simply refers from a symbol in a symbolic calculus to a corresponding word expression in a calculus of words. We are right in demanding an interpretation for 'E', but that will be given indirectly by semantical rules referring to elementary signs together with the formulas connecting them with 'E'. This interpretation enables us to use the laws containing 'E' for the derivation of predictions. Thus we understand 'E', if "understanding" of an expression, a sentence, or a theory means capability of its use for the description of known facts or the prediction of new facts. An "intuitive understanding" or a direct translation of 'E' into terms referring to observable properties is neither necessary nor possible. The situation of the modern physicist is not essentially different. He knows how to use the symbol 'ψ' in the calculus in order to derive predictions which we can test by observations. (If they have the form of probability statements, they are tested by statistical results of observations.) Thus the physicist, although he cannot give us a translation into everyday language, understands the symbol 'ψ' and the laws of quantum mechanics. He possesses that kind of understanding which alone is essential in the field of knowledge and science.

Selected Bibliography

CARNAP, R. *Abriss der Logistik.* Wien, 1929.

———. "Die Mathematik als Zweig der Logik," *Blätter für deutsche Philosophie*, Vol. IV (1930).

———. *Logical Syntax of Language.* (Orig., Wien, 1934.) London and New York, 1937.

———. "Formalwissenschaft und Realwissenschaft," *Erkenntnis*, Vol. V (1935).

———. "Testability and Meaning," *Philosophy of Science*, Vols. III (1936) and IV (1937).

DEDEKIND, R. *Was sind und was sollen die Zahlen?* Braunschweig, 1888.

EINSTEIN, A. *Geometrie and Erfahrung.* Berlin, 1921.

FRAENKEL, A. *Einleitung in die Mengenlehre.* 3d ed. Berlin, 1928.

FRANK, P. *Interpretations and Misinterpretations of Modern Physics.* Paris, 1938.

FREGE, G. *Die Grundlagen der Arithmetik.* Breslau, 1884.

———. *Grundgesetze der Arithmetik*, Vols. I and II. Jena, 1893 and 1903.

HAHN, H. *Logik, Mathematik und Naturerkennen.* Wien, 1933.

HEYTING, A. *Mathematische Grundlagenforschung, Intuitionismus, Beweistheorie.* Berlin, 1934.

HILBERT, D. "Axiomatisches Denken," *Math. Annalen*, Vol. LXXVIII (1918).

HILBERT, D., and ACKERMANN, W. *Grundzüge der theoretischen Logik.* Berlin, 1928. 2d ed., 1938.

HILBERT, D., and BERNAYS, P. *Grundlagen der Mathematik*, Vol. I. Berlin, 1934.

Foundations of Logic and Mathematics

LEWIS, C. I., and LANGFORD, C. H. *Symbolic Logic*. New York and London, 1932.

MENGER, K. "The New Logic," *Philosophy of Science*, Vol. IV (1937).

MORRIS, C. W. *Logical Positivism, Pragmatism, and Scientific Empiricism*. Paris, 1937.

PEIRCE, C. S. *Collected Papers* (esp. Vol II). Cambridge, Mass., 1931 ff.

QUINE, W. V. "Truth by Convention," in *Philosophical Essays for A. N. Whitehead*. London and New York, 1936.

REICHENBACH, H. *Philosophie der Raum-Zeit-Lehre*. Berlin, 1928.

RUSSELL, B. *The Principles of Mathematics*. Cambridge, 1903. 2d ed., 1938.

———. *Introduction to Mathematical Philosophy*. London, 1919. 2d ed., 1920.

SCHOLZ, H. *Geschichte der Logik*. Berlin, 1931.

TARSKI, A. "Der Wahrheitsbegriff in den formalisierten Sprachen," *Studia philosophica*, Vol. I (1935).

———. "Grundlegung der wissenschaftlichen Semantik," in *Actes du congrès international de philosophie scientifique*. Paris, 1936.

———. *Einführung in die mathematische Logik*. Wien, 1937.

WAISMANN, F. *Einführung in das mathematische Denken*. Wien, 1936.

WHITEHEAD, A. N. and RUSSELL, B. *Principia mathematica*, Vols. I, II, and III. Cambridge, (1910) 1925, (1912) 1927, and (1913) 1927.

WITTGENSTEIN, L. *Tractatus logico-philosophicus*. London, 1922.

Index of Terms

[The numbers refer to the sections of this monograph.]

Abstract term, 24
Axiom system, 16

Basic calculus, 16
C- . . . , 8, 9
Calculus, 8
Classification of signs, 5
Conclusion, 9
Consistent, 8
Convention, 12
Customary interpretation, 13

Definition, 13
Derivable, derivation, 9
Descriptive sign, 4
Designatum, 2

Elementary term, 24
Equivalent, 6
Exclude, 6
Explanation, 1, 15
Expression, 4

F- . . . , factual, 7
False, 5
Finite rule, calculus, 10
Formal, 8
Formalism, 20
Formalization, 25
Formula, 9

Geometry, 21

Implicate, 6
Interpretation, 10
Intuitionism, 20

L- . . . , 7
L-exhaustive, 10
Logical calculi, 13, 14

Logical sign, 4
Logicism, 20

Mathematical calculi, 17
Metalanguage, 3

Name, 5
Normal interpretation, 13, 14
Number, 14, 17, 18

Object-language, 3

Physical calculi, 23
Pragmatics, 2, 3
Predicate, 5
Prediction, 1, 15
Premiss, 9, (19)
Primitive sentence, 9, (19)
Proof, provable, 9
Proposition, 5, 9

Rules of formation, 5, 9
Rules of inference, 9
Rules of transformation, 9

Semantics, 2, 4
Sentence, 2
Sentential calculus, 13
Sign, 4
Singular sentence, 15
Specific calculus, sign, 16
Synonymous, 6
Syntax, 2, 8

Transfinite rule, calculus, 10
Translation, 17
True, 5

Understanding, 5, 25
Universal sentence, 15

Linguistic Aspects of Science

Leonard Bloomfield

Linguistic Aspects of Science

Contents:

I. INTRODUCTION

                                        PAGE

  1. Language and Science 218
  2. The Present State of Linguistics 218
  3. Scope of This Essay 220
  4. Terminology 222
  5. Delimitation 223
  6. Writing and Language 224
  7. Language and Handling Activity 226
  8. Apriorism 229
  9. Mentalism 230
  10. Solipsism 231

II. THE FUNCTION OF LANGUAGE

  11. The Speech Community 233
  12. Relayed Speech 234
  13. Verbal Self-stimulation 234
  14. Meaning 235
  15. Acquisition of Meaning 237

III. THE STRUCTURE OF LANGUAGE

  16. Phonemic Structure 239
  17. Grammatical Structure 242

Contents:

PAGE

18. The Sentence 245
19. Constructional Level and Scope 246
20. Varieties of Reference 248
21. Substitution and Determination 251

IV. PRECISION IN NATURAL LANGUAGE
22. Reporting Statements 251
23. Negation 253
24. Abstraction and Approximation 254
25. Counting 256
26. Infinite Classes of Speech-Forms 257

V. SCIENTIFIC LANGUAGE
27. Development of Scientific Language 260
28. General Character of Scientific Language . . 263
29. Publicity and Translatability 264
30. Postulational Form 265
31. Sentence-Forms of Scientific Language . . . 267
32. Syntactic Features 269
33. Special Features 271

VI. SUMMARY
34. The Place of Linguistics in the Scheme of Science 272
35. Relation of Linguistics to Logic and Mathematics 273

SELECTED BIBLIOGRAPHY 275

INDEX OF TECHNICAL TERMS 277

217

Linguistic Aspects of Science

Leonard Bloomfield

I. Introduction

1. Language and Science

Language plays a very important part in science. A typical act of science might consist of the following steps: observation, report of observations, statement of hypotheses, calculation, prediction, testing of predictions by further observation. All but the first and last of these are acts of speech. Moreover, the accumulation of scientific results (the "body" of science) consists of records of speech utterance, such as tables of observed data, a repertoire of predictions, and formulas for convenient calculation.

The use of language in science is specialized and peculiar. In a brief speech the scientist manages to say things which in ordinary language would require a vast amount of talk. His hearers respond with great accuracy and uniformity. The range and exactitude of scientific prediction exceed any cleverness of everyday life: the scientist's use of language is strangely effective and powerful. Along with systematic observation, it is this peculiar use of language which distinguishes science from non-scientific behavior.

The present essay attempts to state briefly such general considerations of linguistics as may throw light upon the procedure of science.

2. The Present State of Linguistics

The ancient Greeks succeeded in describing the main syntactic and inflectional forms of their language. The description, to be sure, was incomplete and contained some serious misconceptions; especially the method of definition was unscientific.

219

With slight modifications, the ancient Greek doctrine has come down into our schools and constitutes today the equipment, as to linguistic knowledge, of the educated man. The ancient doctrine was able to persist because the principal modern languages of Europe are related to ancient Greek and have much the same structure.

Around the beginning of the nineteenth century the Sanskrit grammar of the ancient Hindus became known to European scholars. Hindu grammar described the Sanskrit language completely and in scientific terms, without prepossessions or philosophical intrusions. It was from this model that Western scholars learned, in the course of a few decades, to describe a language in terms of its own structure. The greatest progress in this respect occurred perhaps in connection with the native languages of America. The study of languages of diverse structure broke down parochial misconceptions according to which special features of Indo-European grammar were universal in human speech; it destroyed the pseudo-philosophical dogmas which were built up on these misconceptions. The Greek and medieval notions of linguistics, which still hold sway, it would seem ineradicably, in our schools, have thus been long out of date.

Acquaintance with the Sanskrit grammar and, at the same time, a new mastery of historical perspective brought about, at the beginning of the nineteenth century, the development of comparative and historical linguistics. The method of this study may fairly be called one of the triumphs of nineteenth-century science. In a survey of scientific method it should serve as a model of one type of investigation, since no other historical discipline has equaled it. Historical linguistics, like descriptive, has failed to influence our schools or to enter higher education, except for a haphazard and usually incorrect statement here and there in our schoolbooks. A confusion, strangely enough, of linguistic relationship with the breeding, in races, of domestic animals has found its way into popular sociology.

In the last half-century the study of variations within a language, especially of local dialects, has given us a new insight into the minute processes which make up linguistic change.

Unlike the science of chemistry, which is only a decade or so older, modern linguistics, perhaps for lack of immediate economic use, has failed to put its results into public possession. Not only the man in the street but also the educated public and even the practitioners of closely related branches of science are generally untouched by linguistic knowledge. Under the spell of the traditional dogma which is carried along in the schools the non-linguist responds often enough with incredulous surprise to well-established results that have long ago become commonplace among linguists.

3. Scope of This Essay

In this essay we are concerned only with such phases of linguistics as may throw light upon scientific procedure. Historical-comparative linguistics will not here concern us, since the use of language in science presupposes complete stability in the habits of speech. The linguistic changes which occur within a generation or two are relatively minute; such as they are, they disappear entirely under the agreements, explicit and other, which underlie the scientist's use of language.

As to descriptive linguistics, the vastly greater part of its data also can here be left aside, for we need not enter upon any examination of the widely differing types of structure that appear in different languages. It is enough for us to know that nearly all the structural features of our language which we are inclined to accept as universal—features such as the actor-action sentence, the elaborate part-of-speech system, or the special inflections of our nouns and verbs—are peculiarities of the Indo-European family of languages and are by no means universal in human speech. In part, moreover, the uniformity with which these features seem to appear in the languages that figure in our schooling is an illusion produced by the inaccuracy of the conventional description.

For our purpose we need scarcely touch upon the structural features of any one language; we shall be concerned only with the most general aspects of linguistic structure and with the function and effect of language. Beyond this, we shall need to consider the peculiarities of only one type of specialized dialect, for the language activity of scientists, in so far as it differs from other uses of language, constitutes the jargon or dialect of a craft, comparable to the dialects, say, of fishermen, miners, or carpenters.

This great restriction of our task does not remove the difficulties which are bound to arise. Even simple and well-established statements about language are likely to disturb or baffle the non-linguist, not because of any intrinsic difficulty but because of his prepossession with ancient doctrines that in the course of time have taken on the tenacity of popular superstitions. For the rest, our task demands that we go beyond such simple and well-established facts to matters which are subject to controversy. The writer can only do his utmost to speak plainly, in the hope that the reader will read attentively and, so far as may be, without injecting the prepossessions of our common-sense views about language.

4. Terminology

Linguists naturally have no respect for words. Engaged in describing structural systems of the widest diversity, they are accustomed, at need or at convenience, to invent and to discard all manner of technical terms. The varying terminologies of different linguists will concern only the reader who chooses to go farther afield: within this essay only one set of terms will appear.

On the other hand, the technical terms which are established among linguists often conflict or overlap with the meaning of these same terms as they occur in general scholastic usage and in philosophical discussion. It would be impractical in this essay to avoid them and to introduce the reader, instead, to fanciful innovations. At the time of definition technical terms will be given in italics, and an index will facilitate later reference to these definitions.

Since this essay, moreover, tries to make a specialized and hitherto neglected application of linguistics, we shall be compelled to use some temporary and unaccepted terms, and these, too, are likely to conflict with popular or scholastic usage.

Talking about the forms of speech is not easy. Perhaps only linguists are inured to this kind of discourse; others are likely to find it confusing. This difficulty is superficial and should be overcome without much trouble. Linguists are accustomed in printing to italicize the speech-forms which are under discussion or else to inclose them in square brackets; in this essay, however, conforming to more general usage, we shall place them between single quotation marks (for example: the word 'word').

5. Delimitation

Deep-seated inhibitions block the observation of language. To the unaccustomed observer language seems shapeless and fleeting. One does not talk about language. Among us an avenue of rationalization is provided in the traditional lore of the schoolroom. Where this does not suffice, we are accustomed to several lines of evasion: instead of observing language, one observes or, at any rate, discourses upon such matters as correctness of speech, literary values, or writing.

The notion that there is such a thing as "correct" and "incorrect" speech, that a speaker may "make mistakes" if he speaks "carelessly" in his native language, and that some kind of authoritative ruling may be sought, arose in the eighteenth century as an outgrowth of a peculiar social development. It has a realistic background (from which, however, it departs almost beyond recognition) in the existence, in large modern speech communities, of a standard dialect, spoken by the privileged and educated classes and employed in public speech (school, church, stage, court of law) and in nearly all writing; the standard dialect, common to all of a speech area (English) or to a large part of it (German, Dutch-Flemish), contrasts with local or regional dialects which enjoy a less favored position.

The discussion of literary values—that is, of the artistic use of language by specially gifted individuals—enjoys general favor as a substitute for the observation of language. This exercise, in spite of its prevalence in journalistic and academic spheres, need not here concern us.

6. Writing and Language

When the non-linguist sets out to talk about language, he very often lapses into discourse about writing. In order to avoid this confusion, we shall here use the term *'language'* only of the conventional use of vocal sound ("spoken" language), distinguishing this from *substitutes*, such as writing or drum signals, and from other actions, such as facial mimicry, which may serve in communication. In the present essay the term 'language' will therefore have a more restricted meaning than in other sections of this *Encyclopedia*. Similarly, phrases like 'a language' will here be used only of established ("natural") languages that prevail in communities, and not of restricted systems of symbolism, such as appear in mathematics and logic. For all linguistic study it is of primary importance to know that writing is not the same thing as language. Failure to make this distinction was one of the chief factors that prevented a beginning of linguistic science in the seventeenth and eighteenth centuries. The popular-scholastic contrast of "spoken language" and "written language" is entirely misleading.

In point of time language appears as a characteristic of the human species; men spoke as far back as we know anything about them; we must reckon here with biological reaches of time. To conjecture about the origin of language is to conjecture about the origin of the human species. Writing appears, by contrast, as a modern invention. At the earliest points of use (Mesopotamia, Egypt) it is only a few thousand years old.

Writing is a device for recording language by means of visual marks. By "recording" we mean that the beholder, if he knows the language of the writer and the system of writing, can repeat the speech which the writer uttered, audibly or internally, when he set down the marks.

Writing has been practiced in only few communities, and in these, until recently, almost always as a special skill of a few people. Anything like widespread literacy is a recent development covering only a small part of the world. Languages are quite the same whether their speakers practice writing or not; where there is widespread literacy, writing produces certain sur-face irregularities in the historical development of a language; these are not extensive, but their exact scope has not been de-termined. The best-known effects of this sort are the so-called spelling pronunciations, as when some people pronounce a *d* in 'Wednesday' or a *t* in 'often.'

A literate person finds it easy to talk about writing, because he acquired this skill under explicit verbal instruction, and be-cause the permanent written markings record the movements of the writing hand. His acquisition of language, on the other hand, was unaccompanied, in the nature of the case, by any verbal accounting, and the sounds of language are fleeting and intangible. Hence the normal speaker finds it difficult to talk about language and cannot even describe the movements with which he produces speech. Some of these movements, in fact, though every child learns to make them, have so far baffled the physiologist's attempt at determination. The non-linguist, when he discusses linguistic matters, is likely to mistake writing for language.

A mathematical writer, for example, appreciating the advan-tages of a duodecimal system of numbers, invents two written symbols, say *t* for 'ten' and *e* for 'eleven' (so that *10* now repre-sents 'twelve,' *84* represents 'one hundred,' *10e* 'one hundred and fifty-five,' etc.), and he is disappointed when people find this system too difficult to adopt. He has confused writing with language. Since the number-words of our language are decimal in structure, the proposed written forms are totally out of gear with our language and can be interpreted only by a cumbersome process of calculation.

All this concerns us here because the language of science tends to appear in written notation; in fact, its most characteristic and powerful form, mathematical discourse, can be transmitted only by means of a written record. The branching-out of written no-tation into forms which cannot be matched in actual speech represents a peculiar and highly specialized development of writing. This will concern us in the sequel; here we need say only that, in order to understand the specialized development, we must have a clear and correct picture of the underlying habit.

What applies to writing applies also to less elaborate communicative systems, such as gestures or signals. A period when such systems may have been independent of language, or coordinate with it, would, in any case, lie very far behind us. The conditions which might have prevailed in such a period are quite unknown. In times that are accessible to us and under human conditions as we know them, any system of gestures or signals that goes beyond vague beckoning appears simply as an outgrowth and substitutive reproduction of language.

The confusion, extremely common among literate but linguistically untrained persons, between language and its derivative substitutes often takes the form of inverting this relation or of ascribing some kind of "independence" to the substitutes. Alterations, no matter how fantastic, in a system of writing do not affect the language which is represented. If a new system of writing came into wide use in a community, it might in time produce some very modest linguistic modifications. This must not lead us to suppose that a change, say, from the Arabic to the Latin alphabet, such as has recently been effected for Turkish, has any immediate bearing on the language.

7. Language and Handling Activity

The difference between utterance of speech and all other events is naturally paramount in the study of language. The human acts which are observable under ordinary conditions are thus divided into *language,* the vocal utterance of the conventional type, and *handling activities,* a somewhat narrow name to cover all other normally observable acts, including not only manipulation but also mimicry, gesture, locomotion, acts of observation, etc. The linguist studies sequences in which language mediates between non-linguistic events.

Language creates and exemplifies a twofold value of some human actions. In its *biophysical* aspect language consists of sound-producing movements and of the resultant sound waves and of the vibration of the hearer's eardrums. The *biosocial* aspect of language consists in the fact that the persons in a community have been trained to produce these sounds in certain situations and to respond to them by appropriate actions. The biosocial function of language arises from a uniform, traditional, and arbitrary training of the persons in a certain group. They have been trained to utter conventional sounds as a secondary response to situations and to respond to these slight sounds, in a kind of trigger effect, with all sorts of actions.

In their turn, on a tertiary level of response, handling actions may be subjected biosocially to language: specific handling actions are conventionalized as a response to specific forms of speech, and these handling actions or their products serve as stimuli to call forth these same forms of speech. These substitutes for speech occur scantily in simple communities but play a great part in our civilization. Apart from speculations about a prehuman or semihuman antiquity, in man as we see him, all handling actions which bear elaborate communicative value owe this value to their biosocial subjection to language.

The most familiar example of this is the art of writing: permanent visible marks are produced conventionally as responses to speech and serve as stimuli for the production of speech. For this it is necessary, of course, that the marks be conventionally associated with features of language. These features may be of various kinds: Chinese characters and our numerical digits represent words; the letters of the alphabet represent single typical speech-sounds. The biophysical aspect of writing consists in the materials used for producing the marks, in the shape of the marks, and in the movements by which one makes them. Once we have established the communal habit or convention which attaches them biosocially to speech-forms, the biophysical features may vary within the convention. Thus, the variety of paper, pencil, or ink; the shape, always within our convention, of the letters—script, printed, Roman or italic, small or capital, etc.—and the movements of the writer's hand or the kind of printing press are all indifferent so far as the verbal message is concerned.

The biophysical features of substitutes for language are important enough on their own level. Systems of writing may differ as to the time and labor required for making the marks and as to their legibility. The permanence of the materials is sometimes important. Mechanical devices like the typewriter and the printing press may yield great advantages.

More important for our discussion are the conventions which associate the written marks with forms of language. The alphabet, which associates, in principle, each mark with a typical sound of speech, makes reading and writing relatively easy, since it requires only a few dozen different characters; for the same reason it lends itself easily to mechanical transmission and reproduction. The system of writing, best exemplified in Chinese, where a character is assigned to a word of the language, is much harder to use and to mechanize. On the other hand, a piece of writing can be read in various languages, provided that these agree in structure and that minor divergences are bridged by means of supplementary conventions. Moreover, word-writing has the value of compactness and easy survey. These advantages appear very clearly in the case of our numeral digits. Here, by confining a system of word-writing to a very limited domain of speech-forms, we make it readable, with the aid of supplementary conventions, in any language. Thus, a graph like *71* is read in English as 'seventy-one,' in German as 'one and seventy,' in French as 'sixty-eleven,' in Danish as 'one and half-four-times' (*sc.*, 'twenty'), etc.

Our numeral digits illustrate the advantage of a well-arranged written symbolism. Simple discourses which move entirely in this domain can be carried out by means of a calculating machine. Short of this extreme case, we may cite cable codes, where the written forms are revised in the direction of brevity; or, again, systems of notation such as are used in symbolic logic, where a stock of symbols is so devised as to yield extremely simple rules of discourse. The construction of a calculating machine or the rules of arrangement and substitution in a system of logistics require great care and present great interest: it is decidedly advantageous to study them for their own sake and without reference to the linguistic end points of the sequences in which they operate. Nevertheless, if they are to operate, they must be so planned that, starting with speech-forms, they yield speech-forms in the end. No serious use would be made of a calculating machine, or of a code, or of a system of logistics, which failed to deliver a linguistically significant end result. By way of contrast, the reader may find it not without value to consider such non-linguistic systems as musical notation, say for the piano, or, again, the moves of a game of chess. Persons unaccustomed to the consideration of language are prone to the error of overlooking the linguistic character of notational or mechanical subsidiaries of language and viewing them as "independent" systems; at the same time they may resort to the metaphor of calling these systems "languages." This metaphor is dangerous, since it may lead to the notion that such systems can liberate us from uncertainties or difficulties which inhere in the working of language.

Some students of language, among them the present writer, believe that widespread and deep-seated errors in supposedly scientific views of human behavior rest upon misconceptions of this sort—upon failure to distinguish between linguistic and non-linguistic events, upon confusion of the biophysical and the biosocial aspects of language and its subsidiaries, and, above all, upon a habit of ignoring the linguistic parts of a sequence and then calling upon metaphysical entities to bridge the gap. Several phases of this error will here demand our attention.

8. Apriorism

In the prescientific view of these matters, a term such as "reasoning" covers, on the one hand, observations which cost no great labor and, on the other hand, utterances of speech which are not recognized for what they are. Thus there arises the notion that knowledge may be obtained by a process of "reasoning a priori." Everyday observations, generally human or systematized by tribal tradition, are viewed as innate data of reason, and the ensuing deductions are clothed in a mystic valid-

ity. If the deductions are correctly made, the "a priori" procedure differs from an ordinary act of science only in that the basic
observations are unsystematic and remain untested.

In scientific procedure we mean by *deduction* the purely verbal
part of an act of science which leads from the report of observation and the hypotheses to a prediction. If we replace the report of observation by arbitrarily invented postulates, the discourse makes no pretense to validity in the sphere of handling
actions. Deductive discourse of this kind is produced in logic,
mathematics, or the methodology of science. It is made to fit
some type of observational data, or else it exists for its own sake,
in readiness for the emergence of observational data to which it
may be applied. Until modern times, Euclidean geometry was
viewed as an "a priori" system: the underlying everyday observations about the spatial character of objects were viewed as
inborn and unquestionable truths. Today the same system,
apart from the correction of flaws, is treated as a purely verbal
discourse of deduction from postulates. It is especially useful
because these postulates, by virtue of their historic origin, are
such as to make the discourse applicable to the placing of objects, as this placing is observed in the first approximation that
is customary in everyday handling. We have learned, however,
that astronomical magnitudes make sensible the error in these
postulates, and they accordingly demand a different discourse,
based upon other postulates which, in turn, will be chosen so as
to fit the new observations. One employs postulates which fit
the observed data within the margin of error. The postulates
are chosen so as to yield a simple calculation; the discrepancies
are set aside to await more accurate measurements, which, in
their turn, will make possible a more closely approximative discourse.

9. Mentalism

In the common sense of many peoples, perhaps of all, language is largely ignored, and its effects are explained as owing to
non-physical factors, the action of a "mind," "will," or the like.
These terms, as well as the many others connected with them,

yield service in daily life, in art, and in religion; that they have no place in science is the contention of many scientists. Mentalistic terms do not figure in the procedure of physics, biology, or linguistics, but many students of these subjects employ them in the theoretical parts of their discourse.

An individual may base himself upon a purely practical, an artistic, a religious, or a scientific acceptance of the universe, and that aspect which he takes as basic will transcend and include the others. The choice, at the present state of our knowledge, can be made only by an act of faith, and with this the issue of mentalism should not be confounded.

It is the belief of the present writer that the scientific description of the universe, whatever this description may be worth, requires none of the mentalistic terms, because the gaps which these terms are intended to bridge exist only so long as language is left out of account. If language is taken into account, then we can distinguish science from other phases of human activity by agreeing that science shall deal only with events that are accessible in their time and place to any and all observers (strict *behaviorism*) or only with events that are placed in co-ordinates of time and space (*mechanism*), or that science shall employ only such initial statements and predictions as lead to definite handling operations (*operationalism*), or only terms such as are derivable by rigid definition from a set of everyday terms concerning physical happenings (*physicalism*). These several formulations, independently reached by different scientists, all lead to the same delimitation, and this delimitation does not restrict the subject matter of science but rather characterizes its method. It is clear even now, with science still in a very elementary stage, that, under the method thus characterized, science can account in its own way for human behavior—provided, always, that language be considered as a factor and not replaced by the extra-scientific terms of mentalism.

10. Solipsism

Much of the procedure of science and particularly the end result of each act of science consists of speech in a somewhat specialized form. In order to give an account of science, we must know something of language. Linguistics, however, in any feasible arrangement, occupies a rather late place among the divisions of science, since it comes after biology and all studies of the animal (and not specifically human) circumstances of man.

The fact that we are thus inevitably caught in a circle should not be exploited in favor of a naïve solipsism which is surprisingly prevalent even among otherwise competent men of science. It is true, of course, that science can deal with only such events as have called forth a verbal response on the part of some person—at any rate, of the scientist who undertakes to deal with a given event. More generally, the solipsist says that the universe consists, of necessity, only of his "experiences" or "perceptions." The truth of this axiom cannot be affected by the study of language or of "experiences" or "perceptions," since these studies occupy a late place in science and deal with only a small part of the universe.

If it were strictly applied, the solipsist's axiom would remain irrelevant and harmless. Any predicate that is applied to everything in the world will play no part in science, for science deals only with correlations of events within the universe. The solipsist, having tagged everything, indifferently and without exception, as his "percept" or "experience," goes on to study things exactly as other scientists study them. He does not allow his solipsism to interfere with work in his own field of science. In more general discussion, however, the solipsist commits the fallacy of using his meaningless axiom as a bludgeon to knock over special points of discourse. He is most likely to do this in the face of any attempt at scientific study of human behavior, and especially in the face of any linguistic considerations. It is as though in practical life the solipsist should suddenly recall that an oncoming railway train was only a creature of his perceptions.

II. The Function of Language

11. The Speech Community

Language makes human behavior extremely different from that of animals because it establishes a minute and accurate interaction between individuals. Among animals such interaction, where it exists, concerns only a few gross situations or else is rigidly set in an unvarying pattern. Apart from this, each individual's response to stimuli is limited by the powers of this individual's body. In man these ordinary responses, to which we give the name of 'handling actions,' are paralleled by a second set of responses—the utterance of speech. The speech sounds act as a stimulus upon other persons who may then perform a handling response such as to give a biologically favorable outcome to the speaker's situation. Thus a person who is in need of food but unable by his own bodily power to get it may, by speech, prompt others to get it for him.

Thus, in addition to the normal biological series $S \to R$, man has also the series

$$S \to r \text{——} s \to R \, .$$

Here r——s denotes the act of language; the biologically effective stimulus S and response R are no longer confined to occurrence within one body. Language bridges the gap between the individual nervous systems. It makes possible a minute division of labor and high specialization of individual abilities.

Much as single cells are combined in a many-celled animal, separate persons are combined in a speech community—a higher and more effective type of organization. If the word 'organism' be not confined to denote an individual animal, we may speak here, without metaphor, of a social organism. Primarily, the social organism is the speech community—the community of persons speaking one language—but bilingual or multilingual persons mediate everywhere between these communities: culture areas, such as Europe with her daughter-nations, approach the coherence of a single-speech community; some degree of communication now subsists between all persons on earth.

12. Relayed Speech

The simplest case of speech utterance is that in which the hearer performs a handling action in response to a non-linguistic stimulus which impinges on the speaker. Very often, however, the hearer responds by addressing speech to the first speaker or to other persons. Many relays of speech may intervene between a non-linguistic stimulus and a handling response. The stimuli of various individuals may contribute to the sequence of speech; the sequence itself may be multiple, carried on simultaneously by many chains or a network of speaker-hearer connections; and the handling response may be performed by many persons, co-operatively or in independent actions. The intermediate utterances in such a chain or net exemplify the case where speech is prompted by no immediate non-verbal stimulus and calls forth no immediate handling response but figures merely as a part of a sequence of linguistic and other events in a community. The word 'sumach,' for instance, in its primary and simple use will be uttered in the presence (under the visual stimulus) of a certain kind of tree, but a speaker hitherto ignorant of the word, once this is pointed out to him, will be able, without further explanation, to take part in sequences of action and discourse in which the word 'sumach' occurs in absence of the primary stimulus. Other speech-forms, such as most of those used in this monograph, adhere to no simple stimulus of that sort but occur entirely in complex situations of discourse.

Speech utterance itself, on a different level, may serve as a relatively final response, as in instruction in a foreign language or mathematics, or in a discussion leading to agreement as to the use of some technical term. If we follow a sequence far enough, we expect, of course, to find some modification of non-verbal activity: even poetry or fiction will in the end lead to a more than verbal result.

13. Verbal Self-stimulation

The normal human being, totally uncritical as to language, sees himself not only as a separate body but also as a source of effect upon other persons. However pervasive and at times powerful this effect may be, it is nearly always uncertain. In contrast with this, he is able, with rare exceptions, to co-ordinate his earlier speech ("intention") with later handling actions of his own. This contributes toward the view of the "self" or "ego" as a more than bodily entity.

What here interests us is not merely the relative sureness but also the usefulness of verbal self-stimulation. The utterance of a speech-form is not only a response but serves also as a stimulus to the speaker himself. The utterance can be easily repeated and replaces conveniently an evanescent or remote stimulus. The varied situations of counting (as, say, a flock of sheep before and after a storm) can be adduced to illustrate this.

A child talks to himself at first out loud; then he learns to mumble or whisper; finally, he suppresses all audible and even all visible movements of speech. In the popular view no movement at all is supposed to remain. This is a view which we could adopt only by abandoning the basic assumptions of physics. We must suppose rather that the movements of speech —which, as we shall see, consist of a small number of contrasting units—are replaced by internal movements, at first presumably as mere reductions of the normal movements of speech, but capable, in the course of time, of any degree of substitution. This *inner speech* accounts for the main body of the vaguely bounded system of actions that in everyday parlance goes by the name of "thinking."

14. Meaning

In the sequence $S \to r \text{---} s \to R$ and its complex derivatives, the speech or sequence of speeches has no immediate gross biological effect and serves merely as a "sign" mediating between the more practically important stimuli and responses, which are here represented by the large letters S and R. These, to be sure, may themselves consist of speech, but, in the intention of the formula, this speech will then be on a lower level than the speech represented by the small $r \text{---} s$; an example of this would be discourse about the validity of an earlier speech, or, with another difference of level, discourse about the forms of some language. These special cases need not here concern us: we may speak of the end points S and R as speaker's stimulus and hearer's response, without regard to their verbal or non-verbal character. The two together constitute, in linguistic terminology, the *meaning* of the speech utterance r——s. This holds good even under a mentalistic view: in this view it is merely supposed that the speaker's stimulus and the hearer's response are "ideas," "concepts," or the like, which may be postulated in more or less exact accommodation to the uttered speech-forms and serve to link these to the actually observable stimulus and response.

The term 'meaning,' which is used by all linguists, is necessarily inclusive, since it must embrace all aspects of semiosis that may be distinguished by a philosophical or logical analysis: relation, on various levels, of speech-forms to other speech-forms, relation of speech-forms to non-verbal situations (objects, events, etc.), and relations, again on various levels, to the persons who are participating in the act of communication.

When the correlations of speech-forms with meanings are known, some utterances turn out to be conditioned more immediately by the speaker's stimulus, and others by the hearer's response. Contrast, for example, a report, such as 'It's raining,' with a command, such as 'Come here!' In earlier stages of an infant's language-learning, speaker's meaning and hearer's meaning may be imperfectly correlated, but in the normal use of language both these aspects of meaning are so firmly knit to the speech-form that no distinction is possible. In the description of a language we need not define each form twice over, determining first the situations in which it is uttered and then the responses to which it leads. In general, we define in terms of stimulus because the earlier step in the sequence exhibits less variation. The speaker has been trained, as part of his acquisition of language, to play indifferently either part in the interchange of speech. To be sure, each speaker understands a wider range of speech-forms than he utters, but this does not bring about any discrepancy between the two aspects of meaning: in the adult speaker they are entirely merged in a firm correlation which is guaranteed by the evident working of language.

15. Acquisition of Meaning

Hearing a speech in a language of which we are entirely ignorant, we must include in our first estimate of the meaning the total situation of the speaker and all the ensuing actions of the hearer. To consider meaning from any less inclusive position would lay us open to prepossessions which lurk close by. The more striking features of gesture, situation, and immediate response will often lead us to a hypothesis, to be removed or confirmed by further experience and by the correlations that will be gradually built up. After we have thus acquired the use of a small supply of speech-forms, becoming, to this extent, members of the community, we may induce full-fledged speakers to talk about their language: we ask them the names of things and lead them to define speech-forms in terms of other speech-forms. In the history of travel and adventure all this has many times occurred; the student of language naturally prefers to seek the mediation of a bilingual speaker. For the infant's acquisition of language, however—though little is known of it—we must postulate exactly this process. The elders, of course, facilitate the process by uttering simple speech-forms with strikingly apparent and much repeated practical emphasis upon the relevant features of situation and response. This elementary process has been called *demonstration;* this use of the word is distinct, of course, from its use in logic and mathematics as a synonym of 'proof.' A speech-form has been acquired by demonstration when the learner utters it in the conventionally appropriate situations and responds to it by the conventionally appropriate actions.

Teacher and learner may develop an agreement under which the demonstration of meaning, for certain types of speech-forms, is shortened to a single act, such as holding up an object or unmistakably pointing at it while speaking its name.

Another factor of learning is the occurrence of a speech-form in contexts whose other components are familiar. Certain types of forms, such as the words 'and,' 'or,' 'because,' 'the,' are presumably acquired in this way.

Finally, there is the process of *definition*. For the most part, definition is informal: a speech-form is equated with roughly synonymous forms. Thus we may tell a child that 'rapid' means 'quick' or 'fast,' and we expect the statistical effect of future demonstration and context to train him to use the new word in its more exact conventional setting. Bilingual statements of meaning, such as 'cheval:horse' are in general of this rough type; even apart from differences of structure and function, the meanings in two languages almost never coincide.

We call a definition *formal* when, for some sphere of discourse, the new form and the old one by which it is defined are freely and completely interchangeable. Formal definition occurs in ordinary life only in the case of a very few of the simplest and most abstract speech-forms. The notable instance is furnished by the number words: 'eight' means 'seven and one.' Other instances doubtless occur: 'down' could be defined as 'in the direction in which heavy things fall when we drop them.' The chief place of formal definition, however, is in explicit agreement. This is familiar to us from scientific practice: for a certain discourse, and perhaps for whole types of discourse, participants in some technical procedure agree to use a new term as the exact equivalent of some older speech-form. Thereafter they initiate learners into this agreement. Formal definitions may be bilingual (as 'Kraft:force'), provided abstraction is made from structural and functional differences (for instance, gender of a noun in French or German, as opposed to English).

It will be seen that, even in the case of formal definition, the degree of uniformity among the speakers and even the consistency of any one speaker will be subject to any uncertainty as to the meaning of the defining speech-forms. These may in turn have been formally defined, but in the end we must come to forms which have been only roughly defined and, for each speaker, to forms which he got in infancy from context or, at the very beginning, from demonstration. The meaning of a speech-form in the habit of any one speaker contains factors and fringes of variation; these may be greatly reduced, but probably never entirely eliminated, for the observer by consideration of many speakers, and for the speakers themselves by copious and consistent interchange of speech upon some topic that demands accurate agreement.

III. The Structure of Language

16. Phonemic Structure

In ordinary discussion human situations and responses appear perhaps less flowing and vague than we have here described them in the guise of meanings, for there we are free (as we are not here, in our linguistic discussion) to endow them with some of the permanence and neat outline which they obtain by virtue of their correlation with the forms of language. In language we order and classify the flowing phenomena of our universe; our habit of doing this is so pervasive that we cannot describe things as they may appear to an infant or a speechless animal. The price we pay is a sensible inadequacy of our speech, offset by the privilege of any degree of approximation such as may be seen in some microscopic investigation of science or in the work of the poet. It is the entire task of the linguist to study the ordering and formalization which is language; he thus obtains the privilege of imagining it removed and catching a distant glimpse of the kind of universe which then remains.

The ordering and formalizing effect of language appears, first of all, in the fact that its meaningful forms are all composed of a small number of meaningless elements. We should obtain, in this respect, a parallel to language if, with a dozen or so of different flags, we devised a code in which the exhibition of several flags (in the limiting case, of one flag) in a fixed position and arrangement would constitute a meaningful sign.

The forms of every language are made up out of a small number—ranging perhaps between fifteen and seventy-five—of typical unit sounds which have no meaning but, in certain fixed arrangements, make up the meaningful forms that are uttered. These signals are the *phonemes* of the language. The speakers

239

have the habit of responding to the characteristic features of sound which in their language mark off the various phonemes and of ignoring all other acoustic features of a speech. Thus, a German who has not been specially trained will hear no difference between such English forms as 'bag' and 'back,' because the difference in his language is not phonemic; it is one of the acoustic differences which he has been trained to ignore. In the same way, a speaker of English will hear no difference, until he is trained to do so, between two Chinese words which sound to him, say, like 'man,' and differ as to their scheme of pitch; we fail to hear the difference because in our language such a difference is not connected with a difference of meaning and is consistently ignored whenever it chances to occur. The acoustic features which set off a phoneme from all other phonemes in its language, and from "inarticulate" sound, exhibit some range of variation. It is not required that this range be continuous: acoustically diverse features may be united, by the habit of the speakers, in one phoneme.

The number of phonemes which will be stated as existing in any one language depends in part upon the method of counting. For instance, we shall recognize an English phoneme [j] which appears initially in forms like 'yes,' 'year,' 'young,' and another phoneme [e] in the vowel sound of words like 'egg,' 'ebb,' 'bet.' The longer vowel sound in words like 'aim,' 'say,' and 'bait' may then be counted as another phoneme, or else one may describe it as a combination of the phonemes [e] and [j]. This option would not exist if our language contained a succession of [e] plus [j] which differed in sound, and as to significant forms in which it occurred, from the vowel sound of 'aim,' 'say,' 'bait.' Thus, the English sound [č], which appears in words like 'chin,' 'rich,' 'church,' must be counted as a single phoneme and not as a combination of [t] as in 'ten' and [š] as in 'she,' because in forms like 'it shall' or 'courtship' we have a combination of [t] plus [š] which differs in sound and as to significant forms from [č] in 'itch Al' or 'core-chip.' The fact that these last two forms are unusual or nonsensical does not affect the distinction. The count of phonemes in Standard English will vary, according to economy, from forty-odd to around sixty.

For the most part, the phonemes appear in utterance in a linear order. Where this is not the case, the arrangement is so simple that we can easily put our description into linear order. For instance, the noun 'convict' has a phoneme of stress (loudness) which starts with the beginning of the word and covers the first vowel phoneme; the verb 'convict' differs in that the same stress phoneme is similarly placed upon the beginning of the partial form '-vict.' If we wish to put our description of these forms into linear order, we need only agree upon a convention of aligning a symbol for the stress phonemes, e.g., *'convict* and *con'vict.*

Thus, every speech in a given dialect can be represented by a linear arrangement of a few dozen symbols. The traditional system of writing English, with its twenty-six letters and half-dozen marks of punctuation, does this very imperfectly but sufficiently well for most practical needs.

This rigid simplicity of language contrasts with the continuous variability of non-linguistic stimulation and response. For this reason linguists employ the word *'form'* for any meaningful segment of speech, in contrast with their use of 'meaning' for stimulus and response.

The sound produced in a speech is to all ordinary purposes a continuum. To determine which features are phonemic, we must have some indication of meaning. A German observer, say, who, studying English as a totally unknown language, noticed in a few utterances the acoustic difference between 'bag' and 'back,' could decide that this is a phonemic difference only when he learned that it goes steadily hand in hand with a difference of meaning.

Two utterances, say of the form 'Give me an apple,' no matter how much they may differ in non-phonemic features of sound, are said to consist of the *same* speech-form; utterances which are not same are *different*. The decision of the speakers is practically always absolute and unanimous. This fact is of primary concern to us, since by virtue of it the speakers are able to adhere to strict agreements about speech-forms and to establish all manner of correspondences, orderings, and operations in this realm. To take an everyday instance: anyone can look up a word in a dictionary or a name in a directory.

It would not do to overlook the fact that the phonemes of a language are identifiable only by differences of meaning. For this, however, a relatively small number of gross differences will suffice: once the phonemes are established, any form of the language is completely and rigidly definable (apart from its meaning) as a linear or quasi-linear sequence of phonemes. We do not possess a workable classification of everything in the universe, and, apart from language, we cannot even envisage anything of the sort; the forms of language, on the other hand, thanks to their phonemic structure, can be classified and ordered in all manner of ways and can be subjected to strict agreements of correspondence and operation. For this reason, linguistics classifies speech-forms by form and not by meaning. When a speech-form has been identified, we state, as well as may be, its meaning: our success depends upon the perfection of sciences other than linguistics. The reverse of this would be impossible. For instance, we shall usually seek a given word in a thesaurus of synonyms by looking it up in the alphabetical index. We could not use a telephone directory which arranged the names of the subscribers not in their alphabetical order, but according to some non-verbal characteristic, such as weight, height, or generosity.

17. Grammatical Structure

Some utterances are partly alike in form and meaning; for instance:

> Poor John ran away.
> Our horses ran away.
> Poor John got tired.
> Our horses got tired.

This forces us to recognize meaningful constituent parts, such as 'poor John,' 'our horses,' 'ran away,' 'got tired.'

A form which can be uttered alone with meaning is a *free*

form; all our examples so far are free forms. A form ('form' always means 'meaningful form') which cannot be uttered alone with meaning is a *bound* form. Examples of bound forms are the suffix '-ish' in 'boyish,' 'girlish,' 'childish,' or the suffix '-s' in 'hats,' 'caps,' 'books.'

A free form which does not consist entirely of lesser free forms is a *word*. Thus, 'boy,' which admits of no further analysis into meaningful parts, is a word; 'boyish,' although capable of such analysis, is a word, because one of the constituents, the suffix '-ish,' is a bound form; other words, such as 'receive,' 'perceive,' 'remit,' 'permit,' consist entirely of bound forms.

A free form which consists entirely of lesser free forms is a *phrase;* examples are 'poor John' or 'poor John ran away.'

Sets of words, such as 'perceive: receive: remit' or 'perceive: permit: remit,' establish a parallelism between the extremes, 'perceive' and 'remit.' The habit which is thus revealed is a *morphologic construction*. In the same way, sets of phrases, such as 'John ran: John fell: Bill fell' or 'John ran: Bill ran: Bill fell,' establish a parallelism between the extremes 'John ran' and 'Bill fell,' and illustrate a *syntactic construction*. The parts of a form which exhibits a construction are the *constituents* of the form: the form itself is a *resultant* form.

In the study of an unknown language we proceed as above: partial similarities between forms reveal their complexity, and we progressively recognize constituents and determine, often with some difficulty, whether they are free or bound. In presenting the description of a language, however, we begin with the constituents and describe the constructions in which they appear.

A construction, morphologic or syntactic, consists in the arrangement of the constituents. In addition to the meaning of the constituents, the resultant form bears a constructional meaning, which is common to all forms that exhibit the same construction. Even more than other elements of meaning, constructional meanings are likely to present difficulties of definition, for they are often remote from simple non-linguistic events.

The features of arrangement differ in different languages.

243

Modulation is the use of certain special phonemes, *secondary* phonemes, which mark certain forms in construction. In English, features of stress play a large part as secondary phonemes. We have seen this in the contrast between the verb 'convict' and the noun 'convict.' In syntax it appears in the absence of word-stress on certain forms. Thus, in a phrase like 'the house,' the word 'the' is unstressed; on the other hand, it may receive a sentence stress when it is an important feature of the utterance.

Phonetic modification is the substitution of phonemes in a constituent. For instance, 'duke,' when combined with the suffix '-ess' or '-y' is replaced by 'duch-'; in syntax the words 'do not' are optionally replaced, with a slight difference of meaning, by 'don't.' Neither modulation nor phonetic modification plays any part in the specialized scientific uses of language; it is otherwise with the features of arrangement which we now have to consider.

The *selection* of the constituent forms plays a part apparently in all languages. If we combine the word 'milk' with words like 'fresh,' 'cold,' 'good,' we get designations of special kinds of milk: 'fresh milk,' 'cold milk,' 'good milk'; if we combine it with words like 'drink,' 'fetch,' 'use,' we get designations of acts: 'drink milk,' 'fetch milk,' 'use milk.' The difference in constructional meaning goes hand in hand with the selection of the forms. We describe these habits by saying that the construction has two (or more) *positions* which are *filled* by the constituents. A *function* of a form is its privilege of appearing in a certain position of a certain construction. *The function*, collectively, of a form is the sum total of its functions. Forms which have a function in common constitute a *form-class*. Thus, the forms 'milk,' 'fresh milk,' 'cold water,' 'some fine sand,' etc., are in a common form-class, since all of them combine with forms of the form-class 'drink,' 'don't drink,' 'carefully sift,' etc., in the construction of action-on-object. In syntax, as these examples indicate, words and phrases appear in common form-classes. If words alone are considered, their largest inclusive form-classes are known as *parts of speech*. In many languages, and very strikingly in English, the form-classes of syntax overlap in so complex a fashion that various part-of-speech classifications are possible, according to the functions which one chooses primarily to take into account.

The forms of a form-class contain a common feature of meaning, the *class meaning*. The traditional grammar of our schools gets into hopeless difficulties because it tries to define form-classes by their class meaning and not by the formal features which constitute their function.

The use of *order* as a feature of arrangement is by no means as widespread as the use of selection, but, on account of its simplicity and economy, it plays a great part in the scientific specializations of language. In English the order of the constituents is a feature of nearly all constructions; thus, in 'fresh milk' or 'drink milk' the constituents appear only in this order. In some instances, features of order alone distinguish the positions: contrast, for example, 'John hit Bill' with 'Bill hit John.'

18. The Sentence

In any one utterance a form which, in this utterance, is not a constituent of any larger form is a *sentence*. By definition, any free form and no bound form can occur as a sentence. Various supplementary features are used in different languages to mark the sentence, especially its end. In English, secondary phonemes of pitch are used in this way. In much the same manner as constructions, *sentence types* are distinguished by features of arrangement. The meanings of these types have to do largely with the relation of speaker and hearer ("pragmatic" features of meaning). Thus, pitch and, in part, selection and order determine in English such types as statement ('at four o'clock'), yes-or-no question ('at four o'clock?'), and supplement question ('at what time?').

In many languages, perhaps in all, certain free forms are marked off as especially suited to sentence use. A sentence which consists of such a form is a *full* sentence. In English the favorite sentence forms are phrases which exhibit certain constructions. The most important is the actor-action construction in which a nominative substantive expression is joined with a finite-verb expression: 'Poor John ran away.' 'John ran.' 'I'm coming at four o'clock.' 'Can you hear me?' A sentence which does not consist of a favorite sentence form is a *minor* sentence: 'Yes.' 'Fire!' 'At four o'clock.' 'If you can hear me.'

English and many other languages distinguish clearly a type of sentence whose type-meaning can perhaps be described by the term 'report.' In English the report sentences are full-sentence statements exhibiting the actor-action construction or a co-ordination of several actor-action phrases. A great deal of labor has been spent upon attempts at giving a precise definition of this type-meaning, in disregard of the likelihood of its differing in different languages and in oblivion of the danger that our sociology may not be far enough advanced to yield such a definition. For our purpose, at least a rough outline of this meaning will be needed. In the normal response to a report the hearer behaves henceforth as if his sense organs had been stimulated by the impingement of the reported situation upon the sense organs of the speaker. Since the meaningful speech-forms of the report, however, constitute at bottom a discrete arrangement, the hearer's responses can correspond to the speaker's situation to the extent only that is made possible by the approximative character of the report. Thus, when a speaker has said, 'There are some apples in the pantry,' the hearer behaves as though his sense organs had been stimulated by the impingement of the apples upon the speaker's sense organs—as though the speaker's adventure with the apples, to the extent that it is represented by the meanings of the speech-forms, had been witnessed by the hearer, not visually, but through some sense organ capable of a certain discontinuous range of stimulation.

Irony, jest, mendacity, and the like represent derived types of speech and response; they need not here concern us.

19. Constructional Level and Scope

Constructions are classified, first of all, by the form-class of the resultant form.

If the resultant form differs, as to the big distinctions of form-class, from the constituents, the construction is said to be *exocentric*. For instance, actor-action phrases like 'John ran away' or 'He ran away' differ in form-class from nominative substantive expressions like 'John' or 'he' and from finite-verb expressions like 'ran away.' Similarly, the functions of prepositional phrases, such as 'in the house' or 'with him,' differ from those of a preposition ('in,' 'with') and from those of an objective substantive expression ('the house,' 'him').

If the resultant form agrees as to the major distinctions of form-class with one or more of the constituents, then the construction is said to be *endocentric*. For instance, the phrase 'bread and butter' has much the same function as the words 'bread,' 'butter.'

If, as in this example, two or more of the constituents have the same function as the resultant form, the construction is *coordinative* and these constituents are the *members* of the coordination. If only one constituent agrees in form-class with the resultant form, the construction is *subordinative;* this constituent is the *head* of the subordination, and any other constituent is an *attribute* of this head. Thus, in 'fresh milk,' the head is 'milk' and the attribute is 'fresh'; in 'this fresh milk,' the head is 'fresh milk' and the attribute 'this'; in 'very fresh,' the head is 'fresh' and the attribute 'very'; in 'very fresh milk,' the head is 'milk' and the attribute is 'very fresh.'

The difference of analysis in these two cases is worth observing:

this / fresh milk
very fresh / milk

Although we are unable to give precise definitions of meaning, especially of such ethnically created ranges as constructional meanings and class meanings, yet the mere subsistence of like and unlike sets determines schemes of construction. Only in rare cases does the structure of a language leave us a choice between different orders of description. At each step of analysis we must discover the *immediate* constituents of the form; if we fail in this, our scheme will be contradicted by the constructional meanings of the language.

If a form contains repeated levels of endocentric construction, there will be a word or co-ordinated set of words which serves as the *center* of the entire phrase. Thus, in the phrase 'this very fresh milk,' the word 'milk' is the center.

The formal features of construction—selection of constituent forms, order, phonetic modification, and modulation by means of secondary phonemes—differ greatly in various languages and sometimes lead to very complex structures of word or phrase, but they seem nowhere to permit of an unlimited box-within-box cumulation. Even simple formations may lead to ambiguity because the *scope*—that is, the accompanying constituents on the proper level—of a form may not be marked. For instance, 'an apple and a pear or a peach' may mean exactly two pieces of fruit: then the immediate constituents are 'an apple / and / a pear or a peach,' and the phrase 'a pear or a peach' and the phrase 'an apple' constitute the scope of the form 'and.' On the other hand, the phrase may mean either two pieces of fruit or one piece: then the immediate constituents are 'an apple and a pear / or / a peach,' and the scope of 'and' now consists, on its level, of the phrases 'an apple,' 'a pear.' Similarly, 'three times five less two times two' may mean 26, 18, 11, or 3. These uncertainties are not tolerable in the scientific use of language; it is a striking peculiarity of this use that they are removed only in written notation—as especially by the parentheses and brackets that are used in algebra. The result is a system of writing which cannot be paralleled in actual speech.

20. Varieties of Reference

A thoroughgoing comparison of speech-forms, say in some one language, with features of the non-linguistic world is impossible at the present state of our knowledge. Our system of responses, with its neat discrimination of objects, classes, positions, qualities, movements, etc., results very largely from our use of language. We cannot return to the animal's or the infant's state of speechless response.

In order to find out how much of our world is independent of any one language, we might try to compare the grammars and lexicons of different languages. At present we have reasonably complete data for a few languages only; at some future time, when this task can be undertaken, the results will be of great interest. The forms of any one language could scarcely serve as a frame of reference: we should need, instead, a non-linguistic scale by which to measure.

It is the task of science to provide a system of responses which are independent of the habits of any person or community. These responses are twofold, in accordance with the universal scheme of human behavior: science provides relevant handling responses and clarified speech-forms. In the nature of the case, however, the entire result is transmitted and preserved in a verbal record. If science had completed its task, we could accurately define the meanings of speech-forms.

Even the most favorable type of meanings will show the difficulty of definition. Clusters of stimuli which produce roughly the same elementary responses in all people and, in accordance with this, are not necessarily tied up with communal habits, have been successfully studied: this is the domain of so-called external phenomena, the domain of physical and biological science. Here some of the simpler lexical classifications of language correspond in the main to the classifications of science, as, for instance, in the names of familiar species of plants and animals. However, there is often some gross divergence, as when several species are called by the same name, or one species by several names, and there is a great deal of less manageable vagueness at the borders—species which sometimes are and sometimes are not included in a designation. Even in this simplest sphere, the meaning of many speech-forms involves ethnological features. Here, too, we encounter, on the simplest level, speech-forms which have no extra-linguistic validity, unless it be in the designation of secondarily created artifacts: dragons, griffins, unicorns, etc.

Where science reveals a continuous scale of phenomena, such as color, the segments included under linguistic terms vary greatly in different languages; they overlap and grow vague at the edges; and they are subject to extraneous limitations ('bay,' 'roan').

When we come to meanings which are involved in the habit of communities and individuals, we fall even farther short of accurate definition, since the branches of science which deal with these things are quite undeveloped. In practice we resort here to artistic, practical and ethical, or religious terminologies of definition, and these, however valuable for our subsequent conduct, fail to satisfy the peculiar requirements of science.

In all spheres the structure of languages reveals elements of meaning which are quite remote from the shape of any one situation and are attached rather to constellations which include, often enough, personal or ethnical features. Relatively simple instances are words like 'if,' 'concerning,' 'because,' or the subtle difference, so important in English, between the types 'he ran' : 'he was running' : 'he has run' : 'he has been running.' The difficulty is even greater in the case of bound forms, which cannot be isolated in their language; consider, for instance, the deprecative feature in some of the uses of '-ish' ('mannish').

Constructional meanings and class meanings pervade a language, in part as universally present *categories;* they generally defy our powers of definition. The singular and plural categories of nouns in English are relatively manageable, but include some troublesome features, such as 'wheat' versus 'oats.' Gender-classes, as in French or German, are almost entirely ethnological in character. The normal speaker, without special training, is incapable of talking about these features; they are not reflected in any habits beyond their mere presence in the structure of the language. The major form-classes are remote from any extra-linguistic phenomena. If we assign to the English class of substantives some such meaning as "object," then words like 'fire,' 'wind,' and 'stream' require an ethnologic commentary. The mechanics of a language often require that otherwise similar designations occur in more than one grammatical class. Thus, in English, as a center for the actor in the actor-action construction, we require a noun: hence we have forms like 'height' beside 'high' or 'movement' beside '(it) moves.' Duplications of this kind are not symptoms of any special level of culture but result merely from a rather common grammatical condition.

21. Substitution and Determination

Apparently, all languages save labor by providing *substitute* forms whose meaning rests wholly upon the situation of speaker and hearer, especially upon earlier speech. Since these occur more frequently than specific forms, they are easily uttered and understood; moreover, they are nearly always short and, often enough, bound forms. Thus 'I' and 'you' replace names, and 'this' and 'that' the naming of a thing which may be identified by gesture. The most important type of substitution, for our subject, is *anaphora:* the substitute replaces repetition of a speech-form which has just been uttered or is soon to be uttered. Thus, the set 'he, she, it, they' replaces noun expressions, and the set 'do, does, did' replaces finite-verb expressions ('I'll go by train if John does'). A form competent to fill one position of a construction may suffice for anaphora of a phrase embodying the whole construction: 'Mary dances better than Jane'; here 'Jane' serves as the anaphoric substitute for 'Jane dances.'

Akin to the substitute forms, and very often identical with them, are *determiners*, which indicate a range within the class of phenomena that is designated by an ordinary speech-form: 'this apple,' 'the apple' (anaphora), 'every apple,' 'all apples,' etc. We shall be interested in determiners which leave the specimen entirely unrestricted: 'an apple,' 'some apple,' 'any apple.' If only one specimen is involved, anaphora is easily made ('it,' 'the apple,' 'this apple'); but, where several specimens are involved, English, like other languages, provides very poor means for distinguishing them. To provide for the identification of more than one variable, we must look to other phases of language which contain the germs of a more accurate system of speech.

IV. Precision in Natural Language

22. Reporting Statements

To discuss the meaning of all the varieties of utterance would be equivalent to outlining a complete sociology. We need deal only with a single type, the report (§ 18), since it alone is required for science.

The welfare of a community depends, so far as the actions of people are concerned, most directly upon simple handling activities whose occasion and performance are plainly observable— activities such as the gathering of food, hunting, fishing, construction of dwellings, boats, and containers, manufacture of clothing and tools, etc. These are manipulations of non-human objects, satisfactory in their biophysical aspect. Even where human bodies figure as objects, as in surgery or conflict, these actions suffice in themselves, with a minimum of biosocial significance. The situation in which an act of this sort will succeed does not always present itself in full to the performer; another person may mediate by speech: 'There are berries beyond the cliff'; 'The fish are biting today'; 'My moccasins need patching'; etc. Reports like these concern matters where behavior is uniform: in general, people will agree on the outcome of a test. This is the sphere of ordinary life out of which science grows forth. Natural science grows forth directly; the scientific study of man is hampered by the difficulty of subjecting biosocially conditioned behavior to such simple and testable reports.

If we try for a moment, and with full recognition of inadequacy, to ignore the forms of language, we may perhaps say that a report of this kind conveys, in the first place, the verbal substitute of a stimulus: 'It's raining.' Here the 'it' is an empty and merely formal indication of a point of reference, for which there is in this case no practical need. In most instances, however, some other stimulus, which has already affected the hearer or has been verbally represented to him, serves as a point of reference for the placing of the new stimulus which the hearer has not experienced: 'There are some apples in the pantry'— the last three words may represent nothing new, a familiar complex of stimuli to which the apples are now added. This adherence of a new stimulus to an old one is perhaps the practical background that is formalized in the actor-and-action or subject-and-predicate constructions which appear in the favorite sentence types of many languages: 'The fish | are biting |

today'; Russian 'Iván dóma,' literally 'John | at home.' However, there is no rigid agreement between the structure of the practical stimuli and the formal structure of the utterance.

The behavior of the speakers distinguishes very well between the features of the report which convey the relevant handling stimulus and the features of purely social and personal significance, such as especially the formal structure. Two reports of different structure ('There are berries beyond the cliff' and 'Behind the steep cliff over there you can find some berries') may mediate the same simple handling sequence; it is only the accompanying personal and social adjustments which differ. It is a well-tried hypothesis of linguistics that formally different utterances always differ in meaning; they may be *equivalent*, however, as to some one partial phase of meaning. Of this, the best example is the practical phase of the simple report; out of it there grows forth the equivalence of variously worded statements in scientific discourse. Here, as in the simple reporting of ordinary language, the equivalence covers the phase of meaning which is observable indifferently by all persons.

23. Negation

Every language, apparently, contains inhibitory forms, such as 'Don't!' The report may be thrown into reverse by utterance in a *negative* form: 'There aren't any apples in the pantry.' As this English example shows, the negative version may differ formally in complicated ways from the positive. The human voyage in space and time is linear: response cannot be made simultaneously to a report and its negative. If both are received as sentences, without comment as to biosocial values, the hearer is no better able to act than before; he has heard a *contradiction*.

It is very important, but not always easy, to distinguish, in this matter, between formal grammatical features and features of meaning. Contradiction is a feature of meanings, not of grammatical forms. In the normal case, where the contradictory phrases appear as separate sentences, there is, by definition, no grammatical nexus between them. In the more difficult case, where they appear in the same sentence ('It is green be-cause it is not green'), there is no grammatical incongruity. It is the meaning of the negating speech-forms which is here involved.

If we wished, using the full extent of a language (like English) in its ordinary form, to guard ourselves against contradiction in every discourse, no matter how long, we should have to mas-ter the universe in what would amount to omniscience.

One is likely to be deceived about this because in simple cases the rules for non-contradiction may take on a formal character. For instance, in Old English and in most dialects of modern English, a negated sentence often contains more than one negative unit ('I ain't got none'); in Latin and in present-day Standard English two negatives are separately superadded so as to cancel each other ('I did not have nothing'). What is here involved is the meaning of the negative words. The contradic-tion in a phrase like 'a round square' rests upon the meaning of the terms 'round' and 'square' or upon that of the basic terms which underlie successive definitions.

In miniature linguistic systems, such as are produced in logic and mathematics, the conventions for avoiding contradiction are rules of meaning (semantic rules). Thus, in a system of logistic, it may be agreed that a statement and the same state-ment preceded by the symbol '∼' may not, upon pain of con-tradiction, appear as sentences in any one discourse. This agree-ment defines the meaning of '∼.'

The dichotomy of contradictory sentences inheres in the na-ture of human speech: a sentence in a discourse excludes its contrary. Thus we obtain *implication*. The sentences 'Socrates is a man' and 'All men are mortal' exclude the sentence 'Socrates is not mortal,' and this is the same as saying that they imply the sentence 'Socrates is mortal.' These effects are due to the mean-ings of the words 'all,' 'is,' 'a,' and 'not,' and to the meanings of the grammatical forms.

24. Abstraction and Approximation

A report concerning a simple handling activity covers, in a rough way, only so much of the situation as is useful. If the speaker is prompted, say by an additional question ('What color are the apples?'), he is usually able to extend his report. Moreover, he can often subject himself, in the way of continued observation, to further stimuli of the complex in question ('I'll look and see how many apples there are'). Every apple has a color, and every set of apples a number, but these features are not communicated in the report, 'There are some apples in the pantry.' It is a trick of pseudo-philosophy to postulate a metaphysical "concept" of an apple to account for the imperfect reporting function of the word 'apple.' The obvious fact is simply that a speech does not mention every feature of stimulus. Since the ranges of stimulation and of predisposition are to all practical purposes continuous, and language can provide only a discrete set of forms, this *abstract* character of language is inevitable: not all the features of a situation appear in the report. If we do not consider the extension of an object, we may speak of it as a 'point'; if we speak of one dimension only, we may call it a 'line'; if of two only, we call it a 'surface'; terms like 'straight line,' 'plane,' 'triangle,' etc., add further characteristics, but still leave unmentioned certain simple features which are present in every object. This does not create a world of "concepts."

By lengthening his report, the speaker may tell more: we distinguish degrees of *approximation*. So far as the speaker has observed, or else perhaps only under reservation of irrelevant detail, the linear object if turned in a certain way would occupy the same position as before; hence the speaker says that the 'line' is a 'straight' one. It may be true that no object has been found to fulfill this condition to such an extent as to appear 'straight' under our best observation with mechanical aids. This means merely that the speaker did not employ these aids or else that his report was incomplete in this respect.

Although the meanings of language are discrete, there is no limit to their cumulation. By extending his utterance, the speaker may come closer and closer to a full picture of the situation. This is familiar, for one thing, from the art of fiction. In the realm of handling operation one rarely approximates beyond the features that are useful. In the scientific expansion of this domain, however, one often dwells upon features which have no immediate use in practical life—such features, for instance, as appear in the botanist's systematic classification of families, genera, and species. Accordingly, scientists and specialists in practical operation invent *technical terms*, either by redefining everyday expressions or by borrowing or creating new words. A technical term, then, replaces long phrases, or even a complicated discourse, and its meaning is fixed by an agreement of definition, which, in science, receives explicit formulation and strict adherence.

The useful approximation, in a simple society, will be a rough one; as civilization progresses, usefulness is discovered in closer approximations. Utterances and responses become more variable. This *variability* of response in individuals, and, thanks to manifold specialization of individuals, in the community, may yield a basis for a scientific definition of what we mean in everyday life by such words as 'welfare' or 'happiness.'

25. Counting

The part of ordinary language which most immediately opens into the language-forms of science is the number complex. The equality and inequality of sets of discrete objects are defined in terms of the result of placing the objects of the sets in one-to-one correspondence. Actual placing together of the objects is often inconvenient or impossible. The fingers (and, in suitable climate and surroundings, the toes) serve as an intermediate set. Speech-forms are more convenient. The quinary, decimal, and vigesimal groupings of these forms reflect the habit of counting on the fingers. Some languages provide only a few number-words; many, however, have an indefinitely extensible system. If any number-words are otherwise meaningful, this does not affect their use in counting: all that is needed, and, in most instances, all that is provided, is a set of otherwise meaningless free forms which every child learns to recite in a fixed order.

The set is more easily remembered, and can become indefinitely extensible, if it consists of a short sequence repeated over and over again, with an auxiliary counting of the repetitions. Thus, in the decadic system we have:

one, two, ... , nine, ten;
ten one, ten two, ... , ten nine, two tens;
two tens one, two tens two, ... , two tens nine, three tens;

. . . .

ten tens one, ten tens two, ... , ten tens nine, ten tens ten;
ten tens ten one, ...

The longer forms may be replaced by shorter substitutes, as 'ten tens' by 'one hundred,' 'ten hundreds' by 'one thousand,' etc. Later in the sequence such abbreviative forms may be agreed upon by specialists, as 'billion,' which in England means 'a million millions,' but in the United States 'a thousand millions.'

The system thus outlined appears plainly in Chinese; in most languages it is encumbered, but not changed, by irregularities, such as our 'eleven' for 'ten one,' 'twenty' for 'two tens,' etc.

Evidently there is no point at which the recitation has to end. No matter what number expression be given, one can always cap it by one which belongs later in the sequence. This is what we mean, of course, when we say that the system is indefinitely extensible.

Of all the features of ordinary language, the number expressions enter most directly into the scientific use of language, and they continue there to occupy the principal place. Putting discrete objects in one-to-one correspondence is perhaps as uniform, from person to person, as any operation. Scientists try to reduce their tasks to this shape. Our uniformity in the use of number expressions becomes the basis of complex systems of discourse which confine themselves largely to these expressions and enable us to perform long calculations without leaving this safe ground.

26. Infinite Classes of Speech-Forms

The indefinite extensibility of the system of number expressions of ordinary language leads to the *infinite classes* of mathematics. This linguistic background is worth examining, if only to save mystical aberrations.

Given any set of number expressions, every speaker is able to utter, without doubt or dispute, a number expression that is not in the set. This is what we mean by an *infinite class* of speech-forms.

In mathematics a less direct definition is useful: to say that a class is *infinite* means that it can be put in one-to-one correspondence with a genuine part of itself. Thus, we can put the set of all (positive integer) numbers in one-to-one correspondence with the set of all even (positive integer) numbers by simply assigning to each number its double.

This mathematical definition demands something more than the linguistic definition which we have given above, since we can define a one-to-one correspondence only if there is some order or system in the class. This condition is fulfilled, of course, in the case of the number expressions of ordinary language, and it will be fulfilled in any infinite class of speech-forms that is at all useful in discourse. In practice mathematicians sometimes employ the linguistic version, as, for instance, in Euclid's proof of the theorem that the class of prime numbers is infinite.

The popular or religious use of the word 'infinite' has no place in science, for in this use the word means something which cannot be dealt with, even verbally—something which cannot be grasped or understood—by the powers of a human being. In the sense with which we are concerned, the only infinite classes are classes of speech-forms. One might perhaps devise an indefinitely extensible system of other actions, such as gestures or graphic markings, but our saturation in language would from the very outset force us to assign verbal substitutes to the forms of such a system. Persons not accustomed to the observation of language are likely, in this matter, to see objects where only speech-forms are present. To name or define an infinite class of hypothetical objects or events is merely to adduce a class of speech-forms. Archimedes showed that one may name a number greater than the possible number of grains of sand in the solar system.

Any infinite class of speech-forms can be put into linear order, for each form consists of a linear (or quasi-linear, § 16) sequence of phonemes, and the forms may be alphabetized according to an arbitrary rank assigned to the phonemes. Apart from the imperfections of our traditional system of writing, which do not affect the matter, this is what we do in our everyday alphabetizing. No matter what new word may come into English, or what strange name into a list of telephone subscribers, there is no difficulty as to its place in the alphabetical sequence.

If we make the additional demand that the order be like that of the number expressions in counting, such that every form has an immediate successor (with no intervening forms), then we must call in the number expressions: the speech-forms must be grouped first according to the number of phonemes (or, in traditional writing, of letters) in each one and, then, within each of these groups, alphabetically.

Mathematicians, to be sure, set themselves a much harder task, since they deal with infinite classes each of whose members is an infinite class of speech-forms, as, for instance, with classes of unending decimal expressions. One begins with such forms as 'one-third' or 'the square root of two' or 'pi' (this last defined as the limit sum of a simply constructed infinite series). Then one may demand some special form of expression (such as that of a decimal) whose constituents (such as the digits of the decimal) may require laborious calculation, one by one. Then, further, one may define new classes whose members depend upon these singly calculated forms, as, for example, an unending decimal whose digits depend, according to some stated formula, upon the corresponding digits of the decimal expression of pi. Various types of discourse can be agreed upon, according to the kind of infinite classes that may be admissible. The actually uttered speech-forms, no matter how elaborate, which define or name any class, can always be alphabetized.

The utterance of all members of an infinite class can be required neither in mathematics nor anywhere else.

V. Scientific Language

27. Development of Scientific Language

Persons who carry on a specialized activity develop technical terms and locutions; these shorten speech and make response more accurate. Such are the special vocabularies of fishermen, carpenters, miners, and other craftsmen. These terminologies contribute to the dialectal differentiation which exists in all fairsized speech communities. The special vocabularies and turns of speech which are used in the various branches of science belong in this same general type; only, as scientific observation reaches beyond the interests of ordinary life, the vocabulary of science becomes very large. From timid neologisms it grows to a state where some scheme of word creation stands at the service of every member of the guild. In this way European and American scientists freely coin words by derivation and composition of Latin and ancient Greek stems; such words, with adaptations of grammar and phonetics, are accepted as loan-words in the scientific dialects of the several languages.

The exact responses, and the careful and often complex calculations of science, enforce an unusually meticulous style of speech. The syntactic scope of forms and the domain of substitutes have to be clearly indicated. This, with the elimination of personal factors, produces a general scientific style of utterance. The sentences may extend to great length and may evoke an immediate response only in hearers or readers who are favorably predisposed by training; on the other hand, the message, once grasped, is unmistakable. In pseudo-science the difficulties but not the advantages are imitated; the clinical symptoms are locutions which do not lead to handling response, appeals to personal or ethnical connotation, and, above all, an obscurity which remains even under analysis by a trained recipient.

At an advanced stage the demands for exhibition of data and for complex but unerring calculation lead to speech-forms and especially to written discourses which move outside the sphere of ordinary language. The ancient Greeks carried on mathematical demonstrations largely in ordinary language; it was the development, in the early modern period, of arithmetic and algebra, with its box-within-box markings of scope, that divorced scientific calculation not only from ordinary language but, to all practical purposes, from vocal utterance. People learned to calculate rapidly and accurately by visual reception and graphic manipulation of a small stock of characters in simple arrangements. Thus there arose the plan of conducting, or at least outlining and testing, scientific discourse by means of simple and rigidly manipulated graphic systems.

Without an entirely sharp boundary, we have, then, linguistically, two types of scientific discourse which we shall here distinguish by the names *informal* and *formal*. (We are here using the term 'formal' in its everyday sense and not with the technical meaning which it has in logic.) *Informal* scientific discourse uses ordinary language with the addition of technical words and turns of phrase and with syntactic and stylistic restrictions in favor of uniform response. It is generally capable of reception by a qualified listener. *Formal* scientific discourse uses a rigidly limited vocabulary and syntax and moves from sentence to sentence only within the range of conventional rules. In general, it can be carried on only in writing, mainly because no vocal equivalents have been devised or practiced for the elaborate markings of scope.

Within formal scientific discourse it is customary to distinguish, again without a sharp boundary, between mathematical discourse in general and a special type, symbolic logic, which is devised to establish and test the basic rules of scientific discourse.

Within mathematical discourse there is probably no linguistic difference between applied mathematics—that is, calculations which form part of a scientific discourse—and pure mathematics, where calculation is made for its own interest, with arbitrary axioms replacing the observational data and hypotheses of scientific procedure. Linguistic variety consists primarily in the use or non-use of the number vocabulary. The term 'mathematics' is ordinarily applied only to formal discourse whose basic vocabulary is numerical, and to geometry even when numbers are absent or play a subsidiary part. Other examples of non-numerical mathematical discourse appear, however, in such symbolisms and calculations as those of chemistry and of linguistics. One tries, wherever it is possible, to employ numerical discourse, because of the favorable character of the number-words and because of the stock of ready-made devices and calculations which has been accumulated in the pursuit of pure mathematics.

In all this development we have not left the domain of language. We reach its outskirts in the use of written discourses which will fail of effect if given in vocal form (§ 19). In general, to be sure, the separate written characters have been agreed upon as substitutes for specific words or phrases. In many cases, however, we manage best by ignoring these values and confining ourselves to the manipulation of the written symbols; systems of symbolic logic, especially, may be viewed, in a formal way, as systems of marks and conventions for the arrangement of these marks. The transition to non-linguistic activity could be made if we devised such marks and conventions without any initial or final linguistic interpretation, quite as in children's games that are played with paper and pencil. The marks could then no longer be qualified as writing. Actually, our formal systems serve merely as written or mechanical mediations between utterances of language. On the other hand, important linguistic conditions may be obscured when we speak of a formal system, like a system of symbolic logic, as a "language." The interpretation, initial and final, of the procedure is made in terms of some natural language (such as English), and the system as a whole is meaningless to a reader to whom it has not been interpreted in these terms. It differs even from an "artificial language," such as Esperanto, since the latter supplies a complete set of speech responses for ordinary use, while a mathematical system or a system of symbolic logic supplies only a limited set of responses that mediate between acts of speech.

By way of parenthesis it may be well to add that no artificial language, so far, has reached the function of a natural language. The artificial language is devised by speakers of a natural language, inevitably in terms of this and with little semantic deviation, and it is acquired, in similar terms, by persons who already speak a natural language. It would attain the status of such a language if a group of infants acquired it, by the usual process of demonstration, as their native speech. The only actual event even approaching this has been the implanting of Hebrew upon infants in Palestine: preserved only in written records, Hebrew had for some two thousand years been no person's native language.

Even such a thing as a tabulation of numbers is linguistic; apart from the verbal character of the elements, the arrangement leads to a linguistic interpretation. We leave the domain of language only when we come to a drawing, a geometrical diagram, or a map: here, indeed, we employ a non-verbal object directly as a representation of another non-verbal object.

A formal dialect, such as a system of symbolic logic, may very well be used to state the rules which govern its use. This means simply that the formal dialect suffices for statements about word-order, selection, and substitution. It neither, on the one hand, removes the system from its linguistic status nor, on the other hand, gives it the standing of an independent "language."

28. General Character of Scientific Language

An act of speech is a happening in the world and, as such, an object of science; the branch of science which studies it is linguistics. Scientists, however, are speakers and may agree to utter speech in certain ways; thanks to the simplicity of phonetic structure, they are able quite accurately and uniformly to adhere to these agreements. Accordingly, they treat their own utterances not as an object of science but as a part of scientific procedure. In this sense, and to the extent that social agreements about speech can be maintained, scientists may be said to "control" the forms of their technical dialect, in contrast with the world of meanings—that is, the world of events, including the utterance of speech other than that which forms part of the scientist's own activity. This outside world is reportable and predictable only to the extent that earlier acts of science have mastered it—at best imperfectly.

The scientist may construct a discourse, as in pure mathematics, in which the speech-forms have no meaning beyond that which is created by the scientific agreements governing their use—a type of discourse anticipated by the natural numbers of ordinary speech and, in most instances, based upon them. Such a discourse produces a calculation, made for its own interest, or as a model, or with a view to eventual use; about the outside world it tells nothing. We may be sure of its correctness because it moves only within the verbal agreements upon which it is based.

On the other hand, as soon as we admit meanings of the outside world, we risk error, and our certainty is then only such as may result from earlier acts of science. In a priori reasoning we see a sleight of hand which introduces observations or beliefs of everyday life into a seemingly pure calculation.

It is our task to discover which of our terms are undefined or partially defined or draggled with fringes of connotation, and to catch our hypotheses and exhibit them by clear statements, instead of letting them haunt us in the dark. The mentalist fails to list his undefined terms or to state his hypotheses; he wonders at the obtuseness with which we refuse to accept certain "concepts" which are necessary in the vocabulary of everyday life.

29. Publicity and Translatability

Science deals with the phases of response that are alike for all normal persons. Its observations and predictions can be tested by anyone. Where a temporal feature is present (as in the passage of a meteor), the question concerns the acceptability of a report and is usually settled by agreement of qualified independent observers. Unique personal or communal behavior figures in science as an object, which may be observed like any other; but it does not figure as a part of scientific procedure. This exclusion demands a redefinition of speech-forms; even number words like 'seven' or 'thirteen' have to be stripped of superstitious connotation. In the terminology of physics, the most advanced branch of science, one can see how far this stripping has been carried and surmise how much farther it still must go. Linguistically, as well as in handling, science is a *public* activity. In scientific colloquy recognition is granted neither to the private predispositions of the participants nor to the private connotations which they attach to forms of language; each participant burns his own smoke.

This does not mean that such predispositions, private adventures, or connotations are excluded from the object range of science: as objects they will be studied in psychology, biography, history, aesthetics, etc.

As, by convention and training, the participants in scientific discourse learn consistently to ignore all private factors of meaning, the lexical, grammatical, and stylistic features of their informal discourse become indifferent: each scientist responds to each discourse only with the relevant operations or their linguistic substitutes. Thus, half-a-dozen differently worded treatises, say on elementary mechanics, will produce the same result, so far as science is concerned. In this uniformity the differences between languages (as English, French, German), far-reaching and deep-seated as they are, constitute merely a part of the communicative dross. We say that scientific discourse is *translatable*, and mean by this that not only the difference between languages but, within each language, the difference between operationally equivalent wordings has no scientific effect.

30. Postulational Form

This stripping-down of meanings and exclusion of silent hypotheses has cost mankind much labor and many heartaches, and will cost more. Twists of meaning and of belief that prevail in our community may deceive us all; whoever has detected one or another finds himself misunderstood, since his fellows often relapse into attaching the traditional meaning to his words or inserting the traditional hypothesis into their reading of his calculations.

In order to unmask blind connotations and beliefs and to bring disagreements into the light of operation, where they may be subject to test, scientists resort—not often enough, to be sure—to the method of *postulates*, an explicit statement of what is taken for granted. One lists the undefined terms which one takes from everyday language or, more often, from other branches of science where they have been defined. Physicalism, as we have seen (§ 9), is the hypothesis which supposes that, in the unified vocabulary of science which is thus laid out, the ultimately undefined words will be simple terms in the domain of elementary physics—which, in turn, of course, rest upon physical terms of everyday life. In the same way, alongside the observed data, and on a par with them as starting-points for calculation, one states the hypotheses—the suppositions, generally results of other branches of science, which one accepts as prior to the work in hand. All new terms, after that, are rigidly defined: the new term is to be fully and freely interchangeable with the defining phrase. Only predictions, reached by acceptable calculation, may be added, by way of testing, to the list of hypotheses. If the terms are few enough and the structure of hypothesis and observation is simple enough, a mathematical system may be found or devised for the calculations. A dramatic instance of this was the invention of co-ordinate geometry. At best, of course, the uncertainties of all human response will play into the choice and character of our observation and, indirectly, through the failings of other sciences, into our structure of hypothesis. In the sciences that deal with man there is little enough that we have even learned to observe. It is all the more desirable that we lay bare our situation and our doubts by the frank survey of the postulational method.

It is the task of logic to examine the consistency of our calculations, mathematical or informal. If calculation can be performed in terms of numerical mathematics, its validity stands or falls with the validity of this discipline, and this, so far, has been the main concern of modern logic. If we cannot translate our discourse into an accepted dialect of mathematics, we had better travel with care. Formal rules of calculation would be equivalent to the meaning of our terms and to the content of our hypotheses, but these are often too vague or complex for formal discourse. Most scientific investigations are born with the makeshift help of informal methodology rather than with the professional guidance of formal logic.

31. Sentence-Forms of Scientific Language

The translatability of scientific discourse allows informal discourse to maintain a wide range of variation in vocabulary, grammar, and style; but, by the same token, this variation plays no serious part in the work of science. In principle we could agree upon any number of formal dialects, each capable of a uniform version of the results of science. The character of such a dialect is limited only by the fundamental linguistic type of scientific discourse. However, we have seen that certain speech-forms, most strikingly the number expressions, offer advantages great enough to make our discourse tend always in their direction. Apart from the special vocabularies of the several branches of science—or rather, apart from the great vocabulary of science—our discourse tends always toward the shape that is provided by mathematics and formal logic. Accordingly, we need here outline only the features which seem necessary in a formal dialect that is to serve for the use of science, and we may ignore the possibilities of expansion and variation.

Of sentence types, only the full-sentence statement comes into consideration. Every sentence will consist either of a statement-phrase (§ 18) or of several such in co-ordination. Statements, of course, will figure also as subordinate parts of large sentences.

This restriction would make bound forms of all parts of utterance other than the favored type of phrase: the structure of our statements would be a matter of morphology. In fact, it is possible to devise systems of discourse in which, for example, every phoneme (or, graphically, every letter) will have a meaning.

Such a system will be very compact but inconveniently remote from ordinary language.

It is customary, instead, even in formal systems, to use words and phrases of ordinary language as the smallest meaningful units. This is done because, on the one hand, the morphology of most languages is whimsical and complicated, and, on the other hand, we wish in explanatory discourse, in definition, and in logical discussion to isolate every meaningful form. Hence we speak of forms like 'not,' 'and,' 'plus,' etc., as "words," even though within the system of formal speech they dare not appear as sentences.

Calculation may be viewed as a process of exclusion. A set of initial sentences (report of observation and hypotheses) excludes certain other sentences, including the negatives of certain other sentences; a sentence whose negative is excluded is thereby included (implied) in our discourse. All other sentences are irrelevant. Accordingly, a scientific dialect will contain such forms as 'not,' 'excludes,' 'implies' ('therefore,' 'if . . . . , then . . . .'). One or several such must be undefined; the rest can be defined. The dichotomy of including and excluding statements inheres in the nature of language. A system which contained intermediate values (as, say, of probability) would still have to provide this dichotomy for statements in its discourse.

A report transmits either the mere existence of a stimulus or the accompaniment of a known stimulus by a new one. In either case, the report may give some analysis of the stimuli. Hence we may expect statements of such types as 'There exists . . . .' and '. . . . is. . . . .' Formally, if we make the necessary agreements, the mere presentation of a speech-form or of a sequence of speech-forms in a fixed order might be made to suffice; we approximate ordinary language by using words or phrases of *existence* and of *predication*.

In most languages the constituents of a predication (two-part statement) belong to different form-classes: 'John | ran'; 'This apple | is red.' Senseless locutions like 'Ran | ran' or 'Is red | is red' are grammatically impossible. However, the provision of duplicate forms, such as 'red' and 'redness'—lexically similar expressions in more than one form-class—makes it grammatically possible to say 'Redness | is red.' If this is not apprehended as nonsense, contradictions may arise. Thus, if a class of three members is a "triad," and in our discussion there appear exactly three such classes, *A, B, C*, then we may be tempted to say that the "class" of triads (*T*) "is a member of itself"; then, however, there are four such classes—*A, B, C, T*—and the "class" of triads is not "a member of itself"; but, remove *T*, and there are again three—a hopeless contradiction. Accordingly, if one wishes to use locutions of the type 'Red is a color,' one must distinguish different levels among the forms which enter into predication.

If the stimuli are broken into parts, more than one speech-form may appear in statements of existence, and more than two in predicative statements: 'There exists (a pair of men in the relation of) father-son'; and, George and Edward being known, 'George (and) Edward are father-son.' Here 'father-son' figures as a unit combining with the two units 'George' and 'Edward.'

A third type of sentence will be *equational*, of a form like '. . . . equals . . . .' or '. . . . means. . . . .' It is this type which makes definition possible; accordingly, it must remain undefined. We have seen that, in scientific speech, definition implies complete interchangeability of the new term with the old: if this is not agreed upon, we cannot develop the kind of dialect that serves in scientific use.

Finally, we shall have a form for the exclusion of statements, such as 'not. . . . .' From this most strikingly, but actually not more than from our other fundamental expressions, there arise the rules of calculation. These rules embody such meaning as is granted to the undefined forms of a system. They govern the sequence of sentences: they are not grammatical (morphologic or syntactic) but lexical.

32. Syntactic Features

The type of statement which says that (the known) George is the father of (the known) Edward will present a formal shape something like 'FS (George, Edward).' Here the two names are syntactically co-ordinate but not interchangeable, since the order (father first) has been agreed upon. A commutative construction would appear if we said that Edward and George were brothers, as 'BB (Edward, George).' We require a construction which co-ordinates any number of terms, to begin with, in a linear sequence, with order fixed or free. Such sequences, familiar from Cartesian geometry and from physics, are *vectors*. The definitions of this term which appear in elementary treatises put the cart before the horse, since by appealing to "direction" or specifying conditions for a "change of co-ordinates," they very improperly use an application or a special eventuality of reckoning to define a form of speech. It is only after we have defined a numerical vector as a sequence of numbers in fixed order that it devolves upon the geometer or physicist to show that the vector can be used to state a direction or to discuss how its terms will be altered by a new frame of reference. Otherwise we are exposed to mysticism: the vector may become in the end a "creation of our thoughts" ("ein Gedankending").

. A more complex constellation may require a higher ordering: *matrices* are syntactically of two dimensions, as when we name someone's grandparents, say paternal first and men first:

John	Edna
Thomas	Lilian

The written form allows of two syntactic dimensions; if we used a model in space, say of wires and significant beads, we could construct super-matrices of three syntactic dimensions. It would be easy to formulate problems which required more than three syntactic dimensions, but difficult to devise an appropriate symbolism.

Scientific speech inevitably follows ordinary language in designating sets of similar phenomena by one term, but it then requires a far more systematic determination (§ 21). If several phenomena of the set play different parts in the discourse, the term itself is a *variable*, and some unmistakable identification distinguishes the separate specimens (as, in algebra, 'x_1,' 'x_2,' 'x_3,' etc.). This allows us to dispense with the 'a,' 'any,' 'the' of ordinary language. By means of existence statements one can then define precise terms in the sphere of our 'some,' 'all,' 'no.'

33. Special Features

We can bind ourselves to no limit upon the number of occurrences of any one type of phenomenon which will be studied or upon the minuteness of subdivision to which a phenomenon may be subjected. This is equivalent to a demand that our dialect contain infinite classes of speech-forms. In this matter philosophers and mathematicians have often failed to recognize fundamental linguistic facts. So far as concerns anything accessible to science, the only infinite classes are classes of speech-forms; such situations as are referred to by terms like 'limit,' 'dense,' and 'continuous' arise only from our agreements as to the use of speech-forms and will be sought in vain in that outer world which is studied by science.

In the class of all English sentences that may be uttered, without limit upon length, we have an infinite class of the lowest type, and, if we imagine ourselves incapable of phonetics and ignorant of writing, this class lacks anything like order. By selecting from it and agreeing upon types of order, we define the infinite classes which serve us in scientific discourse. Our ordinary speech furnishes the simplest of these in the shape of the number-words, linearly and discretely ordered, with a first member 'one.'

Linguistic devices like these are entirely the creation of society; their usefulness, but never their operation, depends upon our handling activity. Viewed from the strictly linguistic level, it is an accident that the divisibility of matter goes beyond the power of our finest instruments, and an accident, no less, that our physics postulates molecular structure. Regardless of the latter circumstance, the former brings to us an engineering demand for a densely and linearly ordered infinite class: in the order there shall be a form between any two forms that may be named. This demand is met by means of number-pairs (two-place vectors) in the shape of the rational numbers, with the familiar agreement that, if *ad* precedes *bc*, then among the

271

rational numbers (a, b) shall precede (c, d). However, one could also agree upon an ordering of the natural numbers that would make them dense. From this it follows, that, leaving the decadic syntax of the natural numbers, we could devise a densely ordered class out of repetitions of a single speech-form and, in greater variety, out of phrases built, say, of two speech-forms—witness the decimal expressions consisting of 0's and 1's.

A "continuous" object, such as a piece of wire, presents itself between any two of its points, no matter how we swing the knife. A straight line, in geometry, is met by lines at points which are not designated by rational co-ordinates. Given a linearly and densely ordered class of speech-forms, we require, therefore, that any expression (within the vocabulary and syntax of our dialect) which divides ("cuts") the class in the familiar manner of Dedekind's postulate shall be a member of a new class. This class is then "continuous"; it contains a member to match every member of the old class, since these members make the required division, and it contains, further, any other phrases (within our dialect) which define a cut. Mathematicians sometimes ignore the obvious fact that the possibility of defining the "cuts" depends entirely upon the vocabulary and syntax of our discourse and by no means upon our "perception of time" or upon any mystical realm of "ideas." In the discourse of numbers the agreements of multiplication ('square,' etc.) make possible one kind of cut, and those of limits another. It is for the speakers to decide what expressions they will admit into their discourse.

VI. Summary

34. The Place of Linguistics in the Scheme of Science

The subject matter of linguistics, of course, is human speech. Other activities, such as writing, which serve as substitutes for speech, concern linguistics only in their semiotic aspect, as representations of phonemes or speech-forms. Since the meanings of speech cover everything (designata, including denotata; syntactic relations; pragmatic slants), linguistics, even more than other branches of science, depends for its range and accuracy upon the success of science as a whole. For the most part, our statements of meaning are makeshift. Even if this were not the case, linguistics would still study forms first and then look into their meanings, since language consists in the human response to the flow and variety of the world by simple sequences of a very few typical speech-sounds.

Linguistics is the chief contributor to semiotic. Among the special branches of science, it intervenes between biology, on the one hand, and ethnology, sociology, and psychology, on the other: it stands between physical and cultural anthropology.

Language establishes, by means of sound waves and on the basis of a communal habit, an ever ready connection between the bodies of individuals—a connection between their nervous systems which enables each person to respond to the stimuli that act upon other persons. The division of labor, civilization, and culture arise from this interaction. Popularly and even, to a large extent, academically, we are not accustomed to observing language and its effects: these effects are generally explained instead by the postulation of "mental" factors. In the cosmos, language produces human society, a structure more complex than the individual, related to him somewhat as the many-celled organism is related to the single cell.

35. Relation of Linguistics to Logic and Mathematics

Specialized uses of language involve no great alterations of structure; the specialization consists rather in the way that language is applied. Thus, the study of literature requires that we investigate the institutions and traditions of the community and the psychology (physiology, social status, biography) of the creative individual. In connection with science, language is specialized in the direction of forms which successfully communicate handling responses and lend themselves to elaborate reshaping (calculation). To invent and to employ these forms is to carry on mathematics. The critique and theory of scientific speech is the task of logic. Logic is a branch of science closely related to linguistics, since it observes how people conduct a certain type of discourse. In contrast with this, the invention and skilful manipulation of speech-forms is not a science but a skill, craft, or art; it is as such that we class mathematics. Mathematics appears as a science only so long as we believe that the mathematician is not creating speech-forms and discourses but exploring an unknown realm of "concepts" or "ideas."

Since mathematics is a verbal activity and logic a study of verbal activities, both of these disciplines presuppose linguistics. However, the forms of language which enter into mathematics, and are to be examined by logic, are simple and fairly normal: in principle, neither pursuit requires any technical knowledge of linguistics. In practice, since we labor under a load of traditional and popular misconception about language, a great deal of doubt, error, and dispute will be avoided if mathematicians and logicians acquire enough linguistics to remove these misconceptions. The principal sources of difficulty are twofold. Our popular belief distorts the relation of writing to language, placing the two on a par ("written" and "spoken" language) or even reversing the dependence, so as to suppose that a change in writing equals or prompts a change in language. Our popular belief replaces the function and effects of language by "mental" factors, which, whatever their place in literature or religion, are excluded, as non-physical, from the subject matter and procedure of science.

Selected Bibliography

§ 2. On the doctrine of the ancient Greeks: B. Delbrück, *Einleitung in das Studium der indogermanischen Sprachen* ("Bibliothek indogermanischer Grammatiken," Vol. IV [6th ed.; Leipzig, 1919]).

For the Sanskrit grammar: *Language*, V (1929), 267. The comparative method: A. Meillet, *La Méthode comparative en linguistique historique* ("Instituttet for sammenlignende kulturforskning," Ser. A, No. 2 [Oslo, 1925]). History of modern linguistics: H. Pedersen, *Linguistic Science in the Nineteenth Century* (Cambridge, Mass., 1931). Elementary outline of the subject: L. Bloomfield, *Language* (New York, 1933; London, 1935). Dialect geography: E. C. Roedder in *Germanic Review*, I (1926), 281.

§ 4. Distinction between speech-forms of discourse and speech-forms under discussion: R. Carnap, *Logical Syntax of Language* (Vienna, 1934; London and New York, 1937), sec. 42.

§ 5. On "correctness" in language: C. C. Fries, *The Teaching of the English Language* (New York, 1927); S. A. Leonard, *The Doctrine of Correctness in English Usage, 1700–1800* ("University of Wisconsin Studies in Language and Literature," Vol. XXV [Madison, 1929]); see also *American Speech*, II (1927), 432.

§ 6. Writing: Pedersen, *op. cit.;* E. H. Sturtevant, *Linguistic Change* (Chicago, 1917). Duodecimal notation confused with speech-form: *Atlantic Monthly*, CLIV (1934), 459. Gesture: W. Wundt, *Völkerpsychologie*, Vol. I: *Die Sprache* (3d ed.; Leipzig, 1911), p. 143.

§ 7. Biophysical and biosocial: A. P. Weiss, *A Theoretical Basis of Human Behavior* (2d ed.; Columbus, 1929), p. 84.

§ 9. Mentalistic doctrines of various types: A. P. Weiss in *Psychological Review*, XXIV (1917), 301, 353; in *Journal of Psychology*, XVI (1919), 626. See also *Psychological Review*, XXVI (1919), 327; *ibid.*, XXIX (1922), 329. Behaviorism: Weiss, *Theoretical Basis of Human Behavior;* operationalism: P. W. Bridgman, *The Logic of Modern Physics* (New York, 1927); physicalism: R. Carnap, *The Unity of Science* ("Psyche Miniatures: General Series," No. 63 [London, 1934]); *Philosophy and Logical Syntax* ("Psyche Miniatures: General Series," No. 70 [London, 1935]); O. Neurath in *The Monist*, XLI (1931), 618; see also *Language* XII (1936), 89.

§ 10. A. P. Weiss in *Psychological Review*, XXXVIII (1931), 474.

§ 14. Meaning is discussed with bibliography, on a mentalistic basis, to be sure, by G. Stern, *Meaning and Change of Meaning* ("Göteborg högskolas årsskrift," Vol. XXXVIII, No. 1 [Gothenburg, 1932]). See also C. W. Morris' monograph, *Foundations of the Theory of Signs* ("International Encyclopedia of Unified Science," Vol. I, No. 2 [Chicago, 1938]).

§ 16. The phoneme as a signal: E. Sapir, *Language* (New York, 1921).

§ 18. The sentence: in *Language*, VII (1931), 204. On "pragmatic" features of meaning: Morris, *op. cit.*

§ 19. Levels and ranks of construction: O. Jespersen, *The Philosophy of Grammar* (London and New York, 1924).

§ 20. Categories: F. Boas, "Introduction" in *Handbook of American Indian Languages* (Smithsonian Institution, Bureau of American Ethnology, Bull. 40 [Washington, 1911]), Vol. I.

Linguistic Aspects of Science

§ 24. See in *Philosophy of Science*, II (1935), 499; *Language*, XII (1936), 94. On variability: Weiss, *Theoretical Basis of Human Behavior*, p. 134.

§ 26. Euclid's proof, conveniently given in L. E. Dickson, *Introduction to the Theory of Numbers* (Chicago, 1929), p. 4. *Archimedes*, ed. J. L. Heilig (Leipzig, 1913), II, 219.

§ 31. Calculation: K. Menger in *Philosophy of Science*, IV (1937), 299.

§ 33. Elementary outline: F. Waismann, *Einführung in das mathematische Denken* (Vienna, 1936).

Index of Technical Terms

[Numbers refer to pages of this monograph.]

Abstraction, 255
Actor-action, 245
Anaphora, 251
Approximation, 255
Attribute, 247

Behaviorism, 231
Biophysical, 226
Biosocial, 227
Bound form, 243

Categories, 250
Center, 248
Class meaning, 245
Constituent, 243
Construction, 243
Contradiction, 253
Co-ordinative, 247

Deduction, 230
Definition, 238
Demonstration, 237
Determiner, 251
Different, 241

Endocentric, 247
Equation, 269
Equivalent, 253
Existence, 268
Exocentric, 247

Fill, 244
Form, 241
Formal, 261
Formal definition, 238
Form-class, 244
Free form, 242
Full sentence, 245
Function, 244

Handling, 226
Head, 247

Immediate, 247
Implication, 254
Infinite, 258

Informal, 261
Inner speech, 235

Language, 224

Matrix, 270
Meaning, 236
Mechanism, 231
Member, 247
Mentalism, 230
Minor sentence, 246
Modulation, 244
Morphology, 243

Negation, 253

Operationalism, 231
Order, 245

Parts of speech, 244
Phoneme, 239
Phonetic modification, 244
Phrase, 243
Physicalism, 231
Position, 244
Predication, 268

Resultant, 243

Same, 241
Scope, 248
Secondary phoneme, 244
Selection, 244
Sentence, 245
Sentence type, 245
Subordinative, 247
Substitute, 251
Syntax, 243

Technical term, 256

Variability, 256
Variable, 270
Vector, 270

Word, 243
Writing, 224

277

Procedures of Empirical Science

Victor F. Lenzen

Procedures of Empirical Science

Contents:

I. INTRODUCTION

                                                     PAGE

1. The Problem of Empirical Science 281
2. The Subject Matter of Empirical Science . . 283

II. OBSERVATION

3. Perception 284
4. Counting. 286
5. Measurement of Length 289
6. Measurement of Time 295
7. Measurement of Weight 302
8. Observation through Causality 304
9. Observation of Microphysical Entities . . . 306
10. Partition between Object and Observer . . . 308

III. SYSTEMATIZATION

11. Classification 311
12. The Correlation of Events. 315
13. Successive Approximation 321
14. Successive Definition 324
15. Atomism. 326
16. Statistics in Quantum Theory. 331
17. Atomism in Biology 333

IV. CONCLUSION

18. Unity of Science 335

SELECTED BIBLIOGRAPHY. 338

Procedures of Empirical Science

Victor F. Lenzen

I. Introduction

1. The Problem of Empirical Science

The problem of empirical science is the acquisition and systematization of knowledge concerning the things and phenomena experienced in observation.

The basic procedures of empirical science occur in daily life. Prior to his cultivation of science an individual perceives relatively stable things in space and time, describes them with the aid of symbols which record and communicate the results of observation, and explains perceptible phenomena in terms of causes. Simple experimental techniques are also used in the ordinary conduct of life. The child learns the operations of counting, of measuring length and time, of weighing with a balance. The builder uses tools, the housewife applies heat to produce the chemical reactions of cookery, the farmer cultivates crops. Such procedures are based upon prehistoric discoveries and inventions which have become the heritage of the race. In our historic era empirical science criticizes, augments, and systematizes practical experience. The science of one generation becomes incorporated in the technology of the succeeding one. Science and practice co-operate in the adjustment of man to his environment.

The preceding description of the interdependence of science and practice may be illustrated by a sketch of the origin of science in ancient Greece. It is traditional that the Greeks created European science, but their achievement was founded on knowledge and procedures which had been inherited from earlier Babylonian and Egyptian civilizations. The observations of correlations between the apparent positions of the heavenly bodies by the Babylonians provided a basis for the measurement of time and furnished data for a geometrical picture of celestial motions. The procedures used by the Egyptians in their surveys of land were formulated in propositions whose deductive relations were set forth in Euclidean geometry, which has been the classic model for systems of science. The Egyptians practiced medicine and surgery, and the Greeks under the leadership of Hippocrates developed this field by observation and experiment. Out of elements derived from practice was created the science of biology, which was especially developed by Aristotle, who made observations on animals and invented a system of classification. Machines were employed prior to the creative Greek period; on Egyptian temples the god Osiris is depicted weighing the soul with a balance. The principles implied in such machines are formulated in the science of statics which was founded by Archimedes. Thus the Greeks built upon the particular knowledge and techniques of the Babylonians and Egyptians. They organized the observations and procedures of their predecessors into theories founded upon principles.

The Greek interest in systematization also led to the creation of constructive hypotheses which reduced the diversity of perceptible things to a unitary basis. The theories concerning the nature of things initiated by Thales and his successors eventually led to the creation of the atomic theory, which in the modern era has provided a unified interpretation of physical and chemical phenomena. The Greeks emphasized the rational factor in science at the expense of observation. They were fertile in the creation of theories, but failed to develop adequately the technique of experiment. Greek science lost its creative power as ancient civilization decayed, but with the revival of science during the dawn of the modern era the method of controlled experimentation came to be adequately appreciated. Vesalius' insistence upon dissection for anatomy, Galileo's experiments on falling bodies, and Bacon's attempt to formulate the method of induction recognized the need for observation and experiment in empirical science. The experimental point of view was clearly stated by Newton in his assertion that physical properties are to be derived from experiments. The properties of things are modes of reaction to conditions that are subject to experimental control. Significant experimentation requires the guidance of hypotheses which serve to predict the results of observation. Accordingly, in the well-developed empirical sciences a theory which applies to a selected universe of discourse assumes the form of a hypothetico-deductive system, and scientific procedure consists in deriving predictions which are tested by the results of experiments. This union of the Greek conception of theory with experimental procedure has resulted in the modern development of science, which through technology created the industrial revolution and today is even more radically transforming the practice of daily life. Refinements in technique are providing the instruments for future developments in the theories which systematize the results of observation.

2. The Subject Matter of Empirical Science

In preparation for the analysis of procedures I shall sketch the general character of the subject matter of empirical science. The initial objects of science are the things experienced in perception, and their most general characters are position in space and time. The systematic and ultimately quantitative investigation of space-time order may be called generalized physics. Thus one arrives at the doctrine of physicalism, which asserts that the concepts of empirical science are reducible to those which express the properties of spatio-temporal things.

Carnap has shown that by a method of reduction, of which definition is a special case, the terms of a science are introduced by statements which involve terms designating perceptible things and properties. An outline of physicalist analysis, with especial reference to biology, is given in Volume I, Number 1. I present an analysis, adapted from Carnap,[1] of some biological terms. An organism is a special type of space-time structure, and hence the designation of terms such as 'species,' 'genus,' 'events in organisms,' is always determined on the basis of perceptible criteria. Thus " 'fertilization' is defined as the union of spermatazoon and egg; 'spermatazoon' and 'egg' are defined as cells of specified origin and specified perceptible properties; 'union' as an event consisting of a specified spatial redistribution of parts." Such biological terms as 'metabolism,' 'growth,' 'regeneration,' 'cell division,' etc., are also introduced in terms of perceptible criteria.

In behavioristics the responses of organisms to their environment are investigated, and in sociology the relations of human beings to one another and to the environment. As Neurath has emphasized, sociology deals with space-time structures. The essential difference between mechanics and sociology is that the relative simplicity of mechanical phenomena renders possible the repetition of experiments, whereas the complexity of social processes makes controlled experimentation difficult.

The formation of precise and quantitative concepts for the properties of spatio-temporal things is the first problem of physics, which plays a fundamental role in empirical science. The analysis of physical procedure is therefore basic in this monograph.

II. Observation

3. Perception

The basic procedure of empirical science is observation. Scientific description starts with observation; confirmation of the hypotheses of a theory is attained when phenomena predicted on the theory are observed. The term 'observation' has a meaning which is relative to a scientific situation. In daily life observation consists of perception with the unaided senses, in astronomy observation of the heavenly bodies is mediated by telescopes, in biology observation is through a microscope, in atomic physics observation is theoretical interpretation of experimentally controlled phenomena. The observation of microphysical entities requires the explicit use of physical principles as instruments of interpretation, but all observation involves more or less explicitly the element of hypothesis. In the present part of this monograph I shall describe the different types of observation in the order of increasing explicitness of the hypothetical factor. Let us begin with perception.

The defining characteristic of perception is the occurrence of a sense datum and its interpretation as the aspect of an objective thing. Perception thus involves the hypothesis that there exists an object to which given aspects are referred. The hypothesis that various common things exist is continually being confirmed by the reproduction of perceptions of them. I repeatedly have perceptions which are described as responses to a desk with relatively stable properties. Different types of perception are found to be correlated; thus, upon sight of my desk I expect to touch it. The truth of a perception is confirmed or disconfirmed by testing the predictions derivable from it. The development of the concept of an object is completed by the hypothesis of the identity of the perceptible objects of a society of observers. Thus the concept of objective thing is social; science is tested by social procedure.[2] The scientific criterion of objectivity ultimately rests upon the possibility of occurrence of predicted perceptions to a society of observers.

In daily life perception usually passes unreflectively into action. As I walk along the street, I perceive someone coming toward me and automatically step out of the way. I reach the curb and start to cross the street, but I perceive an automobile and hesitate. Nevertheless, even in the almost unreflective practice of common experience there are flashes of discrimination of qualities and comparisons between them. I notice that the approaching automobile is red; I judge that one man is taller than his companion. A phase of perception, accordingly, is discrimination and comparison of given qualities and relations. The results of such procedure can be recorded and communicated only if the experienced qualities and relations are symbolized. When things are perceived to have aspects that are similar in a specific respect, they are assigned a common quality or relation which is designated by a general name. The designation of the name of a simple quality such as redness can be understood only if one has immediately experienced the quality. General names are instruments of analysis. A qualitative analysis of the data of perception provides the basis for the description of a thing in terms of predicates. For example, an apple may be described as red, round, smooth, sweet, etc. Thus perception is completed by analysis of perceptible properties and the expression of the results of observation in descriptions. This procedure is a first step in the systematization of knowledge. Description of things is a mode of classification which in turn furnishes material for a more general classification. In its descriptive stage, which is based upon perception, empirical science introduces order into cognitions by classification.

On the perceptual level of knowledge the properties of a thing are commonly discovered by manipulation. An object which visually appears to be at hand may be denied reality and called a hallucination because it cannot be grasped.

But things may also be investigated by sight; indeed, for the study of microscopic objects the procedure is to observe them with the aid of optical instruments. In virtue of the capacity of glass to refract light, it is possible to make microscopes and telescopes, so that on looking at an object through an optical instrument its visual aspect is magnified. The phenomenon of magnification renders it possible to infer the existence of objects which do not appreciably affect the unaided senses. Observation with instruments is an extension of perception in which the hypothetical element is more explicit than in the ordinary perceptions of daily life. Yet the hypotheses have been so often confirmed that practical certainty is achieved.

In the development of empirical science qualitative description of things and phenomena is supplemented by quantitative representation. In addition to characterizing a group of objects as many, one may assign to it a number; one may not rest with the description of a rod as long or short, but record that it is so many meters long. Quantitative representation requires procedures for assigning numbers to measurable properties. I shall sketch the methods for the assignment of numbers to collections and to continuous manifolds such as space and time.

4. Counting

The fundamental quantitative procedure of science is counting a set of objects in order to characterize it by a number. A collection may consist of similar things such as the sheep in a flock; or the members of a collection may be as dissimilar as the books, papers, and other things on my desk. For the purpose of counting, differences in characteristics of the members of a collection are ignored, but their distinguishability must be preserved.

The basic operation in counting is illustrated in the following quotation from Conant's *The Number Concept.*[3] "The savage can form no mental concept of what civilized man means by such a word as soul; nor would his idea of the abstract number 5 be much clearer. When he says five, he uses, in many cases at least, the same word that serves him when he wishes to say hand, ... and his only comprehension of the number is, 'these objects are as many as the fingers on my hand.' " This example suggests that counting is an operation of determining similarity between collections. Two collections are defined to be similar if they can be put into one to one correspondence, so that to every member of one collection is correlated a member of the other collection. Similar collections, or classes, are assigned the same number. The null class is characterized by the number zero, unit classes by one, couples by two, trios by three, etc. Indeed, Whitehead and Russell define a cardinal number as the class of all similar classes. Thus we may define the natural numbers which serve to describe collections of objects.

The number of a collection is determined by counting. As one counts, the members are put into one to one correspondence with the natural numbers. If one has used the natural numbers from 1 to n, it is established that the number of objects in the collection is the same as the number of natural numbers from 1 to n, and, since this number is n, the collection counted has n objects. The number of a collection is independent of the order in which the counting occurs.

Given two collections we may combine them to form a new collection. For example, the addition of a collection of three objects to a collection of two objects yields a collection of five objects, and the result is the same if the collection of two is added to the collection of three. The formation of the new collection by addition may be represented by the addition of the numbers of the collections added, thus $3 + 2 = 2 + 3 = 5$. A similar relation holds for all instances of addition, and hence we may infer a general law, $a + b = b + a$, the commutative law for addition.

Laws for multiplication are also derived from observation. If one combines two collections each containing three objects, the resultant collection contains six objects. The combination of three collections each containing two objects also yields six objects. The formation of collections by multiplication may be represented by the numerical expression '$2 \times 3 = 3 \times 2 = 6$.' A similar relation holds for every multiplication of collections, and, hence, one infers the general commutative law for multiplication, $a \times b = b \times a$.

The preceding discussion of the commutative laws of addition and multiplication demonstrates that the fundamental propositions of ordinary algebra are initially derived from observation. Numbers are first discovered as characters of collections; but, once principles expressing their relations are set up, the numbers acquire properties defined in terms of their relations to one another. They become the subject matter of an abstract theory which defines their properties independently of any empirical application. As the number system is extended from the natural numbers to include negative, rational, irrational, and complex numbers, the original connection between numbers and collections is lost. The real numbers find a geometrical representation in the points of a line, and complex numbers in vectors.

The operation of counting is performed by pointing or otherwise indicating the members of a collection to which one assigns the numbers 1, 2, etc., respectively. The physicist, however, has invented mechanical counters so that, for example, the number of revolutions of a wheel is recorded. In contemporary physics the number of microphysical events, such as the passages of cosmic-ray particles through a Geiger counter, can be registered by a mechanical device. In order to determine the number of molecules in a specific volume of gas, however, it is necessary to employ indirect methods based upon theories.

Statistical surveys are important for many problems, and counting is their initial procedure. I have indicated the importance of statistics in physics, but the social sciences are even more dependent upon counting for their materials. Counts are taken of the number of inhabitants in defined areas, of the number of births per year, and similarly of deaths, marriages, homicides, etc. From such statistical data there are calculated birthrates, death-rates, and other results which give useful information about a given society.

5. Measurement of Length

Measurement is the general procedure of assigning numbers to the properties of objects. A measurable property is usually called a magnitude, but the term 'quantity' is also used. A basic measurement is that of length or distance by a procedure which was invented in remote antiquity. In order to measure the length of a rod, one selects a standard of length and counts the number of times that it can be laid off on the rod. I shall explain the general procedure of measurement by analyzing that of length.

The basic principle of measurement is that the space-time coincidence of objective events is perceptible to a community of observers. The most direct exemplification of this principle is offered by the measurement of length, which depends upon the determination of contact of two things from the contact of their aspects in perception. The empirical concept of contact is only approximately precise, but for theory one sets up the postulate that two coincident points are perceived to be coincident. Brunswik has explained how coincidence is the basis of objectivity. Two coincident points are perceived under exactly the same conditions; hence identical spatial perceptions are stimulated by the objects. If, for example, perception is mediated by light which is scattered by the coincident points, light from both points will travel over the same path to the same point on the retina and produce the same perceptual reactions. The objectivity of measurement is based upon the fundamental function of the perception of coincidence.

The physical concept of length is empirically derivable from solid bodies, which in view of the properties to be described may be called practically rigid bodies. Two points on a rigid body determine a stretch which is defined to include its end points. Let us place two bodies adjacent to each other so that two points on the one coincide with two points on the other. Assuming constancy in external conditions, the coincidence of the two sets of end points of the stretches remains constant in time. It is also independent of position in space, for, if the two bodies are displaced together, coincidence is preserved. Accordingly, one characterizes the two stretches as congruent. Congruence is observed to be a symmetrical relation, that is, if A is congruent to B, B is congruent to A; it is also transitive, that is, if A is congruent to B, and B is congruent to C, then A is congruent to C. To the congruent stretches one assigns the same number, the length of each stretch or the distance between its end points.

Thus far I have described the test for congruence of adjacent stretches. The test breaks down if the two stretches are separated in space. Observation shows, however, that, if separated stretches are again brought together, the coincidence of end points is restored. Furthermore, stretches that may successively be exhibited as congruent to a standard stretch may be demonstrated to be congruent when brought together at a distance from the standard. In view of such experimental results, one is led to postulate that stretches are congruent at a distance if they prove themselves to be congruent when adjacent. In effect, we postulate that the distance between two points on a solid body is unchanged in a displacement. In the present context it is meaningless to ask if this convention is really true. Such a question would be significant only if there were a more fundamental definition of length or distance. Accordingly, we may adopt as a standard of length the distance between two marks on some solid body. The international legal standard of length is the meter. It was originally defined as 1/10,000,000 of the distance from either pole to the equator and was approximately exemplified by the distance between two scratches on a bar, the *Mètre des Archives*, at the temperature of melting ice under atmospheric pressure. Later the distance between the two scratches was arbitrarily chosen as the standard. The wavelength of a red spectral line emitted by cadmium has been measured in terms of the standard meter, thus preparing the way for the adoption of this wave-length as the standard of length.

Thus far we have considered congruence of stretches and identity of length of congruent stretches. I have not yet explained the procedure for measuring the length of any stretch in terms of a standard. Let us suppose we are given a line. In nature a line may be realized by a ray of light, by a stretched cord, or by the edge of some solid. One may measure the length of a line by the following procedure: Beginning at one end, place stretches, congruent to the standard stretch, end to end until the other end of the line is reached. The number of stretches is the numerical measure of the length of the line relative to the adopted standard. In practice the congruent stretches are constructed by placing a standard alongside the line and making points on it which coincide with the end points of the stretch. The accuracy of a measurement of length can be indefinitely increased by decreasing the standard of length.

Our description of procedure in measuring length has presupposed constancy of the external environment and of the standard. In practice, however, corrections must be made for a change in conditions. Suppose that we have measured the dimensions of various bodies with a standard rod. If the measurements are repeated with a heated rod, it will be found that the numerical measures are smaller. As Carnap has shown, we have in principle a choice between two methods of explaining the change. We may postulate that the standard rod retains the same length but that the heating of the rod has produced, by an action at a distance, changes in the dimensions of the bodies in the environment. Or we may infer a change in the standard rod. If the standard rod is heated by a flame in contact with it, the second explanation is in accordance with a principle of contiguous causality which states that a physical change is to be attributed to an agency in the neighborhood of the body undergoing the change. This universally accepted kind of explanation allows the standard of length to vary with temperature. Accordingly, the definition of the standard of length requires specification of its temperature. But the definition of temperature presupposes the concept of length; for example, temperature may be defined in terms of the length of a thread of mercury in a glass tube. Thus we seem to have become involved in a vicious circle. In order to measure length, we must correct for the temperature of the rod, but, in order to measure temperature, we need to measure length.

The circle is avoided by the method of successive approximation. Let the standard of length be the distance between two scratches on a particular rod; this constitutes a definition to the first approximation. With the aid of our standard we can construct a thermometer with which we may discover the law for the dependence of length on temperature. Thus the definition of a standard to the first approximation provides a basis for a definition of temperature in terms of which the standard can then be defined to a second approximation. The possibility of this procedure is based upon the fact that a change in temperature produces a relatively small change in length. An error in the measurement of temperature resulting from a small error in the standard of length used to construct the scale of the thermometer gives rise only to errors of the second order of small quantities in the standard of length that is specified in terms of the temperature. The method of successive approximation is also used in controlling the constancy of other physical properties of the rod. The procedure of measuring length thus involves the performance of certain operations under controlled conditions. Our description of the measurement of length illustrates the experimental basis and operational nature of scientific concepts.

The results of measurement need to be appropriately symbolized in order to enter as elements in a mathematical development. The result of a measurement of length is stated by a proposition such as, 'The length of this rod is 2 meters.' Carnap has shown that the foregoing proposition expresses a relation between the number 2 and the object to which it is assigned. The proposition may then be written

2 (length in meters) this rod.

The last proposition is of the form xRy, where the relation R is one-many; there are many rods to which the number has the relation designated by 'length in meters.' Whitehead and Russell have shown that a one-many relation gives rise to a descriptive function; thus from xRy one obtains

$$x = \text{the } R \text{ of } y \ ,$$

or, in symbols,

$$x = R\,{}^{\backprime}y \ .$$

Applying this analysis to the present problem one obtains

$$2 = \text{the length in meters of this rod} \ ,$$
$$2 = \text{the length in meters 'this rod} \ .$$

Relations between measurable properties are represented by functional relations between their numerical measures. This may be illustrated by the law of addition which characterizes length as an extensive quantity. Two stretches A and B on a straight line may be adjoined to form a stretch C. If the lengths of the stretches are measured in terms of a standard, L, the results of measurement may be recorded by '$a = $ the length relative to L of A,' etc. The adjunction of the stretches is then represented by the proposition which expresses a relation between measures, '$a + b = c$.' The goal of exact empirical science is the expression of natural laws as functional relations between numerical values.

The quantitative concept of length is the basis of metrical geometry which expresses the spatial properties of figures. The geometer initially studies the spatial relations of practically rigid bodies, but he also makes figures by stretching cords and by drawing with a ruler and compass. Since the paths of light rays are usually straight, triangles and other figures may be formed of light rays. Indeed, the rectilinear propagation of light furnishes a practical test of straightness; the carpenter tests the straightness of an edge by sighting along it.

The properties of triangles may be used to measure distances by calculation from observations. In the method of triangulation one side of a triangle is the straight line between two stakes, and the other two sides are rays of light from a distant object to the stakes. The surveyor measures the length of the straight line between the stakes with a tape; from measures of the angles that the other two sides make with the base line and the general laws for triangles he calculates the distances of the object from the stakes.

Geometry deals initially with fixed structures at rest in a frame of reference such as the surface of the earth. If geometrical figures are changing, one may specify their instantaneous properties. The motion of figures relative to a given frame of reference introduces new problems. In classical kinematics it was assumed that geometrical figures are independent of their state of motion. Analysis of the operation of measuring the length of a body in motion reveals that the concept of simultaneity is presupposed. Suppose, for example, that a straight rod is moving with uniform velocity with respect to a given frame of reference. The procedure for measuring its length that may be adopted by an observer at rest in that frame is to mark on the frame the simultaneous positions of the end points of the rod. The length of the moving rod relative to the given frame is the distance between the two points marked on the frame. The special theory of relativity determines simultaneity to be relative to the frame of reference, so that the length of a rod relative to a frame in which the rod is moving is different from its length relative to a frame in which the rod is at rest.

The geometrical study of the spatial relations of bodies yields a concept of physical space. In order to increase definiteness, use is made of relatively small bodies or small portions of bodies which are called points. Examples of points are dots, pinholes, knots in cords. For maximum definiteness there is formed the limiting concept of a point which has a definite position. Space may then be described as the system of relative positions of points. The metrical structure of space is defined in terms of the properties of configurations of rigid bodies. In view of relativity, space is defined for a selected frame of reference. The surface of the earth was used by the Greeks as the frame for physical space, but in the modern era the preferred frame has been one with its origin at the center of mass of the solar system, with the axes oriented with respect to the fixed stars. In this frame physical space is Euclidean to at least the first approximation. This means, for example, that it is possible to construct a cubical lattice out of equal rods, that is, a Cartesian coordinate system. The general theory of relativity, however, involves the view that matter determines physical space to be non-Euclidean.

The propositions of geometry are to be viewed as initially generalizations from observations on the spatial properties of structures of practically rigid bodies. In Euclidean geometry the science is based upon a small number of axioms from which the propositions can be deduced as theorems. In abstract geometry the original axioms have been transformed into postulates which define implicitly the fundamental concepts of geometry. From this point of view the postulates of Euclidean geometry are descriptions of the formal spatial properties of rigid bodies. That the concepts of these forms are applicable to the objects of perception is a hypothesis to be confirmed and limited by observation.

6. Measurement of Time

The problem of the measurement of time is to invent a procedure for assigning to events numbers, called the times or dates of events. An event is a process of relatively short duration, the shortness depending on the precision with which one intends to describe phenomena. In theory one forms the concept of instantaneous event as a theoretical construct. The basic temporal relations are simultaneity and succession of events. Two events are objectively simultaneous if they are simultaneously perceptible to a community of observers. A similar criterion may be given for succession. We must distinguish between local time, the time system at a specific place, and extended time, the time system throughout a space.

The procedure of measuring local time is based upon some concrete temporal process, usually a periodic motion. The behavior of a standard clock defines the scale of time. Now, various physical processes can be used for the definition of metrical time: the rotation of the earth about its axis, the revolution of the earth around the sun, the vibration of a pendulum, the revolution of the moon around the earth, etc.

The procedure for measuring time may be illustrated by adopting as the fundamental temporal process the vibration of a pendulum. The fundamental principle is that successive vibrations of the pendulum take equal times. This principle has an empirical basis. We have a qualitative estimate of the duration of processes and can judge, for example, that the time of one vibration of a pendulum is less than the time between sunrise and sunset. Hence we may estimate that two successive vibrations of a pendulum take the same time. Again, all pendulums which are similar in structure vibrate in synchronism, that is, if two pendulums of equal length are released simultaneously, they vibrate together, passing through corresponding points at the same time. But the principle that successive vibrations take the same time, although suggested by experience, is a definition of equal intervals of time. It is a convention concerning which it is meaningless to ask whether it is true or false. The question of truth or falsity would presuppose another standard, for which one would have to assume that each performance of a periodic process requires the same time.

Let us then suppose that we have chosen the vibration of a pendulum as the basis of a metrical structure for time. We may select the beginning of a particular vibration as the origin of time; to the ends of successive vibrations are assigned the numbers 1, 2, 3, etc. The time of any event is expressed by assigning to it the number which is correlated with the vibration, the end of which is simultaneous with the event. If the end of the vibration and the event are not simultaneous, we may imagine clocks with shorter and shorter periods and thus, by interpolation, approximate as accurately as we please to a precise assignment of time to the event. The duration of a process is, then, expressed numerically by the difference of the times of its beginning and end.

The physical process that serves for the definition of a time scale must be subject to constant conditions. The period of a pendulum, for example, is constant only if the conditions of the environment are invariable. For example, the earth's gravitational field must remain constant, the action of electric and magnetic fields must be negligible, the temperature must not change, etc. The control of conditions is accomplished by the method of successive approximation. We have initially a qualitative estimate of constancy of conditions, and therefore we may apply the principle that a specific pendulum keeps the same time on various occasions. For daily life such a definition may be adequate, but for scientific observation the conditions must be quantitatively defined. The definition of time to the first approximation enables one to define 'acceleration,' 'force,' 'electric field,' etc., quantitatively. We may then specify quantitatively the constant conditions for the clock, and thus the time scale can be defined to a second approximation.

In practice it may not be possible to control the conditions to which clocks are subject. If one has a quantitative description of the actual conditions, however, it may be possible to correct the times assigned in terms of the actual clock, and thereby find the times which would be indicated by an undisturbed clock. The earth is the standard clock for astronomical measurements; the definition of time is given by the convention that the angular velocity of the earth about its axis is constant. Now, certain anomalies in the moon's motion can be explained on the hypothesis that the earth is slowing down on account of tidal friction. This explanation presupposes the laws of mechanics, which have been confirmed by experiments in which time was measured by the rotation of the earth. The explanation implies that the earth-clock indicates only an approximate time. One may say, however, that time is to be defined by the rotation of the earth—assuming that there are no frictional forces. The empirically indicated time would then have to be corrected in order to find the time that would be indicated by a frictionless earth. The astronomer does not actually compute the correction from the friction and the laws of mechanics but obtains the correction by comparison with the moon. This illustration, however, shows that corrections may be made for disturbing conditions.

In the historical development different processes have been employed for the measurement of time. The ancients measured time by the motions of the heavenly bodies, but they also measured time by the amount of sand that ran out of an hourglass. The two methods agree approximately on account of the empirically observed correlation between the flow of sand out of the hourglass and celestial motions. We choose the astronomical method as the standard and ascribe the inaccuracy of hourglass time to variable conditions such as differences in the size of grains, smoothness of surfaces, etc. That is, the disagreement between hourglass time and astronomical time is explained by the hypothesis that differences in the grains affect their motions and not the motion of the heavenly bodies. The use of accepted physical principles in the definition of physical quantities is here exemplified.

I supplement the description of procedure by a discussion of the considerations which determine the choice of a standard clock. In the first place, one seeks a process that is as permanent as possible; the rotation of the earth and the motion of the moon especially satisfy this requirement. Furthermore, the clock is to be as free as possible from disturbing influences. The clocks on the earth are subject to disturbances, hence the astronomical clocks are preferable. A pendulum, however, is readily reproducible and is useful as a secondary clock when it has been standardized. Most desirable of all for a definition of time is independence of special properties of matter. A pendulum requires the earth's gravitational field; the mainspring of a watch is dependent on the elastic properties of a particular substance; the heavenly bodies are subject to destruction. The preferred definition is in terms of functional relations between physical quantities. For example, we may define time as the independent variable in the equations of dynamics. The equations which express physical laws then constitute an implicit definition of time. In particular, time may be thought of as defined by the first law of motion which states that a body acted upon by no forces persists in a state of uniform rectilinear motion. By this definition, equal intervals of time are indicated by equal distances passed over by a body under no forces. This definition of time in terms of the first law achieves the dissociation of the time scale from special processes or bodies. The change in definition exemplifies again the method of successive definition. Initially, physical quantities are defined in terms of special operations and provide the basis for the empirical discovery of general laws, which may then be transformed into implicit definitions of the physical quantities involved in them. But it then becomes an experimental fact that the measure of a quantity which is obtained by special operations approximately satisfies the definition of the quantity in terms of the general laws. For example, the measure of time by a pendulum approximately satisfies the definition of time in terms of the first law of motion.

So far, the discussion has been restricted to local time; I now turn to the problem of the time system throughout a space. If P_1 and P_2 are two separated points in space, how are we to define a time scale at P_2 which is the same as that at P_1? We must postulate a law of connection between local times.

In this discussion the points of space will be referred to a definite frame of reference. It is assumed that similar clocks at rest may be placed at any point in the given space. Similar clocks have the property that, if at rest at the same place, they keep the same time. The problem is: Given the time scale at P_1, how shall we extend it to P_2? Suppose that at P_1 there is a set of similar clocks. If one of the clocks is moved slowly to P_2, we may transport its time scale by the postulate that the clock at P_2 is synchronous with the clocks at P_1. Since no immediate perception of the synchronism is possible when the clocks are separated, the postulate is a definition by which the time system at P_1 is extended to the point P_2. There is some empirical foundation for this definition. If two clocks which are synchronous at P_1 are transported to P_2, they will be synchronous at P_2. This experimental result is found even if the two clocks are transported along different paths. Again, if a clock which is synchronous with the clocks at P_1 is transported and then returned to P_1, it will again be synchronous with the clocks at the starting-point. All these experimental results support the convention that a time system is extended throughout space by the slow transport of a clock. This method of extending a scale of time is employed in daily life. Thus, the locomotive engineer extends along the track the time system of the place at which he sets his watch. If a ship carries a chronometer, it extends a time system along its course. In transporting a clock, we must not unduly accelerate and disturb it; but the accelerations to which ordinary clocks are subject do not affect them appreciably.

The preferred method of extending a time system employs light signals. At time t_1 as indicated by the clock at P_1 a signal is sent and arrives at P_2 at the time t_2 as indicated by the clock there. The clocks at the separated points keep the same time if $t_2 = t_1 + l/c$, where l is the distance between the two points and c is the speed of the signal. In effect, the principle of the constancy of the speed of light in a vacuum is postulated as a definition by which a time system is extended throughout space. Recognition of this procedure occurs in the special theory of relativity.

Thus far I have described procedures developed by astronomers and physicists for the exact measurement of time. Estimates of time are also made by the geologist, who investigates the history of the earth. A fundamental problem of geology is the temporal order of strata. If it is possible to assume that strata have not been appreciably displaced from the positions in which they were created, the following principle is applied: The order of superposition is the order in time, the oldest stratum being at the bottom and the newest at the top. This application of the principle is restricted by the circumstance that strata may disappear underground or be continued by a different series. William Smith, however, discovered a method of correlating strata in different regions. From observations upon exposed sections of strata in England, he found that each group of strata contained characteristic fossils, so that the order of superposition of the strata indicated the order of succession of the fossils. This discovery furnished the empirical basis for the procedure of correlating the strata in different regions by means of their characteristic fossils. The hypothesis implied by this procedure is that strata which have been laid down in widely separated periods of time will contain quite different fossils, while those which were formed approximately contemporaneously will contain similar fossils.

The preceding discussion explains the procedure of determining simultaneity and succession of strata. The geologist also seeks to make quantitative estimates of time. In order to determine the age of a sedimentary deposit, he may make an estimate of the present rate of deposition and, from the thickness of the deposit, calculate the time required to produce it. The same procedure is used in estimating the age of the earth from the mass of salt in the ocean and the rate at which it is entering from the rivers. The most accurate method of estimating geologic time is based upon radioactive phenomena. Radioactive elements like uranium and thorium spontaneously disintegrate at a definite rate and give rise to elements of lower atomic weight. The final products are metallic lead, of atomic weight 206 if derived from uranium, and of atomic weight 208 if the final product is of the thorium series. If a rock is rich in uranium, one may measure the masses of uranium and lead which it contains and, from the rate of transformation of uranium into lead, calculate the age of the rock. Since helium is a product of radioactive disintegration, it may also be used to estimate the age of rocks. Chemical analyses of rocks from all parts of the world show that radioactive elements are widely distributed; these radioactive clocks in rocks indicate their ages.

Lecomte du Noüy has defined biological time by the dependence of the rate of reparation of cells upon the age of organisms. He found that the rate of cicatrization of wounds decreases with age as ordinarily determined; in a man of sixty the rate is one-fifth that of a child of ten. Processes that are uniform when measured in terms of astronomical time become nonuniform on the biological time scale.

7. Measurement of Weight

The measurements of length and time are observations in which the results of operations are perceived and interpreted with the aid of principles. In fundamental measurements of length the basic operation consists of laying off the standard on a line. Since a measuring rod is operated by the observer and its state is experimentally controlled, the procedure of measuring length is experimental. Now, it may be contended that there is direct perception of length and time, and hence measurements of them are frequently called direct. The term 'direct' contrasts length and time with properties that are known only as they manifest themselves in experiments which are constructively interpreted. A discussion of some of these experimentally exhibited properties will reveal further the method of experiment and will illustrate Newton's dictum that the properties of things are derived from experiments.

The first example is the physical property, weight. In a sense we perceive length and experience duration, but observation of weight requires constructive interpretation. In order to observe that a material body has weight, I may support the body with my hand. The muscles attached to the hand exert a force which is experienced through kinesthetic sensations. Now, a fundamental principle of statics states that, if a body is at rest, the forces on it must be in equilibrium. This principle of equilibrium accordingly requires that I conceive of the body as having a weight, directed downward, which is balanced by the upward force exerted through the hand. This simple experiment therefore exhibits weight as a force that balances some other force. The weight is assumed to continue in existence when the upward force is removed, so that, if the body is released and falls freely to the earth, the weight of the body is interpreted to be the cause of its acceleration.

The measurement of weight may be based upon a lever such as is used in a beam balance. The empirical foundation for the procedure is the fact that two bodies may be found such that if they are attached respectively to the ends of a level with equal arms, the lever remains horizontal. It is also an empirical fact that if two bodies maintain one lever horizontal, they will maintain all others horizontal, regardless of position, length, material, color, etc. The two bodies behave similarly in different contexts, and hence it is convenient to define them to be equal in weight. If two bodies of equal weight are attached to one end of the lever, a body at the other end which maintains the lever horizontal may be assigned a weight twice that of the first. By this procedure one may build up a set of bodies with weights that are integral multiples of a standard weight. One may also define fractional weights: If two standard weights are balanced by three equal weights, each of the latter is two-thirds of the standard. The scale of weights is constructed so that it satisfies the law of addition for extensive quantities. The present discussion interprets the principle that equal weights maintain an equal arm lever horizontal, which was selected by Archimedes to be a fundamental principle of statics, as an empirically founded definition of equality of weight. Of course, one could use another definition of equality of weight, and then the principle of the lever would be an experimental law. But the alternative definition would presuppose the selection of some other principle of mechanics as a definition.

After weight has been ascribed to a body, it is possible to use this force to discover other forces. If a body is attached to a spring balance, the spring is stretched before equilibrium is attained; it is assumed that the end of the spring is acted upon by two forces in equilibrium, the weight of the body and the tension in the spring which is equal in magnitude but upward in direction. If one attaches bodies of different weight to the spring and measures its extension, one can discover Hooke's law that the stress in a spring is proportional to its extension. Hooke's law may be used as an alternative definition of force in statics. To conclude: Given a single type of force, weight, it is possible to interpret statical experiments as exhibiting weight to be balanced by other forces such as tensions in springs and electric and magnetic forces. Observation of forces consists in constructive interpretation of mechanical experiments in the light of principles which play the role of definitions. The physical concept of force expresses the assignment of numbers in order to describe the conditions of motion. The procedure involves the element of hypothesis. The physical quantities which are assigned by constructive interpretation of experiments must be instrumental in predicting the results of other observations.

8. Observation through Causality

I have thus far analyzed observation of perceptible things and properties. Nonperceptible entities are also inferred to exist as the hypothetical causes of perceptible phenomena. Such inference through causality eventually becomes observation. I use here a crude concept of efficient causality, but in the pages on systematization I analyze it as expressing functional relationship.

The transformation of explanation by constructive hypotheses expressing causation into observation may be illustrated by an analysis of vision. Let us suppose that one may touch a thing but cannot see it. Under appropriate conditions the thing becomes visible, that is, it becomes possible to experience visual aspects which are correlated with the data of tactual perception. For example, if I introduce a lighted candle into a dark room, things become visible; with my eyes open I may see a table. Hence we must think of the visual aspect as produced, at least in part, by the candle. If explanation is to be expressed in terms of the perceptible, one must conceive of this production as an action at a distance, since the candle and the thing perceived are separated in space. The physicist, however, interprets the process as an action by contiguous causality; he defines radiation through the principle that the visual aspect of a thing is functionally dependent on radiation which travels from a source to the thing and is scattered toward the observer.

Thus a principle of contiguous causality yields a definition of radiation, which is assumed as a construct. The definition is justified, since the assumed radiation is instrumental in predicting new phenomena. The definition accordingly plays the role of a hypothesis which is tested by its predictions.

Again, if we keep the illuminated thing constant, as estimated by touch, and change the source of radiation, the visual aspect changes—in color, for example. The color of an aspect is the basis of assigning a specific property of radiation, the wavelength as measured in Young's double-slit experiment. The properties of radiation are thus defined in terms of the visual aspects which it makes possible. Radiation not only produces visual aspects but also affects a photographic plate. On allowing radiation to pass through a prism or grating, it is spread into a spectrum; radiation of a specific wave-length produces a line having a specific position on the plate. To the spectroscopist the perception of a spectral line constitutes an observation of the radiation which produced the line. Experiments on the pressure of light have shown that it exerts a pressure; radiation may therefore be observed through the momentum which it communicates to some directly perceptible object.

A visual aspect is partly produced by radiation, but the aspect also depends upon the thing perceived. If the source of radiation is kept constant and the thing illuminated is varied, the visual aspect changes; for example, the visual aspect of a chair differs from that of a desk in the same light. Hence we must think of the visual aspect as functionally dependent on the radiation and thing conjointly. The visual aspect manifests the reaction of the thing to radiation; in physical language, the visual aspect depends upon the reflection of radiation by the thing. The preceding statement expresses a law which may be viewed as a generalization from observation based upon the definition of a thing in terms of tactual aspects and the principle of contiguous causality. In the development of physics, however, the principle that a thing reflects radiation becomes an essential element in the definition of physical reality. This is the basis of observation by photography. An object reflects radiation which is focused by a system of lenses so that an image is produced on a plate. Perception of the image is the basis of inference to the structure of the object.

9. Observation of Microphysical Entities

Observation through causality, which is exemplified by the assumption of radiation, is the procedure for the study of microphysical entities such as electrons, protons, neutrons, etc. Observation of such entities consists in the perception of macrophysical phenomena that may be interpreted as effects of microphysical objects. In this field, principles become constructive instruments of interpretation, and so observation is more subject to the uncertainties of hypotheses than is perception of common things.

I present some examples of the observation of electrified particles through their effects. If a high-speed alpha particle strikes an appropriate screen, it produces a scintillation which consists of light assumed to be emitted by countless atoms and molecules that have been excited by the particle. Thus, a single particle can be detected by visual perception of the macrophysical scintillation that it produces. If an electrified particle with sufficient energy passes through a cloud chamber, it ionizes the molecules that it strikes, producing ions on which water vapor condenses. Perception of the track of waterdrops completes the observation of the ionizing particle. Again, a Geiger counter is a tube in which a momentary current flows when a particle of sufficient energy passes through and produces ionization. The momentary current is amplified and actuates a mechanical counter which registers the number of particles that pass through the tube. Hence the production of scintillations and condensation tracks, and the actuation of Geiger counters, enable the physicist to detect individual microphysical entities. If many particles impinge on an appropriate plate, a perceptible pattern may be produced. The ultimate elements of physical theory are observed through perceptible macrophysical phenomena.

The physicist determines the properties of elementary particles by investigating their behavior under experimentally controlled conditions. For example, an electron may be subjected to magnetic or electric fields, and from the curvature of the path, observed in a cloud chamber by the condensation track and in a cathode-ray tube by the deflection of the effect on a screen, it is possible to determine the ratio of electric charge to mass as defined by assumed physical principles. In Millikan's measurement of electronic charge, electrons were captured by perceptible oil drops the behavior of which was observed under specified physical conditions. The measurement was completed by a calculation of the charge from the data of the experiment and the assumed physical principles.

An observation in which the element of hypothesis is quite explicit is that of the energy levels of the atom. According to contemporary theory, an atom consists of a positively electrified nucleus to which are bound electrons distributed among concentric shells. The electronic shells are defined by quantum conditions and constitute the basis for a discrete set of stationary states of constant energy of the atom. As electrons jump from one shell to another, the stationary state of the atom changes discontinuously with the absorption or emission of radiation, the frequency of which is determined by the difference between the energies in the initial and final states. Radiation emitted or absorbed produces spectral lines upon a photographic plate, and from the serial order of the lines it is possible to determine the energy levels of the atom. These levels, or quantum states, are the object of a constructive hypothesis which serves to correlate and predict spectral data, but the theory has been so often confirmed that an observation of spectral lines now seems practically a direct observation of the energy levels of the atom. In the development of physical theory during the last two decades there has been a transition from the active construction of hypotheses for atomic theory to the almost unreflective acknowledgment of its objects as real. Indirect observation, consisting of measurements on spectral lines plus the theory from which calculations are made, has achieved the practical certainty of direct observation.

Procedures of Empirical Science

10. Partition between Object and Observer

As one traces the development of observation from perception through measurements of length, time, and weight to the detection of radiation and the electrified particles of microphysics, it is evident that the function of apparatus becomes more and more important. Bohr has emphasized the fact that the observer and his instruments must be presupposed in any investigation, so that the instruments are not part of the phenomenon described but are used. The problem accordingly arises of defining the partition between object and the observer and his apparatus.

I present some examples of the partition between object and observer. Tactual perception is an interaction between a body and end organs such as those in the tip of a finger. If one touches a desk with a finger, the partition is between them. An observer, however, may be extended by mechanical devices. Bohr has cited the following example: If one firmly grasps a long stick in one's hand and touches it to a body, the body touched is the object of observation, and the stick is an apparatus that may be viewed as part of the observer. It is a psychological fact that one locates the tactual aspect at the end of the stick, so that the partition is between the body and the end of the stick. If, however, the stick is held loosely in the hand, the stick becomes the perceived object, and the partition is between stick and hand.

In vision perception is mediated by radiation which travels from the object to the observer. On account of the finite speed of propagation of radiation, the state of the object is prior to, and distant from, the perception of that state. The position of the object is of primary biological significance, and so the perceiver habitually places the partition at the object and unreflectively accepts the radiation which the object emits or reflects as part of the observer. For example, the visual perception of a stick is dependent on the sunlight which is reflected from the stick to the eye. In daily life the object of interest is the stick, and hence the light is treated as an instrument of observation which is an integral part of the observer. But, if the stick is in water and appears to be bent, the explanation of this illusion of vision in terms of the refraction of light by the water presupposes that the partition be placed between the eye and the light.

In perception the partition is determined by the biological needs of the observer and is ordinarily difficult to displace. Brunswik, however, has shown that a change in perceptual attitude is possible; for example, we can compare objects with respect to the size of the pattern on the retina produced by the light from them. In such a case we are to think of the partition as at the retina. In view of the extension of the concept of observation to include interpretation in term of principles, it is desirable to distinguish between cognitive and perceptual partitions. The object may be observed through causation of perceptible effects, so that the two partitions are different. The position of the cognitive partition depends upon the interest of the observer. If a physicist is looking at a pointer on a scale, its status depends on the purpose of the observation. If he is using the instrument to measure an electric current, the pointer is an extension of the observer; the object is the electric current. If the physicist is calibrating his instrument, the pointer is part of the object of observation; the light by which the pointer is seen is then an instrument which belongs to the observer. If he studies the properties of light by photographing spectral lines, the light is the object of observation, and a line is registered on the plate, which plays the role of observer. But if the physicist examines the spectral line, it becomes the object, and the light reflected by it in turn is an instrument.

In the analysis of observation the cognitive partition between object and observer may be displaced in opposite directions. The naïve observer probably assumes that in visual perception he is in direct contact with a distant object; indeed, the child reaches for the moon. In perception the partition is habitually at the object. But it is a hypothesis that our visual perception of a distant object is dependent on the radiation that produces its visual aspects. Physical investigation reveals that the radiation is affected by an intervening medium, and so the cognitive partition is placed at the organism to which the observer is now restricted. The problem of the physicist ends when he reaches the boundary of the organism, but the biologist displaces the partition into the body of the observer. The optometrist places the partition at the retina of the eye, and the physiologist of the nervous system displaces it still farther into the organism. While the biologist displaces the partition farther and farther into the organism, the physicist displaces it farther and farther into the physical object. In macrophysics he observes the properties of bodies that are expressible in terms of perceptible phenomena; in microphysics he penetrates to the molecule, the atom, and now the nucleus. The perceptible effects of microphysical entities, which are not directly perceptible, serve as instruments of observation which thereby become incorporated into the observer. After much experience with the condensation tracks produced by electrons and other elementary particles, one may come to view the perception of the tracks as an observation of the particles. It is almost as if the perceptual partition had been displaced so that it coincides with the cognitive one.

The position of the cognitive partition is of fundamental significance in quantum theory. In an observation of a microphysical quantity there occurs an interaction between object and instrument; the instrument reacts against the object and may produce an unpredictable, finite change in the value of a quantity that is canonically conjugate to the one being observed. In such observations it is not possible to control the action of the measuring instrument upon the object, for the instruments cannot be investigated while serving as means of observation. It is impossible to assign definite canonically conjugate quantities simultaneously, since the experimental arrangements through which the effects of microphysical entities are registered are mutually exclusive. The situation in microphysics is expressed by the concept of complementarity and may be illustrated by the following example. The position of a particle may be observed by allowing it to impinge on a screen, which in order to define position must be rigidly attached to the frame of reference. During the interaction between particle and screen there is an unpredictable exchange of momentum between them. The momentum given to the screen is absorbed by the frame of reference, and its value cannot be substituted in the principle of conservation of momentum to calculate the momentum given to the particle. The procedure for observing position excludes the possibility of using the screen to determine the momentum of the particle. If the screen is mobile in the frame of reference, it may be used to measure momentum, but then specification of position is lost. The observations of microphysics require interpretation in terms of classical concepts, but the fundamentally unpredictable, finite effects of the disturbances by the instruments of observation lead to a restriction in the applicability of classical concepts to microphysical objects. The cognitive partition between object and apparatus is the seat of an indeterminacy which limits theoretical physics to the statistical prediction of the results of classically interpreted experiments. The state of the object on one side of the partition is represented by a wave function that changes in accordance with a differential equation which exemplifies a principle of causality. On the other side of the partition there occurs a sequence of events which proceeds from the measuring apparatus to the observer in accordance with classical laws. The statistical interpretation of the wave function states that it is an instrument for predicting the results of observations registered with classically controlled apparatus and perceived after the manner of common experience. The quantum mechanical theory of microphysics has resulted in a limitation of the possibilities of observation.

III. Systematization

11. Classification

In empirical science particular cognitions are organized into systems of knowledge. The particular cognitions of daily life already involve order: the perception of a thing involves the hypothesis that the actual perception is correlated with possible perceptions; the concept of thing expresses a relatively invariable correlation of properties attributed to it. Science presupposes and continues the systematizing procedures of common experience.

A basic procedure for introducing order into knowledge is classification. Classification is founded on the similarities between things or events; it is based upon the fact that things are similar in specific respects and dissimilar in others. The ordering of things into classes is especially characteristic of description in sciences like botany and zoölogy, but the procedure originates in prescientific experience. The occurrence of general names in language indicates that classification is a primitive mental process. The word 'animal,' for example, shows that there has been formed a class of living things that have the power of self-movement. The employment of names to denote any one of a number of common things is evidence that particular cognitions of things have been systematized by classification.

The classifications of common experience are based upon superficial or striking properties of things, and this dependence on superficial resemblances also characterizes the early phases of science. Scientific classification, however, eventually comes to be based upon essential characters which may be discovered only by careful observation and experimentation. The ancients classified material things in terms of earth, air, fire, and water; this set of substances has been superseded by a system of elements, initially classified in terms of atomic weight but now in terms of atomic number. Physics is not ordinarily thought of as a classificatory science, yet it offers classifications such as those of spectra on the basis of quantum numbers characterizing the energy levels in the atom or molecule.

Classification is a characteristic procedure of the descriptive study of living things. The many plants and animals are classified according to the presence or absence of specific properties, functions, etc. A distinction is frequently made between artificial and natural classification. An artificial classification is based upon some superficial similarity in structure, color, habitat, etc. For example, animals may be classified as terrestrial or aquatic depending upon whether they live on land or in water. A natural classification is based upon the fundamental characters of things. The basic criterion is similarity of structure, the study of which is called morphology or anatomy, but similarity of function and behavior are also used. While classification occurs in daily life, science aims at systems of classification.

The basic classification in biology is the division of the organic world into the plant and animal kingdoms. A kingdom is generally progressively subdivided into phyla, subphyla, classes, orders, families, genera, and species. The scientific name of an organism is compounded of the names designating genus and species. Thus man is called *Homo sapiens.* He belongs to the species *sapiens*, the genus *Homo*, the family Hominidae, the order Primates, the class Mammalia, the subphylum Vertebrata, and the phylum Chordata of the animal kingdom. The natural systems of classifications of biology are designed to exhibit genetic relationships.

The principles of procedure for research on problems of classification have been set forth by Jepson, who has systematically described the flora of California. He stresses the importance of field studies of habits, life-history, soil exposure, and associated species. Field records must be made at the place and time of observation and should be validated by specimens for a herbarium. Observation of plants in their natural environment should be supplemented by experimentation with garden cultures. Jepson transplanted individuals of *Eschscholtzia californica* (California poppy) from the Great Valley to the seacoast. Characters attributed to assumed other species developed, and hence there was justified a reduction in the number of species. For the work of classification he sought the following data: (1) entire life-history of a species; (2) biogeographic status; (3) characters at the limits of the area in which it has its greatest development; (4) structure, character, and presence or absence of plant organs; (5) variation in the organs of a species, especially from one individual or from a series of individuals where these have common parentage; (6) results of multilation of an individual.

The concepts of species and genus are fundamental instruments of classification in biology, and Jepson shows how the definitions of these concepts depend upon the state of knowledge. In his work on the flora of California he lists seven species of the genus Eschscholtzia, while Fedde in *Das Pflanzenbuch* reports one hundred species in California. In Jepson the genus Ptelea is represented by one species, whereas Greene has six species for California. In his discussion of criteria for genera and species Jepson states that the species must consist of individuals bound together by intimate genetic connection as determined by the morphology, detailed structure, life-history, genetic evidence, geographic history, and ecologic status. The species should represent a natural unit, especially from the geographic standpoint, and every effort should be made to give it precise definition. Jepson states that a genus should include all species of close genetic connection which have a marked natural resemblance or are closely bound together by structural peculiarities which indicate a close line of descent or form a compact natural group. Genera so founded are sufficiently large to establish relationships on a recognizable scale and to bring out the intimate relationships which exist between floras of different regions or countries as a result of past migrations. Genera having marked characters should not be subject to a segregation which reduces the generic character to the level of a species character. It is, however, necessary that the limits of genera should with increase of knowledge of their structure be subject to revision and modification. That the characters for classification must be adapted to the specific problem, Jepson illustrates by the genus Arctostaphylos. By morphologic characters, or by biometric measurements, or by other methods determined in advance, there would scarcely be more than five or six species for California. He distinguishes twenty-five species on the basis of differences in reaction to chaparral fires; the responses are constant and fundamentally unlike and are further correlated with geographic and ecologic segregation.

A thing is characterized by a relatively invariable correlation of simultaneously existing properties. Similarity between members of a class signifies that such correlation is exemplified by each member. A classification therefore expresses general laws of correlation which, since they apply to simultaneously existing properties, express uniformities of coexistence. But, since functions are considered in classification, laws expressing correlations of properties involve those expressing correlations of events.

12. The Correlation of Events

The formulation of correlations of events is an important phase of systematization. Indeed, it has been argued that events are constitutive of reality. Examples of events are a flash of lightning, an eclipse of the sun, an earthquake, the birth of a living being. In daily life and qualitative science an event may extend through an appreciable duration, but for precision an event is idealized as the occurrence of properties at an instant. The records of events constitute the raw material for the systematizing activity of science.

A stage in the investigations of correlations of events is the determination of temporal sequences. The history of political events, historical geology, and paleontology are arrangements of events in temporal order. Science in the form of history systematizes observations of events by fitting them into schemes of development of the cosmos, life, and society. Empirical science also formulates laws which express regularities in the correlation of the general characters of events. Such laws provide a basis for prediction; in fact, their confirmation or disconfirmation depends on the occurrence or nonoccurrence of events predicted from them. On the view that events are the basic constituents of reality, laws of coexistence become special cases of laws of correlation of events.

Correlations of events in nature are generally complex, so that in science a phenomenon is analyzed into constituent elements. An example of resolution into concurrent factors is Galileo's analysis of the motion of a projectile. Neglecting the

315

resistance of the air, the projectile describes a parabolic path. Galileo analyzed the process into a superposition of two motions which may be considered independently: a horizontal motion with constant velocity and a vertical motion subject to the acceleration of gravity. The analysis of a phenomenon into successive elements is achieved by differential calculus. Kepler formulated the laws of planetary motion in terms of concepts that characterize the motion as a whole. One of his laws states that the orbits of the planets are ellipses. Newton discovered the differential equation which describes the instantaneous character of the motion. He expressed the acceleration, which is a derivative with respect to time, as determined by a force which is a function of the distances and masses of surrounding bodies. The differential analysis gives an exact formulation of the popular concept of causality.

The concept of causality expresses a relation between phenomena, such that one phenomenon is viewed as the cause of some other phenomenon, the effect. The cause is the condition of the effect; a description of the causes of a phenomenon constitutes an explanation. One may exemplify causality by illustrations drawn from all realms of experience. For example, the action of a force on a body is the cause of its acceleration; the application of a stimulus to an organism is the cause of a response. The concept of causality is best understood by an analysis of the physical concept of causality. Indeed, the hypothesis has been entertained that all natural phenomena are analyzable into physical processes. The typical physical cause is the action of force.

In the life of the individual it is probable that the concept of force is derived from experiences of his own exertions and that of his fellows. The individual exerts a force, for example, with a hand, and observes that it produces effects. A father forces a child to fill the wood box. Bodies are set in motion or brought to rest; forces can act in opposite directions and annul one another. The primitive concept of force is derived from experiences of our own exertions and of our fellows; it expresses production, creation, generation, efficacy.

Since the concept of force was originally derived from the forces exerted by an individual, it is understandable that primitive attempts at explaining natural phenomena were in terms of the theory of animism. Preceding the origin of empirical science men invented animistic explanation of natural phenomena. Natural bodies were interpreted to be the abodes of principles of life; for example, the lodestone, which attracts iron, was interpreted to be the seat of a soul. It is to be recognized that animism was an expression of the demand for causal explanation. Scientific progress began when men directed their attention to the perceptible properties of things, observed correlations of properties and sequences of phenomena, and explained the behavior of things in terms of the causal efficacy of other natural things. Animism did not, however, disappear with the development of science. Thales, the first historical figure of Greek science stated, "All things are full of gods." Empedocles conceived of the forces of attraction and repulsion as instances of love and hate. In modern times Kepler thought of the planets as guided in their orbits by angels, and it may be contended that vitalism in biology is a relic of animism.

Comte has explained in his law of the three stages that every science starts at a theological stage, passes through a metaphysical one, and then achieves final form in a positivist stage. The several sciences are at different stages of development. The operational point of view in physics is recognition that physical science has most completely realized the final form; this may be exemplified by the content of the physical concept of causality. Physical causality may be illustrated by the Newtonian theory of the motion of a body in a gravitational field. In this theory material bodies are conceived to have the physical property, mass. Between two material bodies there is a force of attraction which varies directly as the product of the masses and inversely as the square of the distance between them. Fixing our attention upon one of the bodies, we may say that it is acted upon by a force which is exerted by the other body, or, more precisely, that the first body has an acceleration which depends upon the distances and masses of surrounding bodies. Thus one

317

seeks the causes of a phenomenon pertaining to a given body in the correlated states of other bodies. The cause of the acceleration of the body is described not in terms of a vital principle within the body but in terms of the observable properties of other bodies. In the seventeenth century the exertion of a force was viewed as a state of activity of one body that produced an acceleration in another body. Causality was thus interpreted to be the expression of power, efficiency, production, necessary connection. The concept of causality was then in a metaphysical stage which retained traces of animism.

The concept of efficient causality was criticized by Hume. He analyzed the causal process in the collision of two billiard balls. If a moving ball strikes a ball of equal mass which is at rest, the first ball is brought to rest, while it communicates its state of motion to the second ball. We say that the motion of the first ball is the cause of the motion of the second ball, which is the effect. In observing this process, Hume found only a sequence of phenomena; the state of motion of the one ball was succeeded by the state of motion of the second ball. On close observation one would observe that as the two balls come in contact there is a deformation of the surfaces of contact and then a recovery from the deformation, during which process the first ball is brought to rest and the second ball moves away with the original state of motion of the first ball. At no time while the balls are in contact can one see why the process must occur as it does and not in some other way. Prior to the observation of a collision of two balls one could not predict the outcome of the collision. According to Hume, the concept of efficient causality merely expresses a uniformity of sequence of phenomena.

The concept of causality as correlation was definitely formulated by John Stuart Mill in his canons of induction. Mill's canons are exemplified by the method of concomitant variations which states that whatever varies in any manner whenever another phenomenon varies in some particular manner is either a cause or an effect of that phenomenon, or is connected with it through some fact of causation. Mill's canons offer criteria by which one can determine whether or not there is a causal relation between specific phenomena. The mathematical form of causality purges it of the last vestiges of efficiency. Causal laws are stated as functional relations between numerical measures of variable quantities.

The mathematical expression of causal laws may be illustrated by the Newtonian theory of gravitation. If two bodies having masses m_1 and m_2 are at a distance r, each of these bodies will exert a gravitational force upon the other which is directly proportional to the product of the masses and inversely proportional to the square of the distance between them. Let us fix our attention upon m_1. The physical description of the phenomenon is that m_1 is accelerated by a force which is exerted by m_2. The mathematical expression of the law of motion is given by the differential equation

$$m_1 \frac{dv_1}{dt} = G \frac{m_2 m_1}{r^2} .$$

One should also add that, while m_2 is exerting a force upon m_1, the latter is exerting an equal and opposite force upon the former. The causal process in the system consisting of the two bodies is a mutual action.

A special case is that in which m_2 is the mass of the earth and m_1 is the mass of a freely falling body. For the region in the neighborhood of the surface of the earth we may assume that the factor Gm_2/r^2 is a constant g. Let s be the distance measured downward from a point above the surface of the earth. Since $dv_1/dt = d^2s/dt^2$, the differential equation becomes

$$m_1 \frac{d^2s}{dt^2} = m_1 g \qquad \frac{d^2s}{dt^2} = g .$$

One can integrate this equation and express the coordinate s and the speed as functions of the time and two arbitrary constants. Thus

$$s = s_0 + v_0 (t - t_0) + \tfrac{1}{2} g (t - t_0)^2 .$$

The constants s_0 and v_0 are the values of the distance and the speed at an initial time $t = t_0$. Hence, if one knows the position

319

and speed at an initial time, one can calculate the position and speed at some past or future time.

Hume characterized cause and effect as contiguous in time. If in our example the force exerted by m_2 is the cause and the acceleration of m_1 is the effect, it appears that cause and effect are simultaneous, for the acceleration is simultaneous with the force. We could, however, say that the cause precedes the effect if, after a body were introduced into the vicinity of m_1, it would take a finite time for the gravitational force to begin to act on m_1. But, according to the Newtonian theory, gravitational attraction is propagated with an infinite speed. Hence we can find no meaning for the statement that the cause precedes the effect. Again, contiguity of cause and effect in space means that the body exerting the force is in contact with the one accelerated. On the Newtonian theory of action at a distance spatial contiguity is also lost. In the present example discontinuity in causal action gives rise to an ordinary differential equation for the motion. The contiguity of a causal process in space and time is exemplified by wave motion in an elastic medium. If a portion of the medium is set in vibration, the state of motion is communicated by contiguous action to neighboring parts and travels with a finite speed. The space-time contiguity of the process is represented by a partial differential equation for the displacement as a function of the space coordinates and the time. The causal factor is represented in the differential equation by the dependence of the velocity upon the density and elasticity of the medium.

The search for causes is guided by a principle of causality. In qualitative terms the principle is expressed by the proposition, 'the same cause produces the same effect.' For example, if for a system subject to a definite law of force the same initial conditions are realized at some other time and in some other place, the motion will be the same. Suppose that I drop a body from a certain height above the surface of the earth; a specific motion will occur. The principle of causality predicts that if at some other time and place a body is dropped from the same height, the motion will be similar to the first one.

The principle of causality thus appears to assert the existence of causal laws that are independent of time and place. This would imply that causal laws are significant because the order of nature is one in which certain patterns in phenomena recur. In view of the universal interrelatedness of things, however, it would appear that the state of the whole universe is the condition of every phenomenon. But since the state of the universe apparently never recurs, we could not observe the repetition of a specific process in some other place at another time. At first glance it would appear that the principle of causality is empty.

The answer to the foregoing objection is that it is possible progressively to isolate systems and processes. In a mechanical experiment on the surface of the earth it is possible in practice to ignore the influence of the heavenly bodies. In general, the universe is a set of loosely coupled systems. Physical forces usually vary inversely as the second power of the distance and therefore decrease rapidly with distance. For ordinary accuracy the physicist may view his laboratory as an isolated system. Against a background which may be considered constant, the same relative initial conditions give rise to similar processes. Causal processes can be isolated by controlling the state of the environment.

13. Successive Approximation

The control of the conditions under which a causal law is exemplified proceeds by successive approximation. In preliminary experiments one must presuppose that the conditions are constant and thereby obtain an approximate law. With the aid of approximate laws one can then define the conditions of an experiment more precisely, or correct for the disturbing influences, and thus determine a law to a higher order of approximation. Galileo's first experiments on the acceleration of a ball rolling down an inclined plane were performed with crude apparatus. After the laws of dynamics were discovered from the action of gravity, it was possible to define the conditions of the experiment more accurately and indeed to explain how Galileo's experiments were disturbed by friction.

Examples of the method of successive approximation are offered by contemporary physics. It is incorrect to think that the theory of relativity and the quantum theory have destroyed classical physics. Classical theory has to be assumed to a first approximation in order to define the experimental conditions under which relativity holds to a higher approximation. Thus in his first paper on the special theory Einstein begins the systematic discussion with the statement, "Let us have given a system of coordinates, in which the equations of Newtonian mechanics hold to the first approximation." The formulation of the postulates of the theory which corrects classical dynamics requires a system of coordinates which is defined by the condition that the classical theory holds in it to the first approximation. In order that the foregoing procedure may apply, the more precise theory must contain the first approximation as a limiting case. In the special theory of relativity the length and mass of a body are functions of its speed; for speeds that are small in comparison with the speed of light one obtains the classical assumption that length and mass are independent of speed. The classical dynamics is a limiting case of relativistic dynamics.

According to the general theory of relativity, the path of a ray of light is curved in a gravitational field; as is well known, this prediction has been verified by the measurement of the deflection by the sun of light from a star. The observations and calculations whereby this result is verified presuppose, however, that the law of the rectilinear propagation of light holds in a region near the surface of the earth. The experimental procedure is justified because the earth's gravitational field is relatively small, and in the limiting case of zero gravitational fields relativistic theory reduces to the limiting assumption that light travels in a homogeneous medium in Euclidean straight lines.

Classical physics is a necessary basis for quantum physics. Quantum phenomena occur under macrophysical conditions and are measured by apparatus which registers results that are interpreted in terms of classical concepts. The employment of classical physics in the control of microphysical phenomena is justified by the fact that the classical laws are limiting cases of the quantum laws. Since observations of microphysical phenomena are subject to statistical laws, the causal laws of classical physics render possible the definition of experimental conditions in which causality fails.

On account of the complexity of organisms, it is difficult to control the conditions of a process in a single living system. The biologist therefore uses a procedure, called comparative experimentation by Claude Bernard, in which another system serves as a control. Suppose, for example, that it is desired to determine the effect of modification or removal of a deep-seated organ of an animal. Such an experiment requires an operation which disturbs neighboring organs. In order to distinguish between the effect of the operative procedure and that of disturbing the specific organ, the biologist performs a similar operation on a similar animal, but without disturbing the organ under investigation. A controlled experiment requires comparison of the results upon at least two organisms, under conditions which are the same in each experiment except in one respect. Comparative experimentation may use two animals of the same species, or the same animal at different times or different parts of one animal at the same time.

Controlled experimentation in biology is facilitated by the use of cultures. The problem of plant nutrition is the determination of the substances necessary for the structural composition and metabolism of the higher plants. Experimental procedure consists in placing the roots of the plant in a water culture, which provides a controllable external medium from which the plant can absorb the nutrient substances. The nutritional requirements of all sorts of plants have been discovered by varying the kinds and amounts of salts in the medium. An important procedure for physiology is the cultivation of fragments of tissues and organs of animals outside of the organism. Dr. Carrel deposited small fragments of living tissues in fluid plasma or in an artificial medium. The maintenance of life in such culture mediums renders possible controlled experimentation on living processes.

14. Successive Definition

With the development of science there has been discussed the problem of the genesis of scientific laws. At first sight it appears to be evident that the laws of empirical science are generalizations from observation. But laws serve to define concepts, and observation employs laws as principles of interpretation. In the spirit of Kant, who taught that the category of causality is an a priori condition of the possibility of experience, Dingler holds that the principles of physics are postulates which ought to be chosen in accordance with a principle of simplicity. In the present work a solution of the problem is offered by the theory of successive definition. According to this theory, the status of a natural law may change in the development of science. A law which originates as a generalization from experience may be transformed into a convention that expresses an implicit definition of the concepts it involves. The theory of successive definition has been anticipated in the discussions of arithmetic, geometry, time, and weight. I shall now use classical dynamics for a more detailed explanation. Some concepts must be initially assumed as understood; I shall assume that, in addition to the concepts of geometry and time, we have a statical concept of force. Examples of forces are weight and the force exerted by a stretched spring. If a body is supported at rest by a stretched spring, the interpretation of the phenomenon in statics is that the weight acting downward is balanced by the equal and opposite upward force exerted by the spring. If the body is released, it falls under the action of its weight, which is assumed to be the same as when the body is at rest. Similarly, if a spring is stretched a given distance, it is assumed to exert the same force whether it is accelerating a body or is balanced by some other force.

Let us now apply forces to various bodies. It is found that the velocity increases in the direction of the force. In order to describe the action of force, we may assign to a body in motion the physical quantity momentum, *I,* which may be defined by the postulates that a body at rest has zero momentum and that the momentum communicated to a body by a constant force is proportional to the product of force and time. Experiment shows that for velocities small in comparison with the velocity of light the momentum is directly proportional to the velocity. We may introduce a factor of proportionality *m* and write *I = mv.* The usefulness of our definition of momentum depends on the fact that regardless of the kind of force employed to generate momentum in a given body the factor *m* is the same. Hence the physical quantity *m,* the mass, is viewed as an intrinsic property of the body, which for classical dynamics is independent of velocity, but for relativistic dynamics depends on velocity. The outcome thus far is that we have defined momentum in terms of an impulse equation and have transformed the empirical law of the proportionality between momentum and velocity into a definition of mass.

The foregoing dynamical example illustrates the fact that generalizations from experience become definitions of new concepts. I cite some further examples. The law expressing the dependence of the stress in an elastic body upon the strain becomes a definition of elastic constant. The law which states the functional relation between the length of a wire and its temperature becomes the definition of coefficient of linear expansion. Electrical resistance and the constant in the general gas law are other examples of constants that are defined in terms of empirical laws which have been transformed into definitions. A related example of the transformation of empirical laws into definitions is given by the principle of the conservation of energy. In dynamics it is possible to define energy so that in isolated ideal systems it is constant. On the basis of experiments which showed that a definite amount of heat can be produced by the performance of a definite amount of work, the mechanical principle was extended to a general principle for all natural processes. As Poincaré noted, this principle, which was initially an empirical hypothesis confirmed by experiments demonstrating the mechanical equivalence of heat, has acquired the status of a definition. If in a physical process the total change of known forms of energy is not zero, a new kind of energy is assumed in order to preserve the principle.

Further study of dynamics reveals an alternating procedure in successive definition. We initially adopted force from statics, but, having determined how to assign masses to bodies, we may calculate momentum from $I = mv$ and define force in the more general sense by Newton's equation of motion, which states that force is equal to the rate of change of momentum. The usefulness of this definition depends on the fact that simple laws of force may be found in important applications, for example, the inverse square law for gravitation and electrostatics, and Hooke's law which states that in elastic bodies stress is proportional to strain. Thus a limited concept may provide the basis for a law which is used to define a new concept. Then the law may be employed with the new concept to define a general concept which replaces the original limited one. In the foregoing example the equation of motion was used alternately to define momentum and force. The fundamental laws of dynamics have become conventions which implicitly define the concepts that they involve. With C. I. Lewis we view the principles of fundamental sciences like geometry and dynamics as definitions of concepts which serve to interpret the data of observation. It must be recognized, however, that the suitability and applicability of a conceptual scheme are the subject matter of hypotheses which must be confirmed or disconfirmed by observation.

15. Atomism

Thus far we have studied procedures for the systematization of cognitions of directly observable things and phenomena. I shall now consider systematization from a more detailed and constructive point of view. Perceptible phenomena can be analyzed into nonperceptible processes; empirical laws can be derived from more ultimate laws. In physics the transition is from a macrophysical level to a deeper microphysical one.

The more ultimate point of view may be called atomism in a general sense. Atomic theory vividly illustrates the method of hypothesis.

The fundamental ideas of an atomic theory were created by the early Greek philosophers. As Meyerson has so clearly shown, the mind seeks identities in the natural world. This disposition expressed itself in early Greek science in the endeavor to interpret natural things as modes of a permanent substance. Attempts to provide for change in a theory of substance led to the atomic theory of Leucippus and Democritus. According to atomism, substance consists of countless atoms in the void, and these constitute natural bodies. Natural phenomena consist of changes in the groupings of atoms. Thus the theoretical demand for persistent substance is reconciled with the acknowledgment of the reality of change. It is unfair to state that the ancient atomic theory was only speculative. As is exemplified in the exposition of Lucretius, the Greeks explained many phenomena qualitatively in terms of atomism.

Quantitative development of the atomic hypothesis began when it was used by Dalton to explain the laws of chemical combination in the opening decade of the nineteenth century. Chemical processes are exemplified by the rusting of iron upon exposure to the atmosphere and the decomposition of water into hydrogen and oxygen by electrolysis. Such processes are characterized as composition and decomposition of substances. The criteria of substances are initially qualitative properties like color, odor, hardness, but are ultimately physical properties such as density, melting-point, boiling-point, specific heat, electrical conductivity, spectrum, etc. Thus the procedures for observing physical properties are basic to chemistry. The chemist, furthermore, has characteristic techniques for separating substances in physical mixtures by using differences in boiling-point and solubility, by filtering, etc. He employs methods such for producing chemical changes as heating, mixing solutions, electrolysis. In chemical phenomena the physical properties of the final substances are usually quite different from those of initial ones. Chemistry originated in antiquity with procedures of cooking, the working of metals, the tanning of leather; and the study of chemical phenomena was furthered by the alchemists' attempt to transform baser metals into gold. But the quantitative development of chemistry only began in earnest during the closing decades of the eighteenth century, when Lavoisier demonstrated the conservation of mass in chemical reactions. The use of the chemical balance to weigh materials which are mixed or produced in a reaction renders possible the quantitative formulation of an atomic theory.

The atomic theory of chemistry is based upon three fundamental laws: the law of conservation of mass, the law of definite proportions, and the law of multiple proportions. The foregoing laws, which are derived from chemical experiments, are readily explicable by the atomic theory in which the masses of individual atoms are assumed constant and the same for a given element. The constancy of atomic masses provides a basis for the conservation of mass. In recent years it has been discovered that there exist isotopes, that is, atoms of the same element possessing different atomic weights. It may also be noted that in order to extend the law of conservation to nuclear reactions the relativistic equivalence between mass and energy must be invoked. The hypothesis that the molecules of a substance are always composed of the same kinds of atoms explains the law of definite proportions. The hypothesis that molecules may be formed of different numbers of the same kinds of atoms explains the law of multiple proportions.

Atomism in the theory of matter furnished a basis for an atomic theory of electricity. Faraday discovered the laws of conduction of electricity through electrolytes. An explanation of these laws is immediately given by the hypothesis that in an electrolyte molecules are dissociated into positively and negatively charged ions which carry integral multiples of a unit of electric charge.

Further progress in atomism was made by the kinetic-molecular theory of heat. This theory was based upon the discovery of the mechanical equivalence of heat, which provided the empirical foundation for the hypothesis that the heat content of a body consists of molecular mechanical energy. The kinetic theory of matter has been especially developed for gases. In this theory a homogeneous gas consists of countless molecules, similar in mass, and moving with high speeds. In a collision between molecules there is conservation of momentum and energy. In order to make calculations from the hypotheses, it is necessary to use statistical assumptions and methods. It is the statistically defined quantities, however, that are significant for observable phenomena; an important part of the theory consists of assumptions correlating the statistical quantities with experimentally measurable quantities such as pressure and temperature. For example, the pressure of a gas is assumed to be equal to the time average of the total transport of momentum to unit area per unit time. That is, the pressure indicated by a manometer is the resultant force per unit area produced by the reflection of the molecules by the boundary of the gauge. The temperature of a gas is assumed to be proportional to the time average of the kinetic energy per degree of freedom. On simplifying assumptions it is possible to deduce the general gas law and other laws for gases. Thus the observable, macrophysical properties of gases are explained in terms of the action of microphysical bodies.

Another advance in atomic theory was the molecular explanation of the Brownian movement. The success of the atomic theory in chemistry and in the theory of gases failed to convince notable doubters. The kinetic picture was viewed as a fiction which served only as an instrument for economical thought. Atomic weights were merely combining ratios and not significant of realities. The doubters were practically all silenced, however, by the molecular explanation of the Brownian movement. If colloidal particles suspended in a liquid are viewed through an ultramicroscope, the particles are observed to move in an irregular and random manner. The kinetic explanation is that the particles are sufficiently light so that they are appreciably affected on impact with a molecule. The irregular, zigzag motion of the particles is caused by irregular bombardment by the molecules of the liquid. In observations of the Brownian movement, one therefore observes the effects of a single molecule.

The present illustration shows that a principle of contiguous causality is a factor in the observation of physical objects. The properties of nonperceptible entities are inferred from the behavior of perceptible bodies in collisions which are assumed to satisfy principles of the conservation of momentum and energy. The molecule must be assumed in order to interpret perceptible phenomena in accordance with accepted physical principles. Thus the physical existence of molecules, atoms, electrons, radiation, etc., is like the existence of the physical properties of perceptible bodies. If a colloidal particle exists, then so does the molecule which interacts with it in conformity to physical principles. If a zinc sulphide screen exists, then so does the alpha particle which excites it to scintillate. The functional relations which are expressed by the laws of nature relate the physical properties of perceptible bodies to the physical properties of microphysical entities. The physical world is the object of a hypothetico-deductive system which is assumed in order to interpret the results of observation—initially to interpret a datum of perception as an aspect of a body to which one attributes mass, temperature, electric charge, etc., on the basis of its behavior; further to interpret phenomena such as Brownian movements, scintillations, and condensation tracks as the effects of microphysical entities; and then to express macrophysical processes as statistical resultants of microphysical processes.

The development of atomism is one of increasing detail in its representation of physical objects. Dalton failed to distinguish between atoms and molecules. In simplified kinetic theory the volumes of molecules are neglected and also the forces between them, except in impact. The atoms were initially assumed equal in mass, but isotopes were discovered. Electric phenomena in gases led to the discovery of the electron, and Rutherford's discovery of the nucleus provided a basis for the picture of the atom as a positive nucleus surrounded by shells of electrons. A first clue to the structure of the nucleus was given by radioactivity, which was interpreted as the disintegration of unstable nuclei and the correlated emission of alpha particles, beta particles, and gamma rays. In recent years instruments have been invented that will communicate to electrified particles energies which are high enough to shatter the nucleus. At present the nucleus is assumed to be constituted of protons and neutrons.

The preceding discussion shows that the concept of atom is relative to experimental procedure. The generic meaning of the term is something that cannot be divided, but the concept of indivisibility is relative to the instruments employed. The chemical atom of the nineteenth century was not divisible by ordinary chemical methods, but its outer structure was shattered by electric discharges, and now its inner structure is being smashed by the projectiles from powerful atomic guns.

In the history of physics theories of continuity have competed with atomism. The Cartesian vortices of the seventeenth century, the nineteenth-century ether models of the atom, and recent electromagnetic field theories of the electron testify to the scientific impulse to reduce apparent discontinuity to continuity. In a field theory of matter the fundamental physical quantities are those that characterize the electromagnetic field in space-time. Regions of high value of the field are interpreted to be electrified particles, the inertial mass of which consists of the energy of the surrounding field. Since the field is continuous, the separateness of the electrified particles is only an appearance. Despite attempts to build theories of matter by modifying Maxwell's equations and by generalizing the theory of relativity, theories of continuity are not playing a vital role in contemporary physics.

16. Statistics in Quantum Theory

The discussion of atomism in physics may be supplemented by an exposition of the basic significance of statistical laws in quantum theory. The wave function that represents the state of a microphysical system satisfies a differential equation which expresses determinism. But the results of observations on quantities characterizing a system are subject to statistical laws. Statistical laws, which in classical kinetic theory are founded on determinism, are fundamental in quantum theory.

The statistical laws of quantum theory are exemplified in observations on the state of polarization of light. If ordinary light enters a Nicol prism, the emergent beam consists of plane-polarized light. Since light exhibits corpuscular properties, the beam is conceived to consist of photons which are in a state that is defined by the plane of polarization. An observation on the state of polarization of light is performed by allowing the beam to enter a crystal of tourmaline. If all the light passes through, it has been observed to be plane-polarized in a plane perpendicular to the optic axis of the crystal. If no light passes through, its plane of polarization has been observed to be parallel to the axis. If the fraction of the original beam that passes through is expressed as $\sin^2 a$, the plane of polarization has been observed to be inclined at an angle a to the axis of the crystal. For a constant experimental arrangement the results of an experiment are reproducible. The effects registered in observation are produced by the mass effect of many photons in accordance with the statistical law that a fraction $\sin^2 a$ of the total number passes through and a fraction $\cos^2 a$ is absorbed. The result of an experiment with a single photon, however, is not predictable with certainty.

In order to describe the individual experiment, I shall, in conformity with Dirac, represent the state of polarization perpendicular to the optic axis by ψ_1 and the state parallel to it by ψ_2. The state inclined at an angle a, and represented by ψ_0, may then be expressed as formed by the superposition of states ψ_1 and ψ_2. If c_1 and c_2 designate the weights of the component states, we may write $\psi_0 = c_1\psi_1 + c_2\psi_2$. The action of the tourmaline in an observation on the photon in state ψ_0 is to force it to jump unpredictably into state ψ_1 or ψ_2, and the probabilities are respectively proportional to the squares of the weights. The probability of jumping into state ψ_1 and passing through is $\sin^2 a$, and the probability of jumping into state ψ_2 and being absorbed is $\cos^2 a$. If the photon is initially in state ψ_1, the result of an observation may be predicted with certainty.

The general formulation is as follows. Of any atomic system one may say that it can be prepared so that it is in a determinate state. The preparation is carried out with the aid of deterministic macrophysics. The quantum theory assumes that whenever a system is in a determinate state it can be regarded as formed by the superposition of two or more states. In consequence of this superposition the results of observation are in general not predictable with certainty. There is a finite probability that a result will be observed which is characteristic of one component state, and a finite probability that the result will be characteristic of another component state. If the system is initially in one of the component states, the result of an observation is predictable with certainty.

17. Atomism in Biology

The systematization and explanation introduced by atomism are exemplified in biology. The phenomena of heredity are explained by a theory of genes which is analogous to atomic theory.

The concept of heredity expresses the empirical generalization that organisms are similar to their parents. The scientific study of heredity presupposes an analysis which distinguishes different characters of an organism, so that a complex process may be analyzed into its constituents. In organisms created by sexual union traits arise that are combinations of those of their progenitors.

The fundamental law of heredity was formulated by Mendel as a result of his experiments on plant hybridization. He studied in varieties of garden peas the inheritance of selected characters such as length of stem, whether tall or short; form of ripe seeds, whether round or wrinkled; color of food material within the seeds; etc. Mendel crossed two forms having distinct characters and counted the number of descendants in successive generations possessing one or the other of these characteristics. He first determined that the selected characters were constant for certain varieties, a procedure analogous to the isolation of pure substances in chemistry. He then crossed varieties having different characters and observed the offspring. When a tall variety was crossed with a short one, the offspring in the first filial generation were all tall. If these individuals were bred among themselves, the result was a ratio of three tall to one short. This ratio was a consequence of the fact that in the second generation the proportions of pure tall, impure tall, and pure short were 1:2:1. Correns crossed a white-flowered variety of Mirabilis, the four o'clock, with a red-flowered one and obtained blended, pink hybrids. The offspring of the pink flowers were white, pink, and red in the proportions 1:2:1. The Mendelian law of heredity is analogous to the law of definite proportions in chemistry.

The explanation of Mendelian inheritance is based upon the constitution of the germ cell. The occurrence of pure dominants and pure recessives from hybrid parents is explained by the hypothesis that the germ cells carry only specific unit characters, for example, the factors for red or white flowers. Since all germ cells are pure with respect to a type of character, the hybrid offspring of parents having contrasting characters will produce in equal numbers germ cells bearing the dominant character and those bearing the recessive character. The chance combination of these two classes of male and female cells will yield the typical Mendelian proportion: 1DD, 2D(R), 1RR. The law of heredity therefore has an atomistic basis similar to that for chemistry.

The explanatory stage of systematization in the science of heredity is devoted to the factors in the germ cells which determine traits or characters. Within the nucleus of the cell occurs a substance called chromatin, which takes the form of threads, or chromosomes, preparatory to cell division. The chromosomal theory of heredity is that the determinants of particular traits, called genes, are located in linear series in the chromosomes. By the use of special fixing and staining methods and observation under a microscope it may be observed that the chromosomes contain distinguishable structures called chromomeres. John Belling set forth the parallelism between the serial arrangement, structure, attraction between homologous entities, rate of division, and number of chromomeres, and the corresponding properties of the genes. Belling concluded that correct scientific procedure demands the adoption, as a working hypothesis, of the assumption that chromioles and chromomeres are genes, doubtless with more or less of an envelope.

Confirmation of the preceding hypothesis has been obtained by a study of the large chromosomes in the salivary gland of *Drosophila melanogaster*. Painter and his co-workers separated the elongated threads within the nuclear wall and observed on the chromosomes a great variety of bands, some broad and deeply staining, others narrow or made up of a series of dots. The patterns of bands and lines were characteristic of a given element, and it was possible to recognize the same element in the nuclei of different individuals. Irradiation with X-rays produced changes in the structure of a chromosome; there occurred translocations of elements, deletions or breaks, and these were observed under the microscope. The correlation of a short deletion with the absence of a trait fixed the location of the corresponding gene. By a genetic study of the characters transmitted, and a cytological study of the changes in the structure of the chromosomes, it was possible to correlate traits with position and thereby to locate the genes.

IV. Conclusion

18. Unity of Science

The present monograph on the procedures of empirical science may properly be concluded with an appraisal of the prospects for a unified science.

A basis for unity is the circumstance that the initial subject matter of the several sciences is furnished by common experience. The first objects of empirical science are things and phenomena given in perception, space-time structures whose properties are described in terms of perceptible qualities and relations. The motion of bodies, the behavior of organisms, and the relations between societies are initially described by concepts abstracted from perception. A body whose law of fall is investigated in physics may be characterized as extended, heavy, smooth, and gray. Gold is described as heavy and yellow. Living organisms are characterized by growth, reproduction, and death. Carnap has shown how the symbols for such perceptible things and properties may be used to construct a language for science.

Unity is introduced into science by the fundamental position of physics. In generalized physics there are fashioned precise concepts of space and time which serve as instruments for the description of the general properties of matter and energy in specialized physics. General and specialized physical concepts are instruments for the description of chemical, biological, behavioristic, and social phenomena. The chemist identifies substances by their physical properties, such as density, specific heat, boiling-point, characteristic spectrum. The biologist studies the chemical constitution and reactions of matter constituting a living organism, investigates the exchanges of energy between organism and environment, and interprets phenomena of the nervous system to be electrical. In order to study the responses of an organism to its environment, the behaviorist subjects it to experiments with physically controlled apparatus. It therefore appears that the methods of observation and experimentation of physics constitute a unified procedure for science. Science starts with the perceptions, analyses, and operations of daily life, and progressively applies the measuring rods, clocks, balances, thermometers, and ammeters of the physicist in pursuit of precise data. Claude Bernard especially emphasized the significance of physico-chemical techniques for biology. The procedures of specialized physics, however, must be applied with discrimination in investigations of organisms. New methods must continually be devised to solve new problems in physics. It is therefore to be expected that the special techniques of physics will have to be appropriately modified on application to living processes. It is an open question whether or not all natural laws are reducible to the laws of specialized physics. However, since the phenomena of science are spatio-temporal processes, one must recognize with the doctrine of physicialism that the procedures of generalized physics are the basis of all empirical science. In this unity of procedure resides the unity of science.

One may further ask whether or not it is possible to construct a single deductive theory for all science. Rationalists in the past have sought to unify all knowledge by a single principle, but the logical study of deductive systems has shown that a set of postulates is required for a fruitful theory. It may therefore be doubted that a limited set of principles will systematize both physical and biological laws. In agreement with Bohr, Heisenberg has stated[4] that new phenomena require new concepts and laws for their description and has stressed the relatively closed character of systems for different realms of phenomena.

In considering the possibility of a single system of science, it is instructive to review the progress toward systematic unity in physics. In the seventeenth century Descartes sought to reduce all the phenomena of the material world to matter in motion. A great impetus to a mechanical theory of nature was given when Newton systematized the knowledge of gravitational phenomena, expressed in Galileo's law of falling bodies and Kepler's laws of planetary motion, by the application of a law of gravitation to the laws of dynamics. Huyghens provided a basis for the phenomenon of light by a material ether, while Newton proposed a corpuscular theory. During the nineteenth century elastic solid theories of the ether were developed, and heat was reduced to the mechanical energy of the systems of molecules into which perceptible bodies were resolved. Maxwell created the electromagnetic theory of light with the aid of Faraday's theory of a medium and attempted to explain it in terms of the mechanical properties of an ether. Thus unity in physical theory was sought on the basis of a mechanical theory of nature. But the mechanical theory no longer provides a basis for systematic unity in physics. Electrodynamic theories have been proposed which make electric charges and their electromagnetic fields fundamental. Since the advent of the general theory of relativity, in which the force of gravitation is replaced by space-time curvature, theories have been developed which reduce physical quantities to characteristics of curved space-time. Quantum mechanics had to be developed in order to systematize observations on microphysical processes. In the face of apparent disunity, developments in contemporary physics inspire the hope that quantum mechanics and the theory of relativity may be united in a single theory. And because of the basic function of generalized physics and the ever increasing development and adaptation of the techniques of specialized physics, the progress of physics toward unity augurs well for the unity of all empirical science.

NOTES

1. R. Carnap, *The Unity of Science* (London, 1934).
2. Cf. C. W. Morris, *Logical Positivism, Pragmatism, and Scientific Empiricism* (Paris, 1937), p. 35.
3. Quoted by Harold Chapman Brown, "Intelligence and Mathematics," in *Creative Intelligence*, ed. John Dewey (New York, 1917).
4. W. Heisenberg, *Wandlungen in den Grundlagen der Naturwissenschaft* (Leipzig, 1935).

Selected Bibliography

BELLING, JOHN. *The Ultimate Chromosomes of Lilium and Aloe with Regard to the Numbers of Genes.* "University of California Publications in Botany," Vol. XIV, No. 11. Berkeley, 1928.

BERNARD, CLAUDE. *Experimental Medicine.* Trans. HENRY COPLEY GREENE. New York, 1927.

BOHR, N. *Atomtheorie und Naturbeschreibung.* Berlin, 1931.

BROWN, HAROLD CHAPMAN. "Intelligence and Mathematics" in *Creative Intelligence.* Ed. JOHN DEWEY. New York, 1917.

BRUNSWIK, EGON. *Wahrnehmung und Gegenstandswelt.* Leipzig and Wien, 1934.

CAMPBELL, N. *What Is Science?* London, 1921.

CARNAP, R. *Physikalische Begriffsbildung.* Karlsruhe, 1926.

———. *The Unity of Science.* Trans. M. BLACK. London: Kegan Paul, 1934.

———. "Testability and Meaning," *Philosophy of Science*, October, 1936, and January, 1937.

DIRAC, P. A. M. *Principles of Quantum Mechanics.* Oxford, 1935.

HEISENBERG, W. *Wandlungen in den Grundlagen der Naturwissenschaft.* Leipzig, 1935.

JEPSON, W. L. *A Manual of the Flowering Plants of California.* Berkeley, 1925.

Selected Bibliography

LENZEN, V. F. *The Nature of Physical Theory.* New York, 1931.

———. *Physical Causality.* "University of California Publications in Philosophy," Vol. XV. Berkeley, 1932.

———. *The Schema of Time.* "University of California Publications in Philosophy," Vol. XVIII. Berkeley, 1936.

———. "The Partition between Physical Object and Observer," *American Physics Teacher,* Vol. V (1937).

MACKENSEN, OTTO. "Locating Genes on Salivary Chromosomes," *Journal of Heredity,* Vol. XXVI (1935).

MORE, L. T. *Isaac Newton.* New York, 1934.

MORRIS, C. W. *Logical Positivism, Pragmatism, and Scientific Empiricism.* Paris, 1937.

NEURATH, OTTO. *Empirische Soziologie.* Wien, 1931.

NOÜY, LECOMTE DU. *Biological Time.* London, 1936.

PAINTER, THEOPHILUS S. "Salivary Chromosomes and the Attack on the Gene," *Journal of Heredity,* Vol. XXV (1934).

REICHENBACH, H. *Philosophie der Raum-Zeit Lehre.* Berlin and Leipzig, 1928.

RUSSELL, B. *Introduction to Mathematical Philosophy.* London, 1920.

WHITEHEAD, A. N. *The Principle of Relativity.* Cambridge, 1922.

WOLF, A. *Essentials of Scientific Method.* London, 1928.

Principles of the Theory of Probability

Ernest Nagel

Principles of the Theory of Probability

Contents:

PAGE

I. The Materials for the Study of Probability
1. Introduction 343
2. Development and Applications of the Theory of Probability 347

II. The Calculus of Probability and Its Interpretations
3. Preliminary Distinctions 359
4. Fundamental Ideas of the Frequency Interpretation of Probability 361
5. Fundamental Theorems in the Calculus of Probability 368
6. Nonfrequency Interpretations of Probability Statements 386

III. Unsettled Problems of General Methodology
7. Logical Problems of the Frequency Interpretation of the Probability Calculus 393
8. Probability and Degree of Confirmation or Weight of Evidence 402
9. Concluding Remarks 417

Selected Bibliography 421

Principles of the Theory of Probability
Ernest Nagel

I. The Materials for the Study of Probability

1. Introduction

The daily affairs of men are carried on within a framework of steady habits and confident beliefs, on the one hand, and of unpredictable strokes of fortune and precarious judgments, on the other. Our lives are not filled with constant surprises, and not all our beliefs are betrayed by the course of events; nevertheless, when we examine the grounds even of our most considered actions and beliefs, we do not usually find conclusive evidence for their correctness. We undertake commercial or scientific projects, although we do not know whether illness or death will prevent us from completing them; we plan tomorrow's holiday, although we are uncertain what weather tomorrow will bring; we estimate our budget for next year, although we are not sure whether the consequences of floods, droughts, or wars will not seriously throw it out of balance. In spite of such uncertainties, we manage to order our lives with some measure of satisfaction; and we learn, though not always easily, that, even when the grounds for our beliefs are not conclusive, some beliefs can be better grounded than others. Our claims to knowledge may not be established beyond every possibility of error, but our general experience is warrant for the fact that even inconclusive arguments may differ in their adequacy.

These observations are commonplaces. But they immediately lose their triviality if, by setting them in the context of a penetrating comment of Charles Peirce, we extend them to the procedures and conclusions of the various special sciences. The American logician once remarked that in the exact sciences of measurement, such as astronomy, no self-respecting scientist will now state his conclusions without their coefficient of probable error. He added that, if this practice is not followed in other disciplines, it is because the probable errors in them are too great to be calculated. The ability of a science to indicate the probable errors of its measurements was thus taken by Peirce as a sign of maturity and not of defect. By his remark Peirce therefore wished to indicate that for the propositions in the most developed empirical sciences, no less than for those in the affairs of everyday life, no finality is obtainable, however well they may be supported by the actual evidence at hand.

The temper of mind which is illustrated by such an appraisal is itself the product of modern science and of a preoccupation with its procedures. It is based on the conviction that the methods of the natural sciences are the most reliable instruments men have thus far devised for ascertaining matters of fact, but that withal the conclusions reached by them are only probable because they rest upon evidence which is formally incomplete. The import of such an insistence upon the fallible character of science can be best appreciated by contrasting it with the classic conception of science, formulated in Greek antiquity and perpetuated in a powerful intellectual tradition. This conception of scientific knowledge was modeled upon the ideal of a completely demonstrative and absolutely indubitable natural science, such as Euclidean geometry was believed to be. It was assumed that the subject matter of genuine science was a realm of precise, unalterable laws, and that scientific knowledge, as distinct from belief, opinion, or mere experience, was be be equated with demonstrated knowledge. For such knowledge facts are not contingent, since they must be apprehended through their "reasons" or "causes," and the propositions which express them must therefore be "necessary." Furthermore, it was maintained that the "basic propositions" required as premises for demonstrated knowledge could be grasped by the intellect directly and infallibly and could be seen to be true with even greater assurance than any of the conclusions derived from them. The scientific enterprise was accordingly construed as the progressive apprehension of an eternal order of necessary connections, so that complete certainty was the earmark of genuine knowledge. The changing and the variable could not be subject matter for science; they could at best be the concern of belief and opinion. Variability in the materials studied or in the outcome of measurements was taken to indicate either the obdurateness of subject matter to rational connections or the failure of thought to reach its proper objectives. In a word, experience in the sense of observation and experiment, since it could not yield necessary propositions, could not be the ground for scientific knowledge.

This ideal of science dominated the minds of the great pioneers of modern science and of many of their most illustrious successors; and it is this conception which forms the tacit premiss of many philosophic commentators upon modern science, such as Descartes, Locke, Leibniz, and Kant. It is scarcely possible to exaggerate the significant role which this ideal has played in intellectual history. In proclaiming the ideal of science to be *systematic* knowledge, the rationalist tradition has stimulated research and has led to the development of science as something other than an indigestible miscellany of dubious facts. On the other hand, the great services of classic rationalism cannot hide the fact that its theory of self-evidence rests upon an inadequate analysis of the methods of science, so that it has frequently blocked the progress of inquiry and, though pledged to the ideal of clarity, has not seldom successfully courted obscurantism. Rationalism made complete certitude the theoretical condition for genuine science, but its belief that the latter was obtainable could be maintained only by neglecting or misinterpreting the approximate and contingent character of statements dealing with matters of fact. The long history of science and philosophy is in large measure the history of the progressive emancipation of men's minds from the theory of self-evident truths and from the postulate of complete certainty as the mark of scientific knowledge. Some of the major turning-points in that history consist in radically diminishing the class of statements certifiable simply by a rational insight into their truth. And some of its most dramatic moments have occurred when the approximate and incompletely grounded character of allegedly indubitable propositions was recognized.

The forthright admission of the probable or contingent character of even our most soundly based beliefs and the emphasis upon the general reliability of the *methods* of scientific inquiry rather than upon its conclusions are characteristic of contemporary empiricism. For the traditional empiricism of Locke and Mill, which in intent was a revolt against the exaggerated claims of rationalism, accepted in all essentials the standards and preconceptions of the views it nominally opposed. But that admission of the probable character of our beliefs is not the outcome of a capricious decision: it is not a pronouncement made for the sake of wilfully opposing a historically powerful tradition, nor is it a thesis advanced for the sake of a special set of values and an ulterior conception of nature. That admission and that emphasis have been wrung from students as a consequence of their reflection upon the history of science and of a painstaking examination of its methods. Contemporary empiricists who maintain that our knowledge of matters of fact is "probable" do not thereby maintain that such knowledge is inferior to knowledge of some other kind obtainable by methods different from those the natural sciences employ. On the contrary, they maintain that "probable knowledge" is the only kind of knowledge we can find or exhibit, and that the methods and techniques of the sciences are efficacious and dependable precisely because they make available knowledge of that character.

2. Development and Applications of the Theory of Probability

Although the term 'probable' has been employed several times in the preceding section, no precise sense has been attached to it. It is one of the objects of this essay to assign a clear meaning to sentences which contain the term and its derivatives; but it must be admitted at the outset that an analysis of what is meant by 'probable,' which would meet the unanimous approval of competent students of the subject cannot be given at the present stage of research. In the present and preceding sections the statement 'Knowledge of matters of fact is probable' is to be understood in the rather loose sense that conclusions of factual inquiry are not in principle incorrigible, because the formal conditions for assuring the logical validity of those conclusions are not completely realized, and because statements having factual content are not logically necessary.

The doctrine that knowledge of matters of fact is only probable is one of the central theses of contemporary analysis of scientific method. The implementation of this doctrine with modern logical and mathematical techniques is relatively recent. But even during the heyday of classic rationalism the status of beliefs which fell short of its ideal of scientific knowledge was frequently and vigorously discussed. Out of the permanent needs which generated such discussions have grown the modern calculi of probability and the diverse interpretations and applications which the term 'probable' has received. The possible equivocality and unquestionable vagueness of the term are therefore in part due to the history of empirical science. The brief survey, to which we now turn, of some of the contexts in which the term is and has been applied aims to achieve three things: to emphasize the intimate connection between the development of empirical science and the growing need for a theory of probability; to indicate the great range of applications of the term 'probable' and so to provide the materials for a discussion of its meaning; and to serve as a convenient introduction to the issues and techniques under contemporary discussion.

a) Aristotle's logical writings formulate the rationalist ideal of science, but his biological works exhibit less exacting standards of scientific adequacy. His evaluation of the then extant theories of sexual reproduction is characteristically judicious; evidence, some of it observational, is presented in opposition to the Hippocratean doctrines and in support of his own views, but there is not even a pretense that the question is settled beyond further debate. His examination of the facts of heredity show him to be familiar with at least the crude elements of a statistical explanation of the similarities and differences between ancestors and descendants. The mechanism which he suggested as an explanation for the observed facts was in essentials that of a shuffling and recombination of characters, so that only certain traits would normally recur. Even before Aristotle the principle of natural selection was advanced by Empedocles, though Aristotle had little use for it. No ancient mathematician developed a technique for handling statistical aggregates, and it is possible that the prevalent view of chance as an agent was an insurmountable impediment to a consistent working-out of a statistical view of nature which the theory of natural selection suggests. Nevertheless, passages in Aristotle and in the writings of the Ionians, Democritus, and Hippocrates which could be cited indicate that such a view was not foreign to the ancient mind.

b) Occasions for dealing with evidence which is not conclusive, but which nevertheless carries some weight, presented themselves in the legal and social transactions of both Athens and Rome. For example, there was a rule in Athenian courts excluding hearsay evidence, on the ground of the general untrustworthiness of reported statements as compared with the evidence of eyewitnesses. The courts of Rome took pride in deciding cases before them upon the basis of reason and the evidence of fact rather than caprice, and complicated safeguards were instituted to assure the adequacy of the evidence presented. Curious and distorted survivals of these appear in the formalistic rules of evidence of the Middle Ages. For example, two witnesses were required for a "full proof," the testimony of a single reliable witness counted as "half-proof," while a doubtful witness counted for "less than half." The object apparently aimed at was to convert the process of rendering a decision into a calculation of the "resultant force" of the testimony submitted. There was thus some basis in fact for Rabelais' portrait of Judge Bridlegoose, who made his decisions which were "correct" in the long run by throwing appropriately loaded dice. Years after, Leibniz was again intrigued by the possibility of a calculus of evidence, and the ideal of a quantitative science of proof has frequently hovered before students of probability. When the calculus of probability was finally developed, many of its great masters, like Laplace, Poisson, and their followers, attempted to turn it to such a use, though with singularly poor success.

c) Although no individual knows the exact date of his death, he can reasonably expect a definite span of life. His expectations are based on statistical regularities manifesting themselves in large groups of men. The use of such statistical uniformities for predicting individual behavior illustrates a common type of "uncertain inference," which in recent times has become an exacting and important discipline; and it was exemplified in ancient practices as well. Various forms of commercial insurance existed in Babylonia, Greece, and Rome, and the Romans were no strangers to life insurance. Just how and on the basis of what kind of statistical information the various rates were estimated is now unknown, although it is fairly clear that the estimates were not arbitrary. For example, the rates on bottomry and marine insurance depended on the destination of the vessel and the season during which it sailed; and, although the careful gathering of vital statistics is a modern phenomenon, a census of populations was frequently made in antiquity for military and taxation purposes. While therefore the practice of insurance was not placed upon a sound basis until the end of the eighteenth century, it was built on a large body of factual information; and, even though the beliefs which rested on this information fell short of the classic ideal of science, they made possible the planning and execution of important policies.

During the Middle Ages the Italian cities saw the beginning of commercial insurance as a profit-making enterprise; by 1700 the business of insurance was rapidly developing in western Europe, with life insurance in regular demand a century later. These enterprises required to be supported by adequate statistical techniques, and in fairly rapid succession there appeared a number of important statistical studies. For example, in 1662 John Graunt showed how to employ the register of deaths, which began to be kept in London during the Black Death, to make forecasts on population trends; during the same century John De Witt, grand pensioner of Holland, and Halley, the English astronomer, concerned themselves with annuity problems. Halley laid the basis for a correct theory of the subject, and he showed how to calculate from the mortality tables which he constructed the value of an annuity on the life of a person of given age. While scientific knowledge in accordance with the rationalist ideal was not obtainable for these domains, probable knowledge was, and it became the guide to life.

d) The entire subject of statistical inference now called for a theoretical foundation. The need was supplied from an unexpected quarter—the theory of games of chance. Dice games played with ankle bones were popular in antiquity, and the ancients distinguished between the "likelihoods" of certain combinations of throws. They did not, however, develop any technique for assigning numerical measures to the different "degrees of likelihood." The quantitative study of games of chance begins with the modern period and was cultivated by a brilliant succession of mathematicians.

Solutions of special problems in the division of stakes and the placing of wagers were first given by Cardan and Galileo (sixteenth century); but the general attack on the theory which was involved in their analyses began with Pascal and Fermat (seventeenth century), who showed that all the special problems under consideration could be reduced to problems in the mathematical theory of permutations and combinations. Upon this basis a convenient calculus was developed, which was subsequently applied to many different fields of inquiry. Huygens, the Bernoullis, Montmort, De Moivre, and Bayes are the most prominent figures in the early history of the subject. Their work was systematized and completed in the great treatise of Laplace (early nineteenth century), and the point of view from which they conducted their analyses remained until quite recently the basis for the interpretation and extension of the mathematical theory. The principle upon which Laplace assigned numerical values to probabilities was that of analyzing the possible outcome of a situation into a set of alternatives which could be judged as "equally possible." Accordingly, although we might be ignorant of which one of these alternatives would occur, a method was provided by the aid of which an appropriate "degree of rational belief" could be assigned to propositions about "chance events." In brief, fortuitous events which had heretofore been denied the status of genuine objects of scientific knowledge could now be handled in an expert manner with the help of probability theory. The intellectual instrument was thus forged for developing what is now known as the statistical view of nature, and for exhibiting important continuities in techniques and methods in different scientific disciplines.

e) The theoretical foundations of the probability calculus as formulated by Laplace still had their roots in traditional rationalism. On the one hand, probability judgments were understood to betoken ignorance: Laplace maintained that all events are regulated by "the great laws of nature" which a sufficiently powerful intelligence could use to foretell the future in the most minute way. On the other hand, judgments of equipossibility were made to rest on a nonexperimental basis. A critique and reformulation of these foundations were not to come for several decades. Nevertheless, these rationalistic preconceptions were conveniently overlooked in the application of probability theory. One of the earliest and most successful of these applications was to the systematization of measurements and observations in the experimental sciences. Astronomy was the first to employ the theory of probability for this purpose. Justly regarded for a long time as the most exact science of measurement, it nevertheless was patent to everyone that the measurements actually performed did not yield identical numerical values for what was presumably the same magnitude, however carefully gross disturbing factors were eliminated. In consequence, the measurable predictions calculated from astronomical theory were not in precise agreement with the numbers obtained by direct measurement. Given the climate of opinion within which astronomical theory was developed, it was congenial to interpret these fluctuations as deviations or "errors" from the "true values" of magnitudes, and to attribute the "inexactitude" of actual measurements to human failing.

Nevertheless, there was a pressing need for techniques to estimate the "true values" from the actual measurements and to measure the degree to which the latter "approximate" to the former.

This situation is not local to astronomy. As Boyle once explained, "You will meet with several observations and experiments which, though communicated for true by candid authors or undistrusted eyewitnesses, disappoint your expectations, either not at all succeeding constantly or at least varying much from what you expected." Indeed, to test any theory, empirically specified initial conditions must be given, and the consequences logically derived from them with the help of the theory must be compared with the outcome of further observational procedures. Thus, two series of actual measurements or observations must be instituted to test a theory; and, for both series, we find that as a matter of fact there are groups of discordant statements reporting the issue of our measurements. Whether the theory is in accordance with the "facts" cannot therefore be decided without some further hypothesis on the actual measurements we make.

The study of this problem in terms of the theory of probability constitutes what is known as the theory of errors. It was begun in the eighteenth century by Boscovitch, Lambert, Euler, and Thomas Simpson, and was continued by Daniel Bernoulli, Legendre, Gauss, and Laplace. Gauss showed that if we assume that the deviations from the "true magnitude" are produced by a large number of hypothetical "elementary errors" acting independently of one another, the form of the law of distribution of the actual measurements can be deduced, and an approximation to the "true value" can be calculated from the data. The Gaussian "Law of Error" and the Method of Least Squares for systematizing discordant observations have played an important role in subsequent researches in the theory of measurement and statistics. Recent critical work on the foundations of probability shows that Gauss's arguments for the law rest on assumptions which cannot always be made legitimately. In consequence, alternative laws for the distribution of errors have been proposed, notably by Poisson, Pearson, Gram, and Charlier, each suitable for different circumstances.

f) The expanding national economies following the breakup of the feudal system required the gathering of extensive factual information in order to guide the formulation of financial, military, and political policies. The earliest attempts to tie up the mathematical theory of probability with the analysis of such descriptive statistics were made by De Moivre, Nicholas and Daniel Bernoulli, Euler, and D'Alembert. Under the influence of the ideas of the French Encyclopedists, who sought a rational basis for monetary undertakings, public-health administration, judicial procedure, and even the conduct of elections, Condorcet tried to apply on a comprehensive scale the new mathematical instrument of probability to all such matters. Like Laplace and Poisson after him, he achieved only a modicum of success. It was characteristic of this group of writers to misunderstand and consequently to overrate the function of the probability calculus; their procedure frequently seemed to rest on the assumption, as one commentator remarked, that valuable results can be obtained from unreliable and insufficiently analyzed data by employing a sufficient number of signs of integration. However, it is to the great credit of these men to have insisted on the fusion of statistical methods with the theory of probability. Interest in this fusion was further stimulated by the Belgian astronomer Quetelet, who saw in the theory of probability the appropriate tool for developing a reliable social science. Poisson had enunciated in a somewhat confused form a "law" which he called "the law of great numbers"; according to this law large aggregates of elements exhibit definite properties with a stable relative frequency, even though these properties occur quite fortuitously within the aggregates. Quetelet popularized this idea in the context of the social disciplines. He regarded the "average man," as computed from the extensive statistics he gathered, as the analogue in social matters of the center of gravity in mechanics; and he saw in the statistical regularities with which certain human actions occur the operation of comprehensive laws of social development. He thus found it easy to believe that determinate laws could be formulated to connect the different social averages—determinate laws modeled upon those recorded in Laplace's *Celestial Mechanics.* However, Quetelet was uncritical both in gathering his statistical material and in interpreting it; he was never really clear as to the meaning of statistical averages, and never appreciated the limitations of the probability calculus. His influence, great at first, rapidly waned, and for a time so did the interest in applying theoretical statistics to the social sciences.

When interest in the subject was once more revived, it was supported by research needs in biology, psychology, and theoretical physics. Statistical methods subsequently developed on the basis of probability theory were then applied to matters as remote and different as the calculation of the density of telephone traffic and the maintenance of manufactured products at a certain standard of quality. The determination of the character of an indefinitely large population on the basis of samples drawn from it is a problem common to many disciplines and many daily occupations. The elements of an adequate theory of sampling within the framework of a theory of probability were first laid down by Lexis, and further developed by Bortkiewicz, Tschuprow, Markoff, and others. They showed that the sheer number of instances in a sample is no guaranty of its representative character, criticized statistical practice which relied upon the accumulation of unanalyzed numerical data, and developed a technique for obtaining trustworthy statistical coefficients from data grouped carefully according to the variety, homogeneity, and number of the instances. More recently, R. A. Fisher and his school have approached the problem from a different point of view, and, in addition to devising important criteria for the adequacy of statistical coefficients, he has called needed attention to the serious limitations of many of the Laplacian formulas. Other distinct contributions to the theory of sampling have been made by Fechner, Bruns, Galton, Thiele, Pearson, and Neyman. In consequence of these researches, the theory of errors, the theory of sampling, the theory of curve-fitting, now all fall within a comprehensive theory of probability.

g) Although the importance of the main ideas of the mathematical theory of probability for systematizing measurements was quickly recognized in the sciences, the theory of probability was for a long time usually regarded as simply ancillary to the theoretical disciplines. Thus, it was commonly assumed in physics that its laws are statable in "deterministic" form, such that the positions and velocities of elementary particles at one time are connected in precise ways with the positions and velocities at any other time. It is today a commonplace, however, that some of the most fruitful applications of the theory of probability occur *within* the theoretical framework of various sciences. The ancient idea that the apparently permanent objects around us as well as the regularities in their behavior could be viewed as aggregate effects of a large number of hypothetical elements undergoing random changes has frequently attracted the creative minds in science. Thus, Kepler played with it to explain the appearance of a new star in 1604; Boyle had a corpuscular theory for the states of aggregation of bodies; Huygens even formulated a corpuscular theory of gravitation; and Daniel Bernoulli's interpretation of Boyle's law for gases in terms of the kinetic theory of matter is well known.

Apparently, the first man to work out such theories with sufficient quantitative detail to make possible an empirical evaluation of the magnitudes associated with the hypothetical elements was Joule. He computed the average velocities of hydrogen molecules on the basis of statistical considerations and showed that, in order to produce the observed effects, the velocities must lie in specified intervals. The statistical explanation of thermal phenomena was carried to much greater lengths by Maxwell: he showed that, if certain assumptions are made concerning the probabilities with which the particles of a gas acquired different positions and velocities, the familiar gas laws could be deduced. But perhaps the greatest triumph of probability theory within the framework of nineteenth-century physics was Boltzmann's interpretation of the irreversibility of thermal processes; this he was able to do in terms of the most probable distribution of the energies of the molecules of a gas.

In consequence, the second law of thermodynamics can be formulated as a theorem in probability, and irreversible processes turn out to be statistical phenomena.

Thermodynamics is not an isolated instance of the use of statistical concepts within theoretical formulations. Even before Maxwell employed probability theory for the study of gases, G. G. Stokes used it to analyze the effects of polarized light coming from different sources as the average or most probable effects. Again, statistical mechanics, which consistently employs the theorems and the point of view of probability theory, has been fruitfully applied in the study of the history and distribution of the stars. And more recently the entire theory of radiation has been developed to include systematically within itself hitherto unrelated phenomena, on the basis of a profound and radical application of the theory of probability.

But physics is not the only science with has profited from using statistical concepts in its theories. Democritus tried to explain the resemblances and dissimilarities between parents and children in terms of a shuffling of the atoms coming from the ancestors of a child; Aristotle employed related notions in discussing similar problems. And ever since Darwin called attention to the importance of the facts of variation for any adequate biological theory, students of biology have been developing a statistical treatment of the subject. It is obviously essential to distinguish between variations due to heredity and those due to environment; this phase of the subject has been explored by Pearson and his school with the help of the mathematical theory of probability. But attempts such as those of Galton to formulate the laws of heredity in terms of average contributions from the ancestors of a given set of progeny are now known to be unsatisfactory; and Galton's mistakes indicate some of the limitations of statistical methods in general. The theoretical basis for modern experimental genetics was supplied by Mendel. The theory of the mechanism of heredity he proposed, which involved the transmission, segregation, and combinations of unit characters in various proportions, obviously lends itself to be exploited in terms of the fundamental ideas of mathematical probability; with its help, the artificial selection of plants and animals has been brought to a high stage of perfection. The mathematical theory of natural selection for those groups in which a Mendelian analysis can be made has been worked out mainly by R. A. Fisher, J. B. S. Haldane, and S. Wright.

In general, therefore, the introduction of probability notions into the theoretical structure of physics and biology has been most fruitful. It has made possible the prediction of the relative frequency with which definite characters occur in groups of individuals, even when it is not feasible to predict the occurrence of such characters for a given individual.

h) The developments which have thus far been surveyed have gradually tended to undermine the authority of classic rationalism in science. For, as points of view borrowed from the theory of probability and statistics assume central roles, both within the theoretical framework of the sciences as well as in the procedures of applying theories to matters of fact, it becomes progressively more difficult to assume that the principles of a science are self-evident or necessary. This change in the climate of opinion has been further supported by a general logical criticism of the assumptions of the classic view which began in antiquity. The Epicureans, as well as Skeptics like Carneades, developed conceptions of the logic of inquiry which made allowances for the formally incomplete character of the evidence for empirical statements. In modern times it was Hume's discussion of causality which put the rationalist notion of necessity on the defensive, and since then every variety of empiricism has had its day in court.

Nevertheless, although the Humean analysis was a powerful dissolvent of ancient preconceptions, it was not so powerful as some of the internal technical developments within the special sciences. A few of these have already been indicated. But perhaps the most significant single technical achievement, from the point of view of its general effect upon the philosophy of science, has been the definitive refutation of the thesis that Euclidean geometry is the apodeictic science of space. For the discovery of the non-Euclidean geometries exhibited logically possible alternatives to conceptions previously regarded as indubitable, while the recognition of a distinction between pure and applied mathematics cut the ground from under the claims of traditional rationalism. Whatever doubt still lingered as to the possibility of alternative conceptions of "space" was finally removed when the Newtonian physics and its Euclidean framework for mechanics were displaced by relativity physics and its framework of Riemannian geometry. And perhaps the final coup de grâce to the claim that physical principles are indubitable and necessary was supplied when the familiar physics of continuous action was found to be inadequate for vast ranges of phenomena, and made way for the contemporary physics of quanta.

There are thus both historical and analytic grounds for the view central to empiricism that there is no a priori knowledge of matters of fact, and there are similar grounds for the thesis of contemporary empiricism that no amount of empirical evidence can establish propositions about matters of fact beyond every possibility of doubt or error. On the other hand, the recognition of this state of affairs raises an important problem. Although our beliefs cannot be established with absolute finality, we do, as we must, differentiate between them on the ground of the character of the evidence which supports them. We regard it more probable that Napoleon was a historical character than that he is a solar myth. We believe that the prognoses of a modern physician are more reliable than those made a century ago. A chemist accepts Lavoisier's theory of combustion as better founded than Stahl's phlogiston theory, and a physicist will urge that the quantum theory of radiation is today more securely based than it was twenty years ago. There is clearly an obvious need for canons to evaluate the evidence supporting any proposition, and for the formulation of the principles we employ in deciding that one statement is better grounded than another. Judging by the success of past attempts to supply them, it may be suspected that every proposed list of such canons and formulations will be incomplete and will require emendation with the progress of inquiry. The need, however, is a permanent one; and the attempts to satisfy it constitute the broader setting and the larger theme in contemporary discussions of probability.

II. The Calculus of Probability and Its Interpretations

3. Preliminary Distinctions

The vast range of material which has just been outlined has been traditionally regarded as constituting the subject matter for a theory of probability. It has been frequently assumed that a precise meaning can be found for the term 'probable' which is common to its use in each of the contexts indicated. Upon this point competent students are not in agreement. Without prejudging the issues involved, it is possible to distinguish between two groups of statements in which the term 'probable' or its derivatives occur. The first group contains statements such as the following:

'The probability that a man of thirty will survive his thirty-first birthday is .955'; 'The probability that a normal coin will present a head after being tossed is $\frac{1}{2}$'; 'The probability that on the basis of the evidence in 1938 the electronic charge e has a value in the interval $(4.770 \pm .005) \times 10^{-10}$ electrostatic units is .67'; 'The probability that a molecule of hydrogen has a velocity in the interval $v - dv$ and $v + dv$ is p'; 'The probability of a 10° deflection of an a-ray passing through a film is $\frac{1}{4}$'; 'The intensity of a spectral line is determined by the probability of the corresponding quantum transition'; and 'A snowstorm in New York during January is more probable than during November.'

The second group contains such statements as:

'Relative to our present evidence the theory of light quanta has a probability which is greater than its probability relative to the evidence available in 1920'; 'The evidence makes it highly improbable that Aristotle composed all the works attributed to him'; 'The theory of evolution has a higher probability on the evidence than the theory of special creation'; 'It is probable that, had Cleopatra's nose been a half-inch longer, the course of the Roman Empire would have been different'; and 'It is not probable that Christ was a descendant of King David.'

Statements in the first group employ the term 'probable' in a sense which, by practically unanimous consent, is subject to the rules of the calculus of probability; indeed, the calculus has been explicitly devised to handle such "probabilities." Statements in the second group apparently employ the term to indicate the "degree" of the adequacy of the evidence supporting the proposition; students are not agreed whether the mathematical calculus of probability is applicable to such "probabilities." In the present section we shall for the most part confine ourselves to such statements which clearly fall into the first set; the discussion of the second group of statements is reserved for Section III. But nothing said in the present section will exclude the possibility that both classes of statements, in spite of apparent differences between them, are subject to the same interpretation.

Even though we have restricted the scope of the present section in the indicated way, it is still not possible to specify a sense of 'probable' or even a formulation of the calculus of probability, upon which reasonably complete agreement is obtainable. There are, in fact, three major interpretations of the term. According to the first, a degree of probability measures our subjective expectation or strength of belief, and the calculus of probability is a branch of combinatorial analysis; this is the classical view of the subject, which was held by Laplace and is still professed by many mathematicians. It is not always clear whether by 'expectation' proponents of this view understand *actual* expectations or *reasonable* expectations. According to the second, probability is a unique logical relation between propositions, analogous to the relation of deducibility; its most prominent contemporary supporter is the economist Keynes. According to the third, a degree of probability is the measure of the relative frequency with which a property occurs in a specified class of elements; this view already appears in Aristotle, was proposed by Bolzano and Cournot during the last century and further developed by Ellis, Venn, and Peirce, and was finally made the basis for a subtle mathematical treatment of the subject by von Mises and other contemporary writers. We shall begin with the exposition of the frequency interpretation of probability and its calculus; subsequently, the other two views will be briefly considered, and it will be argued that the frequency view is the one most suitable for the first of the foregoing two classes of statements; in Section III we shall finally examine some important methodological problems which cluster around the frequency view.[1]

4. Fundamental Ideas of the Frequency Interpretation of Probability

The basic ideas of the frequency conception of probability emerge upon an examination of such a statement as 'The probability that a person of thirty residing in the United States survives his thirty-first birthday is .945.' The meaning of such a statement can be ascertained by examining how it is established. That procedure, greatly simplified, is somewhat as follows. Suppose that during a period of years there is no migration to or from the United States, and that during these years exact counts are made of its inhabitants who fall into definite age groups. Thus, suppose that in 1900 there are 2,000,000 persons who have just reached their thirtieth birthday, and that exactly one year later there are 1,890,000 persons who have just reached their thirty-first birthday; that is, of the thirty-year-olds in 1900 a ratio of .9450 survive at least another year. We imagine that similar figures are obtained for the four succeeding years, and that the ratios of thirty-year-olds who survive their thirty-first birthday are .9452, .9456, .9451, and .9454, respectively. We notice that, although these ratios are not constant, the differences do not appear until the fourth decimal is reached. We may say, therefore, that during these five years approximately 945 out of a thousand thirty-year-old residents of the United States live for at least another year; and we may make the further *assumption* that for an *indefinite number of future years* the corresponding ratios of survivals remain in the neighborhood of .945. Accordingly, the statement 'The probability that a thirty-year-old resident of the United States survives his thirty-first birthday is .945' means that in the long run the relative frequency with which thirty-year-olds in the United States survive for at least one year is approximately .945.

The following points must be noted in this example. In the first place, the probability statement supplies no information about any *individual* resident of the United States; the information is relevant to the individual Tom Brown only in so far as he belongs to the *class* of thirty-year-old residents. Second, the statement supplies information about no property of this class of residents other than the one explicitly specified, namely, the property of surviving at least one year. Third, the statement supplies a numerical value—the value of a *relative frequency*. Fourth, the statement does not mean that in *every* thousand thirty-year-olds 945 will live for at least another year. And, finally, this numerical value is intended to specify the relative frequency of survivals during an indefinite number of years, or "in the long run," and not only during the years for which an actual count has been made; that is to say, the statement makes a prognosis.

We now turn to the general definition of probability statements. But at once difficulties arise. A proposed definition must be precise and unambiguous and at the same time should be modeled as closely as possible upon the procedures which the foregoing example illustrates. On the other hand, those procedures have been described with the help of terms which are not precise; in particular, the expressions 'in the long run' and 'approximately' are highly vague, and it is not easy to develop a mathematical theory in terms of them. Accordingly, the definition to be proposed will replace these expressions by more precise ones, which are appropriate for developing a *calculus* of probability. Hence, although the definition will be modeled upon the illustration, it will employ precise mathematical concepts to which there cannot easily be assigned a simple empirical meaning. The methodological problems which are a consequence of this procedure will have to be considered subsequently.

Let R be a non-empty class of elements (i.e., it contains at least one member), to be known as the *reference class*; for reasons which will be soon apparent, the elements of R will be supposed to be serially ordered. Let A be some property which the elements of R may exhibit. Suppose R contains n elements, and let nu(A and R) be the number of elements in R which have the property A. We may now define the expression 'the relative frequency with which elements in R have the property A,' which we abbreviate into 'fr$_n(A, R)$', as follows:

$$\text{'fr}_n(A, R)\text{' is short for } \frac{\text{'nu}(A \text{ and } R)\text{'}}{n}$$

It is evident that a relative frequency is a proper fraction. Suppose now that the number of elements in R increases. In general, the fraction fr$_n(A, R)$ will be different for different values of n. It may happen, however, that these fractions will crowd around some fixed number p, and will differ from it by a small positive magnitude ϵ which diminishes as n increases: in familiar language, fr$_n(A, R)$ will tend toward p in the long run. The mathematically precise way of rendering this possibility is to say that fr$_n(A, R)$ approaches p as a limit with increasing n; that is,

$$p = \lim_{n \to \infty} \text{fr}_n(A, R) .$$

What mathematicians understand by 'limit' is illustrated by the following. Consider the infinite series of fractions $\frac{1}{2}, \frac{2}{3}, \frac{3}{4}, \frac{4}{5}, \ldots$; its limit is 1. Suppose we have the infinite series of numbers $x_1, x_2, x_3, \ldots x_n, \ldots$, where the subscripts indicate the ordinal position of the numbers in the series. To say that p is the limit of this series means that, however we may select a positive number ϵ, there is a number N such that for every n, if $n > N$, then the absolute difference between x_n and p (i.e., neglecting signs) is less than ϵ. The reason for requiring R to be serially ordered is now clear. If R contains only a finite number of elements, $fr_n(A, R)$ is unaffected by the order in which the elements are counted; but the limit of $fr_n(A, R)$ when R is not finite does depend on the order in which the elements of R (and therefore the relative frequencies) are arranged.

It is very convenient in developing the calculus of probability to define 'probability' as 'the limit of relative frequency.' If we abbreviate statements of the form "The probability that an element has the property A if it is a member of R is p' into 'prob$(A, R) = p$,' the definition takes the following form:

$$\text{'prob}(A, R) = p\text{' is short for '}p = \lim_{n \to \infty} \text{fr}_n(A, R)\text{'}$$

'prob$(A, R) = p$' may be read conveniently as 'the probability of A in R is p.'

Expressions like 'prob(A, R)' which describe numbers are called *numerical expressions* and consist of the *functor* 'prob' together with its arguments; 'nu$(A$ and $R)$' is also a numerical expression, and 'nu' another functor. The expression 'prob(A, R)' is the fundamental numerical expression in the mathematical theory of probability developed on a frequency basis; it describes a real number, which may be irrational, in the interval 0 to 1 inclusive. Within the calculus of probability the statement 'The probability that a thirty-year-old resident of the United States survives his thirty-first birthday is .945' must now be taken as equivalent to 'The limit of the relative frequency with which the property of surviving at least one year occurs in the ordered class of thirty-year-old residents of the United States is .945.'

It has already been pointed out that the foregoing definition of 'probability' has been proposed for the sake of its great convenience in calculations. It employs the notions of infinite ordered classes and of limiting values of relative frequencies in such classes. It is obvious, of course, that in empirical procedures we are occupied with finite classes which may or may not be ordered, and with relative frequencies rather than limits of relative frequencies. Some writers (e.g., Copeland and Popper) have proposed to use as the definition of 'probability' *not* 'the limit of relative frequencies,' but 'the *condensation point* of relative frequencies.' p is said to be a condensation point of the series $x_1, x_2, x_3, \ldots x_n, \ldots$, if for every positive number ϵ and every N there is an n such that $n > N$ and the absolute value of the difference between x_n and p is less than ϵ. Such a definition has the merit that a proof can be given that there is at least one condensation point for relative frequencies in an infinite reference class, even though no limit exists; it suffers from the disadvantage that according to it a property may have more than one probability in a given class, so that the calculus of probability becomes more complicated.[2]

It is essential to note the following points in connection with probability statements interpreted in terms of relative frequencies:

a) No meaning can be attached to any expression which, taken literally, assigns a probability to a single individual as having a specified property. Statements of probability predicate something of an individual (e.g., Tom Brown) only in so far as he is an element in a specified reference class. Probability statements which do not do so *explicitly* must be regarded as incomplete if they are to be significant: they must be understood as making an implicit specification of the reference class within which the designated property occurs with a certain relative frequency.

b) Every probability statement of the form thus far considered is a factual statement, into whose determination empirical investigations of some sort must always enter. Probability statements are on par with statements which specify the density of a substance; they are not formulations of the degree of our ignorance or uncertainty. To assert that the probability of a normal coin presenting head after being tossed is $\frac{1}{2}$, is to ascribe a physical property to a coin which is manifested under determinate conditions.

c) Since probability statements require the specification of a reference class with respect to which a given property has some degree of probability, a given property can be associated with different degrees of probability, according to the reference class which is specified. The probability of surviving at least one year may be .945 with respect to the reference class of thirty-year-old residents of the United States; it may be .734 with respect to the reference class of sixty-year-old men; and it may be .345 with respect to the class of domesticated cats.

d) Since the explicit definition of probability statements is in terms of relative frequencies, the *direct* evidence for them is of a *statistical* nature. Thus, waiving difficulties to be mentioned, the direct evidence for the probability of a coin falling head is obtained by counting the frequency with which it falls head. However, probability statements do not always occur singly and are often part of a more or less inclusive *system* of statements or a *theory*. In such cases the estimation of the numerical values of the probabilities and the subsequent testing of such values may be made on the basis of *indirect* evidence which in some cases may even be nonstatistical. This point will receive further attention in Section III.

e) Since a probability has been defined as the limit of a relative frequency (or, even more loosely, as the relative frequency in the long run), every probability statement is a hypothesis; such a hypothesis cannot be completely confirmed or finally verified by the (necessarily) finite amount of evidence actually at hand at any given time. It is thus quite possible that the numerical value estimated for a probability on given evidence is not correct, so that revisions of the estimate may have to be made repeatedly. It is partly for this reason that in the history of the subject discussions of probability have run parallel with discussions of the problem of induction. The situation with respect to probability statements is indeed more serious than has been just indicated. For not only cannot probability statements be completely confirmed; they cannot even be completely disconfirmed by any actual evidence. The issues involved will receive further attention below.

f) Finally, it is a mistake to suppose that the successful use of probability statements depends in any way upon the issues of what is popularly known as "determinism." Because current microscopic physics employs theories involving in an essential way probability considerations, many thinkers, including reputable scientists, have been persuaded into supposing that the general breakdown of "mechanistic" explanations has been demonstrated, that processes in nature are "noncausal," and that contemporary physics supplies evidence for the existence of human "freedom" and for a "spiritualistic" world-view. Such suppositions feed upon mistaken or misleading formulations of the actual issues in modern physics, as has been pointed out repeatedly, among others, by Venn, Peirce, Philipp Frank, and Henry Margenau. It is perhaps sufficient to note that the use of probability statements requires no commitment, even by implication, to any wholesale "deterministic" or "indeterministic" world-view; they can be used successfully in such contexts in which specified properties occur with stable relative frequencies in specified classes of elements.

One of the main difficulties in most debates on causality is that the term is not explained with sufficient precision to make discussion fruitful. (As a matter of fact, specific contributions to the sciences of nature rarely if ever contain the term.) Without entering into detailed analyses of the issues sometimes raised, the following observations may help clarify some of them.

(*i*) Questions of causality can be significantly discussed only if they are directed to the *theories* or *formulations* of a science and not to its subject matter. No clear sense can be given to most pronouncements that the world or any segment of it is a causal process. On the other hand, in discussing the causal or noncausal character of a given theory, two factors must be examined: the *state* (or system of properties) in terms of which the physical system under discussion is described and the *laws* (or system of equations) which connect the states at different times and places. The state of a system is sometimes specified with the help of properties belonging to what are taken as "individual elements," sometimes with the help of the properties of a field, and sometimes in statistical terms involving the properties of aggregates of individuals. The laws also can differ markedly in form: they may establish a unique correspondence between states at different times or they may have the form of probability statements; they may be explicit functions of the time variable or they may not, etc. No universally accepted criterion has been formulated for judging whether a theory is "causal." Classical mechanics is frequently considered as the example par excellence of such a theory; the states considered by it are the positions and momenta of material particles, and its laws are certain differential equations of the second order not containing the time variable explicitly. It is often assumed that, in order to be a causal theory, the states employed by the theory must be those of classical mechanics. In that case, however, neither classical electromagnetics nor modern quantum mechanics are causal theories, although the former is usually so regarded. In some cases, on the other hand, the distinction between noncausal and causal theories is made on the basis of whether the states are specified in statistical terms or not, so that classical statistical mechanics and modern quantum mechanics would both be classified as noncausal theories. The main point to be borne in mind is that both factors, specification of state and form of law, are relevant to the discussion. Even theories which employ statistically specified states have been said to be causal because their laws establish a unique correspondence between its states at different times—although with respect to certain properties of individuals in the system the theories have been classified as noncausal, because the equations supply only probability statements concerning the occurrence of properties of individuals.

(ii) Because probability statements supply no information about any individual member of the reference class, it has been imagined that a physical theory involving probability considerations precludes a "causal" explanation of the phenomena under consideration. Now such a theory will usually specify the state in statistical terms; and, as a consequence, the predictions of the theory may have the form of probability statements concerning the properties of individuals. In some cases, however, it is also possible to describe the situation in terms of nonstatistical states, so that laws of a "causal" type may connect these new states. Whether it is possible or convenient to do so is obviously a matter to be decided for each case by experiment and scientific policy. It so happens that for the phenomena studied by classical statistical mechanics it is possible to do this; and, as a consequence, the "indeterminism" of classical statistical mechanics has been usually regarded as eliminable or inessential. Such an elimination is not possible for modern quantum mechanics within the framework of its procedures, and marks an important difference between classical and recent physics. In any case, nothing more than a very technical scientific difference is involved; and at least some physicists are of the opinion that future research may remove this difference. It should also be noted, moreover, that if the Ψ-function in modern quantum mechanics is taken to specify the state of the system, without seeking to interpret this function statistically, quantum theory may also be regarded as a "causal" theory, for its laws have the form of equations usually regarded as of the causal type: they establish a unique correspondence between states at different spatio-temporal regions.[3]

5. Fundamental Theorems in the Calculus of Probability

1. *The function of the calculus.*—It should now be clear that probability statements cannot in general be certified on purely formal grounds, so that pure mathematics and logic are not in the position to assert probability statements of the form considered thus far. What then, it may be asked, are the function and nature of the mathematical calculus of probability? To readers of the preceding monographs in this *Encyclopedia* the answer will be familiar. The calculus of probability has the same general function as a demonstrative geometry or a demonstrative arithmetic: given certain initial probabilities, the calculus of probability makes it possible to calculate the probabilities of certain properties which are related to the initial ones in various ways. Thus, arithmetic cannot tell us how many people live in either China or Japan; but, if the population of China and the population of Japan are given, we can compute the

combined population of these countries. The calculus of probability functions in the same way. It is important to recognize that the propositions asserted *in the calculus* are not factual or empirical statements: they are all certifiable on formal grounds alone, and are analytic of the definitions and rules initially laid down. The proposal to establish the theorems of the calculus of probability by experimentation, which has sometimes been made, is as ill-considered as would be the proposal to prove experimentally that $3^2 + 4^2 = 5^2$. The function of the probability calculus, like that of other calculi, is to make possible the *transformation* of probability statements in order that their theoretical content be made evident. The calculus thus has an *instrumental function* in the context of empirical investigations. It permits us to derive the relative frequencies with which certain properties occur from initial probability statements which do not explicitly mention those frequencies; in this way the calculus makes possible a more adequate testing of the probability statements which we entertain by making explicit the predictions they involve.

The detailed discussion of the calculus of probability can be undertaken only with the help of the technical apparatus of mathematical analysis. Some familiarity with at least the elementary theorems of the calculus is, however, essential for a just appraisal of its function and limitations. In the present section we shall accordingly state a few standard theorems of the calculus and, incidentally, obtain important material for evaluating the claims of standpoints in the philosophy of science which do not subscribe to an empirical outlook.

2. *Elementary theorems of the calculus.*—Suppose we wished to obtain the probability that children of white parents are both blue-eyed and blond. The reference class R consists of children born to white parents; the problem requires for its answer the (limit of the) relative frequency with which the properties A (being blue-eyed) and B (being blond) jointly occur in R. This number, prob(A and B, R), could be estimated directly. It may, however, be *calculated* from the following two numbers: the probability of A in R; and the probability of B in the refer-

ence class consisting of blue-eyed children of white parents; i.e., from prob(A, R) and prob(B, A and R). The following theorem, known as the General Product Theorem, can be easily demonstrated: The probability of A and B in R is equal to the probability of A in R, multiplied by the probability of B in A and R. Using familiar mathematical symbolism this can be stated as follows:

$$\text{prob}(A \text{ and } B, R) = \text{prob}(A, R) \times \text{prob}(B, A \text{ and } R) . \quad (1.1)$$

We happen to know that the relative frequency of blond hair in the class of blue-eyed children of white parents is not equal to the relative frequency of blond hair among children of white parents in general. In some cases, however, the probability of B in R does equal the probability of B in the narrower reference class A and R. The properties A and B are then said to be "independent" of each other with respect to R. In such cases we obtain the Special Product Theorem:

$$\text{prob}(A \text{ and } B, R) = \text{prob}(A, R) \times \text{prob}(B, R) . \quad (1.2)$$

Theorem 1.1 may itself be generalized for the joint occurrence of n properties $A_1, A_2, \ldots A_n$.

Suppose now that we required the probability that children of white parents are either blond or black-haired. The properties A (being blond) and B (being black-haired) cannot as a matter of fact jointly occur in the class R; they are said to be "exclusive" with respect to R. For such exclusive properties the following Special Addition Theorem can be easily proved: The probability of A or B in R is equal to the probability of A in R plus the probability of B in R. Again employing mathematical symbolism we obtain:

$$\text{prob}(A \text{ or } B, R) = \text{prob}(A, R) + \text{prob}(B, R) . \quad (2.1)$$

Let us next obtain the probability that children of white parents are either male or female. Properties such as male and female are called "contradictory properties" in the class of human births because they are both exclusive and exhaustive. It is obvious that the probability of being male or female in the class of children of white parents must be equal to 1. In particular, we can demonstrate the following theorem:

$$\text{prob}(A \text{ or not-}A, R) = 1 , \tag{2.2}$$

and with the help of theorem 2.1 we also obtain

$$\text{prob}(A, R) + \text{prob}(\text{not-}A, R) = 1 . \tag{2.3}$$

Thus if the probability of a male birth among humans is .51, the probability of a female birth in that class must be .49.

Theorems 1.1 and 2.1 are fundamental in the elementary calculus of probability. From them a large number of important consequences can be derived by applying the ordinary rules of logic and arithmetic. A few of them will be mentioned because of their practical and methodological importance.

There is clearly no difference between the probability of A and B in R and the probability of B and A in R. Accordingly,

$$\text{prob}(A \text{ and } B, R) = \text{prob}(A, R) \times \text{prob}(B, A \text{ and } R)$$
$$= \text{prob}(B, R) \times \text{prob}(A, B \text{ and } R) ,$$

from which we obtain the Division Theorem:

$$\text{prob}(B, A \text{ and } R) = \frac{\text{prob}(B, R) \times \text{prob}(A, B \text{ and } R)}{\text{prob}(A, R)} , \tag{3.1}$$

which can be given the following more convenient form:

$$\text{prob}(B, A \text{ and } R) =$$
$$\frac{\text{prob}(B, R) \times \text{prob}(A, B \text{ and } R)}{\text{prob}(B, R) \times \text{prob}(A, B \text{ and } R) + \text{prob}(\text{not-}B, R) \times \text{prob}(A, \text{not-}B \text{ and } R)} . \tag{3.2}$$

Theorem 3.2 is one form of what is known as Bayes's theorem. A more general form is the following: Let $B_1, B_2, \ldots B_n$ be a set of mutually exclusive and exhaustive properties with respect to R, and let B_i be any one of them. Then

$$\text{prob}(B_1, A \text{ and } R) = \frac{\text{prob}(B_i, R) \times \text{prob}(A, B_i \text{ and } R)}{\sum_1^n \text{prob}(B_i, R) \times \text{prob}(A, B_i \text{ and } R)} , \tag{3.3}$$

where, as usual, 'Σ' is the sign of summation.

Bayes's theorem and the consequences which have been drawn from it have played important roles in discussions of the foundations of probability, induction, and scientific method. It is therefore important to illustrate how it may be employed, especially since the limitations of its use have not always been clearly understood or remembered. Let R be the very numerous class of shots fired at a certain target; let A be the property of a shot hitting the bull's eye; and, finally, let B_1 be the property of a shot that it is fired from Rifle 1, B_2 from Rifle 2, and B_3 from Rifle 3. All the shots are supposed to be fired from these rifles. The (limiting) relative frequency of shots from Rifle 1 is $\frac{3}{8}$, from Rifle 2 is $\frac{1}{8}$, and from Rifle 3 is $\frac{4}{8}$; furthermore, the probability that a shot fired from Rifle 1 hits the bull's eye is $\frac{1}{5}$, while from Rifle 2 it is $\frac{2}{5}$, and from Rifle 3 it is $\frac{1}{5}$. What is the probability that a shot which hits the bull's eye is fired from Rifle 2? The question asks for the value of prob(B_2, A and R); it is obtainable from theorem 3.3 if we remember than $n = 3$, prob(B_1, R) $= \frac{3}{8}$, prob(B_2, R) $= \frac{1}{8}$, prob(B_3, R) $= \frac{4}{8}$, prob(A, B_1 and R) $= \frac{1}{5}$, prob(A, B_2 and R) $= \frac{2}{5}$, and prob(A, B_3 and R) $= \frac{1}{5}$. A simple calculation shows that the required probability is $\frac{2}{9}$.

Bayes's theorem is frequently referred to as a theorem in "inverse probabilities," and it has been traditionally regarded as the instrument for discovering the probability of "causes" or "hypotheses" from known "effects" or "consequences." The reason for this terminology is perhaps evident from the illustration: the probability which is sought is that of the "cause" (namely, of a shot being fired from Rifle 2), on the assumption that certain "effects" have set in (namely, of the shot hitting the bull's eye). But although Bayes's theorem can be demonstrated in the calculus of probability, it can be employed to determine the probability of "causes" only if *all* the probability coefficients in the right-hand side of the formula are given. Of special importance are the probabilities of the form 'prob(B_i, R)' which are sometimes designated as the "antecedent probabilities of the causes." Now it has been often assumed that, if we possess no information to the contrary, these antecedent probabilities are equal to one another. This assumption has been supported by what is known as the Principle of Indifference. With the help of this principle it has been supposed that probabilities could be determined a priori—that is, without recourse to empirical, and more particularly to statistical, investigations. Consequently, this assumption proceeds from a different conception of probability than the one developed in §4; and for a relative frequency conception of probability the equating of probabilities to one another simply on the ground that we know no reason why they should be unequal is a major error. Proceeding within this different conception of probability, Laplace deduced from Bayes's theorem the so-called Rule of Succession, which for a long time was accepted by eminent thinkers as the basis for reliable scientific predictions. According to this rule, if n events of a certain kind have been observed in succession, then the probability of its recurrence is $(n + 1)/(n + 2)$. Following Laplace, Quetelet declared that, "after having seen the sea rise periodically ten successive times at an interval of about twelve hours and a half, the probability that it will rise again for the eleventh time would be $\frac{11}{12}$." But it also follows from the rule that, if the tide has not been observed to rise at all, the probability of its rising is $\frac{1}{2}$; and such a consequence is a *reductio ad absurdum* of the rule and of its premises for any view of probability which defines it in terms of relative frequencies.

In most problems it is not practically or theoretically possible to assign values to the antecedent probabilities in Bayes's theorem which could have any empirical significance. For this reason Bayes's theorem has only a limited use, and few writers today take it seriously as a means for determining the probability of a given hypothesis on the basis of given evidence.

3. *Theorems depending on irregularity in the reference class.—* The theorems which have been mentioned thus far can be demonstrated on the sole assumption that the relative frequency of a property in its reference class has a limit. But many theorems in the calculus which are of greatest importance in practice require that other conditions are satisfied as well.

Suppose that in the class R (e.g., tosses with a coin, where the tosses may be imagined as temporally ordered) the property H (head falling uppermost) occurs as follows, where T (tail falling uppermost) and H are exclusive and exhaustive properties with respect to R:

$$H\ T\ H\ T\ H\ T\ H\ T\ H\ T\ H\ T\ \ldots \tag{i}$$

That is to say, we suppose that every other toss yields a head, so that the probability of a coin falling head is $\frac{1}{2}$. Here the property H occurs with an obvious regularity; and, if such were indeed the case for actual throws with a coin, we would very likely not employ probability considerations with respect to it. In fact, however, in actual cases heads and tails occur in no such regular order, but with an irregularity somewhat as follows:

$$HHTTTHHTHTTHHTTHHHHTHTHHTTT \ldots \tag{ii}$$

In this finite segment of a hypothetically infinite series the relative frequency of H is .51; but we may imagine that the limiting value of this ratio is also $\frac{1}{2}$. The second series is like the first in having $\frac{1}{2}$ as the limit for the relative frequency of H; it is unlike the first in that H occurs in it irregularly or at random. Various theorems in the calculus of probability depend upon the assumption that the reference classes involved possess such a random character.

It is, however, not easy to give a precise sense to what we mean by 'at random,' and an extensive technical literature now exists which deals with the problems of defining 'irregularity' in a manner suitable for mathematical purposes. The first one to have called attention to the importance of conditions of irregularity and to have worked out systematically a mathematical theory of probability with them in mind is von Mises. His procedure takes its point of departure from the following observation: If in the first of the foregoing series we select the (nonfinite) subseries R' by including in it only the odd terms of R, the probability of H in R' is no longer $\frac{1}{2}$ but is 1. On the other hand, if we select the subseries R' from a random series R (such as the second series above is supposed to be) in the same way as before, the probability of H in R' is still $\frac{1}{2}$; that is, prob(H, R) = prob(H, R'). Now let S be any nonfinite subseries of R, subject to the sole condition that the elements of S

are not selected on the basis of their possessing or not possessing H. If for *every selection* of such a subseries S from R, prob(H, R) = prob(H, S), the reference class R is said by von Mises to be irregular. He believes that this definition makes precise our intuitive notion of irregularity and that it formulates the conditions found in games of chance and other fortuitous events. Moreover, he maintains that in order to demonstrate many of the standard theorems in the calculus of probability his condition for irregularity must be assumed. (It is well to bear in mind, however, that considerations such as these which involve infinite classes or classes having certain types of order are pertinent primarily to the *calculus* of probability. They are introduced for the sake of constructing a consistent and powerful instrument of symbolic transformations.)

However, many students have found von Mises' definition unsatisfactory. It can be shown that, if a reference class satisfies von Mises' condition of irregularity, the order in which the specified property occurs in it cannot be formulated by any mathematical function; and doubts have therefore been raised as to the *logical* possibility of a reference class which is to satisfy so stringent a condition of irregularity. Indeed, if the phrase 'every selection of such a subseries S' in the definition is taken seriously, a contradiction can be exhibited in the notion of an irregular reference class. Various attempts have accordingly been made by a number of writers to overcome such difficulties (e.g., by Doerge, Kamke, Tornier, Reichenbach, Popper, Copeland) by distinguishing between different types of irregularity and by proposing conditions of irregularity whose consistency can be established. None of these substitutes, however, is sufficiently strong logically for demonstrating the standard theorems in question in their full generality. But more recently it has been shown by A. Wald that by suitably relativizing the selection of subseries in von Mises' definition to certain very general classes of selections, the logical difficulties can be obviated, while at the same time the consequent restrictions upon those theorems do not seriously impair their general validity.[4]

We shall assume that a mathematically satisfactory definition of irregularity can be given, and proceed to mention a few important theorems which may be demonstrated for reference classes satisfying it. Let R be such a reference class (e.g., throws with a coin) in which the property H (head uppermost) has the probability p while the property T (tail uppermost) has the probability $1 - p$. We now suppose the elements of R to be grouped into sets of n successive elements each, and ask for the probability that exactly r elements in a set(where $r \leq n$) have the property H while the remaining $n - r$ elements have

T. The numerical value of this probability can be shown to be equal to

$$\frac{n!}{r!(n-r)!}\, p^r(1-p)^{n-r} \tag{4.1}$$

where $r! = 1 \times 2 \times 3 \times \ldots \times (r-1) \times r$, with $0! = 1$.

It is of some importance to understand clearly what this number signifies. Suppose that H and T occur in R as in series (ii) on page 32 above, and suppose that R is broken up into sets of four successive elements each. The following sequence of sets then results:

$$(HHTT)\ \ (HTTT)\ \ (TTTH)\ \ (TTHH)\ \ (THHT)$$
$$(HHTH)\ \ (HTHT)\ \ (THTT)\ldots \tag{iii}$$

Some of these sets, such as the first and second, are overlapping, in the sense that they contain common terms from $R;$ others, such as the first and fifth, are nonoverlapping. If we let $r = 1$, the number given by theorem 4.1 is the limit of the relative frequency with which these sets contain one H and three T's; that is, $4p(1-p)^3$.

Suppose now that $p = \frac{1}{2}$. The probability that in sets of four successive elements from R there is just one element with the property H and three with T, is then $\frac{1}{4}$. But the probability that in such sets there are just two heads and two tails (here $r = 2$) is $\frac{3}{8}$. Hence, when n and p are fixed, the number determined by theorem 4.1 will vary with r. What value of r will yield a maximum value for this number? It can be shown that r must satisfy the condition

$$pn + p \geq r \geq pn + p - 1. \tag{4.2}$$

When n is very large, the value for which r yields a maximum may be taken to be pn. This means that the probability of sets with n successive elements containing just r elements with the property H is a *maximum*, when r is approximately equal to $pn;$ that is to say, the most probable value occurs for the case when the relative frequency of H in a set of n elements is approximately equal to the *limit* of the relative frequency of H in R.

A very important consequence, known as Bernoulli's theorem, can now be derived, which plays a central role in the practical use of the probability calculus. It can be stated as follows:

Let R be a reference class which is irregular with respect to a property H, and let prob(H, R) = p. Let R be broken up into sets of n successive elements each, and let ϵ be any positive number no matter how small. The probability that H will occur in these sets with *frequencies* lying in the interval $pn \pm \epsilon n$ (or with *relative frequencies* lying in the interval $p \pm \epsilon$) approaches 1 as a limit as n increases (4.3).

The following will illustrate the theorem: R is the irregular class of throws with a coin, and the probability of getting a head is taken to be $\frac{1}{2}$. We ask for the probability that in sets of n successive throws each, the frequency of heads will differ from $n/2$ by not more than $n/10$ (or that the relative frequency of heads will differ from $\frac{1}{2}$ by not more than $\frac{1}{10}$). According to Bernoulli's theorem, this probability tends to 1 as the value of n is increased. Thus, the probability that in ten successive throws there will be anywhere from four to six heads (i.e., that the relative frequency of heads will lie in the interval $\frac{1}{2} \pm \frac{1}{10}$ or $\frac{4}{10}$ to $\frac{6}{10}$) is .47; the probability that in thirty successive throws there will be anywhere from twelve to eighteen heads is .73; the probability that in fifty throws there will be anywhere from twenty to thirty heads is .84; the probability that in one hundred successive throws there will be anywhere from forty to sixty heads is .95; the probability that in five hundred throws there will be anywhere from two hundred to three hundred heads is .99, etc.

These numerical values are calculated with the help of mathematical techniques explained in treatises on probability. Of particular importance in the application of the probability calculus is the analytic formula

$$\phi(t) = \frac{2}{\sqrt{\pi}} \int_0^t e^{-x^2} dx \, ,$$

which is obtained from theorem 4.1 by a series of approximations. $\phi(t)$ is the probability that in sets of n successive elements of R which is irregular with respect to H, H occurs with a frequency lying in the interval

$$pn \pm t \sqrt{2p(1 - p)n} \, ,$$

377

where $\text{prob}(H, R) = p$; and tables of values of '$\phi(t)$' for different values of 't' have been constructed. In many problems the elements of R are assumed to take on any one of an infinite set of properties. Thus, in measuring the length of objects we may suppose that the measurements are carried out with great precision, and we may accordingly find it convenient to assume that the possible values of the length are real numbers (in the strict mathematical sense). The problems arising in such cases lead to the theory of continuous or geometrical probability.

The theorem of Bernoulli has been generalized by Poisson, and more recently Cantelli and Polya have given an important extension for it. A more general theorem than that of Bernoulli has been established by Tchebycheff, which was further elaborated by Markoff.

There is also an inverse of Bernoulli's theorem, which is sometimes referred to as Bayes's theorem and is obtained with the help of theorem 3.3; it has played an important part in the theory of statistics. But this theorem in inverse probability, with whose help the probability of a statistical hypothesis is to be established on the basis of the samples that have been drawn, suffers from the serious limitations and difficulties already pointed out in connection with theorem 3.3. Critical statisticians no longer make use of it. Statisticians have now developed more suitable procedures for handling the sort of problems Bayes's theorem was intended to solve; the method of maximum likelihood, recently proposed by R. A. Fisher, is a valuable and interesting contribution to this phase of theoretical statistics.

As already indicated, many theorems of the probability calculus are demonstrable only on the assumption that the reference classes are irregular—or in easily understood intuitive terms, that there is a general independence between the occurrence of a property on one occasion and its occurrence on another. However, in many fields of research (e.g., the behavior of gases) such independence cannot, on physical grounds, be assumed to exist. Nonetheless, it has been shown that the calculus of probability may be applied even to such domains with consistency and success.[5]

It is worth while mentioning a seemingly fatal criticism of the definition of 'probability' as 'the limit of relative frequencies.' Let R be irregular with respect to H, and let $f_1, f_2, \ldots . f_n, \ldots .$, be the series of relative frequencies of H in R after the first, second, nth terms. (Thus, in the series [ii] of p. 32, the relative frequencies are: $\frac{1}{1}, \frac{2}{2}, \frac{2}{3}, \frac{2}{4}, \frac{2}{5}, \frac{3}{6}, \ldots .$) Suppose that p is the limit of these frequencies. Then, once a number ϵ has been selected, there must be an N such that for every n greater than N the difference between f_n and p is less than ϵ; and this means that after the Nth term in R, the relative frequency of H in R will have to remain close to p. But according to theorem 4.1 there is a probability, which though small is not zero, that a very long run of successive H's will occur; and, according to the criticism being considered, there is this probability that even after the Nth term in R such a long run of

H's will set in. However, the criticism continues, a sufficiently long run of *H*'s will make some of the f_n's (with $n > N$) differ from *p* by *more than* ϵ. A contradiction is thus alleged in the calculus of probability developed on a limit basis. (Thus, suppose $p = .50$, $\epsilon = .01$, and that *N* is taken to be 100; and suppose that beginning with the one hundred and first throw a run of two hundred heads sets in. If *x* is the number of heads which have appeared in the first one hundred throws, $f_{300} = (x + 200)/300$ which differs from .50 by more than .01.)

However, the allegation of contradiction itself rests on a blunder and proceeds from a conception of probability according to which it is significant to ascribe a probability to a single occurrence. The probability specified by theorem 4.1 does *not* permit us to infer a long run of *H*'s starting with an *assigned* term in *R*, for example, with the *N* + 1th. That probability has for its reference class *R**, the class having as *its* elements *sets* of *n* successive elements from *R;* while *p* has *R* for its reference class. (It is possible and significant to ask for the probability that a definite run of *H*'s begins at some assigned term; but the answer to it is *not* given by theorem 4.1. Such a question involves the consideration of a *series* of reference classes such as *R*. An examination of this more complicated problem shows that the objection being considered confuses convergence in *R* with *uniform convergence* in a series of *R*'s.) There is thus no incompatibility between the statement that there is a nonvanishing probability of *H* occurring with a relative frequency *different* from *p*, in sets of *n* elements each (here the reference class is *R**); and the statement that the probability of *H* occurring in *R* is *equal* to *p*. It is true, of course, that in assigning a certain value to *N* we may be committing an error, because for a time the relative frequencies of *H* may diverge from *p*. But this does not establish a *contradiction* in the limit definition of probability; it simply testifies to the difficulty in fixing a value for *N*. That definition does not supply us with an effective method for obtaining a value for *N* either by calculation or in some other way; it merely asserts the existence of such a number *N*. For this reason it has been subjected to various criticisms by finitists, some of which will be considered below.[6]

4. *Formalization of the calculus of probability.*—Two points should be noted in the foregoing presentation of theorems in the calculus. In the first place, the theorems were formulated and explained in terms of an explicit definition of 'probability' as 'the limit of relative frequencies.' And, second, no primitive propositions were specified from which the theorems of the calculus may be derived with the help of the rules of logic. From the standpoint of a formal mathematical discussion, as well as from the point of view of modern methodology, these are defects, and they require a brief discussion.

If the functor 'prob' is introduced as the defined equivalent of 'the limit of relative frequencies,' every proposition of the calculus is simply a transcription of a theorem in the theory of limits; and every proposition is an analytic statement which can be certified on formal grounds alone. When the calculus is developed in this way, there is no need to supply a special set of primitive sentences: the primitive sentences sufficient for the theory of real numbers are also sufficient to establish every theorem in the calculus.

However, while it may be an advantage to have every theorem of the calculus an analytic sentence of arithmetic, the frequency interpretation of the functor 'prob' is not the only one that is possible. The state of affairs here is strictly analogous to what obtains in geometry. As geometry is employed in physics, the terms 'point,' 'line,' 'plane,' etc., which occur in Euclid, designate certain physical configurations; consequently, the propositions of geometry (such as that the angle-sum of a triangle equals two right angles) formulate measurable relations between physical configurations in exactly the same way as do the propositions of mechanics. But the derivation of geometric theorems from the primitive propositions of Euclid does not depend upon the correlations which happen to be established between terms like 'point' and 'line' and determinate physical configurations. Indeed, formal or demonstrative geometry is not a branch of physics: its theorems cannot be significantly characterized as empirically true or false, because the nonlogical terms in them (e.g., 'point') are uninterpreted. Only after *semantical rules* have been introduced (sometimes also called *co-ordinating definitions*), which correlate such uninterpreted terms with terms employed to designate empirical subject matter, is a formal geometry transformed into a part of natural science. By distinguishing between pure and physical geometry, not only do we avoid confusing questions of formal validity with questions of empirical fact but we also increase the applicability of pure geometry. Alternative co-ordinating definitions may be introduced, so that qualitatively different subject matters may be explored in terms of the same formal system. On the other hand, we may also find that, of the many distinct pure geometries which are logically possible, one system is a more effective means than another for organizing the materials of an empirical subject matter.

Similarly, it is not necessary to interpret 'probability' in terms of frequencies in order to develop a formal calculus of probability. The formalization of the probability calculus is of special importance because of the conflicting interpretations which have been given to the term 'probable,' as well as because of the wide range of opinion concerning the conditions under which probability statements are to be regarded as significant. As in the case of geometry, the probability calculus can be formalized in different ways, depending on what terms are selected as primitive, on the mathematical apparatus which is to be employed in developing it, and also upon the use to which it is to be put subsequently. Only one condition is usually observed in formalizing the calculus of probability: it is required that theorems which have been traditionally regarded as standard ones in the subject (such as the addition theorem or Bernoulli's theorem) be derivable from the primitives of the system.

Only a brief mention is here possible of some of the points of view from which the calculus may be formalized. To understand some of them, the distinctions (made in Vol. I, No. 3) will have to be recalled between the language of a science itself, the syntax language whose object-language is the language of science, and the semantic language of the language of science. Statements in the first language refer to what is commonly called the subject matter of the science, statements in the second refer to the order and possible arrangements of the expressions in the object-language, while statements in the third refer to the relations between an expression in the object-language and its subject matter. One difference between probability calculi arises from the fact that probability statements have been formulated in each of these three languages; another difference is due to the fact that some probability statements are metricized while others are not; and a third difference is due to several attempts to incorporate the probability calculus into a general logic which would include both necessary and probable inference.

Two broad classes of calculi of probability may be distinguished: those which provide a metric for the fundamental functor 'prob' and those which do not. Nonmetrical probability calculi may be further distinguished according as they introduce a definite serial order for probabilities or not. The motivation for the construction of nonmetrical calculi has usually been the desire to interpret probability statements in a nonfrequency sense. Such interpretations are often used by writers who have their eyes on the problems of induction and the estimation of evidence in history and legal procedures. A nonmetrical calculus has been developed by Keynes, but the subject is still in a very unsatisfactory and primitive state.

It is possible to formulate a frequency theory of probability both in the object-language and in the semantic language of a science, the choice between these alternatives being largely a matter of convenience. The probability statements of physics occur in its object-language, and most writers who approach the problems of probability from the natural sciences prefer an object-language formulation. Calculi in the object-language usually associate a number p with a probability, such that $0 \leq p \leq 1$. Some writers restrict the values of 'p' to rational numbers; others permit it to vary in the field of real numbers. A formalized calculus may be developed by taking 'prob(A, R)' as an uninterpreted two-place numerical expression; the logical properties of the expression are then determined by a set of postulates from which, with the help of the usual rules of logic, the standard theorems may be derived. These postulates are abstract in the sense that no restrictions are imposed on the possible interpretations of the functor other than the trivial one that every such interpretation satisfy these postulates. Abstract sets of postulates for probability have been given by Borel, Cantelli, Kolmogoroff, Popper, Reichenbach, and several other writers. (It is also possible to formalize the calculus by taking a one-place numerical functor as primitive, and subsequently defining a two-place functor in its terms; and there are other possibilities as well.) From an abstract mathematical point of view, the probability calculus is a chapter in the general theory of measurable functions, so that the mathematical theory of probability is intimately allied with abstract point-set theory. This aspect of the subject is under active investigation and has been especially cultivated by Borel, Fréchet, and a large number of French, Italian, Polish, and Russian mathematicians. In object-language calculi, the arguments '*A*' and '*R*' to the numerical expression 'prob(*A*, *R*)' are usually predicates or predicate variables; in semantic and syntactical calculi the arguments are usually names of sentences or variable designations of sentences. Postulates for metricized semantic calculi are similar to metricized object-language calculi except for the difference in the kind of arguments the functors take. Such semantic postulates have been given by Mazurkiewicz, Popper, and others.[7]

The possibility of interpreting a formal calculus of probability in different ways can be illustrated by the following list: (i) The functor 'prob' may be interpreted as the limit of relative frequencies in an infinite reference class; the postulates are then transformed into analytic propositions in the theory of real numbers. (ii) The functor may be defined as a relative frequency in a finite reference class; some of the postulates then become analytic propositions in the elementary arithmetic of rational numbers, while others must be suppressed. (iii) The functor may be interpreted, as in the classical Laplacian formulation, as the ratio of the cardinality of two sets of alternatives; the postulates are again converted into analytic propositions in elementary arithmetic. (iv) The functor may be interpreted, as by F. P. Ramsey, as a measure of "partial beliefs," where a degree of probability is a measure of the extent to which a man is prepared to act on a belief. (v) The functor may be interpreted as the ratio of two areas; the postulates become statements in some system of geometry. (vi) A proposal has been made by C. G. Hempel to introduce co-ordinating definitions for the functor in such a way that, while it will refer to relative frequencies in a class, the postulates are converted into synthetic statements of physics. (vii) A semantical interpretation has been given to the functor (Reichenbach), according to which it designates the truth-frequency of sentences in certain ordered classes of sentences. (viii) According to another interpretation, which also appears to be semantical, the functor denotes the "degree of falsifiability" of a theory (K. Popper). (ix) The functor has been interpreted as referring to the degree of a unique relation between a "proposition" and a set of premises. (It is not clear, however, how this view is to be understood. The language in which it is proposed sometimes suggests that the relation is a syntactical one holding between sentences, sometimes that it holds between the "possible facts" which the sentences designate, and sometimes that it is a semantical relation.) Some writers who take this interpretation do not regard the functor as a numerical one (Keynes), while others explicitly do so (H. Jeffreys).[8]

There is another standpoint from which the formalization of the calculus has been undertaken. Leibniz was one of the earliest writers to broach the possibility of a general formal logic in which the calculus of probability would occupy a central place. According to such a project, the standard relations of deducibility between propositions are to be regarded as limiting cases of a more inclusive relation of "probability implication." Many writers after Leibniz, including Boole and Peirce, kept the ideal of such a general logic a live one; and Clerk-Maxwell went to the extent of declaring that "the true logic for this world is the calculus of probability." However, little was done to actualize this possibility until the very recent development of polyvalent logical calculi. The fusion of familiar formal logic and the calculus of probability into one compendent formal system is now actively investigated. But, although much important work has been already done, there is at present still no satisfactory system of such a general logic.

The calculi of n-valued logics of sentences (with n a finite integer) were first developed by J. Lukasiewicz and E. Post. These calculi reduce to the standard sentential calculus (e.g., of *Principia mathematica*) when $n = 2$; that is, when a sentence is permitted to take just two "truth-values," namely, truth and falsity. There are certain partial analogies between the theorems of polyvalent logics and theorems of the probability calculus when the latter is suitably formulated; and a number of writers, including Mazurkiewicz, Reichenbach, and Zawirski, have been exploiting these analogies, with the intent of formalizing the calculus of probability as a polyvalent sentential calculus. Reichenbach's method, stated in outline, consists in interpreting an infinite-valued sentential calculus (with values lying in the interval 0 to 1 inclusive) so that each truth-value is the limit of the relative frequency with which the members of definite sequences of propositions are true. He has urged, moreover, that such an infinite-valued "probability logic" is the one most appropriate for science—on the ground that no empirical statement can be completely verified and can therefore be associated with a "truth-value" which in general is different from 0 (falsity) and 1 (truth). Reichenbach's proposal is not free from technical difficulties, and most students are not convinced that he has achieved a fusion of the probability calculus and a general logic of propositions. For example, Reichenbach's polyvalent "probability logic" contains expressions which apparently are subject to the rules of the ordinary two-valued logic; and it therefore seems that his probability logic is constructed upon a basic two-valued schema. Again, his probability logic is nonextensional, in the strict sense of this term in standard use, while the general system of logic commonly employed in mathematics and physics is extensional; it is therefore not easy to see how the latter can be a specialization of the former.[9]

Interest in the fusion of formal logic and the calculus of probability into one comprehensive system has also been exhibited by physicists impressed by the part which probability statements play in modern quantum theory. In that theory certain noncommutative operators occur, as a consequence of some of the fundamental physical assumptions of the theory; and it is possible to regard such operators as a species of logical multiplication upon propositions dealing with subatomic phenomena. However, instead of superimposing such noncommutative multiplications upon the general framework of a logic of propositions in which multiplication (i.e., and-connection) is commutative, proposals have been made to revise the general logic of propositions. According to some of these proposals, a multiplication which is noncommutative will be governed by the formal rules of the logic of propositions and will not be introduced simply as a consequence of a special physical theory. Attempts to re-write quantum mechanics upon the basis of an altered sentential calculus have been made by J. von Neumann, G. Birkhoff, M. Strauss, and others. But researches in this field have not yet gone far enough to permit a judgment on the feasibility and convenience of the proposed emendations.[10]

6. Nonfrequency Interpretations of Probability Statements

We must now briefly consider other interpretations of probability statements than the frequency view proposed earlier in the present section.

1. *The classic conception of probability.*—As already noted, the mathematical theory of probability was first developed in connection with games of chance, and the point of view from which it was cultivated received its classic formulation in the treatise of Laplace. According to the Laplacian view, all our knowledge has a probable character, simply because we lack the requisite skill and information to forecast the future and know the past accurately. A degree of probability is therefore a measure of the amount of certainty associated with a belief: "I consider the word *probability*," De Morgan explained, "as meaning the state of the mind with respect to an assertion, a coming event, or any other matter on which absolute knowledge does not exist." What is required for a mathematical treatment of probability, however, is an exact statement of how this measure is defined; and the classical account is as follows.

Judgments of probability are a function of our partial ignorance and our partial knowledge. We may know that in a given situation the process studied will have an issue which will exhibit one out of a definite number of alternative properties; thus, in tossing a die any one of the six faces may turn up. (These alternative properties have been called the "possible events.") On the other hand, we may have no reason to suppose that one of these events will be realized rather than another, so that, as Laplace remarked, "in this state of indecision it is impossible for us to announce their occurrence with certainty." But a measure of the appropriate degree of belief in a specific outcome of the process can be obtained. We need simply analyze the possible outcome into a set of "equipossible alternatives," and then count the number of alternatives which are favorable to the event whose probability is sought. This measure, the probability of the event, is a fraction whose numerator is the number of favorable alternatives, and whose denominator is the total number of possible alternatives, provided that all the alternatives in question are equipossible. Thus, the probability of obtaining six points with a pair of dice is $\frac{5}{36}$, because the dice can fall in any one of thirty-six equally possible ways, five of which are favorable to the occurrence of six points in all. On the basis of this definition, the probability calculus was developed as an application of the theory of permutations and combinations.

Almost all writers on probability in the nineteenth century (e.g., Poisson, Quetelet, De Morgan, Boole, Stumpf), and many contemporary mathematicians (e.g., Borel, De Finetti, Cantelli, Castelnuovo), follow Laplace with only relatively minor variations. Because of its historical role, as well as because of its contemporary influence, we shall briefly examine this view.[11]

a) According to the Laplacian definition, a probability statement can be made only in such cases as are analyzable into a set of equipossible alternatives. But, while in some cases it seems possible to do this, in most cases where probability statements are made this is not possible. Thus suppose that a biased coin is assigned the probability of .63 that it presents a head when tossed; there is no clear way in which this number can be interpreted as the ratio of equipossible alternatives. This is perhaps even more evident for statements like 'The probability that a thirty-year-old man will live at least another year is .945.' It is absurd to interpret such a statement as meaning that there are a thousand possible eventuations to a man's career, 945 of which are favorable to his surviving at least another year. Moreover, the Laplacian definition requires a probability coefficient to be a rational number. But irrational numbers frequently occur as values for such coefficients, and there is no way of interpreting them as ratios of a number of alternatives. Thus, on the basis of certain assumptions, it can be calculated that the probability that two integers picked at random are relatively prime is $6/\pi^2$. This number cannot be made to mean that there are π^2 equally possible ways in which pairs of integers can be picked, six of which are favorable to getting relative primes.

b) Writers on the subject have not always been clear as to whether they regarded a probability as the measure of a (psychological) belief, or whether they regarded it as a measure of the degree of belief one *ought* to entertain as reasonable. If a probability coefficient is the measure of a degree of actual certainty or the strength of a belief, the addition and multiplication of probabilities require that we determine procedures for combining certainties or beliefs in some corresponding manner. There are, however, no known methods for adding beliefs to one another, and indeed it is difficult to know what could be meant by saying that beliefs are additive. The proposals of Ramsey and De Finetti, to measure strength of beliefs by the relative size of the bets a man is willing to place, are based on a dubious psychological theory; and at least Ramsey's proposal leads directly to a definition of probability in terms of relative frequencies of actions. On the other hand, if probability is a measure of the amount of confidence one ought to have in a given situation, the Laplacian view offers no explanation of the source of the imperative. It is possible, finally, that a probability coefficient is simply a conventional measure of a degree of belief; in that case, however, probability statements turn out to be bare tautologies.

c) According to the Laplacian definition, the alternatives counted must be equally possible. But if 'equipossible' is synonymous with 'equiprobable,' the definition is circular, unless 'equiprobable' can be defined independently of 'probable.' To meet this difficulty, a rule known as the Principle of Indifference (also as the Principle of Insufficient Reason and as the Principle of the Equal Distribution of Ignorance) has been invoked for deciding when alternatives are to be regarded as equiprobable. According to one standard formulation of the rule, two events are equiprobable if there is no known reason for supposing that one of them will occur rather than the other.

It can be shown, however, that, when this form of the rule is applied, incompatible numerical values can be strictly deduced for the probability of an event. An emended form of the rule has been therefore proposed, according to which our *relevant* evidence must be *symmetrical* with respect to the alternatives, which must not, moreover, be divisible into further alternatives on the given evidence. This formulation seriously restricts the application of the Principle of Indifference. Apart from this, however, two points should be noted: A coin which is known to be symmetrically constructed (so that according to the principle its two faces are to be judged as equiprobable) may nevertheless present the head more frequently than the tail on being tossed; for the relative frequency of heads is a function not only of the physical construction of the coin, but also of the conditions under which it is tossed. Second, no evidence is perfectly symmetrical with respect to a set of alternatives. Thus, the two faces of a coin are differently marked, they do not lie symmetrically with respect to the earth's center at the instant before the coin rises into the air, etc. The emended rule therefore provides that it is only the relevant evidence which is to be considered. But if 'relevance' is defined in terms of 'probable,' the circle in the Laplacian definition is once more patent; while, if judgments of relevance are based on definite empirical knowledge, the ground is cut from under the basic assumption of the Laplacian point of view.

d) It is usually assumed that the ratio of the number of favorable alternatives to the number of possible ones (all being equipossible) is also a clue to the relative frequency with which an event occurs. There is, however, no obvious connection between the 'probability of obtaining a head on tossing a coin' as defined on the classical view, and 'the relative frequency with which heads turn up.' For there is in fact no *logical* relation between *the number of alternative ways* in which a coin can fall and *the frequency* with which these alternatives in fact occur. It has, however, often been supposed that Bernoulli's theorem demonstrates such a connection. For as already explained, according to that theorem if the probability of head is $\frac{1}{2}$, then the probability approaches 1 that in n tosses there are approximately $n/2$ heads as n increases. But the supposition that Bernoulli's theorem establishes a relation between a priori (i.e., determined in accordance with the classical definition) and a posteriori probabilities (i.e., determined on the basis of relative frequencies of occurrence) is a serious error. It commits those who make it to a form of a priori rationalism. For within the framework of the classical interpretation of the calculus, Bernoulli's theorem simply specifies the relative number of certain types of equiprobable alternatives, each consisting of n tosses; it is no more than a theorem in arithmetic and does *not* permit us to conclude that these alternatives will *occur equally often*. That is to say, only if the expression 'The probability of heads is $\frac{1}{2}$' designates a relative frequency of occurrence, can the phrase 'The probability approaches 1' be legitimately interpreted as designating relative frequencies of occurrences.

2. *Probability as a unique logical relation.*—A number of modern writers, conscious of the difficulties in the classical view of probability as a measure of strength of belief, have advanced the view that probability is an objective logical relation between propositions analogous to the relation of deducibility or entailment. According to this version, a degree of probability measures what is often called "the logical distance" between a conclusion and its premises. The evaluation of a degree of probability therefore depends upon recognizing the inclusion, exclusion, or overlapping of logical ranges of possible facts. Though varying considerably among themselves, something like this view (which has had its forerunners in Leibniz and Bolzano) is central to von Kries, Keynes, J. Nicod, F. Waismann, and several other writers. Only the standpoint of Keynes will be examined here.[12]

For Keynes, probability is a unique, unanalyzable relation between two propositions. No proposition as such is probable; it has a degree of probability only with respect to specified evidence. This relation of probability is not a degree of subjective expectation; on the contrary, it is only when we have perceived this relation between evidence and conclusion that we can attach some degree of "rational belief" to the latter. (As already noted, Keynes's formulation of his view is not unambiguous. His occasional language to the contrary notwithstanding, it does not seem likely that he regards his probability relation as a syntactical one. The present writer is inclined to the opinion that it is a semantical relation.) It is characteristic of Keynes's standpoint that the *secondary* proposition, which asserts that a proposition p has the probability relation of degree a to the proposition h, can and must be known to be true "with the highest degree of rational certainty." Such a highest degree of rational certainty is obtainable, according to Keynes, when we see that the conclusion of a syllogism follows from its premises, as well as when we see that a conclusion "nearly follows" from its premises with degree a of probability. However, degrees of probability are not quantitative and are not in general capable of measurement; indeed, according to Keynes, probabilities cannot in general be even ordered serially, although in some cases they are comparable. The comparison of probabilities, whenever this is possible, is effected with the help of the modified Principle of Indifference mentioned above; and the judgments of relevance which the principle presupposes are themselves direct judgments of degrees of probability. In terms of such an apparatus of concepts, Keynes develops a calculus which formulates the relations between comparable probabilities, and finally explains how and under what limited circumstances numerical values may be assigned to degrees of probability.

Although Keynes avoids some of the difficulties of the classical view of probability, his general standpoint has difficulties of its own. Omitting all discussion of the technical difficulties in his calculus, we shall confine ourselves to a brief mention of three central issues.

a) On Keynes's view we must have a "logical intuition" of the probable relations between propositions. However, few if any students can be found who claim for themselves such an intuitive power; and no way has been proposed to check and control the alleged deliverances of such direct perceptions in cases where students claim it. Moreover, the possession or lack of this power is wholly irrelevant in the actual estimation of probabilities by the various sciences. No physicist will seriously propose to decide whether two quantum transitions are equi-probable by appealing to a direct perception of probability relations; and, as N. R. Campbell remarked, "anyone who proposed to attribute to the chances of a given deflection of an a-ray in passing through a given film any sense other than that determined by frequency could convince us of nothing but his ignorance of physics."

b) Since on Keynes's view numerical probabilities can be introduced only when equiprobable alternatives are present, he cannot account for the use of numerical probabilities when such an analysis is not possible. Moreover, like the classic interpretation, Keynes cannot establish any connection between numerical probabilities and relative frequencies of occurrences. His theory, when strictly interpreted, is incapable of application to the problems discussed in physics and statistics, and at least from this point of view remains a vestal virgin.

c) On Keynes's view it is significant to assign a probability, with respect to given evidence, to a proposition dealing with a single occasion. For example, it is permissible to declare that on given evidence the probability of a given coin falling head uppermost *on the next toss* is $\frac{1}{2}$. However, the coin, after it is thrown and comes to rest, will show a head or it will show a tail; and no matter what the issue of the given throw is, the probability of obtaining a head on the initial evidence is and remains $\frac{1}{2}$. No empirical evidence is therefore relevant either for the confirmation or for the disconfirmation of that probability judgment, unless we invoke indirectly a relative frequency in a group of statements—which would be contrary to Keynes's intent. But this is to fly in the face of every rule of sound scientific procedure. A conception of probability according to which we cannot in principle control by experiment and observation the probability statements we make is not a conception which recommends itself as germane to scientific inquiry.

Except for matters to be discussed in Section III, the difficulties which have been pointed out for the classical and the logical interpretations of probability do not embarrass the frequency view. For this negative reason, but especially because it is in accord with scientific practice, the frequency interpretation of probability is the one most suitable for the first class of statements which was specified at the beginning of this section.

III. Unsettled Problems of General Methodology

7. Logical Problems of the Frequency Interpretation of the Probability Calculus

It was shown in Section II that the definition of 'probability' as 'the limit of relative frequency' is *suggested* by common practice in assigning probability coefficients. It has been argued that a probability of $\frac{1}{2}$ for head turning up when a coin is tossed means, roughly, that in half the cases of flipping a coin the head is presented. However, such a statement does not mean that in every two tosses a head turns up just once, for in that case it would be absurd to apply it to an odd number of throws; and we would not regard the statement as erroneous if, after getting a tail, we did not get a head on the next succeeding throw. Accordingly, a less misleading explanation of what a probability of $\frac{1}{2}$ signifies is that in a long run of throws the relative frequency of heads is approximately $\frac{1}{2}$. But it has also been pointed out that a definition of 'probability' as 'the approximate ratio of frequencies in the long run' is not precise and is not suitable for mathematical purposes. A definition in terms of limits, on the other hand, has the requisite precision, and a logically consistent calculus can be developed on such a basis. The convenience and fruitfulness of such a definition for the purposes of a calculus of probability are indeed beyond question.

However, from the point of view of the application of the calculus to empirical matters, it would be of little profit to have a precise mathematical definition of 'probability' if as a consequence every probability statement would acquire a theoretical content which cannot be controlled by acknowledged empirical methods. But an examination of the form of probability statements, when these are interpreted in terms of limits of relative frequencies, seems to indicate that such is indeed the case. This may be seen concretely in the following way. Suppose we test the hypothesis that the probability of heads is $\frac{1}{2}$ by flipping the coin a thousand times, and suppose we get a run of a thousand heads. We might be inclined to conclude that the hypothesis has been definitely proved erroneous. However, on that very hypothesis such a run of heads is not excluded, since that hypothesis asserts something about the limiting ratio of heads in an *infinite* class and not in a *finite* one. In general, that hypothesis is compatible with *any* results obtained in any finite number of throws; and, conversely, a given result within a finite class of throws is compatible with any hypothesis about the numerical values of the probability. In short, it seems that no direct statistical evidence obtainable from actual trials (which must obviously be finite in number) can establish or refute a probability statement.

(It should be observed, moreover, that this difficulty is not obviated, as some writers have thought, by employing a less precise definition for 'probability.' For example, if we define it in terms of approximate ratios in long runs, a finite number of observations on the direct evidence for a probability statement will still not suffice to establish or refute it completely and unambiguously.)

The formal argument is as follows. If $f_1, f_2, \ldots f_n, \ldots$, is the series of relative frequencies of heads and ϵ a positive number, to say that the probability of getting a head is $\frac{1}{2}$ is to say that $\frac{1}{2}$ is the limit of these ratios. And this means that for every ϵ there is an N, such that for every n, if $n > N$, then the absolute difference of f_n and $\frac{1}{2}$ is less than ϵ. Or, in the notation of modern logic,

$$(\epsilon) \, (\exists \, N) \, (n) \, [(n > N) \supset (|f_n - \tfrac{1}{2}| < \epsilon)] \, .$$

This statement contains three quantifiers, the two universal quantifiers 'for every ϵ' and 'for every n,' and the existential quantifier 'there is an N.' Because of the presence of the universal quantifiers, this statement cannot be established by examining a finite number of ϵ's and n's; or, in the language proposed by Carnap, the statement is not completely confirmable.

This situation is familiar throughout science. For example, the statement 'All bodies attract each other inversely as the square of their mutual distances' is not completely confirmable either. But, according to strict logic and textbook scientific method, this latter statement is capable of *complete disproof* by *one negative instance;* and it is usually said, therefore, that, although we can never be in the position to assert the truth of universal statements, we may be in the position to assert their falsity.

However, probability statements do not fall under this dictum. For, in order to completely disprove such a statement, its formal contradictory would have to be completely confirmed. But the formal contradictory of the specimen probability statement is: There is an ϵ, such that for every N, there is an n, such that $n > N$, and the difference between f_n and $\frac{1}{2}$ is not less than ϵ. In symbolic notation

$$(\exists \epsilon)\ (N)\ (\exists n)\ [(n > N) \cdot (\,|f_n - \tfrac{1}{2}\,| \geq \epsilon)]\,.$$

However, this statement also contains a universal quantifier, namely, 'for every N,' so that it cannot be completely confirmed.

In sum, therefore, a probability statement can be neither completely confirmed nor completely disconfirmed.

Many writers have therefore concluded that probability statements interpreted in terms of relative frequencies are devoid of empirical meaning because what they assert cannot be controlled by determinate empirical procedures. Such a conclusion, if it were warranted by the facts, would be fatal to a frequency interpretation of probability. For it is a cardinal requirement of modern science that its statements be subject to the criticism of empirical findings. This simply means that not every state of affairs can be confirmatory evidence for a given statement and that observable states of affairs must be specifiable which would be acknowledged as incompatible with its truth. On the other hand, such a conclusion is paradoxical because in actual practice probability statements interpreted in terms of frequencies *are* accepted or rejected on the basis of empirical evidence; and no one seriously doubts that we order affairs of everyday living, of industry, and of science with their help.

What is required, therefore, is a specification of the semantical and pragmatic *rules* in accordance with which probability statements are accepted and rejected on the basis of empirical findings. Although a complete set of rules cannot be given at present, so that the problem is in a very unsettled condition, it is believed that the following observations will be found relevant to the issue raised.

a) An objection often made to the limit definition of probability is that limits, in the strict sense of the term, do not exist for empirically determined relative frequencies and that in actual statistical material the ratios of frequencies fluctuate more or less widely. Such an objection, however, should in all consistency be made also to the use of general mathematical analysis in the natural sciences. For the limit concept is employed not only in probability but elsewhere also. For example, the masses or centers of gravity of bodies are frequently calculated with the help of the integral calculus, and the integrations are performed on the assumption that the mathematical functions which specify the density of the bodies are continuous; the calculation of these quantities thus involves limits at several places. Moreover, the assumption of a continuous density distribution is not warranted by our present theories of matter as discontinuous. We do not, however, reject the powerful tools of analysis for these reasons. An even simpler illustration of the use of limits occurs in measurement, against which no one seems to raise difficulties of the sort indicated. Every actual measurement, for example, of the length of the diagonal of a square, yields a *rational* number; nonetheless, in theoretical work we frequently employ irrational numbers, such as $\sqrt{2}$, for specifying lengths; and irrational numbers involve limit notions. The reason for employing terms involving limits in probability theory, as elsewhere, is the same: we thereby obtain powerful and economical methods in making *mathematical transformations*. And the reason why the use of such "calculus terms" and the procedures requiring them is countenanced in the natural sciences (even when direct empirical evidence and theoretical considerations indicate that the conditions for their use are not fully satisfied) is that we know how to *correlate* with them *groups* of directly measured magnitudes lying in certain intervals.

b) It is indeed a naïve conception of scientific method according to which the statements of science (whether singular or general) are to be rejected on the ground of a single negative instance. It was pointed out in Section I that, even in the exact sciences of measurement, the numerical values of magnitudes as predicted by a theory are not in precise agreement with the numerical values obtained by actual measurement and observation; the theory of errors had its genesis in the study of just such situations. A theory is not in general dismissed as false or worthless because the confirmation of its predictions by observation is only approximate—even though *formally* every deviation from a predicted value of a magnitude is a negative instance for a theory. The amount of allowable deviation between predicted and observed values is not specified by the theory itself, and even a "large" deviation may not be decisive against the theory. The reasons for this are twofold: An empirically testable consequence of a theory does not follow from the theory *alone*, but from it conjoined with statements reporting matters of observation and possibly other theories. Consequently, an apparent negative instance for a theory may be argued to be incompatible not with *it* but only with some of the other premises of the argument; and by a suitable alteration in the assumptions from which the testable consequences are drawn, the theory itself may be retained as in accordance with the "facts." Second, the amount of allowable deviation between predicted and observed values of a magnitude may be a function of a number of variable factors, such as the number of observations made, the purposes for which the inquiry is conducted, the kind of activity which the theory is intended to coordinate and foretell, or the character of the instruments by means of which the testing is carried on. These factors cannot in general be completely enumerated or specified in detail, although those who conduct researches have been trained to make allowance for them in the concrete cases before them.

A crude illustration of this second point, for the case when *direct statistical evidence* for a probability statement is evaluated, can be constructed as follows. Suppose the hypothesis that $\frac{1}{2}$ is the probability of obtaining a head with a coin is to be tested, by tossing it one hundred times. According to the hypothesis, we may expect approximately fifty heads. If heads turned up forty-nine times, we would regard this as confirming the hypothesis; if heads turned up forty-five times, this may still be regarded as confirmatory; but if heads turned up only twenty times, we might suspect that the coin is loaded and doubtless propose a different value for the probability of getting a head. That is to say, somewhere between getting twenty and getting fifty heads in one hundred throws, we might fix a value such that a frequency less than it is to be taken as disconfirming the hypothesis of $\frac{1}{2}$. In other words, the actual hypothesis which would be tested under these circumstances is that the relative frequency of heads lies in an *interval* $\frac{1}{2} \pm \delta$, where the positive number δ is not fixed once for all but varies with circumstances. Now the probability of obtaining deviations of specified magnitudes from $\frac{1}{2}$ (on the assumption that *sets* of such trials are repeated indefinitely) can be calculated with the help of Bernoulli's theorem; and *this* probability depends upon the initial hypothesis that p is the probability of getting a head as well as upon the number n of throws which are made. Hence δ will often be a function of p and n. But it may be a function of other factors as well: e.g., of our knowledge of the physical construction of the coin and of the circumstances under which it is thrown, of the size of our fortune if we are gambling, etc. The definition of 'probability' in terms of 'limit' is therefore important for the purpose of constructing a consistent and powerful *calculus*. The calculus itself is instrumental in effecting transitions from one set of empirically controllable statements to other such sets. Provided that appropriate semantical and pragmatic rules are instituted for applying the calculus, it is not a serious objection to it that some of its terms cannot be taken as *descriptive* of the subject matter of science.

In modern theoretical statistics various methods have been devised for evaluating the goodness of an estimate of parameters, such as p, which characterize a hypothetical infinite population. According to the older methods of Lexis, the *aggregate sample* on the basis of which the estimate is made requires to be analyzed into sets of elements which are similar in certain relevant respects; the stability or fluctuation of the estimate in these various groups is then studied. In the more recent methods of R. A. Fisher, J. Neyman, and others, "measures of credibility" are introduced, some of which are carefully distinguished from probabilities. According to these methods, the values assigned to the hypothetical probabilities must meet explicitly stipulated conditions of stability under repeated samplings, and must also make these measures of credibility a maximum. It is not possible at this place to enter into this subject in greater detail.

In many cases, no determinate numerical value can be assigned to a probability, not because a frequency interpretation of probability statements is not relevant but because relevant statistical information is lacking. For example, the proposition is often asserted that when the barometer falls it is highly probable that it will rain, although no numerical value is usually specified for this "high probability." Such a statement clearly means that the relative frequency of rain within a few hours, in the class of cases where the barometer falls, is greater than $\frac{1}{2}$ and possibly close to 1. But lacking precise statistical information, the high probability is assigned and confirmed on the basis of general impressions as to the behavior of the weather. In still other cases, such as that involved in estimating the probability of a witness speaking the truth, the statistical data may be even more meager, and the general impressions upon which we base our estimates may be highly unreliable and even worthless.

We have no final assurance that a hypothesis as to the numerical value of a probability is a correct one. However, the method of inquiry we employ is a self-corrective one, and in general we place greater reliance upon our rules of procedure and their *net* results than upon particular conclusions obtained. We are not in a position to assert with finality that the empirical frequencies we obtain do converge to a limiting value. But as Peirce and more recently Reichenbach have pointed out, if these ratios do tend to remain within certain narrow intervals, we can discover what those intervals are by a repeated and systematic correction of the estimates which are suggested by the samples we continue to draw.

c) Thus far, only the *direct statistical* evidence for a probability statement has been considered. But it was explained in Section II that, whenever such a statement is part of an inclusive system of statements, the evidence may be indirect and even of a nonstatistical character. There are, indeed, the following possibilities: Let S be a probability statement of the form '$\text{prob}(A, R) = p$'; and let Σ be a class of statements which in general will contain *singular* statements reporting matters of observation (e.g., statements which ascribe a property to a definite space-time region), as well as general or theoretical statements some of which may have the form of probability statements.

(i) The value of p in S may be estimated directly from statistical evidence concerning the frequency of A in R; this case has already been considered.

(ii) From S and Σ another probability statement S_1 may be derived which may be tested by direct statistical evidence for S_1. Thus, if S ascribes the probability of $\frac{1}{2}$ to a coin falling head uppermost, S_1 may ascribe the probability of $\frac{1}{4}$ to the coin falling heads up twice in succession.

(iii) From S and Σ a statement S_2 may be derived which is nonstatistical. Thus, let S ascribe the probability of $\frac{1}{3}$ to an atom in a state with a magnetic moment of one suffering a transition into a state with a magnetic moment of two when a deflecting field is introduced; then S_2 may assert that the intensity of the ionic current across the path of the molecular beam is of a specified magnitude. In this case, no problems arise in connection with the empirical control of probability statements which do not arise in connection with other statements of science.

(iv) The value of p in S may be deduced from Σ. Thus, if Σ contains the Schrödinger equation together with a number of boundary conditions, we can calculate the numerical value of the probability that an atom in a given space-time region will be a in a specified state.

Although some of the formal logical problems in connection with cases (iii) and (iv) have not been thoroughly worked out, such cases do occur. And it is evident from them that the correctness of a given hypothesis as to the numerical value of a probability may be controlled in much the same way as the more familiar nonstatistical hypotheses of science are controlled.

In recent years, following a suggestion of Poincaré, what is sometimes known as "a causal theory of probability" has been developed by G. D. Birkhoff, E. Hopf, and others. The main idea of these researches is the deduction of a probability value (e.g., the probability of the ball in roulette coming to rest in a red sector) from underlying dynamical assumptions governing the average values of certain quantities with increase of time. It is incorrect to maintain, as some have done, that a probability value can be deduced from a dynamical theory which contains no material assumptions about the distribution of frequencies or average values. Nevertheless, these researches, apart from their technical interest, emphasize one very important point: An estimate of a probability which is made simply on the basis of unanalyzed samples or trials is not likely to be a safe basis for prediction. If nothing is known concerning the mechanism of a situation under investigation, the relative frequencies obtained from samples may be poor guides to the character of the indefinitely large population from which they are drawn. Thus, because we know very little about the mechanism of historical changes in human societies, it would be unsafe to use the life-probabilities computed in the first quarter of the present century as a basis for conducting a life insurance business in America two centuries hence. On the other hand, because we know something about the mechanism of biological heredity, a relatively few observations on the number and types of descendants of a plant may suffice to confirm hypotheses about the probability of certain types recurring. Again, we assign a value to the probability of getting heads on a freshly minted coin with great assurance, even before making any actual trials with it, because the homogeneity of the products of national mints, as well as of the conditions under which the coin would be thrown, are fairly well established. In general, therefore, the amount and kind of evidence required for probability statements depend on their interconnections with the body of our knowledge and theories at a given time.[13]

(v) Some writers, notably Reichenbach, have maintained that, while probability statements are incapable of complete confirmation or disconfirmation, nevertheless a degree of probability (in the frequency sense) can be attached to them. Such a proposal, it turns out, involves a hierarchy of probabilities, in which every probability statement on one level is subject matter for probability statements on a higher level; it is a conception which has stimulated the development of a "probability logic" referred to in Section II. If such a proposal could be implemented with an unambiguous and convenient method for assigning probabilities to probability statements, it would go a long way to solving definitively the logical problem to which the present section has been devoted. Reichenbach's writings make important contributions toward formulating such a method. However, a probability statement is a *general* statement, as was explained on page 52; and we reserve the discussion of the probability of general statements, hypotheses, or theories for § 8.

8. Probability and Degree of Confirmation or Weight of Evidence

At the outset of Section II two classes of statements containing the term 'probable' were distinguished. The members of the first class have now been shown to require a frequency interpretation, and the statements in it are subject to the rules of the calculus of probability. We shall now inquire whether the second class is similar to the first in these respects.

A common objection to the frequency theory of probability is that, although probability statements concerning single occasions or single propositions are often asserted and debated, it is meaningless to assert such statements in terms of the frequency theory. For example, writers like Keynes have urged that such statements as 'It is probable on the evidence that Caesar visited Britain' and 'The evidence makes it improbable that all crows are black' cannot be analyzed in terms of relative frequencies; and they have concluded that a conception of probability is involved in them which is different from, and "wider" than, the frequency view. Frequentists have retorted, quite rightly, that such statements *are* without meaning, if they *literally* attribute a probability in the frequency sense to a single proposition; but frequentists have also urged that such statements do have significance if they are understood as *elliptic* formulations.

There is little doubt that many probability statements which are apparently about single propositions are incomplete formulations and that, when they are suitably expanded, they conform to the conditions required by the frequency theory. On the question, however, whether *all* probability statements about single propositions are to be analyzed in this way there is considerable difference of opinion. This disagreement not only divides frequentists from nonfrequentists like Keynes but it also represents a division among those who subscribe to a frequency interpretation for the first class of statements previously mentioned.

This difference of opinion concerning the range of applicability of the calculus of probability has a long history. Earlier writers on the subject believed that the calculus was the long-sought-for instrument for solving all problems connected with estimating the adequacy of evidence. In particular, it was maintained that the problems associated with establishing general laws on the basis of examined instances and with obtaining some measure for the reliability of predictions (the traditional problems of induction) were part of the subject matter of the mathematical theory of probability. Bayes's theorem and the Rule of Succession were commonly employed for these purposes, and Jevons explicitly regarded induction as a problem in inverse probabilities. On the other hand, writers like Cournot and Venn, two of the earliest writers to propose a frequency interpretation of the probability calculus, were equally convinced, though for different reasons, that the calculus was not relevant to the problems of induction. More recently, Keynes and Reichenbach, arguing from diametrically opposite standpoints, agree on the point that the term 'probable' can be given a consistently univocal meaning; and Reichenbach has given the most complete account at present available of how to extend the frequency view to the consideration of the probability of scientific theories. But other contemporary frequentists, such as Carnap, von Mises, Neurath, and Popper, though supporting the frequency interpretation for a very large class of probability statements, do not believe such an interpretation is appropriate for every statement which contains the word 'probable.' This latter group of writers rejects the notion of a "logical probability" as developed by Keynes and others; but it distinguishes between 'probable' employed in the sense of 'relative frequency' and 'probable' employed in the sense of 'degree of confirmation' or 'weight of evidence.'

It is possible, therefore, to distinguish writers on probability according to the following schema: (1) Writers who interpret 'probable' in a *univocal* sense; such writers differ among themselves according as they accept the classical view, the view of probability as a unique logical relation, or the frequency view. (2) Writers who do not believe that the term 'probable' can be interpreted in precisely the same manner in every one of the contexts in which it occurs.

The present state of research, therefore, leaves the issue unsettled as to the scope of the frequency theory of probability. We shall examine the points at issue, but our conclusion will of necessity have to be highly tentative. We shall concern ourselves explicitly with statements ascribing a probability to a theory, because of lack of space; but the discussion will apply without essential qualifications to probability statements about singular statements like 'Caesar visited Britain,' whenever such probability statements are not analyzable as elliptic formulations involving relative frequencies. By 'theory' will be understood any statement of whatever degree of complexity which contains one or more universal quantifiers, or a set of such statements.

1. *The probability of theories.*—We begin with examining the proposal to interpret probability statements about theories in terms of relative frequencies; and, since Reichenbach has expounded this proposal more fully than anyone else, we shall examine his views. Reichenbach has given two distinct but

allied methods for defining "the probability of a theory." The first of these methods has received an improved formulation by C. G. Hempel, which avoids serious difficulties present in Reichenbach's own version. It should be noted that the definitions given by both methods are semantic ones.

a) Let T be some theory, for example, the Newtonian theory of gravitation. Let C_n be a class of n singular statements, each of which specifies an initial state of a system. (For from T alone, without the specification of initial conditions, no empirically controllable consequences can be obtained; thus, the mass, initial position, and velocity of a planet must be assigned before a future state of the planet can be predicted.) From every such statement with the help of T, other statements may be derived, some of which are empirically controllable by an appropriate observation. Therefore, let E_n be the class of n such singular statements derived from C_n with the help of T. We suppose that a one-to-one correspondence is established between the elements of C_n and E_n; and without loss of generality we shall suppose that every statement in C_n is true. (From a single statement in C an indefinite number of statements belonging to E may be derived; but we can simply *repeat* a statement in C for every one of the distinct consequences drawn from it.) Let $\mathrm{nu}(E_n)$ be the number of statements in E_n which are true. The relative frequency with which a statement in E_n is true when its corresponding statement in C_n is true is given by $\mathrm{nu}(E_n)/n$. Suppose now that n increases indefinitely, so that C_n will include all possible true initial conditions for T, while E_n will include all the possible predictions which are made from them with the help of T. The numerical expression

$$\mathrm{prob}(E,\, C) = \lim_{n \to \infty} \frac{\mathrm{nu}(E_n)}{n}$$

will then be the probability that the consequences, obtained with the help of T from appropriate initial conditions, are true. This, in essence, is Reichenbach's first method of assigning a probability to a theory T.

Although the foregoing exposition requires supplementation

in several ways, there seems to be little question that a precise definition for 'the probability of a theory' can be given on a relative frequency basis. It is, however, by no means evident that such a definition formulates the concept people seem to be employing when they discuss the probability of theories.

(i) On the foregoing definition the probability of a theory is the limiting value of relative frequencies in an infinite ordered class E. This value is therefore independent of the *absolute number* of true instances in E, and is also independent of the absolute or relative number of instances in E which we know to be true *at a given time*. However, we often do say that on the basis of *definite evidence* a theory has some "degree of probability." Thus, a familiar use of this phrase permits us to say that, because of the accumulated evidence obtained since 1900, the quantum theory of energy is more probable today than it was thirty years ago. The foregoing definition is not suitable for this use of the phrase.

(ii) Because the probability of a theory is defined as the limit of relative frequencies, the probability of a theory may be 1, although the class E of its empirically confirmable consequences contains an infinite number of statements which are in fact false. This conclusion could follow even if some of these exceptions to the theory are ruled out as not being genuine negative instances (see the discussion of this point in § 7). But, according to the familiar usage of 'probability of a theory' already referred to, if a theory did have an infinite number of exceptions, not only would not a "high degree of probability" be assigned to it: it would be simply rejected.

(iii) It is difficult to know how even the approximate value of the probability of a theory, in Reichenbach's first sense, is to be determined. The situation here is not quite the same as for the probability statements which occur *within* a natural science and which have been already discussed in §7. In the present case it does not seem possible to obtain other than direct statistical evidence for an assigned numerical value; for it is not apparent how a statement about the probability of theories can be part of an inclusive system, so that the statement might possibly be confirmed indirectly, perhaps even by nonstatistical evidence. Reichenbach's proposal of a hierarchy of probabilities, according to which the probability of a probability statement may be estimated, postpones this problem by referring it to a higher level of probabilities; but postponing a problem does not solve it.

b) The second method proposed by Reichenbach for assigning a probability to a theory in a frequency sense depends upon the first method. The theory *T* under consideration will now be regarded as an element in an infinite class *K* of theories. These theories are supposed to be alike in some respects and unlike in others; and the theory *T* will share with a number of others in *K* a certain definite property *P*. (The following crude illustration may help fix our ideas: Suppose *T* is the Newtonian theory, and *K* the class of possible theories dealing with the physical behavior of macroscopic bodies. *P* may then be the property that the force functions in the theory are functions of the coordinates alone.) The probability of the theory *T* is then defined as the limit of the relative frequency with which theories in *K*, possessing the property *P*, have a probability in Reichenbach's first sense which is not less than a specified number *q*.

We can comment only briefly on this proposal.

(i) Although it is easy to introduce the reference class *K* and the property *P* in the formal definition, in practice it is by no means easy to specify them. The class *K* must not be selected too widely or arbitrarily, but no way is known for unambiguously grouping together a set of allegedly "relevant" theories. The difficulty is even greater in specifying the property *P* for a concrete case. We might wish to say, for example, that the theory of relativity is more probable than the Newtonian theory. But just what is the property *P* in this case on the basis of which they are to be distinguished?

(ii) We do not at present possess a sufficiently extensive collection of theories, so that appropriate statistical inquiries cannot be made with respect to them in accordance with this proposal. This proposal therefore completely lacks practical relevance. Indeed, there is some ground for suspicion that the proposal would be feasible only if, as Peirce suggested, "universes were as plentiful as blackberries"; only in such a case could we determine the relative frequency with which these different universes exhibit the traits formulated by a theory under consideration.

(iii) If we could assign a probability value to a theory according to the first of Reichenbach's two proposals, there would be little need for estimating its probability by the second method. It is consistent with these proposals that a theory which has a probability of 1 on the first method, has the probability of only 0 on the second method. But since we are, by hypothesis, interested in that *one* theory, of what particular significance is it to know that theories of such a type have almost all their instances in conformity with the facts with only a vanishingly small relative frequency? This second proposal, like the first, does not therefore formulate the sense of those statements which assign a "degree of probability" to a theory on the basis of *given* finite evidence. For this second proposal does not permit us to talk *literally* about the degree of probability which *one definite theory* has on the evidence at hand; and it is just this which is intended when the evidence for a theory at one time is compared with the evidence at another time.

2. *Degree of confirmation or weight of evidence.*—These difficulties with the two proposals for assigning a probability to a theory, in the relative frequency sense of the term, are serious enough to have led competent students to seek a different interpretation for such statements. Guided by the actual procedure of the sciences, a long line of writers have urged that a different concept is involved in such statements from the one specified by the frequency theory of probability. This concept has been designated as "degree of confirmation" or "weight of evidence," in order to distinguish it from the various interpretations given to the term 'probable.' We shall briefly explain what is meant by 'degree of confirmation' and discuss some of the problems which center around its use.

The initial task which must be performed before a satisfactory account of 'degree of confirmation' can be given is a careful analysis of the logical structure of a theory in order to make precise the conditions under which a theory may be confirmed by suitable experiments. This has been partially done by Carnap with considerable detail and refinement. We shall, however, not reproduce the results of his analyses, and shall employ distinctions inexactly formulated but which are familiar in the literature of scientific method. In particular, we shall take for granted the following, of which use has already been made: No theory (or for that matter no singular statement) can be established completely and finally by any finite class of observations. But a theory can be tested by examining its instances, that is, the singular sentences *E* derived with the help of the theory from the sentences *C* stating the initial conditions for the application of the theory. Both *C* and *E* may increase in number; but, while theoretically there are an infinite number of instances of a theory, no more than a finite number will have been tested at any given time. Indeed, a theory is said to be capable of being confirmed or verified only incompletely, just because no more than a finite number of its instances can be actually tested. The instances may be confirmed by observation, in which case they are called the *positive instances* for the theory; or they may be in disaccord with the outcome of observations, in which case they are called the *negative instances*.

We shall assume for the sake of simplicity that there are no negative instances for a given theory *T*. Then as we continue the process of testing *T*, the number of positive instances will usually increase. Now it is generally admitted that, by increasing the positive instances, the theory becomes more securely established. What is known as 'the weight of evidence' for the theory is thus taken to be a function of the number of positive instances. And we may accordingly state as a preliminary explanation of what is meant by 'the degree of confirmation' for a theory that the degree of confirmation increases with the number of the positive instances for *T*.[14]

This explanation is, of course, far from precise; but at present no precise definition for the term is available. As matters stand, the term is used in a more or less intuitive fashion in the actual procedures of testing theories. It would obviously be highly desirable to have carefully formulated semantical rules for employing the term; but there is no early prospect that the rules for weighing the evidence for a theory will be reduced to a formal schema. The following observations, however, indicate some of the conditions under which the weighing of evidence is carried on, and will contribute something to making more precise the meaning of 'degree of confirmation.'

a) It does not seem possible to assign a quantitative value to the degree of confirmation of a theory. Thus, at one stage of investigation a theory T may have twenty positive instances in its favor, while at a later stage it may have forty such instances. While the degree of confirmation of T at the second stage would in general be acknowledged as greater than at the first stage, it is nevertheless not appropriate to say one degree of confirmation is twice the other. The reason for this inappropriateness is that, if degrees of confirmation could be quantized, all degrees of confirmation would be comparable and be capable therefore of a linear ordering. That this does not seem to be the case is suggested by the following hypothetical situation.

Suppose that the positive instances for T can be analyzed into two nonoverlapping classes K_1 and K_2, such that the instances in K_1 come from one field of inquiry and those in K_2 from another field. For example, if T is the Newtonian theory, K_1 may be the confirmatory instances for it from the study of planetary motions, while K_2 may be those coming from the study of capillarity phenomena; each set of instances is in an obvious sense qualitatively dissimilar from the other. Now imagine the following possibilities as to the number of instances in K_1 and K_2:

	P_1	P_2	P_3	P_4	P_5	P_6	P_7	P_8	P_9
K_1......	50	50	100	101	99	100	200	100	198
K_2......	0	50	0	49	52	90	0	100	2
E......	50	100	100	150	151	190	200	200	200

The last row of figures gives the total number of positive instances for T. These nine possibilities are arranged in order of increasing number of positive instances. Would we say, however, that this order also represents the order of increasing degrees of confirmation?

It would generally be granted that for both P_2 and P_3 the degree of confirmation is greater than for P_1, simply because of the total number of positive instances. On the other hand, many scientists would be inclined to assign a greater degree of confirmation to P_2 than to P_3, even though the total number of positive instances is the same in these cases. And the reason they would give is that in P_2 there are *different kinds* of instances, while in P_3 there is only one kind. For this reason also P_6 would be assigned a higher degree of confirmation than P_7, even though the total number of positive instances in the former case is less than in the latter case. Again, P_4 and P_5 would often be assigned the *same* degree of confirmation, even though the total number of instances is different in these cases, because the relative number of instances of each kind is approximately the same. Finally, P_8 and P_9 would often be regarded as *incomparable* with respect to their degrees of confirmation, because of the disparity in the relative number of different kinds of instances.

Variety in the kinds of positive instances for a theory is a generally acknowledged factor in estimating the weight of the evidence. The reason for this is that experiments which are conducted in qualitatively different domains make it easier to control features of the theory whose relevance in *any* of the domains may be in question. Hence, by increasing the possibility of eliminating what may be simply accidental successes of a theory under special or unanalyzed circumstances, the possibility of finding negative instances for the theory is increased. In this way of conducting experiments, the theory is subjected to a more searching examination than if all the positive instances were drawn from just one domain. A large increase in the number of positive instances of one kind may therefore count for less, in the judgment of skilled experimenters, than a small increase in the number of positive instances of another kind. It follows, however, that the degree of confirmation for a theory seems to be a function not only of the absolute number of positive instances but also of the kinds of instances and of the relative number in each kind. It is not in general possible, therefore, to order degrees of confirmation in a linear order, because the evidence for theories may not be comparable in accordance with a simple linear schema; and a fortiori degrees of confirmation cannot, in general, be quantized.

Indeed, the foregoing hypothetical situation is only a highly simplified outline of the considerations which are usually taken to be relevant in estimating the weight of the evidence for a theory. Among other factors usually considered is the precision with which the confirmable consequences of a theory are in agreement with experimental findings. Although, as has been repeatedly explained, a theory is not rejected simply because perfect agreement between predicted and experimentally determined magnitudes does not occur, the more closely the observed values center around the theoretically expected magnitudes, the greater weight is usually attached to the supporting observations for a theory. Furthermore, evidence for a theory often consists not only of its own positive instances but also of the positive instances for *another* theory, related to the first within a more inclusive theoretical system. The number of direct positive instances may in such cases be regarded as of small importance, in comparison with the fact that support is given to the theory by the accumulated positive instances for the inclusive system.

b) How large must the number and kinds of positive instances be in order that a theory can be taken as adequately established? No general answer can be given to such a question, since the answer involves practical decisions on the part of those who conduct a scientific inquiry. There is an ineradicable conventional element among the factors which lead to the acceptance of a theory on the basis of actual evidence at hand. It is always theoretically possible to demand further evidence before agreement is reached that a theory has been sufficiently well tested. However, the practical decision is in part a function of the contemporary scientific situation. The estimation of the evidence for one theory is usually conducted in terms of the bearing of that evidence upon alternative theories for the same subject matter. When there are several competing theories, a decision between them may be postponed indefinitely, if the evidence supports them all with approximately the same precision. Furthermore, the general line of research pursued at a given time may also determine how the decision for a theory will turn out. For example, at a time when a conception of discontinuous matter is the common background for physical research, a theory for a special domain of research formulated in accordance with the dominant leading idea may require little direct evidence for it; on the other hand, a theory based on a continuous notion of matter for that domain may receive little consideration even if direct empirical evidence supports it as well as, or even better than, it does the alternative theory.

In particular, the acceptance of definite numerical values for probabilities also involves practical decision, for which no general rules can be given. As already explained, such numerical values are often computed on the basis of more or less comprehensive theoretical systems, and the confidence which we have in the correctness of those values depends on the confidence we have in those systems. It may happen that we can determine the value of a probability with only small accuracy by a theory which has a relatively high degree of confirmation, while a different value may be computed with great precision by an alternative theory with an inferior degree of confirmation. The supposition that in such a case the dilemma can be resolved by a clear-cut method neglects the human and accidental factors which determine the history of science. Certainly no mathematical or logical formula can be given which would mechanically supply a coefficient of weight for the correctness of the decisions which are made in many analogous cases.

c) Assuming that these desultory observations are based on the study of actual scientific procedure, it may be asked why it is that we seem to feel that theories with a greater degree of confirmation deserve our confidence on logical grounds more than those with less—whenever such comparisons can be made.

Why, in other words, should a theory be regarded as "better established" if we increase the number and kinds of its positive instances?

Perhaps a simple example will help suggest an answer. Suppose a cargo of coffee is to be examined for the quality of the beans. We cannot practically examine every coffee bean, and so we obtain some sample beans. We do not, however, sample the cargo by taking a very large number of beans from just one part of the hold; we take many relatively small samples from very many different parts of the ship. Why do we proceed in this way? The answer seems to be that our general experience is such that, when we conduct our samplings in this manner, we approximate to the distribution of qualities in the entire hold; and, in general, the larger our individual samples and the more diversified our choice of the parts of the ship from which they are taken, the more reliable (as judged by subsequent experience) are the estimates we form. It is at least a plausible view that in testing a theory we are making a series of samplings from the class of its possible instances. A theory is "better established" when we increase the number and kinds of its positive instances, because the *method* we thereby employ is one which our general experience confirms as leading to conclusions which are stable or which provide satisfactory solutions to the specific problems of inquiry. At any rate, this was the answer which Charles Peirce proposed to the so-called "problem of induction," and which has been independently advanced in various forms by many contemporary students of scientific method (e.g., M. R. Cohen, J. Dewey, H. Feigl, O. Neurath, and many others). As Peirce succinctly put the matter, "Synthetic inferences are founded upon the classification of facts, not according to their characters, but *according to the manner of obtaining them*. Its rule is that a number of facts obtained in a given way will in general more or less resemble other facts obtained in the same way; or, *experiences whose conditions are the same will have the same general characters*." A degree of confirmation is thus a rough indication of the extent to which our general *method of procedure* has been put into operation. While no probability in a frequency sense can be significantly assigned to any formulation of our method (because it is that very method which is involved in estimating and testing such probabilities), scientific inquiry is based upon the assumption, which is supported by our general experience, that the method of science leads to a proportionately greater number of successful terminations of inquiry than any alternative method yet proposed.[15]

Attempts to find a systematic answer to "the problem of induction" within the framework of a theory of probability, though often made, have not in general been regarded as successful. The *process* of induction has been usually conceived as the search for more or less stable and pervasive relations between properties of objects; and the *problem* of induction has been taken to be the discovery of a principle (the principle of induction) which would "justify" the various conclusions of that process. Stated in this way, it is rather difficult to know just how the "problem" is to be conceived in empirical terms. On the face of it, the "problem" seems to involve a futile infinite regress; and indeed the Achilles heel of attempted solutions of it has usually been the status of the proposed principle of induction: how is the principle itself to be "justified"? The number of different types of answers which have been given to this last question is relatively small; among them are the following: the inductive principle is a synthetic a priori proposition concerning the nature of things in general, it is an a priori proposition concerning the fundamental constitution of the human mind, it is a generalization from experience, and it is a "presupposition" or "postulate" of scientific procedure. It would take too long to examine these answers in detail. It is perhaps sufficient to note that the first two involve positions incompatible with the conclusions of modern logical research; that the third commits a *petitio principii;* and that the fourth, assuming it to have a clear meaning, cannot make of the proposed inductive principle a "justification" of the procedure of science or of its conclusions, since according to this answer the principle is simply an *instrument* of scientific procedure. The position taken in the present monograph is that no antecedent principle is required to justify the procedure of science, that the sole justification of that procedure lies in the specific solutions it offers to the problems which set it into motion, and that a *general* problem of induction in its usual formulation does not exist. Since the notion of the probability of theories (in the specific senses discussed above) has been found to involve serious difficulties, and since the degree of confirmation for a theory has been argued to indicate the extent to which the theory has been tested by the procedure of science, the problem of induction which the present writer recognizes as genuine is the formulation of the general features of scientific method—of the method which, in short, leads to a proportionately greater number of successful terminations of inquiry than the number which other methods may have to their credit.

One brief final remark: It has been customary in the traditional discussions of scientific theories to seek grounds for our knowledge of their *truth* or at least of their *probability* (in some one of the many senses previously discussed). Omitting more than mention of those students (e.g., Wittgenstein and Schlick) who have dismissed such discussions as meaningless because, according to them, theories are not "genuine" propositions since they are not completely verifiable, reference must be made to another group of writers. According to this group, the traditional discussions have not fruitfully illuminated the character of scientific inquiry because those who take part in them neglect the *function* which theories have in inquiry. When this function is examined, it has been urged, it turns out that questions of the *truth* of theories (in the sense in which theories of truth have been traditionally discussed) are of little concern to those who actually use theories. Reflective inquiry is instituted for the sake of settling a *specific* problem, whether it be practical or theoretical, and inquiry terminates when a resolution of the problem is obtained. The various procedures distinguishable in inquiry (such as observation, operation upon subject matter including the manipulation of instruments, symbolic representation of properties of subject matter, symbolic transformation and calculation, etc.) are to be viewed as instrumental to its end product. The use of theories is one patent factor in reflective inquiry. They function primarily as means for effecting transitions from one set of statements to other sets, with the intent of controlling natural changes and of supplying predictions capable of being checked through manipulating directly experienceable subject matter. Accordingly, in their actual use in science, theories serve as *instruments* in specific contexts, and in this capacity are to be characterized as good or bad, effective or ineffective, rather than as true or false or probable. Those who stress the instrumental function of theories are not necessarily committed to identifying truth with effectiveness and falsity with uselessness. Their major insight does not consist in denying the meaningfulness of certain types of inquiries into the truth of theories but in calling attention to the way theories function and to the safeguards and conditions of their effectiveness. A theory is confirmed to the degree that it performs its specific instrumental function. From this point of view, which has been developed with much detail by Dewey, the degree of confirmation for a theory may be interpreted as a mark of its proved effectiveness as an intellectual tool for the purposes for which it has been instituted.

9. Concluding Remarks

In consonance with the discussion and terminology of the theory of signs in Volume I, Number 2, it is convenient to classify the problems connected with probability into three distinct though connected groups. *Syntactical problems:* these are concerned primarily with perfecting the calculus of probability, making more precise its assumptions, simplifying its procedure, establishing its consistency, developing alternative formal techniques, and indicating its relation to other branches of formal mathematics. Some of these matters were considered in Section II. *Semantical problems:* these are concerned with establishing and formulating appropriate rules for applying the calculus to various existential affairs, by indicating under what conditions certain complexes of signs in the calculus are to be co-ordinated with experimentally controllable situations. *Pragmatic problems:* these are concerned with formulating the procedures and conditions involved in the acceptance of probability statements, and with evaluating the efficacy of the calculus in solving the problems set for it in scientific inquiry. Semantical and pragmatic problems were outlined in the present section.

In recent years a growing number of mathematicians and logicians have devoted themselves to the solution of the syntactical problems of probability. Although there are still a number of outstanding difficulties, these are being attacked with the most subtle instruments of modern mathematics. In any case, the calculus has been refined and generalized to an extent undreamed of a century ago. The discussion of the semantical problems of probability is perhaps still in its infancy, though important spade work has already been done. The very recognition of the existence of such problems bodes well for the future, since the classical discussions of probability have been shown to be inadequate largely because semantical problems were not clearly distinguished from syntactical ones. The discussion of pragmatic problems has been carried on in the United States for many years. The most obvious fruits of this activity are the number of substantial contributions to an objective psychology dealing with scientific inquiry. And the present co-operative attack upon this group of problems by biologically oriented thinkers and those trained in the mathematical sciences gives a bright promise that, perhaps for the first time in the modern period, an adequate account of human behavior in the context of getting knowledge will soon be available.

The present section has stressed problems associated with the discussion of probability which are still largely unsettled. An unsettled situation in an intellectual discipline has often been seized upon by those hostile to free inquiry as an opportunity to cry out the "bankruptcy of science," to charge it with "confusion," to preach a wholesale skepticism with respect to its findings, and to invoke dogmatically "perennial truths" in the interest of private and institutionalized vested interests. However, unsettled situations in science usually mark important departures from traditional modes of analysis and are concomitants of active research; and the present state of probability discussions is typical of such situations. Disagreement among competent students certainly indicates that the last word upon the topic under discussion has not been said; but it may also indicate that a community of workers is co-operatively engaged in contributing to the solution of complicated issues. Such in fact is patently the case in current discussions of probability. Even where sharp disagreements occur, those engaged in the discussion have been drawing upon one another's insights, have been influencing one another to state their proposed solutions with greater precision, have been led to recognize alternative possibilities in solutions, and have consequently guarded themselves against a premature commitment to theses which may block the course of further inquiry. What is essential for the future development of probability considerations, as for the development of science in general, is that trained minds play upon its problems freely and that those engaged in discussing them illustrate in their own procedure the characteristic temper of scientific inquiry—to claim no infallibility and to exempt no proposed solution of a problem from intense criticism. Such a policy has borne precious fruit in the past, and it is reasonable to expect that it will continue to do so. In the history of the study of probability it has brought into existence a perfected calculus of probability; it has led to an extension of its range of application to many diverse domains; and it has contributed to showing that the various sciences, however distinct their specific subject matters may be, employ a common logic and common procedures, are faced with common logical problems, and are mutually indebted to one another for indispensable tools of inquiry.

NOTES

These very limited bibliographical notes aim to do no more than suggest further reading on some of the topics mentioned in the text.

1. For the classical view of probability consult Laplace, *Essai philosophique sur les probabilités* (Paris, 1814); A. De Morgan, *An Essay on Probability* (London, 1838). For the logical view see J. M. Keynes, *Treatise on Probability* (London, 1921); J. von Kries, *Die Principien der Wahrscheinlichkeitsrechnung* (Tübingen, 1886); F. Waismann, "Analyse des Wahrscheinlichkeitsbegriffs," *Erkenntnis*, Vol. I. For the frequency view see J. Venn, *Logic of Chance* (London, 1886); Charles S. Peirce, *Collected Papers*, Vol. II (Cambridge, Mass., 1932); R. von Mises, *Wahrscheinlichkeit, Statistik, und Wahrheit* (Vienna, 1936) and *Wahrscheinlichkeitsrechnung* (Leipzig, 1931); H. Reichenbach, *Wahrscheinlichkeitslehre* (Leiden, 1935).

2. For these alternative definitions consult K. Popper, *Logik der Forschung* (Vienna, 1935), and A. H. Copeland, "Admissible Numbers in the Theory of Probability," *American Journal of Mathematics*, Vol. L, as well as his "Predictions and Probabilities," *Erkenntnis*, Vol. VI.

3. Peirce's comments on these matters are scattered throughout his writings, especially in Vols. II and VI of his *Collected Papers*. Philipp Frank has written many monographs on this subject, but the fullest account will be found in *Das Kausalgesetz und seine Grenzen* (Vienna, 1932). Henry Margenau develops his point of view in several articles in *Philosophy of Science* and in R. B. Lindsay and H. Margenau, *Foundations of Physics* (New York, 1936). See also E. Cassirer, *Determinismus und Indeterminismus in der moderne Physik* (Göteborg, 1937).

4. Reichenbach's book cited in n. 1 contains fairly full references to these discussions. A. Wald's paper, "Die Wiederspruchsfreiheit des Kollektivbegriffes der Wahrscheinlichkeitsrechnung," appeared in K. Menger's *Ergebnisse eines mathematischen Kolloquiums*, Heft 8.

5. The technical details referred to will be found in any book on the mathematical theory of probability; for example, the books of von Mises cited in n. 1. Fisher's writings are scattered in many periodicals but are summarized in his *Statistical Methods for Research Workers* (Edinburgh and London, 1925), which also contains a list of his papers, and also in his *The Design of Experiments* (Edinburgh and London, 1935).

6. This criticism has been made by a number of writers, e.g., F. Cantelli, "Considération sur la convergence dans le calcul des probabilities," *Annales de l'Institut Henri Poincaré*, Vol. V; T. C. Fry, *Probability and Its Engineering Uses* (New York, 1928); and R. B. Lindsay and H. Margenau in their *Foundations of Physics*.

7. Consult A. Kolmogoroff, "Grundbegriffe der Wahrscheinlichkeitsrechnung," *Ergebnisse der Mathematik*, Vol. II, and the important series of works edited by E. Borel entitled *Traité du calcul des probabilités et de ses applications*.

8. These interpretations will be found in F. P. Ramsey, *Foundations of Mathematics* (London and New York, 1931); C. G. Hempel, "Über den Gehalt von Wahrscheinlichkeitsaussagen," *Erkenntnis*, Vol. V; H. Jeffreys, *Scientific Inference* (Cambridge, 1931); S. Mazurkiewicz, "Über die Grundlagen der Wahrscheinlichkeitsrechnung," *Monatshefte für Mathem. u. Physik*, Vol. XLI; and other works by authors already cited in previous notes.

9. Reichenbach's views have been stated by him in their most complete form in the work cited in n. 1 and in his *Experience and Prediction* (Chicago, 1938).

10. Consult G. Birkhoff and J. von Neumann, "The Logic of Quantum Mechanics," *Annals of Mathematics*, Vol. XXXVII; M. Strauss, "Zur Begruendung der statistischen Transformationstheorie der Quantenmechanik," *Sitzungsber. der preuss. Akad. d. Wiss.* (1936); Paulette Février, "Les Relations d'incertitude de Heisenberg et la logique," *Comptes rendus des sciences*, Vol. CCIV.

11. For a recent exposition of the subjective view consult B. De Finetti, "La Prevision: ses lois logiques, ses sources subjectives," in *Annales de l'Institut Henri Poincaré*, Vol. VII. Criticisms of the classic view will be found in the writings of Peirce, Venn, von Kries, and Keynes already referred to.

12. In addition to the works cited in n. 1, consult J. Nicod, *Le Probleme logique de l'induction* (Paris, 1923).

13. Poincaré's method is explained in his *Calcul des probabilités* (Paris, 1896) and also by Reichenbach in his *Wahrscheinlichkeitslehre*. For the work of E. Hopf see "On Causality, Statistics and Probability," *Journal of Mathematics and Physics*, Vol. XIII; see also G. D. Birkhoff and D. C. Lewis, "Stability in Causal Systems," *Philosophy of Science*, Vol. II.

14. Carnap's discussion is contained in his "Testability and Meaning," *Philosophy of Science*, Vols. III and IV.

15. For further discussion of these matters consult M. R. Cohen, *Reason and Nature* (New York, 1931); John Dewey, *Essays in Experimental Logic* (Chicago, 1916), *Quest for Certainty* (New York, 1929), and *Logic: The Theory of Inquiry* (New York, 1938); H. Feigl, "The Logical Character of the Principle of Induction," *Philosophy of Science*, Vol. I; Otto Neurath, "Pseudorationalismus der Falsifikation," *Erkenntnis*, Vol. V.

Selected Bibliography

BIRKHOFF, G. D., and LEWIS, D. C. "Stability in Causal Systems," *Philosophy of Science*, Vol. II (1935).

BIRKHOFF, G., and NEUMANN, J. VON. "The Logic of Quantum Mechanics," *Annals of Mathematics*, Vol. XXXVII (1936).

BOREL, E. *Traité du calcul des probabilités et de ses applications.* Paris, 1925.

CARNAP, R. "Testability and Meaning," *Philosophy of Science*, Vols. III (1936) and IV (1937).

CASSIRER, E. *Determinismus und Indeterminismus in der moderne Physik.* Göteborg, 1937.

COHEN, M. R. *Reason and Nature.* New York, 1931.

COPELAND, A. H. "Admissible Numbers in the Theory of Probability," *American Journal of Mathematics*, Vol. L (1900).

———. "Predictions and Probabilities," *Erkenntnis*, Vol. VI (1936).

DE FINETTI, B. "La Prevision: ses lois logiques, ses sources subjectives," *Annales de l'Institut Henri Poincaré*, Vol. VII (1937).

De MORGAN, A. *An Essay on Probability.* London, 1838.

DEWEY, JOHN. *Essays in Experimental Logic.* Chicago, 1916.

———. *Logic: The Theory of Inquiry.* New York, 1938.

FEIGL, H. "The Logical Character of the Principle of Induction," *Philosophy of Science*, Vol. I (1934).

FISHER, R. A. *Statistical Methods for Research Workers.* Edinburgh and London, 1925.

———. *The Design of Experiments.* Edinburgh and London, 1935.

FRANK, PHILIPP. *Das Kausalgesetz und seine Grenzen.* Vienna, 1932.

FRY, T. C. *Probability and Its Engineering Uses.* New York, 1928.

HEMPEL, C. G. "Über den Gehalt von Wahrscheinlichkeitsaussagen," *Erkenntnis*, Vol. V (1935).

HOPF, E. "On Causality, Statistics and Probability," *Journal of Mathematics and Physics*, Vol. XIII (1934).

JEFFREYS, H. *Scientific Inference.* Cambridge, 1931.

KEYNES, J. M. *Treatise on Probability.* London, 1921.

KOLMOGAROFF, A. "Grundbegriffe der Wahrscheinlichkeitsrechnung," *Ergebnisse der Mathematik*, Vol. II (1933).

KRIES, J. VON. *Die Principien der Wahrscheinlichkeitsrechnung.* Tübingen, 1886.

LAPLACE, P. *Essai philosophique sur les probabilités.* Paris, 1814.

LINDSAY, R. B., and MARGENAU, H. *Foundations of Physics.* New York, 1936.

MISES, R. VON. *Wahrscheinlichkeit, Statistik und Wahrheit.* Vienna, 1928.

———. *Wahrscheinlichkeitsrechnung.* Leipzig, 1931.

NEURATH, OTTO. "Pseudorationalismus der Falsifikation," *Erkenntnis*, Vol. V (1935).

Principles of the Theory of Probability

NEYMAN, J. *Lectures and Conferences on Mathematical Statistics*. Washington, D.C., 1938.

NICOD, J. *Le Probleme logique de l'induction*. Paris, 1913.

PEIRCE, CHARLES S. *Collected Papers*, Vols. II and VI. Cambridge, Mass., 1932 and 1935.

POPPER, K. *Logik der Forschung*. Vienna, 1935.

RAMSEY, F. P. *Foundations of Mathematics*. London and New York, 1931.

REICHENBACH, HANS. *Wahrscheinlichkeitslehre*. Leiden, 1935.

———. *Experience and Prediction*. Chicago, 1938.

STRAUSS, M. "Zur Begruendung der statistischen Transformationstheorie der Quantenmechanik," *Sitzungsber. der preuss. Akad. d. Wiss.* (1936).

VENN, J. *Logic of Chance*. London, 1886.

WAISMANN, F. "Analyse des Wahrscheinlichkeitsbegriffs," *Erkenntnis*, Vol. I. (1930).

WALD, A. "Die Wiederspruchsfreiheit des Kollektivsbegriffes der Wahrscheinlichkeitsrechnung," *Ergebnisse eines mathem. Kolloquiums*, Heft 8.

Foundations of Physics

Philipp Frank

Foundations of Physics

Contents:

I. Introduction 427

II. The Logical Structure of Physical Theories 429
  1. Equations, Logical and Semantical Rules 429
  2. Operational Meaning and Validity of a Theory 430
  3. Are the General Laws of Nature Pure Conventions? . . . 431
  4. Misuse of the Vernacular of Physics 433
  5. Mathematical Description or Physical Explanation? . . . 434
  6. The Metaphysical Basis of "Explanation" 435
  7. Pictorial Theories 436

III. Classical Newtonian Mechanics 437
  8. Domain of Newtonian Mechanics; Mass 437
  9. Field of Force; Gravitational and Electromagnetic Fields 438
  10. Newton's First Law of Motion (Law of Inertia) 439
  11. Heavy Mass and Inert Mass 440
  12. Newton's Second Law of Motion 440
  13. Inertial Force and Centrifugal Force 441
  14. The Law of Force Must Be "Simple". 442
  15. Do Forces "Exist"? 443

IV. Heat, Irreversibility, and Statistics 445
  16. Recoverable and Irrecoverable Changes of State . . . 445
  17. The Second Principle of Thermodynamics and Entropy . . 447
  18. Cosmological Implications 448
  19. The Kinetic Theory of Heat 449
  20. Mechanics Cannot Account for Irrecoverable Changes of State 449
  21. Statistics and Irrecoverability · 450
  22. Statistical Hypotheses Are Assertions about Physical Facts . 452

Contents

V. THE THEORY OF RELATIVITY 453

  23. Einstein's Two Basic Principles 453
  24. Newton's Laws of Motion Not Universally Valid 454
  25. The Operational Definition of 'Mass' Has To Be Modified . 455
  26. The Operational Definition of 'Time Distance' Becomes Ambiguous 456
  27. The Relativity of Time 457
  28. The Speed of Light Has the Same Value in All Systems of Reference 458
  29. The Traveling Twins 459
  30. Conversion of Mass into Energy and Vice Versa 459
  31. Creation and Annihilation of Mass 460

VI. LIGHT 462

  32. The Crucial Experiments on Behalf of the Wave Theory . . 462
  33. Conversion of Light into Kinetic Energy; the Photon . . . 463
  34. The Paths of Photons 464
  35. Laws of Motion of a Photon 465
  36. Mechanical Momentum of a Photon 466
  37. The Interpretation of Diffraction in Terms of the Mechanics of the Photon 467

VII. MECHANICS OF SMALL MASSES (WAVE MECHANICS) 468

  38. Failure of Newton's Mechanics in the Domain of Very Small Particles 468
  39. Bohr's Theory of Emission of Light by the Atom 469
  40. The Wave Theory of Matter 470
  41. The New Laws of Motion for Small Particles 471
  42. Diffraction of Material Particles 472
  43. The Uncertainty Relation 472
  44. Misunderstandings about the Relation of Uncertainty . . . 474
  45. Bohr's Idea of Complementarity 475
  46. By What Law Does Wave Mechanics Replace Newton's Second Law? 476
  47. Physical Reality and Causality 478
  48. Object and Subject in Wave Mechanics 480
  49. Metaphysical Interpretations of Wave Mechanics 481

VIII. STRUCTURE OF MATTER 483

  50. Continuity or Discontinuity of Matter 483
  51. Operational Meaning of 'Matter' 483

425

Contents

52. Electromagnetic Mass 484
53. The Number of Molecules in a Gas 485
54. The Problem of Chemical Binding 486
55. The Structure of the Hydrogen Atom 486
56. Electron, Proton, and Neutron 488
57. The Forces of Chemical Binding 489
58. The Structure of the Atoms of Different Chemical Elements 490
59. Chemical Properties and Atomic Weights 491
60. The Nuclear Forces 492
61. The Phenomena of Radioactivity 493
62. Production of Positrons, Electrons, and Photons by Nuclear
Reactions 494
63. The Structure of the Nucleus 494
64. Atomic Power 496

IX. CONCLUSION 498

NOTES 502

SELECTED BIBLIOGRAPHY 503

I. Introduction

In the present monograph an attempt is made to present physics in such a way that it is fit to become part of unified science. The first step is to bring it into such a shape that it becomes clear which statements tell something about observable facts and which are statements about the choice of symbols. This means that we must discuss the operational meaning of all symbols used and the kind of relations which exists between these symbols.

We are not going to set up a complete system of symbols and operational definitions from which one could derive all facts of physics. Such a systematic presentation would be a very hard job, and at that I suspect that only a very few scientists would read it. Most of this presentation would be a dull routine work, and the reader would not know how to find the points to which he should direct his attention. The scientist is interested in logical analysis if, and only if, this analysis is not trivial or commonplace. One does not need to occupy each square foot of a territory in order to control it. What matters is only the control of some key positions. In this presentation of physics we shall attempt to bring under control only a few key positions in the huge symbolic structure by which the scientists have been "mapping" the wide domain of physical phenomena.

It is not difficult to find out these key positions. We must look only for the starting-points of current "interpretations" of physics on behalf of some philosophical creed, usually idealistic, or even spiritualistic. If we examine these "interpretations", we notice almost regularly that they are not interpretations of physics itself but interpretations of the symbolic structure, ignoring the operational meaning of these symbols. This means that they ignore the link between symbols and observable phenomena. Examples of this type of interpretation are obvious. The "energetic" aspect of physics is interpreted as a "refutation of materialism" and the four-dimensional presentation of relativity theory as establishing the "existence" of a four-dimensional world.

It would be a great mistake to say that the philosophers are chiefly responsible for these misinterpretations. As a matter of fact, each one of these "interpretations" has had its origin in a confused presentation of physics given by a physicist. Some of the latter have been honestly surprised at the fruits of their work. They have blamed the confused way of presenting their subject on the present "crisis" of physics, which leaves this field in a state of transition. One physicist suggested even putting up a sign telling all nonphysicists and in particular

Introduction

philosophers: "Keep out, while repairs are going on!" This attitude of "isolationism" would frustrate all attempts toward the unification of science and is, therefore, directly opposed to the aims of this *Encyclopedia* in general and of this monograph in particular.

Physics has been for centuries the spearhead of advance in human thought. In a unified science it should keep its role as a description of the physical universe and should not deteriorate into an incoherent, and somehow mysterious, agglomeration of symbols, rules, and recipes. In this situation quite a few physicists have tried to avoid all these difficulties by "sticking strictly to the facts" and by keeping away from the dangerous enterprise of logical and critical analysis. There is no doubt that this attempt is doomed to failure. Scientists who looked at the world from such different angles as Ernst Mach and A. N. Whitehead have agreed on one point: a physicist who dodges all logical analysis and tries to be a "physicist and only a physicist" will imbue the presentation of his subject with some "chance philosophy", usually a very obsolete one.[1] A careful examination of the presentation of physics given by physicists who pretend to keep strictly to the "facts and only to the facts" will reveal soon that these presentations are written in a spirit of medieval scholasticism. We can notice it even by a perfunctory look at textbooks for beginners. Quite a few make a generous use of words, like "entity", which are borrowed from medieval philosophy and have no status in science.

Obsolete philosophical views in physics are mostly obsolete physical theories in a state of petrification.[2] Aristotle's philosophy of physics is a petrification of a physical theory which covered the experiences of Greek and oriental artisans about physical phenomena. Kant's *Metaphysical Principles of Natural Science* is a petrification of Newton's physics. Eddington's *Philosophy of Physical Science* is a petrification of Einstein's relativity and Bohr's quantum theory.

Remembering these facts, we can easily understand why a certain type of confusion in the presentation of physics gives rise to philosophical "interpretations". The lack of coherence in the presentation of a new physical theory often has its origin in the incoherence between the languages of old and new physical theories. The old language slips easily into the presentations of recent physics. For it pretends to express a certain philosophy which is based upon intuition or good sense and cannot be disregarded by physicists. Actually this language has been invented in order to present older physical theories. Its use means, therefore, confusing the language of recent physics with the language of older and abandoned theories. We can now understand why the key points in the structure of physics are the points where philosophical misinterpretations slip in easily. A coherent presentation of these key points and their vicinity will lead easily to a coherent presentation of the whole domain of physics.

The elucidation of the relations between symbols and facts in physics owes much to the works of men like Ernst Mach, Henri Poincaré, C. S. Peirce, and P. W. Bridgman. The schools of thought which have been called Positivism, Pragmatism, and Operationalism have done a fine job. The present monograph follows their lines in general but does not make any attempt to build up three "self-consistent" systems of Positivism, Pragmatism, and Operationalism. The spirit of all these schools of thought is one and the same. But the theses which we would obtain if we tried to make a clear-cut distinction between them would result in three new branches of metaphysics: positivistic, pragmatistic, and operationalistic metaphysics.[3]

II. The Logical Structure of Physical Theories

1. Equations, Logical and Semantical Rules

Every physical theory consists of three essential parts.[4] There are, first, the equations of the theory, e.g., in mechanics, Newton's equation of motion; in electromagnetism, Maxwell's equations of the electromagnetic field; etc. These equations contain terms like 'co-ordinate', 'time', 'force', 'magnetic field intensity', 'electric conductibility', etc. By themselves they cannot be checked to see whether they are in agreement with the physical facts or not. Nor can we check their logical consistency. Carnap calls these equations the "calculus" of the field of physics.[5] Newton's equations of motion are, e.g., the calculus of mechanics. These equations must have as their basis, as Carnap expresses it, a second calculus by which we can learn what transformation of our equations are allowed without altering their meaning. Mechanics needs as a basis the calculus of algebra and geometry. Only by the application of this "second" calculus can the consistency of the first one be checked.

However, the system of both equations and logical rules (e.g., Newton's equations of motion plus the laws of algebra and geometry) is by no means a system of physical laws. We have to add, as a third part, statements which define the physical meaning of words like 'distance', 'time', 'moment', 'simultaneously', 'force', 'mass', etc. The statements which give these definitions are called by Carnap "semantical rules".[6] Besides the terms of the first calculus ('force', 'mass', etc.) and the terms of the second part, the logical rules ('plus'/'minus', 'and'/'or', etc.), the third part, the semantical rules, contains words like 'iron cube', 'wooden bar', 'warm water', 'one inch long', etc.

These latter words are to be understood in the sense in which they are used in our everyday language. Without this loan taken from the language of our daily life, the language of science would not be shareable knowledge. The semantical rules connect our equations or our calculus with the words of the language of our daily life, at least in physics and, perhaps, elsewhere too. But, in order to avoid any ambiguity, we have to be sure never to apply these words in a wider domain than in the range of their applications in our daily life. In this case this language is understood without controversy by everybody. But it is, e.g., illegitimate to apply the expression 'iron cube' to a body of a size of one million miles through, for we do not even know whether the existence of an iron cube of these dimensions would not violate the laws of physics. The same precaution has to be taken in using words like 'distance' or 'simultaneous'. As Carnap points out, only by adding these semantical rules to the equations do the latter become physical laws which can be checked by experiments.[7]

2. Operational Meaning and Validity of a Theory

In physics these semantical rules consist in the description of physical operations. Bridgman has studied the way the rules have been set up and used by the physicists long before the term 'semantics' had been used in logic.[8] To emphasize their character in physics, Bridgman calls these rules "operational definitions". When we know these rules or definitions, we

know "the operational meaning" of a term. We must, therefore, always be aware of the fact that, e.g., Newton's equations of motions by themselves do not contain the theory of motion. Newton's equations with their operational meaning provide us with the theory of motion. There is no way of checking whether the equations of motion by themselves are valid or true or correct. They are only a part of the theory. We can check only whether a system of equations plus the operational definitions is confirmed by a certain experiment or not.

If there is no disagreement with the experiment, we can say only that the system as a whole is "confirmed". But we cannot say that the equations by themselves or the operational definitions by themselves are confirmed. It is possible to alter the equations and the operational definitions in such a way that some conclusions drawn from the whole system remain unaltered.

Speaking exactly, we have to remember that experiments confirm a system which consists not only of two but of three kinds of statements: equations, plus operational definitions, plus logical and mathematical rules. Therefore, it is theoretically also possible to alter the traditional rules of logic without altering the confirmed conclusions.[9] We have only to introduce suitable alterations of the equations and the operational definitions. By an experimental confirmation we cannot demonstrate that the new logical rules are better than the old ones.

We demonstrate only that this new logic is a part of a system of statements which is as a whole in agreement with the experimental facts.

These remarks have their basis in a very simple argument of elementary logic which Bertrand Russell once elucidated by a drastic example: We start from the assumptions that "bread is made of stone" and that "stone is nourishing". Then it follows logically that "bread is nourishing". This statement can be confirmed by experiments. If someone claims for this reason that we have confirmed our assumptions, he would certainly be ridiculous. According to Russell, a great many confirmations of physical hypotheses are of this type.[10]

Now what is the meaning of the assertion that Newton's equations or Maxwell's equations themselves are confirmed by experiments? This assertion means, speaking exactly: It is possible to add to these equations in a fairly simple way such operational definitions that the whole system is in agreement with the observed facts. This confirmation also reveals, to be sure, a property of the equations themselves. If, e.g., Newton's equations had not this property, it would be impossible to find handy operational definitions which, added to these equations, would turn them into confirmed physical statements. We would need operational definitions which are complicated and hard to handle. But if such an addition is possible in a fairly simple way, we would say that Newton's equations them-

selves are confirmed by experiments. If we speak a little loosely, we may say that in this sense the equations themselves are physical laws.

But, even so, the operational definitions of general terms like 'energy', 'entropy', etc., are of an immense complexity. Bridgman has described the difficulties involved in an elaborate way. According to his analysis, "paper and pencil operations" play a great part. He has pointed out that only under very special conditions is it possible to distinguish by reasonable physical operations even between expressions which have apparently as simple a meaning as 'heat conduction' and 'heat radiation' or 'flux of heat' and 'flux of mechanical energy'.[11]

3. Are the General Laws of Nature Pure Conventions?

There has been a school of thought which has assumed that eventually one could find for any set of equations a set of operational definitions which might be added to these equations and turn them into confirmed physical laws. In this case an experimental confirmation would not reveal any property of the equations at all. The equations by themselves would not say anything about the physical world. However, if operational definitions of some of the symbols in the equations were given, the equations would become operational definitions of the rest of the symbols. Newton's equations of motion, e.g., would no longer be laws of motion. If we added to them the operational definitions of 'acceleration' and 'mass', the equations would become operational definitions of 'force' or a convention how to use the term 'force'.

The school of thought which has ascribed this property to all the fundamental equations of physics has been given the name "conventionalism". The French mathematician Henri Poincaré has been quoted as its most prominent spokesman. Actually, Poincaré emphasized that an experimental confirmation of equations plus operational definitions does not confirm the equations and that, by admitting any imaginable operational definition, we can turn every equation into a confirmed one. This statement can, from the angle of formal logic, hardly be refuted. It is logically equivalent to the statements of section 2. The equations, by themselves, are said to be "valid" or confirmable by experiments only if, by substituting "simple and practical" operational definitions, they become confirmed physical laws. This does not exclude that, by admitting all imaginable operational definitions, almost any system of equations could be converted into confirmed laws, provided that the system is not self-contradictory. If we consistently make the distinction between "simple and practical" operational definitions and arbitrary definitions which may be "complicated and impractical", it becomes clear in what sense the general laws of physics are purely conventional and in what sense they are valid assertions about facts.[12]

We understand this distinction easily if we examine the way in which operational definitions are actually used in physics. A specific operational definition can practically be used only if some specific physical laws are assumed to be valid. What is, e.g., the operational definition of the time distance "one hour"? We may say that this is the time during which the big hand of our pocket watch traverses an angle of 360 degrees. However, we are not interested in the reading of a particular watch. We mean to say that any pocket watch can be used. This means assuming that the hands of all pocket watches proceed with one and the same angular velocity. But this is a statement of a physical law about the behavior of watch springs. Moreover, if we say that a certain phenomenon lasts one hour, we do not mean to say that we can use only a spring watch in order to check the statement. We may as well use a pendulum clock and define one hour as a duration of a certain number of oscillations of this pendulum. But this is only possible if we assume the validity of a physical law: The unwinding of a spring as an effect of its elasticity proceeds at a rate which is proportional to the frequency of the pendulum as an effect of gravity. Since these and similar laws are valid, the expression 'one hour' defined by spring watches or pendulum clocks is helpful in giving a simple description of the actual motions of material bodies. Briefly, an operational definition is "simple and practical" if there are physical laws according to which the numerical result of this definition is identical with the result of other independent operations. Or, in other words, an operational definition is "practical" if it plays an important role in the simplest formulation of valid physical laws.

We must appreciate this fact if we wish to appreciate justly the merits of "conventionalism" in physics and in science in general. Let us consider an example which has been used over and over again for stating the case of conventionalism: the principle of the conservation of energy. We start from the case that only mechanical and heat energy play a role. If we consider an isolated system, we can state: The sum of heat energy H and mechanical energy M remains constant through all interchanges. Mathematically: $H + M$ = Constant. This is certainly no longer true if electromagnetic phenomena also come into play. Then we have to introduce also an electromagnetic energy E, and the principle of conservation of energy would say $H + M + E$ = Constant.

The advocates of conventionalism argue: If the last equation is not confirmed by experiment, we can always add a term U in such a way that the equation $H + M + E + U$ = Constant is confirmed by our experiments. Then this last equation is the operational definition of the term U (unknown kind of energy). However, we note that the electromagnetic energy E has not been defined by our equation of conservation only. There is also a different operational definition of E.

It can be calculated from the electric and magnetic field intensity. Thus we have two operational definitions of E. To say that both these definitions render one and the same result means to assert the validity of a certain physical law. This law gives a practical value to the introduction of the symbol E. The introduction of the unknown energy U would be practical only if we knew a second operational definition of U which is independent of the conservation equation $H + M + E + U$ = Constant. Then the principle of the conservation of energy would no longer be purely conventional but a statement about facts.[13]

4. Misuse of the Vernacular of Physics

In some philosophical discussions the concept of energy has been used in a rather loose way. Speaking of biological or psychological phenomena, quite a few authors have argued as follows: If we examine the phenomena in living organisms, it may be that the sum of the physical and chemical forms of energy remains constant during the living process. If this were the case, it would confirm the hypothesis that life is a physico-chemical phenomenon. Let us now discuss the possibility that the physico-chemical forms of energy would not fulfil the law of the conservation of energy, that, e.g., a living organism would do more work than corresponds to the energy it has absorbed. Even in this case one could uphold the principle that the sum of

433

all forms of energy remains constant. One has only to introduce a new term L (energy of life or vital energy) into the law of conservation. If we denote the sum of all physical and chemical energies by P, we confirm by experiments that P does not remain constant during the phenomena of life. But if we choose L in a suitable way, we can always achieve that $P + L =$ Constant. In this case life could not properly be called a physico-chemical phenomenon. But nonetheless it could be treated by an exact science, a "vitalistic biology", which would contain a law of conservation of energy: The sum of physico-chemical energy and vital energy remains constant.

But in this vitalistic or organismic biology the operational meaning of L (vital energy) is defined by the conservation law $(P + L = $ Constant). This law would state a "convention" as to how to use the word "energy of life L" and would be by no means a law about the physical world. The crucial test of the vitalistic biology is whether we can introduce, besides the state variables of physics and chemistry (mass, electric charge, etc.), a small number of biological state variables from which the energy of life (L) can be computed by a simple formula. Certainly we do not know of any attempt to do so. We have no "practical" operation to define the energy of life.[14]

By comparing the concept of energy in physics with this useless concept of vital energy, we learn to understand the real scientific meaning of the energy law in physics. We know that in physics the conservation law is not the only operational definition of electric energy; or, in other words, there are other equations in the calculus of physics besides the energy equation which contains the symbol E. In addition to it, the number of state variables upon which the energy law depends is a small one. We shall take up this last argument again in the discussion of the concept of force (sec. 15).

5. Mathematical Description or Physical Explanation?

Frequently a distinction has been made between two types of physical theories. One type starts as we did from a set of equations. This type of theory, we have been told, describes the observable phenomena. It is called, therefore, a phenomenological theory or a mathematically descriptive theory. But there is also, according to the traditional distinction, a second type of theory. Here the physical facts are not only "described" but also "explained". If we treat, e.g., the phenomena of light by a theory of the second type, we present a mechanical model of the ether which transmits light and a mechanical model of the atom which emits light. In this way we have a "causal" theory; we are given the causes of the light phenomena, not only a description. This type of theory is sometimes called a theory which uses pictures instead of equations—a pictorial theory. We have also been told frequently that only this type of theory provides an understanding of nature and gives real satisfaction to the mind of the physicist. We have also been told that only this type of theory is really helpful in finding new facts, while the phenomenological theory is only helpful in recording facts which are already known.

As a matter of fact, the whole distinction between the phenomenological and the causal theory has been much overworked and overrated. If we are given a mechanical model of the ether, we have to formulate the equations of this model. And, possessing these equations, we have to proceed exactly as we did in the mathematically descriptive theory. The only thing which would clearly distinguish the second type of theory would be the requirement that these equations are to be of a particular form. They ought to be derived from the equations of motion, as formulated by Newton to handle the motions in the planetary system. But these equations themselves constitute the phenomenological or "descriptive" theory of a particular domain of physical phenomena, the motions of medium-sized material bodies. Therefore, the requirement that all physical phenomena should be explained by mechanical models would mean assuming that the motion of the smallest particles is covered by the same equations that have been extracted from the motions of medium-sized bodies. This assumption is, of course, far from being self-evident and, for that matter, has turned

out definitely to be an oversimplification. We know very well that the motion of subatomic particles cannot be covered by Newton's laws of motion and therefore by no mechanical model. The statement that the laws of "modern" physics (e.g., relativity theory) are not "causal" laws but only "descriptive" ones has a definite operational meaning only if we use the phrase "causal explanation" in the sense of "derivable from Newton's laws of motion".[15]

6. The Metaphysical Basis of "Explanation"

This definition of "causal explanation" would make sense only if we ascribed to Newton's laws of motion a particular logical status. But doing so would mean a discrimination against any attempt to set up a new foundation of physics. The expression "a physical phenomenon is explained" means in the everyday language of the physicist: "The statements about this phenomenon can be derived from a set of equations and practical operational definitions from which we can also derive the bulk of the physical phenomena which are actually known." To single out a particular type of equations as the only legitimate basis of explanation is not justified by science. All attempts at such a justification have been based upon metaphysics. They have had their roots in the belief that the validity of some equations (e.g., Newton's equations) can be established "philosophically" or "epistemologically" without reference to the observable facts which have to be derived. Otherwise the derivations from Newton's equations would not be more of an "explanation" than the derivation from any other set of statements.[16]

Therefore, the distinction between a description and a causal theory is a purely metaphysical distinction. If we look back into the history of science, we notice that Francis Bacon, the leader of empirical philosophy, accused the Copernican system of being purely descriptive while, according to him, the old Ptolemaic system provided understanding.

Philipp Lenard and quite a few "empirically minded" scientists of the twentieth century accused Einstein's theory of relativity of being purely "descriptive", while Newton's theory was supposed to be causal and explanatory. It is instructive to take note of this similarity to the attitude which traditional philosophers took toward the advance in science. The only apparent difference is a shift in the content of the metaphysical creed. In the sixteenth century a physical phenomenon was said to be "explained" if it could be derived from Aristotle's philosophy, while in the nineteenth century only those phenomena were regarded as "explained" which could be derived from Newton's philosophy. This belief in the exceptional logical status of a particular type of equation had become, by-and-by, so firmly established that its metaphysical foundation was forgotten or driven into the subconscious. But the content of the creed remained alive and was more and more credited to "common sense" or "immediate intuition of nature". The empiricist Francis Bacon did condemn Copernicus, and the physicist Lenard did condemn Einstein, not in the name of metaphysics, but in the name of good sense or "unbiased and strictly empirical description of nature." Even a slight glance at the history of scientific thought shows us that the content of yesterday's metaphysics is today's common sense and tomorrow's nonsense.[17]

7. Pictorial Theories

It has been maintained frequently that the theories using the mechanical models are more "visualizable" or "pictorial" or "intuitive" than the theories starting from equations. I think that this claim has a very weak foundation. Speaking from a purely mathematical point of view, every description by an equation is equivalent to a description by a visualizable diagram or picture. But it seems that the philosophers and scientists who discriminate in favor of pictorial theories mean by this word "similar to a familiar group of sense experiences". As a matter of fact, the mechanical interaction of medium-sized material bodies is a phenomenon which we observe again and again in our daily life. These phenomena are covered by Newton's equations in a very satisfactory way. Every theory which describes new thoughts by comparing them with these familiar phenomena strikes a friendly sounding chord in our minds. Such a theory seems to us intelligible and visualizable for this psychological reason. But to require that every theory should be set up according to this pattern would mean to assume that the laws governing this particular domain of phenomena are sufficient to cover all phenomena of physics, even the motion of the smallest particles with the greatest speeds. If this assumption is once rejected, what would be the point in calling a theory "not visualizable"? It would mean that the theory contradicts statements which have been dropped anyway because of their inadequacy.

A particular difficulty in understanding the meaning of physical theories has arisen from the confusing of a theory which is intuitive or visualizable with a theory which is supposed to be supported by our "inner intuition". Quite a few philosophers have asserted and quite a few scientists have agreed that, e.g., the axioms of geometry can be recognized as true by looking into our own minds. In order to give to this ambiguity the appearance of unambiguity, a particular word has been borrowed from German philosophy and introduced into the English vernacular of philosophy, the ambiguous word "anschaulich", which means "conceivable by our sense observation", but also "conceivable by the efforts of our minds even if not observable at all". These two meanings contradict each other in the application of that word to every actual problem.

III. Classical Newtonian Mechanics

8. Domain of Newtonian Mechanics; Mass

Classical mechanics or Newtonian mechanics is still the cornerstone of all physics. By this theory a group of phenomena is covered around which the research work of science was centered in the period when Galileo started the revolution which gave rise to modern physics. These phenomena are the motions of "medium-sized" material bodies, such as the motion of running cars, of machines, of falling and rolling stones. Newton set up a pattern of description which was adjusted as well to the motion of planets as to the motion of these medium-sized bodies. If we understand Newton's laws of motion which define this pattern, we have made the first step toward understanding physics. A great deal of misunderstanding of the recent physical theories (relativity and quantum theory) is due to the fact that Newtonian physics has been frequently misrepresented. There has been, in particular, a tendency in the teaching of physics to suggest that Newton's laws are almost self-evident. In quite a few textbooks the authors express their bewilderment that such obvious truths have been ignored for such a long time. This kind of presentation is a serious obstacle in the way of a real understanding of physics. One has to emphasize, rather, how great a power of imagination was necessary to set up these laws, how long and strenuous the road has been which has led scientists from the observed facts to Newton's laws. The more self-evident these laws appear to us, the less do we understand them.

For medium-sized material bodies we can tentatively define the "mass" of a body by determining the number of grams which we obtain by reading its weight on a spring-balance at sealevel. This operation is unambiguous, as the weight does not change if we move the balance to any point at sealevel. We note here again that the applicability of an operational definition is based on the physical fact that several operations render an identical numerical result. In the domain of the phenomena which are covered by Newton's laws of motion the following statement holds:

If we impart to a body of a given mass in a well-defined environment a certain initial velocity, its future motion is determined. If the size of the body is small, this motion can be described approximately by a curve. Let us now increase the mass, but not the size, of our body more and more without altering any of the other circumstances and examine how the orbit is altered by this fact.[18]

9. Field of Force; Gravitational and Electromagnetic Fields

The environment which has an influence upon the orbit is called the "field of force". It can be of two kinds. In the first case, the orbit is not at all altered by the increase of mass. An example of this case is the orbit of a projectile launched inside a vacuum. Then we call the environment "static gravitational field of force." By 'static' we mean that the field is due to masses which are all at rest (relative to the fixed stars). In the example of the projectile, for instance, the motion of the earth has to be neglected.

A typical example of the second case is the static electric or magnetic field of force. Its influence on the orbit of a charged particle becomes less and less the greater the mass becomes. We may consider, for instance, the path of an electrically charged particle moving through an electrostatic field which has a direction which is different from the initial velocity of the particle. If the charge remains constant while the mass of the particle increases, the particle will be less and less deviated by the field from its original direction. If the mass becomes "infinitely great," the particle will travel along a rectilinear path with constant speed whatever the direction of the field may be. The motion of an infinitely great mass is called an "inertial" motion, for this path is a result of mere inertia and independent of the field of force.

It is obvious that a rectilinear motion relative to a system of reference S cannot be a rectilinear motion relative to other systems S' which are accelerated or rotational in motion relative to S. To be unambiguous, we have to qualify our statement about the path of infinitely great masses. Only if we describe its motion relative to a particular system of reference S is the path of a very great mass in a field of force of the electromagnetic type a straight line and its speed constant. Such a particular system S is called an "inertial system". Our first superficial experience makes us believe that our earth is an inertial system. But more thorough investigations (of the kind of the Foucault pendulum) show us that the earth is in rotation relative to the particular system S and that rather the system of fixed stars should be taken as an inertial system, at least approximately. If we describe the path of an infinitely great mass relative to the rotating earth, it would be a kind of helix. For each system of reference it would be a characteristic curve. The inertial system is recommended only by the simplicity of this curve (straight line).

10. Newton's First Law of Motion (Law of Inertia)

These statements about the path of infinitely great masses are the factual content of what is known as the "Law of Inertia". Obviously, this law is empirical and cannot be derived from any self-evident principle. The only element in it which looks self-evident is the assertion that the straight line is the "simplest" curve. But actually it is an observation statement, too, except that it describes psychological facts, not physical ones.

The Law of Inertia has an operational meaning only if the inertial system is described by a physical operation. As a first approximation we can say that S is identical with the system of fixed stars. Since a particular physical object (the fixed stars) enters into this law, it cannot be a self-evident statement. It is even clear that the Law of Inertia, if given this operational meaning, cannot be exactly true. For the fixed stars have their "peculiar" motions and do not form a rigid frame of reference. Therefore, the Law of Inertia in this form can be only approximately true. Einstein's general theory of relativity and gravitation attempted to replace the old Law of Inertia by a law which takes into account these peculiar motions of the fixed stars. Instead of stating that a mass travels along a straight line with constant speed, we can also state that its "acceleration" is zero. ('Acceleration' means change of speed or direction.)

If we consider a finite mass, we notice by experiments that in the same environment the smaller the mass, the greater the acceleration. An electrically charged particle moving in an electric field is the more deviated from its original course the smaller its mass. (The charge belongs to the environment.) We can state this experimental fact quantitatively by stating that the product mass times acceleration remains constant if the environment does not change. Mathematically: If we denote the mass by 'm' and the acceleration by 'a' and put the product ma equal to f, the quantity f depends upon only the distances of the moving mass from the bodies of its environment.

The law $ma = f$ is known as "Newton's Second Law of Motion". We call f the "force acting upon the body m" and exerted by the bodies of the environment. If all this is true, we can derive a new operational definition of mass: The ratio of two masses is inversely proportionate to the accelerations which they get from one and the same force. This definition is only unambiguous if we obtain the same value of mass whatever force we apply. Again a physical law has to be regarded as valid in order to insure the identical result of the operations which define mass.

11. Heavy Mass and Inert Mass

The values of masses obtained from our original definition (by weighing) and our new one (by the ratio of accelerations) are identical. This is an important physical fact, too. We call "heavy mass" the result of weighing and "inert mass" the ratio of accelerations. Then we can state the law of the identity of heavy and inert mass. If we use this law, we can apply the relation $ma = f$ also to motions in gravitational fields, although it would seem that in this case the acceleration is independent of the mass. Here we have to assume that the force of gravita-

tion is proportional to the heavy mass m. The force f is equal to mg, where g is independent of m. Then $ma = f$ becomes $ma = mg$ and $a = g$. The acceleration a is independent of the mass m and is not altered if the mass becomes infinite. In this way Newton treated the gravitational field as an example of the general field of force. He advanced the hypothesis that the gravitational force is proportional to the inert mass m. This hypothesis is another way of stating the identity of heavy and inert mass. However, we must not fail to understand that the operational meaning of the terms 'inertia' and 'force' in the gravitational field is a particular one. An electromagnetic force produces an acceleration; this means a departure from the inertial motion, while in the static gravitational field the inertial motion (the motion of an infinitely great mass) itself is accelerated. The definition of force by "acceleration" and by "departure from the inertial motion" give now different results. A launched projectile is not deviated by a force if a "force" is defined as causing its departure from the motion of an infinitely great mass.

12. Newton's Second Law of Motion

Discussing Newton's Second Law of Motion ($ma = f$), the question has been raised frequently whether this law states physical facts which can be checked by experiments or whether it is only a definition of the term 'force'. Evidently, whatever may happen in the world of our experience, we can always introduce a force f which obeys the relation $ma = f$. Whenever there is an acceleration a (which is an observable quantity), we may assume the existence of a force f which cannot be observed itself but has to be calculated from the definition $ma = f$. Therefore, this "law" seems to be only a definition of the term 'force' and can be neither confirmed nor refuted by experiments. However, it is obvious that certain physical effects must be confirmed to make $ma = f$ an unambiguous operational definition of f. The product ma has to be independent of the mass m and has to depend only on the situation of the moving body in its environment. Then the law of motion says that equal masses have, under equal relations to the environment, equal accelerations. But the factual content of this law is very poor unless an operational definition of the term 'equal relations' is given which does not contain the acceleration produced.

The most famous example which decided the success of Newton's laws of motion was his theory of gravitation. The relation to the environment is determined in this case by the masses M of this environment and the distances r of the moving mass m from these masses. Every mass M contributes the term $f = Mm/r^2$ to the force. If we substitute this expression in the law $ma = f$, we obtain a law which contains only observable quantities. By using the "calculus" (Newton's equations of motion), the motion of our mass m relative to the inertial system

can be calculated. This theory has succeeded in giving a very good account of the observed motions in the planetary system.

However, this success cannot be interpreted as a confirmation of the law $ma = f$ by itself but is a confirmation of this law plus the law of gravitation Mm/r^2. The meaning ascribed by Newton and the Newtonian school to the law $ma = f$ itself has been the following: Since the substitution of Mm/r^2 turned out to be such a success, it seemed probable that every motion in the universe could be described by substituting for f in $ma = f$ a formula which is in a certain way analogous to Mm/r^2. But it may contain a different power of r or even a different function of r or even the velocity of m besides the distance. The main point was that f should be expressed by a simple function of the variables which determined the relation of m to its environment.

13. Inertial Force and Centrifugal Force

Newton's Second Law of Motion, $ma = f$, where f is the gravitational or electromagnetic force, holds only if a is the acceleration with respect to an inertial system (e.g., the fixed stars). If we want to calculate the acceleration a of a mass m relative to a system which has an acceleration or a rotation relative to the fixed stars, we have to use a modification of Newton's laws. We may call the rotating or accelerated system the "vehicle" (comparing it to a railroad car). The acceleration of our mass m relative to the vehicle may be denoted by 'a_{rel}' and the acceleration of the vehicle (speaking exactly, of the point of the vehicle coinciding with the mass m) by 'a_{veh}'. If both these accelerations have the same direction as the acceleration a of our mass m relative to the fixed stars, we have obviously the equation: $a_{rel} + a_{veh} = a$. Newton's equation becomes then: $ma_{rel} = f - ma_{veh}$.

This means: The acceleration relative to the vehicle can be calculated as if it were at rest (relative to the fixed stars), except that we have to add to the electromagnetic or gravitational force f a new kind of force $f_{in} = -ma_{veh}$ which we call "inertial force". It increases proportionately to the acceleration of the vehicle but has the opposite direction. The simplest example is the shock which we feel when a railroad car is starting or stopping. a_{veh} is the acceleration of the car. If the vehicle is rotating, we can no longer use our simple relations between the accelerations.

But these relations still hold for the components of the accelerations in a certain direction. If we choose a direction perpendicular to the axis of rotation, $f_{in} = -ma_{veh}$ becomes what one calls "centrifugal force". It follows that the acceleration a_{rel}, perpendicular to the axis of rotation, can be calculated by adding to the radial component of f (electromagnetic or gravitational force) the centrifugal force. There is sometimes a confusion about whether the centrifugal force is really

a "force". The answer is simple: according to the original Newtonian definition ($ma = f$), only forces of the type of gravitational or electromagnetic force can be substituted for f. The centrifugal force does not produce any acceleration with respect to the fixed stars. But if we want to describe motions relative to a rotating system, we have to add the centrifugal force f_{in} to the gravitational or electromagnetic force f.

14. The Law of Force Must Be "Simple"

If the force f is not a simple function of the distances, masses, etc., the law $ma = f$ loses its factual content. This becomes obvious if we consider what has been the most outstanding achievement of Newton's laws. We remember that Copernicus described the planetary orbits by circles, Kepler by ellipses. But if we examine the departures from the earth's elliptic orbit produced by the attraction of other planets, by Jupiter, Mars, etc., we obtain, as the effect of these perturbations, curves of an extreme complexity. Newton, however, succeeded in describing these same orbits by the simple statement that the acceleration with respect to the fixed stars is a sum of terms each of which has a simple form Mm/r^2.[19]

This means: A very complicated curve or a very complicated mathematical function can be derived from the extremely simple function Mm/r^2. If this expression of force were as complicated as the equation of the curves performed by the planets, there would be no point in replacing the geometrical description by the dynamical description. The hope of the Newtonian school has been to find for every type of motion a particular law of force of a simplicity similar to the law of gravitation—a law for the forces of cohesion, of chemical affinity, etc. This hope has been partly disappointed but partly also more than fulfilled. Since the turn of the century (1900) it has become more and more evident that the motion of the smallest particles (electrons and nuclei) cannot be described adequately by the Newtonian pattern. In the mechanics of subatomic particles (quantum mechanics) the term 'acceleration of a mass at a certain point in space' does not occur at all. Therefore, the concept of "force" loses its original operational meaning.

But, on the other hand, within the domain in which Newton's pattern of description can be applied at all (medium-sized bodies) not so many different laws of forces have been needed as Newton's school had expected. Besides the gravitational force, which is not even a force in the full operational meaning of this word, only electromagnetic forces have been used in twentieth-century physics— in the whole domain of phenomena which are covered by Newton's laws of motion. There are no special forces which are responsible for the phenomena of cohesion or of chemical valence. The force f in Newton's law $ma = f$

can always be calculated from the laws of electromagnetism and has therefore also an operational meaning which is independent of its definition by $ma = f$. Partly, however, these phenomena of chemical affinity and physical cohesion have to be treated by laws which are different from Newton's laws of motion. We are going to learn later that in these parts of physics "forces" are not defined by acceleration. This holds, in particular, for the phenomena of nuclear physics.

In engineering physics some laws of force are used which are "rules of thumb". In aerodynamic engineering, for instance, a law is used which says that the resistance of air which slows down the speed of an airplane is proportional to the square of its speed. But the physicists believe that this law can be derived from the laws of collisions between the molecules of the air and the molecules of the plane. The forces which are responsible for the effects of these collisions are the electric attractions and repulsions between the electric charges of which ultimately the atoms of gases and solid bodies consist.

15. Do Forces "Exist"?

These considerations will help us a little toward appreciating some attempts which have been made to apply Newton's laws of motion to living organisms by a kind of short cut. Quite a few authors have suggested introducing "vital forces" or even "spiritual forces". This would mean

443

substituting these forces for the symbol 'f' in Newton's laws. We have seen that Newton's laws of motion have been a success in physics only because it has been possible to substitute for 'f' an expression which is simple and has a certain similarity to the law of gravitation. If one had to introduce "entelechies" or "holistic tendencies" as state variables instead of distances and masses, the whole meaning of Newton's laws would be lost. We would be faced by a completely new type of laws. To apply Newton's laws in their real scientific meaning to the motions of living organisms means to assume that the forces determining these motions are not very different from the electromagnetic forces if we remain in the domain of bodies of such a size that terms like 'acceleration' and 'velocity' have an operational meaning. A spiritual theory of the living organism perhaps does not contradict Newton's laws but would make them a definition of the term 'force' and turn them into tautologies. The opinion that such vital or spiritual forces can be introduced into Newton's laws has its source in some connotations of the term 'force'. Since this word is used in our everyday language as a term of psychology, the impression has been produced that by virtue of this term a "psychical" element has a legitimate place in physics.

As a remainder of the organismic philosophy of science which was prevailing in the Middle Ages there has been a widespread reluctance to accept the operational definition of force by which force is defined in terms of motion.

It has been argued that in a stretched piece of rubber there is a force, a tension, even if the rubber is at rest. Therefore, we are told, 'force' cannot be defined by motion and does exist as an entity independent of motion. But the statement, "Force can exist without acceleration", is a very misleading way of stating the facts. It is in flagrant contradiction to Newton's definition of force. What really happens in the stretched rubber is a state of equilibrium between several forces. The algebraic sum of the forces acting upon any particle of rubber is zero. To say that forces "exist" in this rubber would be as correct as to say that in the number zero the number five "exists" because five minus five is equal to zero. The existence of a tension or force within the rubber means in terms of operations that, by removing a part of the rubber, the equilibrium is disturbed. Acceleration of the rubber particles is produced and the force reveals itself. By saying that this force has existed before and was only balanced by other forces, we do not add anything to the operational meaning of our description of facts.

The only logical, sound way of putting the problem of the "existence of force" is again to avoid what Carnap calls the "material mode" of speaking and to stick to the "formal mode." We have to ask: If we set up a system of axioms from which Newtonian mechanics can be derived, is it necessary to introduce 'force' as an undefined term? The answer is clearly in the negative. We can introduce 'force' as a term defined by acceleration if we assume the validity of some statements about physical facts.

The insistence upon the use of the term 'force' as an "entity of its own" has its sources only in some psychological connotations of this word. We can quote the example of the Nazi philosophy in which the word 'force' is regarded as dear to the mind of the Nordic race. Every attempt to introduce a definition of 'force' by motion is branded as an act of the enemies of the Nordic race.

However, it is possible to give an operational meaning to the statement that "forces are physical realities". According to P. W. Bridgman, a term of physics describes a physical reality if several independent operational definitions of this term can be given which render one and the same numerical result.[20] If in a specific case a force cannot only be defined by acceleration but also as a function of masses, distances, etc., we have two independent operational definitions of force. We can say in this case that the term 'force' refers to a physical reality. This is obviously the case for the gravitational and electromagnetic forces which can be calculated from masses, electric charges, etc. But it is obviously not the case for spiritual and vital forces which cannot be calculated from observable data.

IV. Heat, Irreversibility, and Statistics

16. Recoverable and Irrecoverable Changes of State

If a small part of the physical world passes from a state s_0 into a state s_1, the following question arises: Is it possible that in this part of the world the original state s_0 be restored without changing anything in the rest of the world? If this restoration is possible, we call the transition from s_0 to s_1 a 'recoverable transition'. If not, we speak of an 'irrecoverable (irreversible) change of state'. To understand the role of irrecoverable phenomena and in particular the phenomena of heat, we are going to consider a simple example.[21]

The part of the world which we consider may be a perfect gas which is isolated from any supply or loss of heat. But we can change the volume of this gas by moving a piston closing the container of the gas. The initial state s_0 is a certain volume V_0 and a certain temperature T_0 of our gas, e.g., a volume of 2 liters and a temperature of 0° centigrade. The second state s_1 may be the result of an increasing pressure upon the piston. The final volume may be V_1 (e.g., 1 liter). We perform the compression by putting only enough weight on the piston to make the pressure exerted just a little greater than the expansion pressure of the gas. Then the transition from the initial volume V_0 to the final volume V_1 will occur infinitely slowly. Since the gas is, during the whole transition, almost in equilibrium, we speak of a quasistatic change of state. This means that the process is almost static. As no heat is lost, the temperature increases by the compression. When the gas has reached the final volume V_1, the temperature has increased to a value T_1, which is greater than T_0. According to an easy calculation, this temperature can be calculated by the formula: $T_1 V_1^{k-1} = T_0 V_0^{k-1}$ (k has the value 1.4 for a diatomic gas like oxygen).

Outside our gas the effect of the process has been the lowering of the weight which has been pressing upon the piston. Hence, the effect on the whole world has been lowering of a weight, increase in a temperature, and decrease in the volume of our gas. Is this change of state from s_0 to s_1 recoverable? Can the state s_0 be restored? Certainly, yes. We have only to reduce slowly the pressure upon the piston. Then the gas will expand again infinitely slowly because the pressure on the piston is just a little smaller than the internal pressure of the gas. When the original volume V_0 is restored by expansion, the weight on the piston is in its original position and the initial low temperature T_0 is also restored. This means that the change of state from s_0 to s_1 is really recoverable.

Let us now modify our change of state. We start again from s_0 but produce a change of state in a new way. The gas may remain isolated against

heat supply from outside. Now we increase its temperature. But we do it at constant volume V_0. We can, e.g., like in the old Joule experiment, have a paddle wheel rotating in the gas. The rotation can be produced by the fall of a weight which is connected with the wheel by a cord over a pulley. The gas will be heated by friction. The state s_0 will be changed into a state s_2 which has the same volume V_0 but a temperature T_2 which is greater than T_0. Is the change of state from s_0 to s_2 now also recoverable? Can a change of state occur which restores s_0 if no heat is supplied from outside? This change would mean that the weight which produced the friction by falling has been lifted again, while the temperature of the gas which had increased by friction has dropped again. The final result of the change of state from s_2 to s_0 would be the decrease in temperature of the gas and the lifting of a weight outside the gas, while nothing else has changed in the world.

If we generalize the results of all our experiences on the conversion of heat into work, we notice that a weight can never be lifted just by using the heat supplied by the drop in the temperature of one single reservoir. This generalization has become a basic hypothesis of physics. It is usually called the "Second Principle of thermodynamics" (the First Principle states the conservation of energy). From this Second Principle it follows that the change of state from s_0 to s_2 is not recoverable. This means: The increase of temperature by friction at constant volume is not recoverable. It is obvious that by this change of state from s_0 to s_2 (friction) the function TV^{k-1} increases as V remains constant and T increases.

17. The Second Principle of Thermodynamics and Entropy

The function $F = TV^{k-1}$ increases by the process of friction and would decrease if a weight were lifted and nothing else happened than the decrease in the temperature of one body (e.g., our gas) at constant volume. The function $F = TV^{k-1}$ may, provisionally, be called "entropy". This entropy remains obviously unaltered as long as the process is quasistatic and no heat is supplied. The entropy increases if a change of state takes place which is irrecoverable. The entropy would decrease during a change of state which is forbidden by the Second Principle of thermodynamics. This principle can therefore be formulated for a perfect gas as follows: There is a function F of the state variables V, T with this property: if no heat is supplied, F can only increase. It remains constant during a quasistatic process. But this is only a limit which can never be reached by an actual experiment.

If we pass from a perfect gas to a general system, we can formulate the Second Principle in an analogous way. The whole system consists of several bodies with the volumes V_1, V_2, etc., and the temperatures T_1, T_2, etc. Then there is again a function $F(T_1, T_2, \ldots, V_1, V_2, \ldots)$ of the state variables which behaves in a simple

way during changes of state which are not accompanied by an exchange of heat between these bodies and the outside world. The function F remains constant during quasistatic processes. These changes of state are recoverable with the approximation with which a quasistatic process can be performed at all. All changes of state which entail an increase of F are irrecoverable, while changes of state with decreasing F are impossible.

As a matter of fact, every function of F which is increasing with increasing F has the same properties as the "entropy". It turns out to be convenient not to introduce $F = TV^{k-1}$ but $S(V, T) = \log F$ as the "entropy" of a perfect gas. For if we consider two gases with the common temperature T and the volumes V_1 and V_2, we can show easily that during a quasistatic process

$$F(V_1, T) + F(V_2, T) = TV_1^{k-1} + TV_2^{k-1}$$

does not remain constant while

$$\log TV_1^{k-1} + \log TV_2^{k-1}$$

does. If we introduce the function $S = \log F$,

$$S(V, T) = \log TV^{k-1} = \log T + (k-1)\log V,$$

and call it definitely "entropy", we can assert a simple law. If S_1 and S_2 are the entropies of two gases which are in temperature equilibrium, the entropy of the whole system is $S_1 + S_2$. This means that $S_1 + S_2$ remains constant during quasistatic processes of the whole system.

The most popular example of increasing entropy is the passage of a system in which there are great differences in temperature into a system in which these differences are smaller. In the case that no heat is supplied from outside, one can show by an easy calculation that the entropy increases by the disappearance of these temperature differences. Therefore, a process of leveling temperature intensity in an isolated system is irrecoverable. Only in the borderline case of a quasistatic process does the entropy of the isolated system remain constant. The system performs recoverable changes of state. Therefore, the quasistatic processes are also called 'reversible processes'.

18. Cosmological Implications

All these facts and generalizations have been used to predict a very gloomy future for our world. The Second Principle of thermodynamics has been formulated as follows: The entropy of the universe tends toward a maximum, and this maximum is attained if the whole world is of constant temperature. Since the world is an isolated system, every change of state is connected with an increase of entropy. When the world has reached the state of greatest entropy, change is no longer possible; the universe will "die".

In a similar way it has been argued that only a finite number of years have passed since the creation of the world. If the actual laws of nature had been valid through an infinite number of years, the maximum of entropy would have already been attained. Since this is obviously not the case and the world is still "alive", the laws of nature and our world itself cannot have existed for an infinite number of years. It must have been created a finite number of years ago.

Since all processes which are covered by Newton's laws of motion can be reversed, this tendency of the world toward the ceasing of all changes has been interpreted as a feature of nature which cannot be explained by mechanics. A kind of striving toward an end has been envisaged by many philosophers. It has been cheered as a spiritual factor. This running toward the death of the world has been interpreted as being in contrast to the evolution of organisms which shows a tendency toward greater and greater differentiation. Two conflicting tendencies must therefore be recognized in the universe both of which cannot be explained by mechanical laws. Sometimes this conflict has been interpreted as a scientific background for the eternal struggle between God and the devil.[22]

As a matter of fact, the term 'entropy of the universe' is, as P. W. Bridgman says correctly, a pure "paper and pencil affair". We can, of course, say that the entropy of the universe is the sum of the entropies of all parts of the universe. But speaking exactly we do not know whether the expression 'entropy of the Universe' has any operational meaning. For the sum of the entropies of the parts of a system can be regarded only under very restricted conditions as the entropy of the whole system. We do not even know whether the sum is convergent from the purely mathematical viewpoint. Therefore, all these conclusions concerning the creation and death of the world are results of a loose way of thinking.[23]

19. The Kinetic Theory of Heat

We shall understand the whole problem of recoverability much better if we attack the question from a different angle. We ask whether "irrecoverability" is compatible with the hypothesis that heat is a movement of small particles. Then we shall learn that irrecoverability is the result of a rather shortsighted view of the universe, while a more penetrating view would reveal to us a universal recoverability. We must never forget that the results of thermodynamics have operational meaning only under conditions which can be described by the concepts of thermodynamics: temperature, mechanical work, entropy, etc. But there are conditions which cannot be described by what P. W. Bridgman calls the "universe of operations" of thermodynamics. This fact is obvious from the simple consideration that in no statement of thermodynamics is the expression 'velocity' or 'time' used. Therefore, we are given not the slightest estimation of how long it may take to establish a certain state of the world. If a theory does not contain any statement about time, it has no operational meaning to say that the

theory predicts a certain future of the world. For the time may just as well be "never."

There are certainly phenomena of heat which are not covered by the laws of thermodynamics. In this category belong all phenomena of fluctuation like the Brownian movement, the spontaneous change of density in the atmosphere which accounts for the blue sky, etc. For this reason we are in need of another method to deal with the phenomena of heat. This method is provided by the "kinetic" or "statistical" or "molecular" theory of heat. Its basic hypothesis asserts that, besides the observable motions of bodies, there is an irregular zigzag movement of its microscopic and submicroscopic particles which on the average does not contribute anything to the observable velocity of medium-sized bodies. Moreover, according to this hypothesis, the average kinetic energy of this molecular motion is proportional to the absolute temperature of the body.

This hypothesis which accounts for a great many facts connected with the conversion of heat into mechanical work and vice versa faces a great difficulty if we apply it to the existence of irrecoverable changes in the universe.

20. Mechanics Cannot Account for Irrecoverable Changes of State

We consider a simple case. There may be a gas inclosed in a container. The gas molecules are performing rectilinear motions and are only deflected if they hit each other or the walls of the container. We start from a state s_0 in which all molecules are assembled in a small corner of the container while most of its volume is empty. This means in terms of thermodynamics that the density of the gas in this corner is great while the density in the rest of the volume is zero. In the second state, s_1, the molecules may fill up the whole container with equal density. According to the rules for calculating the entropy, this physical quantity has in the state s_1 a much greater value than in the state s_0 of our gas. As in an isolated system the entropy can only increase, the change of state from s_0 to s_1 is, according to the laws of thermodynamics, irrecoverable.

According to these laws, it can never happen that a gas which fills up a volume with constant density can undergo changes of density by which eventually the whole mass is assembled in one corner. However, according to the kinetic theory, we cannot understand why such a thing should be impossible. Let us assume that the state s_1 (equal density) has just been reached. Then we reverse the direction of the velocities of all particles. According to Newton's laws of motion, they must traverse now the same orbits as before but in an opposite direction, until they reach finally again the state s_0 and assemble in one corner. Therefore, the kinetic theory seems to be incompatible with the existence of irrecoverable changes of state. This means that the kinetic theory would be incompatible with the Second Principle of thermodynamics. For from mechanics it

seems to follow that a homogeneous gas can spontaneously move into a small corner.

This argument is conclusive. Newton's mechanics as the general Law of the Universe and the existence of irrecoverable changes are incompatible. The kinetic theory of heat cannot be based on mere mechanics. It needs, in addition to Newton's laws of motion, still a different type of hypothesis. The kinetic theory makes use of *statistical hypotheses*. L. Boltzmann was the first to point out how, by combining mechanical and statistical hypotheses, it can be predicted that the initial state of leveling processes like heat conduction and diffusion is "irrecoverable", if we take this word in its practical sense. However, it can also be derived from statistical hypotheses that this irrecoverability is only a superficial aspect. If we observe the phenomena more in detail and over long periods of time, we can bet that the initial state will once reappear.

21. Statistics and Irrecoverability

The statistical hypotheses, in contrast to the mechanical hypotheses, do not make an assumption about how a specific particle behaves at a specific time in a given field of force. Statistical assumptions say, in a certain sense, more and, in a certain sense, less. We explain their meaning by a simple example. We may divide the whole volume of our container of gas into a great many small cells. Our hypothesis says: If we pursue the path of a particle over a long time (years or centuries), the time during which it dwells in a specific cell is a specific fraction of the whole time considered. The volume of the cell is in the same ratio to the whole volume of the container as the dwelling time in the cell is to the whole time considered. This means that this ratio becomes more and more independent of the time of observation. This ratio may be denoted by 'p'. If there are, e.g., 100,000 cells, p is a very small fraction (e.g., $p = 1/100,000$) and may be called the "dwelling time" in a specific cell. If we consider a second gas molecule moving independently of the first one, it spends again the $1/100,000$ part of any given time stretch in our specific cell. If T is the whole time of observation and T' is the dwelling time of the first particle in our cell and T'' the time during which both the first and the second particle are dwelling in our cell, we have

$$T' = pT \qquad T'' = pT' = p^2T .$$

If we want to know the time during which N particles would all dwell in one specific cell, we would find that it is equal to $p^N T$. If we assume, e.g., that N is equal to one million (10^6), the dwelling time of all N particles in one specific cell would be

$$T \cdot p^{1,000,000} = T \left(\tfrac{1}{100,000} \right)^{1,000,000}$$

$$= T \frac{1}{10^{5 \times 10^6}} = T \frac{1}{10^{5,000,000}} .$$

The denominator is a number in which the digit 'one' is followed by five million zeros. Therefore, if we make observations during a billion years ($T = $ a billion) the average "dwelling time" of our particles in one particular corner will be much less than the billionth of the billionth part of a second. Moreover, if we are the lucky observers of such an event, we can bet that it will soon disappear and not reappear for a billion generations to come.

This is the statistical aspect of irrecoverability: There are rare states of a system and they disappear very soon. Hence, we can bet in what direction the system will develop. But if we start from a very frequent state (e.g., from the state where the particles are equally distributed over the whole container), we can bet that it will hardly happen in a billion generations that the rare state will be restored in which all particles assemble in a particular corner.

The sequence "rare state–frequent state" happens as often as "frequent state–rare state". The illusion of irrecoverability in the realm of our observations results from the fact that we start every experiment from a "rare" state of our system and our time T is always short. Then we can bet that the future will bring states which are not so rare. As a result, essentially there is no asymmetry of sequence but a distinction between rare and frequent states of a system. Starting from a rare state, s_0, the next state will usually be a frequent one, while, starting from a frequent state, the next one will usually be again a frequent one.

If we pass from the special case of a gas to a general case, we may say that in every system we have to set up a statistical hypothesis which allows us to compare the different states of a system with respect to their relative dwelling times.

22. Statistical Hypotheses Are Assertions about Physical Facts

The kinetic theory of heat is based upon statistical hypotheses which make predictions about motions of particles as the mechanical hypotheses do. They could, therefore, contradict Newton's laws of motion. In setting up these statistical hypotheses, their compatibility with Newton's mechanics has to be assumed as a special hypothesis. Every statistical hypothesis assumes a specific regularity about the dwelling time of particles in a certain domain. The essential in "Boltzmann's statistics" is the assumption that the dwelling time of a particle in any specific domain is independent of how many other particles are present in this domain. If we drop this hypothesis, we alter substantially the conclusions drawn from the kinetic theory. In the modern wave mechanics, which has replaced Newton's mechanics, Boltzmann statistics have been replaced by other statistical hypotheses. For each particle a particular statistical hypothesis has been set up. For some particles (e.g., photons) the dwelling time of the particle is increased by the presence of other particles (Bose statistics). In

other cases (e.g., electrons) the dwelling time is unfavorably influenced by the presence of other particles (Fermi statistics).

We must not be too puzzled by the fact that irrecoverability only comes into the picture if we start from a rare or improbable state of the system. But we may ask: How does it happen that we are living in a period and on a place of the universe where the entropy is increasing? We must always remember that the statistical approach to the happenings in the universe starts from the assumption that these happenings are in agreement with our basic statistical hypotheses. This means that the universe has already traversed a great many cycles. All the rare states have appeared and disappeared and will reappear again. This universe is, in the sense of ancient philosophy, an Epicurean universe. The origin of the sun, the earth, the elements, and even of our own human race is due to the law of chance. The appearance of men in the universe is a very improbable event from this viewpoint. In the Aristotelian universe, on the other hand, everything has developed according to plan toward a certain end of higher perfection. However, if we start from the assumption that our universe is an Epicurean one, our whole existence is due to chance, and we, the human race, are a very improbable event. Therefore, it is obvious that the universe will mostly go back to a state which appears more frequently according to the fundamental statistical hypothesis. But all the rare states have their proper chance to reappear.

These considerations are pertinent for the attitude of philosophers toward the principles of thermodynamics. A great many philosophic interpretations of physics have made use of the principle of increasing entropy to bolster up an anti-mechanistic teleological view of the universe which invokes a tendency, a direction toward a certain end, instead of a causal chain of events. Unfortunately, the trend of

events, if we take thermodynamics for granted, tends toward destruction of the universe. If we, on the other hand, replace pure thermodynamics by the kinetic theory of heat (laws of motion + statistical hypotheses) we drop implicitly the Aristotelian theory of the universe and accept the Epicurean view that every tendency toward an end is an illusion and the real actor in the evolution of the world is a play of chances and the survival of the fittest. There is no real irrecoverability.

V. The Theory of Relativity

23. Einstein's Two Basic Principles

Newton's mechanics was applied to a hypothetical medium, the "ether", which was supposed to fill up the whole world-space. The wave motion in this ether was interpreted as being responsible for what appears to our sense observations as phenomena of light. The "ether theory of light" treats these waves according to Newton's laws of motion. The first result of this theory is that in empty space the waves of light are propagated with a speed $c = 3 \times 10^{10}$ cm/sec with respect to the ether. The speed of the source of light has no influence upon this speed of propagation. This result can also be formulated without speaking of the ether. One proceeds in a way which is similar to the replacing of Newton's absolute space by the "inertial system" (sec. 9). We say that in empty space the waves of

light are propagated with the speed c relative to a certain "fundamental system S".

Let us now examine phenomena due to the simultaneous motion of material bodies and propagation of light waves. Then Newtonian mechanics and optics lead to results which are not in agreement with the experimental facts and, moreover, are by themselves rather awkward. If we stick to the belief that Newton's laws of motion are the only legitimate basis of physics and that light is in particular a wave motion in a medium which follows these laws, we have to resort to complicated additional hypotheses.

In 1905 Einstein dropped both assumptions: Newtonian mechanics and the ether theory of light. His new theory of light and motion retained Newton's laws only for a limited class of motions which obviously follow these laws. These are the motions with "small velocities." This means velocities which are small in comparison to the velocity of light. Troubles arise only if the speed of material bodies approaches the speed of light. Further, Einstein retained two results of the ether theory of light which are made plausible by experiments. But he regarded these results as straight generalizations of experimental facts without troubling about whether they could be derived from the ether theory of light or from Newton's laws of motion.

Einstein's two hypotheses (or principles) are: first, the speed of light relative to any room is independent of the speed of the source of light relative to this room (principle of the constancy of light velocity); second, the propagation of light relative to the walls of a moving room is determined by the initial conditions relative to this room provided that the motion of this room is a rectilinear and uniform motion relative to the fundamental system S (principle of relativity).

This principle can also be formulated as a statement about the impossibility of performing a certain set of operations. If our room has the speed v relative to S, it says that there is no physical experiment by which we can measure this quantity v, i.e., the expression 'speed of a room relative to the fundamental system' has no operational meaning. If we substitute into Einstein's two principles the operational definition of 'speed of light relative to a system', 'speed of the

light source', etc., they become physical hypotheses about the interactions between the motion of bodies and the propagation of light. They are to be confirmed as any physical hypotheses by the agreement of their results with direct observations. These results are in conspicuous contradiction to the results derived from Newton's laws of motion in connection with the traditional ether theory of light. For this traditional theory allows for a particular operation by which this speed v can be determined (Michelson's experiment).

24. Newton's Laws of Motion Not Universally Valid

Actually, none of the observable phenomena derived from Einstein's two principles has been in disagreement with the results of observation. In this sense all statements derived from these principles are "true". Then it follows easily that some immediate results of Newton's laws of motion cannot be true. According to these laws, the increment of velocity produced by a given force is independent of the actual (initial) velocity Therefore, by the continuous action of a constant force, any increment of velocity could be produced. We could, e.g., obtain in this manner a velocity v of a body greater than the velocity c of light in a vacuum. But we can show easily that this result is in flagrant contradiction to Einstein's principle of relativity. Assuming the universal validity of Newton's laws, we could consider a room which is moving with the speed of light relative to S. By a light source at rest in this room light is emitted through the vacuum toward one of the walls. This light will never reach this wall which itself moves with the speed of light. Therefore, no reflection can occur in this direction. By the absence of reflection we could make sure that our room has the speed c relative to S. But the determination of this speed is impossible, according to the principle of relativity.

Therefore the assumption that Einstein's principle is valid implies the negation of Newton's laws. We must assume that in contrast to Newton's mechanics the increment of velocity produced by a force is dependent upon the actual velocity. We can calculate exactly this dependence from Einstein's principles and obtain the result that this increment becomes the smaller the greater is the actual velocity, and tends toward zero if the actual velocity approaches the speed of light (the force being constant). This modification of Newton's laws can be checked by experiments. One can observe the increment of velocities produced by electric and magnetic forces if one examines fast-moving electrons, as cathode rays or the rays emitted by radioactive substances. These rays travel with a speed which is comparable to the speed of light. According to Einstein's results, the dependence of the increment of velocity upon these great initial velocities is great enough to be observable.

25. The Operational Definition of 'Mass' Has To Be Modified

This dependence has been confirmed by experiments. Hence some operations which are to render an identical result according to Newton's laws no longer do so. Let us consider the operational meaning of 'the mass of a particle'. If we assume that the field of force is known (e.g., given by Coulomb's law), we can get the mass by measuring the acceleration a (increment of velocity per unit of time). The mass m is defined by $m = f/a$. According to Newton's mechanics, the result is independent of whether the initial velocity was small or great. But if Einstein's principles are right, this operational definition becomes ambiguous. The acceleration a (and, therefore, m) depends actually upon from what initial velocity we start the experiment. In order to obtain an unambiguous result, we have to specify the operations involved, in particular the initial velocity v. If we require that the initial velocity be zero relative to S, the acceleration becomes unambiguously determined. We must therefore use a modified operational definition of 'mass'. We can either make the specification that the initial velocity relative to S is zero; then we define a concept which is called "rest mass" m_0. Or we can include the initial velocity v in the description of the operation. Then acceleration and mass themselves become dependent upon v. We obtain a physical quantity which is no longer a constant but a function of v. This quantity is called "mass" m in the new mechanics. By using this definition, we can formulate the laws of motion in the simple form: mass times acceleration equals force $(ma = f)$. But the mass m is now a function of v.

26. The Operational Definition of 'Time Distance' Becomes Ambiguous

If we assume that all observable results of Einstein's two principles are true, we have to assume that all results which can be logically derived from them are also true, even if they are not observable directly. Among these conclusions are several which have been exciting to a great many people. They have seemed to contradict common sense, in general, and the glorified common sense called "philosophical insight", in particular.

We can conclude, e.g., from Einstein's two principles that a clock which travels with the speed v relative to S loses time compared with the clocks at rest in S. If we recall the operational definition of a clock, we notice again that some operations which rendered, according to Newton's laws, identical results no longer do so if Einstein's principles are assumed to be true. The operations by which the time distance between two events was defined did not mention the speed of the clock relative to any system of reference. For, according to Newton's physics, this speed is without influence upon the march of the clock. If Einstein's principles are true, this operational definition of the time distance between two events becomes

ambiguous. We must specify the speed of the clocks used in this measurement. In order to obtain an unambiguous result of our defining operation, we must no longer say that "between the events A and B there is a time distance of 10 seconds" but that "there is a time distance of 10 seconds if we use clocks which are at rest in a particular system S'". We can express this statement a little more briefly by saying that "this time distance is 10 seconds relative to the system S'". The velocity of S' relative to S must be specifically given. We use again a "relativized language" in order to make the description and the operations unambiguous.

By starting from Einstein's principles, one can derive new laws of motion which differ for high speed from Newton's laws. We can in the same way derive laws for the propagation of light through material bodies which could be derived from Newton's mechanics and traditional wave optics only by complicated additional hypotheses.

The system of all statements which can be derived from Einstein's principles is called "the Theory of Relativity". This theory is so called because its characteristic and basic hypothesis is a principle of relativity. Before Einstein the validity of Newton's laws of motion and of the theorem of relativity for optical phenomena seemed to be incompatible. Einstein's master-idea was to drop Newton's laws of motion as well as the ether theory of light and to generalize

Newton's theorem of relativity into a general hypothesis which should be valid in the whole domain of the motion of material bodies and of light propagation.

27. The Relativity of Time

By people who are interested in the philosophical aspects of relativity theory the question has been raised: Has science really proved that time is relative and not absolute? The answer can be given quite directly.

The statement that "time is relative" can have two different meanings. First, it may mean that the march of a clock is altered by a rectilinear motion with constant speed v. This assertion is a statement about facts which can be checked directly or indirectly by experiments. It is confirmed indirectly by every experiment which confirms an observable result drawn from Einstein's principles. It is checked more directly by experiments which examine how the frequency of the light emitted by a sodium atom is altered by the motion of this atom (atomic clock, Ives's experiment).

But this physical meaning of the relativity of time is not the one which has puzzled the philosophers and the great public. 'Relativity of time' has a second meaning—a logical or philosophical one. From the theoretical and experimental results of Einstein's principles it becomes evident that the statement that "there is a time distance of 10 seconds between two given events" has no operational meaning.

We must always add explicitly a particular system of reference. Therefore, Einstein suggested dropping entirely sentences of the form "the time distance is 10 seconds" from the language of physics and of admitting only sentences of the form "the time distance is 10 seconds relative to the system of reference S". This suggestion is not a statement about physical facts which can be confirmed or refuted by experiments. It is a suggestion for using a language which is recommended as being well adjusted to our experience about physical facts.

Nobody can be persuaded to accept this recommendation if he does not like to. The adjustment of our language to the physical facts may not be the primary motive of our rules of language. Perhaps we want to adjust our language to tasks which are beyond our desire for a clear description of facts. If we wish, we can, of course, use the expression 'real time distance between two events' without referring to a particular system of reference. But we must know how to give to such a statement an unambiguous operational meaning. Either we concede that we do not know which of the arbitrary systems of reference is the "real" one and leave the discovery of this system to a superior spirit, or we single out some convenient system (e.g., the system of fixed stars) and give to the time distance relative to this particular system the honorary title "real or absolute time distance" while we call the time distance relative to other systems "relative or apparent time distance". Nobody can prove that this way of speaking is "false". If we believe, e.g., that our moral values should be described by using the term 'absolute', one could find it convenient to use in the description of the physical world as far as possible the same terminology as in describing the goals of human behavior. Then it would be advisable for believers in "absolute values" to use also in physics the expression 'absolute time distance.'

28. The Speed of Light Has the Same Value in All Systems of Reference

Among the results of Einstein's principles which cannot be checked directly by experiments, the one which has been regarded frequently as its "central absurdity" is perhaps the following: Any light ray has, relative to S, the same speed as relative to a system S' which moves with an arbitrary speed v uniformly relative to S. Some philosophers have taxed this statement as being absurd or even self-contradictory. They argue: The speed of light relative to S is c. If S' moves with the speed v (which is smaller than c) in the direction of the light ray, it advances more slowly than the light. Therefore, the speed of light relative to S' is only $c-v$, which is obviously smaller than c. To assert that the speed of our light ray relative to S' is also c means to say that $c-v$ is equal to c while v is different from zero. This is obviously absurd. However, through

all this argument the velocities are measured by instruments at rest in S. Only if this operational definition of speed is used, can it be proved that the speed of light relative to S' is $c-v$. But in Einstein's statement 'velocity relative to S'' means the velocity measured by instruments which are at rest in S'. Then it cannot be proved that this speed of light relative to S' is $c-v$ but rather that it is c. The confusion comes from the failure to distinguish between mathematical symbols 'v' and 'c' and their operational meaning. Without the addition of the operational definitions, physical statements do not say anything about physical facts and cannot be checked by experiments.

If we have this in mind, we can understand that the statement that "no material body can move with a speed which is equal to or greater than the speed of light" is not a statement about absolute motion. It means, of course, that no material body can move with the speed c relative to S, but it means just as well that no material body can move with the speed c relative to S' (which moves in turn with the speed v relative to S). The real operational meaning of our statement about the speed of light as the upper limit of all speeds of masses is the following: If we have a set of clocks and yardsticks adjusted to a system S, no material system S' can move with the speed c relative to S if we measure c by the instruments in S.

29. The Traveling Twins

Another startling result is the story about the traveling twins. We learned that a clock which is moving with the speed v relative to S loses time compared with the clocks which are at rest in S. A human being is a kind of clock. The chief functions of the human organism, in particular the heart-beats, are achieved by periodical mechanisms. According to Einstein's principles, every periodical mechanism behaves like a clock. If we have twins, one at rest in S and one moving with the speed v relative to S, the traveling one will experience a smaller number of heart-beats. This can be seen particularly clearly if the traveling brother traverses a large circle which eventually leads him back to his stationary brother. Then the traveler must have remained younger than his brother at home. This conclusion is a necessary consequence of Einstein's principles. This result can, of course, neither be confirmed nor refuted by direct experiments. We never have to do with organisms traveling at a speed which is near to the speed of light without any violent disturbance produced by an acceleration. The possibility of this phenomenon can perhaps be understood if we consider the recent physiological experiment concerning the conservation of organisms at low temperatures. For a bath at a temperature well below the freezing-point brings about a slowing-down of the periodical processes inside the organism. It follows certainly from Einstein's two principles that the twin brother who has been at rest in the fundamental system S becomes older than his brother who performed a circular motion relative to this system.[24]

30. Conversion of Mass into Energy and Vice Versa

According to Newtonian physics, the sum of masses cannot be changed by any interaction of material bodies. According to the theory of relativity, the masses are dependent upon speed. However, the question arises whether the sum of the rest masses can be altered by interaction. From Einstein's principles it can be derived that the sum of the rest masses remains unaltered only if the sum of the kinetic energies is not altered by the interaction considered.

But if we consider phenomena which are connected with the conversion of kinetic energy into other forms of energy, e.g., into heat, the matter is different. As the simplest case we consider two bodies of equal rest masses which move with equal speeds relative to S in opposite directions. A collision takes place. If the bodies are perfectly rigid (unelastic), both bodies will come to rest in S and their whole kinetic energy will disappear and be converted into heat energy. It can be strictly derived from relativity theory that by this collision the sum of the rest masses of the colliding bodies is changed. Before the collision the rest masses of each body were m_0; the rest mass of the whole system, $2m_0$. After the collision we have one larger body consisting of both small ones. Its rest mass will be greater than $2m_0$. From the theory of relativity it can be strictly derived that the increment of rest mass will be equal to the loss of kinetic energy divided by c^2 (the square of the speed of light). Since the loss of kinetic energy is equal to the produced heat energy H, we can also say that the rest mass after the collision is $2m_0 + H/c^2$. This is a special case of a more general theorem which follows from the theory of relativity: If by an interaction of material bodies kinetic energy E disappears, the sum of the participating rest masses increases by E/c^2. If kinetic energy E is produced by the interaction, the sum of the rest masses would decrease by E/c^2.

We can verbalize this fact by saying that rest mass is partly "converted" into kinetic energy. If the rest mass decreases by Δm_0, the kinetic energy $E = c^2 \Delta m_0$ is produced. The same thing is true if we replace production of kinetic energy E by production of radiant energy E. Continuing this line of argument, one can envisage the possibility that the whole rest mass m of a body could be converted into energy. Then the energy $E = m_0 c^2$ would be produced and the whole rest mass of the body would disappear. There are phenomena in nuclear physics which seem to lend themselves easily to an interpretation of a conversion of matter into energy (such as in the atomic bomb).

A well-known phenomenon of this kind is the so-called "mass defect" of atomic nuclei. The nucleus of the helium atom, e.g., consists of four particles, each of which has the rest mass m_0 of the hydrogen nucleus (proton). However, the mass of the helium nucleus is smaller than four times the rest mass of the hydrogen nucleus. If we assume that the particles in the nucleus are kept together by attracting forces, the particles which are packed in the helium nucleus have a smaller potential energy than they had when separated from one another. The formation of the helium nucleus is connected with a loss of potential energy and therefore with a production of kinetic or radiant energy. If the rest mass of the hydrogen nucleus is m_0 and the rest mass of the helium nucleus $4m_0 - D$, the drop of potential energy accompanying the formation of one helium nucleus is Dc^2. This energy can be regarded as the "binding energy" which keeps the helium nucleus together. There are a great many cases of disintegration of an atomic nucleus where the binding energy can be measured directly. We find a good agreement with the energy calculated from the mass defect.

31. Creation and Annihilation of Mass

Another phenomenon of this type is the creation or annihilation of an electron-positron pair. As we shall learn in Part VIII, the "electron" has a much smaller rest mass than the proton and a negative electric charge.

The "positron" has the same mass as the electron, but the charge is positive. The magnitude of the charges of electron and positron is equal. If they combine, the charges disappear of course. Moreover, there is sufficient evidence to assume that the rest masses disappear also and are converted into radiant energy. This assumption is confirmed particularly by the measurement of the wave-lengths of the radiation which accompanies the disappearance of an electron-positron pair. For (according to sec. 39) radiation consists of elementary portions of radiant energy called photons. The wave-lengths λ of such a photon can be easily calculated from the energy E by the formula $E = hc/\lambda$. Roughly speaking, the smaller the energy, the greater the wave-lengths. If we calculate the loss of energy E produced by the disappearance of the rest mass of an electron-positron pair ($E = 2m_0c^2$), we obtain the wave-lengths of this radiation from $2m_0c^2 = hc/\lambda$. It is a very hard gamma radiation and is really found if one observes the phenomena connected with the disappearance of positrons.

We have learned now how the conversion of mass into energy or the annihilation of mass can be described very distinctly and clearly by describing performable physical operations. The annihilation of mass has been interpreted occasionally as a refutation of materialism and support of spiritualism. This can be done only if one uses this language without having in mind the operational mean-

ing of the words and sentences. The statement, "Matter can be annihilated and converted into energy", sounds strange to the man of average school training and smacks at least of spiritualism. For he understands tacitly the words 'matter' and 'energy' the way they are used in everyday language. In this language by 'matter' is meant something like rock or ocean. By 'energy' is meant something like a soul or a spirit. The dualistic view of the contrast between body and mind is deeply entrenched in our everyday language. In this language, 'matter' has the operational meaning of some hard and impenetrable stuff, while 'energy' seems to be defined by operations which belong in the field of psychology rather than in the field of physics.

The physicist, in particular the twentieth-century physicist, means by 'matter' a system consisting of a great number of particles like electrons, protons, positrons, etc. The operational definitions of these particles are very different from the operations by which we used to test the presence of some hard or impenetrable stuff like the "matter" of our everyday language. If we take into account the meaning of the words 'electron' or 'proton', which are used in the wave theory of matter, we are very far from operations testing hardness or impenetrability. On the other hand, radiant energy is regarded by the physicist as consisting of photons. The presence of photons is tested by methods which are essentially of the same kind as the methods by which the presence of electrons is tested. Therefore, if we introduce the operational meaning of the terms 'matter' and 'energy', the annihilation of matter and its conversion into energy is not at all obscure and has nothing to do with the experiments of the spiritualists which demonstrate "dematerialization".

VI. Light

32. The Crucial Experiments on Behalf of the Wave Theory

A century ago there were two theories of light both of which were based upon Newtonian mechanics. According to the "corpuscular" hypothesis, light consists of small corpuscles which obey Newton's laws of motion. They perform rectilinear motions with constant speed. If passing from vacuum or air into water, they are attracted by the particles of water and deviated from their original path. The deviating forces are similar to Newton's force of gravitation. From this hypothesis it can be derived that between the angle of incidence α and the angle of refraction β there is the relation $\sin \alpha / \sin \beta = c'/c$, where c is the speed of light in vacuum (or in air) and c' is the speed of light in water. Since by the attraction the light corpuscles are accelerated, it follows that c' is greater than c; the speed of light in water is greater than in air ($c' > c$). But Foucault performed in 1850 an experiment by which he proved that the speed of light in water is smaller than in air in contrast to the claim of the corpuscular theory. Therefore, the second of the conflicting hypotheses about light propagation seemed to be confirmed: the "wave theory" of light. According to this theory, light consists of waves propagated through an elastic medium (ether) which follows Newton's laws of motion From this hypothesis Huyghens could derive that the speed of light in water is smaller than in air ($c' < c$). This result seemed to be confirmed by the Foucault experiment.

The Foucault experiment has been quoted over and over again as the outstanding example of a crucial experiment. It was supposed to have sealed the death sentence of the corpuscular theory. But if we take a strictly logical viewpoint, we can say only that the experiment decided against a corpuscular theory of light which assumed that the corpuscles follow Newton's laws of motion. Therefore, we can speak only of a decision against any corpuscular theory of light if we mean by 'corpuscle' a small mass in the Newtonian sense of this word. It is obviously not excluded by the Foucault experiment that light consists of corpuscles in a wider sense which move according to laws which differ from Newton's laws. We have to understand that an experiment can only be crucial if only one clear-cut alternative exists. An experiment can never decide definitely between wave theory and corpuscular theory but at most between Newtonian waves and Newtonian corpuscles. By Foucault's experiment Newtonian corpuscles are outruled, but this does not mean a decision in favor of Newtonian waves, unless we make the assumption that only this alternative exists. Perhaps Newtonian mechanics has to be dropped outright.

33. Conversion of Light into Kinetic Energy; the Photon

Half a century after Foucault's crucial experiment it had become more and more clear that the wave theory of light in its original form did not suffice to cover the whole realm of newly discovered facts. The new facts which have been greatly responsible for the dropping of the full-fledged wave theory of light are the phenomena of "photoelectric effects." If light waves hit the surface of zinc, this surface obtains a positive electric charge. Electrons (negative electric charges) of a mass m and the charge e are emitted by the zinc atoms. The absorbed energy E of the incoming radiation is used partly to overcome the attraction of the surface (work P) and partly to impart to the electron a kinetic energy $mv^2/2$. According to the law of the conservation of energy, this means $E = mv^2/2 + P$ or $mv^2/2 = E - P$.

If the wave theory of light were correct, the intensity of radiation would decrease inversely proportionate to the square of distance from the source of light. If we choose this distance great enough, the energy absorbed by a square inch of our zinc plate tends toward zero. However, experiments show that the kinetic energy $mv^2/2$ of the electrons which originates from the conversion of the incoming light energy into kinetic energy is not dependent at all on the distance of our zinc from the light source. The kinetic energy $mv^2/2$ and the speed v depend only upon the frequency (color) of the incoming radiation.

Einstein interpreted (1905) this result as follows: The light energy is not homogeneously distributed over the wave surface. Therefore, this energy is not thinned out at great distances, as the wave theory implies. On the contrary, according to Einstein's hypothesis, the light energy is concentrated in small packages of energy called "photons". To emit one electron, just one photon of light energy has to be absorbed by the zinc surface. The energy of a photon is proportional to the frequency $E = h\nu$, where h is a universal constant and ν the frequency of radiation. 'Frequency' means "number of vibrations per second". Therefore, the greater the frequency, the larger the photons and the greater the speed of the emitted photoelectric electrons. However, the greater the distance of the zinc from the source of light, the smaller the number of electrons which is emitted per second. But the speed of these electrons depends only upon the frequency of light and not upon the distance. This hypothesis is formulated mathematically by $E = h\nu$ and therefore $mv^2/2 = h\nu - P$. By checking this equation experimentally, one can determine the constant h. The first elaborate measurement was performed by Millikan. He found the value $h = 6.56 \times 10^{-27}$ erg. sec.

34. The Paths of Photons

If light consists of photons, we must assume that the brilliancy of illumination of a surface is proportional to the number of photons hitting a square inch. If we consider a phenomenon of interference or diffraction, the dark regions on a screen are regions where few photons hit the screen, while the diffraction maxima are regions where a great many photons come in. By this argument one understands easily that the motion of photons does not follow Newton's laws of mechanics but a very different type of law. One has to make use of the superposition of waves to find the distribution of photons. The single photon is not mentioned in these laws at all. The intensity of the light (the square of the wave amplitude) in a certain region is proportional to the number of photons in this region. 'The path of a photon' has no operational meaning. As a matter of fact, this path is not spoken of in any law of physics. We say: Our sun emits photons which heat and illuminate our earth. If we put a screen or a shutter in the way of these photons, we can predict the heat and luminosity effect of the radiation which hits or passes these devices. Since the laws of

wave optics allow us by the superposition of waves to calculate the resultant amplitudes, we know in every region the average number of photons. But the laws of optics do not allow us to describe the path of a single photon on its way from the sun to the earth.

We can also understand more directly why 'the path of a photon' is an expression without operational meaning. If such a path could be physically produced, e.g., a rectilinear path, we could produce a light ray which travels along a geometrical straight line and passes through a definite point in a definite direction. However, the physical existence of such a light ray is incompatible with the laws of wave optics. In order to assure that a particular light ray passes through a definite point in space, we must make it pass through a very small hole in a screen. The smaller the hole, the more precisely the ray passes by a definite point. However, if the diameter of the hole becomes comparable in size with the wave-length of the light ray, the phenomenon of diffraction takes place. The ray passes through a definite point but does not continue in its original direction after passing the hole. If we want to have the ray continuing in its original direction, we have to make sure that the hole is much larger than the wave-length. But in this case the beam of light is of a considerable thickness and does not pass precisely through a definite point.

Therefore, if we want to produce a light ray by a physical operation, we cannot achieve an all-round approximation. Either we (1) bring it about that the ray passes approximately through a given point: then there will be no good approximation of direction; or we (2) achieve a good approximation in direction: then the precise location of the point is poorly defined. This state of affairs has been the basis of the "Principle of Complementarity" in physics.

We must have in mind that a "photon" cannot be thought of as a geometrical point. A package of light energy of the wave-lengths λ must contain a great number (e.g., N) of crests of waves in order to have the characteristics of a wave with the wave-length λ. Therefore, it must have at least the size $N\lambda$. If the photon is of a size which is small in comparison to $N\lambda$, it cannot have a distinct wave-length or frequency at all. 'The path of a photon' of the frequency $\nu = c/\lambda$ has, strictly speaking, no operational meaning, since only a "point" can traverse a "path".

We are accustomed to speak loosely about a light ray in empty space and to imagine it as a straight line. But here again this way of speaking has an operational meaning only if we assume some physical laws to be true. These laws are, roughly speaking, the independence of the light ray of its environment, in particular of the width of the opening through which it passes. But the validity of these laws means exactly the absence of the phenomena of diffraction, the absence of the wave properties of light. Since we know that these phenomena exist, 'path of a light ray', without including the environment of the light ray in the description, is an incomplete expression and has no operational meaning.

35. Laws of Motion of a Photon

We can easily give a quantitative description of this state of affairs. We may have a slit of the width A in a dark screen. A beam of light rays comes in perpendicularly to the screen and passes the slit. According to a simple calculation which we find in any elementary textbook of physics, only a fraction of the light energy moves behind the slit perpendicular to the screen. The balance is deflected. The bulk of it is deviated by an angle ψ (the first diffraction maximum) connected with A and the wave-length λ by the equation $A \sin \psi = \lambda$. If there were no diffraction, a light ray passing through a slit which has approximately the shape of a point ($A = 0$) would keep approximately its direction perpendicular to the screen. This means $\psi = 0$ approximately. But, because of the diffraction, A and ψ cannot be simultaneously zero or approximately zero if the wave-length λ is given. To produce an approximately rectilinear light ray, we have to make A and ψ both as small as possible. However, the smaller A becomes, the greater becomes ψ and vice versa. To achieve an approximately rectilinear light ray, we have therefore to compromise and to make both, A and $\sin \psi$, fairly small, e.g., $A = \sin \psi = \sqrt{\lambda}$. If we use light of a wave-length $\lambda = 10^{-5}$ cm. (ultraviolet), we would obtain $A = 0.003$ cm. and $\psi = 12$ angle minutes approximately. We can describe the light phenomenon by a rectilinear ray if we allow for a margin of 0.003 cm. in the starting-point and of 12 angle minutes in the direction. This means practically a straight line through a point and perpendicular to the screen. The smaller λ, the better is the approximation to a rectilinear path which can be made. If we use this "compromise" path instead of the theory of diffraction, we can predict within the margin described where our light ray will hit a screen put up behind the slit.

36. Mechanical Momentum of a Photon

If light radiation hits a material body, it exerts a force upon it—the pressure of light. This can be stated by saying that radiation imparts a certain momentum upon this body. We can interpret this fact by ascribing to every photon a mechanical momentum exactly as to a moving mass. The action of radiation to a body can be treated as collision between masses. Each mass m has a momentum mv, if the speed is v. The sum of the momenta is not changed by the collision, according to Newton's Third Law of motion. If one of the bodies loses momentum, the other one must gain it. A photon has an energy $E = h\nu$. According to the theory of the electromagnetic field, a package of radiant energy E carries a momentum $M = E/c$. This means: If radiant energy E hits a material body, it imparts the same momentum as if a mass m with the speed v hit the body, mv being determined by $mv = E/c$. A photon in particular has the momentum $M = h\nu/c$. Since every photon travels with the speed of light c, this momentum does not depend upon speed but on the frequency or wave-length. Since $\nu = c/\lambda$, we have

$$M = \frac{h\nu}{c} = \frac{h}{\lambda}.$$

This momentum can be directly checked by the Compton effect, which we observe when we examine the collision of a photon with an electron. By this collision the momentum of the electron as well as that of the photon is affected. The alteration of the momentum of the photon means an alteration of the wave-length. If the electron has originally the momentum of zero, the momentum of the photon can only decrease by the collision. This means that the wave-length λ must increase and that the frequency ν must decrease. From the hypothesis that the momentum of the photon is $M = h/\lambda$, one can derive that the wave-length of a photon which is moving after the collision at a right angle to its original direction increases by h/mc, where m is the mass of the electron. $h/mc = 2.42 \times 10^{-10}$ cm. is called the 'Compton wave-length'. This alteration of wave-length can, of course, be observed only if the original wave-length is very small. Compton carried out this experiment by using X-rays the photons of which have very short wave-lengths. He could confirm the alteration of the wavelengths of X-rays which hit some material bodies which are rich in electrons. This experiment is the most direct measurement of the mechanical momentum h/λ of the photon.

37. The Interpretation of Diffraction in Terms of the Mechanics of the Photon

If we keep in mind that the photon carries a mechanical momentum, we can give to the diffraction experiment also a purely mechanical interpretation. Instead of attempting to produce an exactly rectilinear light ray, we can attempt to produce a photon which has a certain definite position and a momentum in a definite direction. We determine the position again by having the photon pass a slit of a size A in a screen. If no diffraction existed, the whole momentum would be perpendicular to the screen; the momentum parallel to the screen would be zero. By decreasing A more and more, we could achieve with an arbitrary accuracy that the photon passes through a given point and has a momentum exactly perpendicular to the screen. But, because of the diffraction, the momentum of the photon has also a component parallel to the screen. Since the momentum itself has the magnitude h/λ, its component parallel to the screen is $h/\lambda \sin \psi$. We denote this component by 'p'. From $A \sin \psi = \lambda$ (sec. 35) follows $A \sin \psi = h/v$ $\sin \psi$ or $Ap = h$. We see again that it is impossible to produce a photon which has simultaneously an exact

position and an exact direction of momentum (perpendicular to the screen). For in this ideal case we would have $A = 0$ and $p = 0$, which would render $Ap = h = 0$, which is impossible, since h is different from 0. To achieve an exact position ($A = 0$), we use an infinitely small slit. In this case p would increase infinitely. We could not say anything about the direction of the momentum. If we want an exact direction of momentum (p very small), we need a very great A. This means that the position of the photon becomes undefined. We can again compromise by making A and p both fairly small. Then we have a photon with a fairly exact position and fairly exact direction of momentum.

We understand very clearly by these considerations that the photon is not a particle with a certain position and momentum which both exist but cannot be measured simultaneously. Actually, an experimental arrangement which produces a photon with an exact position (very small A) frustrates the achievement of an exactly perpendicular momentum (p very small). 'A photon with an exact position and an exact momentum' is an expression which does not enter into any description of physical facts; it does not enter into the formulation of any physical law. Therefore, this expression has no operational meaning and has to be dropped from the vocabulary of physics.

If we assume the x-axis parallel to the screen, we may introduce the

467

symbols $A = \Delta x$, $p = \Delta p_x$, where Δx means the inexactitude of the position and Δp_x the inexactitude of the momentum, both in the x-direction. If we describe a state of affairs by saying that our photon has a position $x = 0$ and the momentum $p_x = 0$ parallel to the screen, this description is only correct if we allow the "uncertainty" Δx in the position and Δp_x in the momentum. Since $Ap = h$, we

have the relation $\Delta x \times \Delta p_x = h$ between these "uncertainties".

If we use the description $x = 0$, $p_x = 0$ for the present state of the photon, we can predict the future with the "indeterminacy" Δx and Δp_x by assuming a rectilinear path perpendicular to the screen. These remarks are the roots of W. Heisenberg's relation of "uncertainty" or "indeterminacy" (sec. 43).

VII. Mechanics of Small Masses (Wave Mechanics)

38. Failure of Newton's Mechanics in the Domain of Very Small Particles

Newton's physics assumed that the laws which govern the motion of the planets govern also the motions of the smallest particles of matter—the atoms and electrons. One concluded that the orbits of the electrons revolving around the atomic nucleus would follow the same law as the orbits of the planets around the sun, except that the gravitational forces in the solar system had to be replaced by electrostatic attractions in the atom. However, both types of forces followed one and the same law: The force is inversely proportional to the square of the distance. But in quite a few cases, and in crucial ones at that, the Newtonian pattern failed to render an adequate method of describing and predicting the conditions of atomic equilibrium and motion.

In spite of the general belief that the

action of chemical valences can be described by the Newtonian concept of force, it has been actually impossible to substitute into the Newtonian pattern '$ma = f$' an expression for 'f' which would describe the phenomenon of saturation of a valence. It is easy to understand that two particles of unlike electric charges (positive and negative ions) can attract each other. But how can two neutral particles form a compound? However, just this seems to happen in the case of very simple chemical compounds (homopolar compounds). The hydrogen molecule H_2 consists of two neutral hydrogen atoms. The chemist says that the valences of these atoms saturate each other. But from the point of view of the mechanics of electrostatic forces it is hard to understand how this can happen and how a third hydrogen atom can be prevented from being attracted. The phenomenon of saturation of valences remains unexplained in Newton's mechanics.

Still more obvious is the failure of Newton's pattern to describe the motion of electrons in the atoms which must occur in order to explain the emission of the spectral lines. According to the theory of Niels Bohr, the emission of spectral lines from heated hydrogen gas is a kind of reversed photoelectric effect. If a photon hits the surface of a metal, the radiant energy of the photon is converted into the kinetic energy of an electron leaving the surface (sec. 33). Similarly, if the energy (kinetic and potential) of the moving electrons within the atom decreases, the mechanical energy lost by the atom is converted into the radiant energy of the photons which leave the atom. These photons constitute light of a specific wave-length which reveals itself by the spectral lines of the hydrogen gas.

39. Bohr's Theory of the Emission of Light by the Atom

When the atom passes from the energy E_n to the smaller energy E_1, according to Bohr's hypothesis exactly one photon of the energy $h\nu$ is produced. If we accept Einstein's fundamental hypothesis about the conversion of mechanical into radiant energy and vice versa, we have the equation $E_n - E_1 = h\nu$. If we know the initial and final energy of the atom (E_1 and E_n), the frequency ν of the emitted radiation is determined. We know from the examination of the frequencies of spectral lines that there are specific frequencies in the hydrogen spectrum. Therefore, there can be

only specific values of energy in the hydrogen atom.

If we consider the simplest case, we can assume that the negatively charged electron is moving along a circle around the positively charged nucleus. According to Newton's laws of motion and Coulomb's law of electrostatic forces, the attraction e^2/r^2 must be balanced by the centrifugal force mv^2/r, where e is the charge of the electron and of the nucleus, m the mass of the electron, v the linear speed, and r the radius of the circular orbit. From $e^2/r^2 = mv^2/r$ it follows that $v^2r = e^2/m$. This means that for every radius r a circular motion is possible if we choose only the speed v accordingly. The value of the energy E of our circular orbit is given by

$$E = -e^2/r \text{ (potential energy)}$$
$$+ mv^2/2 \text{ (kinetic energy)} = -e^2/2r .$$

Since there is an orbit for any value of r, there is also an orbit for any value of energy E. Therefore, Newton's mechanics cannot help us to derive the specific values of energy (E_n and E_1) which render specific values of frequency. Bohr, following some suggestions of Planck's theory of radiation, assumed that the orbits of electrons have to obey, besides Newton's laws of motion, a still different kind of law, called "quantum laws", by which most of the orbits allowed by Newton's mechanics are excluded.

These quantum laws require that the angular momentum of the revolving electron has to be equal to $h/2\pi$ (where h is the same constant as in secs. 33–37) or an integer multiple of $h/2\pi$. This means that mrv is equal to $h/2\pi$ or $nh/2\pi$, where n is an integer number. Then it follows that the smallest possible value of the radius r can be derived from $2\pi mrv = h$ and $mv^2r = e^2$. By eliminating v, it follows that $4\pi^2m^2r^2v^2 = h^2$, $4v^2mr = h^2/e^2$, and, finally, $r = h^2/4\pi^2me^2$. This value of r provides us with the smallest possible size of an electronic orbit. The charge e of the electron is known from Millikan's experiment on the ionization of oil droplets; m is known from J. J. Thomson's experiments on the deflection of cathode rays in magnetic and electric fields; h is known from the photoelectric effect (sec. 33). From the known values of h, m, and e we obtain that r is approximately 10^{-8} cm. This value agrees with the size of the atoms obtained by other methods, e.g., Brownian motion, diffraction of X-rays by crystal lattices, etc.

Knowing r, we can calculate the energy E_1 of the smallest orbit ($n = 1$). We obtain $E_1 = -2\pi^2e^4m/h^2$. To obtain the energy of the nth orbit, we have to replace h by nh and obtain $E_n = -2\pi^2e^4m/h^2n^2$. This formula allows us, by using the Einstein-Bohr formula $E_n - E_1 = h\nu$, to calculate the frequencies of the spectral lines of the hydrogen atom in agreement with the experiments.

40. The Wave Theory of Matter

This was a great success of Bohr's spectral theory, but we must not forget that it was achieved by a partial abandonment of Newton's mechanics without replacing it by a new mechanics. The crucial formula, $4\pi^2m^2r^2v^2 = h^2$, was obtained by a hypothesis (quantum law) which was superimposed upon Newton's laws of motion, leaving these laws themselves substantially unaltered. In this way mechanics and optics became an incoherent patchwork which gave rise to many pseudo-problems.

The French physicist, Louis de Broglie, looked at this situation from a new point of view. He did not believe in patching up Newton's laws by the quantum laws but in altering Newton's laws themselves. He took his cue from a comparison between optics and mechanics. The optical phenomena can be described and presented in terms of light rays (straight lines in vacuum) in every case where diffraction can be neglected. If diffraction comes in, we have to pass from ray optics (geometrical optics) to wave optics. Under what conditions must diffraction be considered? Obviously, diffraction comes into the picture if light passes through small openings or around small obstacles, the word 'small' meaning 'comparable in size with the wave-lengths of light'. This means that the pure ray optics loses its applicability if the rays have a great curvature. The word 'great' means here again a radius of curvature which is comparable with the wave-lengths of light. Perhaps, argued De Broglie, the orbits calculated from Newton's mechanics play only the role of the light rays in optics. Perhaps Newton's mechanics loses its applicability also if the orbits have a very large curvature or a very small radius of curvature. This is probably the case if we consider the curvature of the electronic orbits around the nucleus of the hydrogen atom. The radius of curvature of the smallest orbit is of the order of magnitude of 10^{-8} cm., much smaller than the wave-lengths of visible light, which is about 10^{-5} cm.

Under such circumstances, De Broglie argued, we have to replace Newton's mechanics by a new mechanics. This generalized mechanics would be in such a relation to Newton's mechanics as wave optics is to ray optics. For small curvature we can regard the optical rays as paths of the photon. But, if diffraction comes in, the behavior of photons can no longer be described by paths (secs. 34, 35).

Louis de Broglie advanced the idea that there is a new type of waves called by him "waves of matter" and later, after him, "Broglie waves" which determine the behavior of material particles and electrons in a way similar to the way in which the light waves determine the behavior of photons. The most natural hypothesis was to assume that the relation between wave-length and mechanical momentum is for the material particle the same as for the photon. We know that for photons the momentum p obeys the formula $p = h/\lambda$. If we know the λ of the kind of light used, we can calculate from this formula the momentum p. In the case of material particles, however, the momentum is known from mechanics $p = mv$ (v be-

ing the speed of the particle), while the wave-lengths of the De Broglie waves must be calculated from the formula $\lambda = h/p = h/mv$.

41. The New Laws of Motion for Small Particles

One notices immediately that if the mass m of a particle tends toward infinity, the Broglie wave-lengths λ tend toward zero. This means that the phenomena of diffraction can be neglected. The motion of great masses can be treated without using the waves. We have to do with orbits of masses which obey Newton's laws of motion. But if we have to do with very small particles, the Broglie wave-lengths become comparable with the radius of curvature of the orbit, as in the case of the electronic orbit in the hydrogen atom, and we have to apply the laws of "wave mechanics". According to these laws the term 'path of a particle' has no more operational meaning than 'path of a photon' in ordinary optics. What we describe according to Newtonian mechanics as a stream of electrons emitted by some source (heated wire or the electrodes of a vacuum tube) must now be described as the emission of "Broglie waves". Exactly as the intensity of the electromagnetic waves describes the number of photons per unit of volume in a certain region, the intensity of the De Broglie waves describes the number of material particles per unit of volume—the density of mass.

The first confirmation of this daring hypothesis was De Broglie's derivation of Bohr's "atomic radius" without using the quantum laws. The wave propagation along the circular orbit of the electron around the nucleus must not destroy itself by superposition and interference. Therefore, the perimeter of the smallest orbit must be equal to one wave-length. This means, according to De Broglie, $2\pi r = \lambda = h/mv$. This result is obviously equivalent to Bohr's quantum law $mrv = h/2\pi$ (in sec. 39).

42. Diffraction of Material Particles

The most direct confirmation of De Broglie's hypothesis, however, would be to show that material particles (e.g., electrons) which pass a small hole produce on a screen a diffraction pattern similar to the diffraction pattern of light. There would be alternatively dark and bright rings. The bright rings are the regions of the screen where a great many particles hit, while the dark rings show us the regions where no, or very few, particles hit. The size of the hole must be comparable to the wave-lengths of the De Broglie waves. The American physicists Davison and Germer used the intervals between the atoms of a metallic foil as slits to produce a diffraction pattern by having electrons pass this foil.

By these and similar experiments it has become evident that very small material particles follow laws of motion which are very different from the laws of Newtonian mechanics. The particles of a beam which pass through the hole in a direction perpendicular to

the screen do not continue moving in this direction behind the screen as they are expected to do according to Newton's law of inertia. Only a certain percentage of the incoming particles do so. They hit the center of the receiving screen and form the central maximum of diffraction. The greatest percentage of the remaining particles is diffracted toward the first diffraction maximum by the diffraction angle ψ, where $A \sin \psi = \lambda$. If we know that, e.g., a thousand particles enter the hole, we cannot predict what every individual particle will do behind the slit. But we can safely predict that a certain percentage will hit the central maximum and a certain percentage the first diffraction maximum ψ, etc. Here again, as in the case of the photon, we can denote the size of the slit by Δx (this means $A = \Delta x$) and the momentum of the particle parallel to the screen by Δp_x. Since the formulas are the same as in the case of the photon, we have again $\Delta x \times \Delta p_x = h$. The product: the "uncertainty of x" times the "uncertainty of p_x" equals a constant h.

43. The Uncertainty Relation

The meaning of this relation in the case of material particles seems to be much more paradoxical than in the case of the photon. For this relation is incompatible with the basic statement of Newtonian mechanics that every mass point has at every moment of time a definite coordinate x and a definite momentum $p_x = mv_x$ in the direction of x. If we know these values for an instant of time $t = 0$, Newton's laws of motion allow us to predict these values for any future instant of time. If we examine carefully our result obtained in the case of photons and apply it to the case of small material particles, we learn that there is no law of mechanics which contains the expression 'position and momentum of a particle at a certain instant of time'. There are experimental arrangements which we can describe by saying that the particle passes the diaphragm at a position $x = 0$ with a margin of error Δx; by narrowing the slit, we can make this margin Δx as small as we wish. But then the equation $\Delta x \Delta p_x = h$ shows us that Δp_x becomes very great. This means that we only know that behind the slit the momentum p_x of the particle is zero with the allowance of a wide margin of error. If we consider a great number of particles passing the slit of the width Δx, most of them have a momentum $p_x = h/\Delta x$ parallel to the screen. But also all smaller and greater momenta will occur with a certain frequency according to the Gaussian distribution of errors. If we fix the position of the diaphragm and of the small hole at $x = 0$ relative to the earth, we obtain diffraction rings on a screen which is at rest behind the slit. We can predict the position of these rings exactly from knowing only the position of the screen relative to the slit. Our information which allows us to predict the behavior of the particles behind the slit is based entirely upon the measurements (in the ordinary sense of the word) which we

can perform in the experimental set-up. In this way we know $A = \Delta x$ (width of the slit) and the distance of the screen from the slit. This information allows us to predict the position of the diffraction maxima on the screen. The momentum Δp_x of an individual electron while passing the slit remains unpredictable. But obviously it is possible to "measure" the momentum of an individual electron in a certain sense. Since the component Δp_x of this momentum is directed parallel to the diaphragm, the particle will, in passing the slit, impart a momentum upon the diaphragm parallel to the diaphragm itself. If the diaphragm is not fixed with respect to the earth, but movable, one can find the momentum of the particle by measuring the momentum of the diaphragm, which means practically its speed obtained by the impact of the particle. But if the screen is moving while the electrons pass the slit, the position of the particle relative to the earth is not predictable even if the slit Δx is very small.

We must always decide in what result we are interested: in the position of the particle passing the diaphragm, or in the momentum of this particle. In each case we can make a prediction. In the first case we can predict the diffraction pattern on the screen. In the second case we know the momentum which the diaphragm gets from the particle, and we can make predictions by means of Newton's mechanics. We can predict the motion of bodies which are hit by the diaphragm. We could, e.g., predict the elongation of a ballistic pendulum.

44. Misunderstandings about the Relation of Uncertainty

We must carefully avoid the misunderstanding which has been caused by the way some physicists have discussed the relation of uncertainty. One hears sometimes the statement: "It is impossible to measure simultaneously the position and the momentum of a small particle". This sounds as if there would be small particles which possess certain positions and certain momenta. We are told that we can measure either of them but that nature is so diabolic as to prevent us from measuring both simultaneously. This statement is rather misleading. The expression 'a particle with a certain position and a certain momentum' has no operational meaning if De Broglie's hypothesis is accepted. However, we can set up an experimental arrangement which gives to the expression 'particle with a certain position' a kind of approximate operational meaning. We can make some predictions from such statements. They determine the future within a certain margin of "indeterminacy". There are also arrangements which allow us the use of the expression 'particles with a certain momentum'. They lead to predictions which are also reliable within a certain margin. The relation of "uncertainty" involves the relation between these two margins of prediction and is called, for this reason, the "relation of indeterminacy". But "a particle with

a certain position and a certain momentum" is not an object of the physical reality as far as very small particles are concerned. Speaking exactly, a particle by itself without the description of the whole experimental setup is not a physical reality. But "a particle passing a diaphragm at a certain point" is a description of a physical reality and "a particle imparting to a diaphragm a certain momentum" is a description of a physical reality too.

As a matter of fact, we would find all these statements less paradoxical if we dropped the word 'particle' altogether from the language of physics. We could say that the mechanics of the smallest particles, wave mechanics, allows us from the knowledge of the initial experimental conditions to predict the future observable phenomena. Among these phenomena are in particular "point events", e.g., scintillation produced at a point of a screen, and "impulse events", e.g., a momentum imparted to an observable material body. Then nobody would be puzzled by learning that scintillations on a screen do not follow an orbit which is determined by Newton's laws of motion. Hans Reichenbach pointed out that small particles with definite positions and definite momenta can be introduced into the language of physics and that their introduction can be justified in a certain sense. The logical coherence of the system of physics would not be violated, but the motions of these particles would follow very awkward laws. Reichenbach speaks of "causal anomalies". He calls this way of describing the subatomic phenomena the "exhaustive interpretation", while he calls the presentation in Bohr's complementary language as in the present monograph the "restrictive interpretation". The latter sticks strictly to experience and to simple operational definitions but sacrifices the traditional language of mechanics. The exhaustive description sticks strictly to the language of Newtonian mechanics but requires very involved operational definitions. It gives a certain satisfaction to the lover of logic but does not give the simplest possible description of the subatomic phenomena.

45. Bohr's Idea of Complementarity

As Niels Bohr has pointed out repeatedly, the physicist feels at ease when he can keep to the language of the mechanics of everyday life as far as possible. However, if we stick to "simple and practical" operational definitions there is no adequate means of describing the subatomic phenomena in terms of "full-fledged particles" with position and momentum. In this situation the physicist takes some satisfaction in speaking, at least, of "particles which have only position" and of "particles which have only momentum". In this way we use, at least partially, the language of corpuscles to which we are accustomed in Newton's mechanics.

If we admit any operational definition without regard to whether it is "simple and practical", we can, of course, measure the exact position and velocity of a particle and give an "exhaustive interpretation" of its state. We use two diaphragms with small slits and observe a particle which passes both slits. Then the velocity can be calculated from the distance of the slits. But the velocity calculated in this way is not the velocity which helps us to predict the future course of the particle. For after the passing of the second diaphragm we must again take into account diffraction, and the future course cannot at all be calculated from the velocity. The "exhaustive interpretation" of the state of a particle is no basis for predicting its future states by simple laws. Therefore, the correct statement of the "relation of uncertainty" is not that "position and velocity cannot be measured simultaneously" but that "there is no law of prediction which contains reference to the simultaneous position and velocity of a particle". If a statement does not contain a rule of prediction, it is not a physical law but purely tautological. It is a definition. Our statement about measuring velocity by using two diaphragms is certainly an operational definition of position and velocity. But this definition does not enter into any law of physics which would help us to predict the future course of events. As we emphasized in section 2, an operational definition is only "helpful" or "practical" if it helps us to formulate physical laws.

The position and the momentum of a particle are two physical properties which reveal themselves on different occasions which exclude each other. The position can be obviously observed as a "point event" and the momentum as an "impulse event"—they have nothing to do with each other. Only if the particle has such a great mass that Newton's mechanics can be applied, can we speak of a particle having a position and momentum and can we calculate the momentum from two positions by the formula $p = mdx/dt$. According to Bohr, the physical world cannot be described by one coherent language. There are two languages which "complement" each other. Under certain circumstances the language of "positions of particles" or "point events" must be used; in other circumstances, excluding the ones above mentioned, we speak of "momenta of particles" or "impulse events". If we make use of all possible information about the present state of the world, we must use both languages. Then we can predict all events which our actual science enables us to predict. This aspect of the world is what Bohr has called the "aspect of complementarity".[25]

The indeterminacy relation $\Delta x \Delta p_x = h$ can also be written $\Delta x \cdot \Delta v_x = h/m$ if we introduce $p = mv_x$. If the mass m increases, h/m tends toward zero. It becomes more and more meaningful to assume that Δx as well as Δv_x are 0 and to speak of a mass with a definite x and a definite v_x.

We understand now very well what prediction can be made if we send particles through a very small slit in a diaphragm. We consider, e.g., a beam of electrons directed perpendicularly to the diaphragm. We proceed as we did in optics (sec. 35) in order to obtain a "compromise beam". This means to make the size of the slit rather small but not so small as to make the momentum undefined. Then we obtain what we call in experimental physics a "beam". According to Newton's mechanics, it would follow from the law of inertia that all electrons, after passing the slit, continue in their course perpendicular to the diaphragm. They strike a screen which is set up parallel to our diaphragm within a region which is nearly congruent to the slit. However, if diffraction plays a certain role, there is for every region of the screen a certain possibility of being hit by a particle. Instead of the rigid law that one limited region is hit and all the rest of the screen remains untouched, we have to say that the frequency of a hit is distributed over the whole screen. One says sometimes, and it means the same thing, that the law of inertia is replaced by a law predicting the statistical distribution of hits. We can only say that in the region where, according to Newton, all particles should hit there is merely an overwhelmingly great frequency of hits.

46. By What Law Does Wave Mechanics Replace Newton's Second Law?

We assume now that the beam of electrons passes a "field of force". We consider only the simplest case where

By What Law Does Wave Mechanics Replace Newton's Second Law?

the force has the direction of velocity. The force f is defined by the formula $f = ma$, where 'a' means the acceleration of a "great mass" m to which Newton's law can be applied. We assume, e.g., that this beam is launched upward against the force of gravity. The acceleration of gravity at this point of the earth may be g. The particles may start with an initial speed v. According to classical mechanics, they would move upward by a distance s which is given by the equation $mv^2/2 = mgs$ or $s = v^2/2g$ (decrease of kinetic energy is equal to the work done against the force of gravity). If the mass m is "small", Newton's mechanics can no longer be applied. By "small" we mean that there is a considerable drop of potential between two points with a distance of one "Broglie wave length": $\lambda = h/mv$. We find now from the laws of wave mechanics that we can affirm only that most of the particles reach the height of $s = v^2/2g$. But there is for every height s a certain percentage of particles which reach it. Newton's law of motion ($ma = f$, where f is the force of gravity) is replaced by a formula which gives us the percentage of particles which reach a certain height s.

This reshaped law of motion becomes particularly important if we replace the force of gravity mg by an electric field which acts upon the charge of the electron. If the potential difference between the diaphragm and the final position of the electron is V (voltage), the force in the Newtonian sense is eV/s. If the mass m is so great

that Newton's mechanics can be applied, the distance s which an electron traverses is again given by $mv^2/2 = eVs$, and therefore $s = v^2m/2Ve$. If the mass is "small", we can no longer predict that all particles will reach s but say only that most of the particles reach the distance indicated by the last formula. However, some particles go farther. There is for every distance s a certain percentage of electrons which reach it.

This result of the new mechanics turned out to be particularly important for the understanding of the phenomena of radioactivity. The alpha radiation of radioactive substances (like radium) consists in the emission of positively charged particles (alpha particles) from the nucleus of the atom. However, if we measure the speed v of the emitted particles, we find that, according to Newton's laws of mechanics, they must have traversed a considerable electrical potential difference in order to acquire this speed. But in this case the nucleus must be surrounded by a very high "potential barrier". This means that there is also a considerable potential difference between the center of the nucleus and its surface. But then it can be calculated from Newton's laws and the observed speed v that the charged alpha particles will not succeed in "ascending" this potential difference and in leaving the atom, as our v would be too small. The Russian physicist Gamow pointed out that, according to wave mechanics there is for any initial value of v a certain percentage of particles which will succeed in reaching the surface of the nucleus and be emitted into the open. In this way the new mechanics finds a wide field of application in the motion of the subatomic particles. According to the new mechanics of small particles, we can start from the observable initial conditions, e.g., the initial electron beam, and predict future observable phenomena. We are not able to predict every single point event, but, if the "force" is given, we are able to predict the statistical distribution of the future point events.[26]

47. Physical Reality and Causality

One must not exaggerate the gap between the new mechanics of small masses and the Newtonian mechanics. According to Newton's mechanics, the future positions and velocities of mass points can be predicted with certainty. This is, of course, only true if we speak in the language of the "calculus" (sec. 1), the language of "physical quantities". However, if we substitute for these quantities the operational meaning, some statistical component always enters. Speaking in terms of sense data, the distinction between prediction of a single event and statistical prediction loses even its clear meaning. For the question always remains arbitrary to a certain degree of what we are to regard as a single sense datum and what as an average of a great many sense data. The difference between a statistical theory and a strictly causal theory is not as much in the realm of sense data as in the realm of physical quantities.

Newton's mechanics allows a strictly causal prediction of the positions and velocities of mass points which are regarded as the physical reality described by this theory. But what is the physical reality described by the new mechanics? If we regard the percentage of particles producing point events in a certain region (the wave function) as this reality, the new theory is causal too. It only loses its causal character if we regard the positions and momenta of particles as the state variables which describe the physical reality. But these variables are in the new mechanics not state variables in the sense in which this word is used in Newtonian physics. There is no description of the world at a certain instant of time which contains both position and momentum of a mass point.

Moreover, while in Newton's mechanics the state of the mass point is described without an experimental setup to which it belongs such a "mass point stripped of its environment" does not occur at all in the new description of the world. This new description which contains, e.g., the term 'position of a mass point' gives a description of the present state of the world by describing actually objects like "a mass point in a particular experimental set-up." These objects have, obviously, operational meaning. And there is a "complementary" description of a second setup which allows us to ascribe a "momentum" to the particles. But there is no setup which would allow us to use both the terms 'position' and 'momentum of a particle'. Obviously, we can make only predictions, starting from statements which have operational meaning. Therefore, we can make predictions starting from a statement about the "position of a particle". But these predictions are not so precise as the predictions in Newton's physics which start from statements about both "position and momentum of a particle". To say, therefore, that in wave mechanics the law of causality is not as strictly valid as in Newton's mechanics would be a very inadequate description of the character of the new mechanics. For these two systems of mechanics do not use the same state variables to describe the physical reality. Therefore, we cannot compare them as to the greater or smaller validity of the law of causality. Such a comparison would make sense only if the old and the new description made use of one and the same set of state variables. But this could be achieved only by reducing both descriptions to the language of sense data or to the "thing-language". But in this language the distinction between a "causal" and a "statistical" theory becomes vague.

The new mechanics, we are often told, does not describe physical reality at all. For the state of a photon or of a mass particle is never described objectively but always in connection with an experimental setup or an instrument of measurement with which it is in interaction. We cannot, e.g., describe the state of a photon on its way from the sun. We can describe it

only at the instant of time when it strikes a screen or comes otherwise in interaction with a material body which has so great a mass that its behavior can be described by Newtonian mechanics. Even a revolutionary mind like Einstein's has for a long time doubted whether it is advisable to drop the traditional conception of objective physical reality, according to which "physical reality" is ascribed to a material particle without regard to its environment. Einstein has suggested not to stop efforts to find a theory of subatomic phenomena which describes "real things" in the traditional sense. Obviously, there is no convincing reason which could prevent us from sticking to the Newtonian concept of physical reality and from regarding Bohr's "Principle of Complementarity" as a provisional state of science. But one may as well admit a different type of world description in which the expression 'position of a particle' cannot be applied without describing the instrument by which this position is measured.[27] If this description is "simple and practical", we can as well ascribe "physical reality" to the objects of our new mechanics, provided we mean "reality" in the operational and not the metaphysical sense.

If we examine the whole question of causality more thoroughly, we soon notice that this law in its whole generality cannot be stated exactly if the state variables by which the world is described are not mentioned specifically. Otherwise, the general formulation of the law of causality would have no operational meaning. For the more general formulation says: let a state A of the world be succeeded by a second state B. If A occurs again, B follows again too. If no specific state variables are used for the description, we cannot check whether the state A has recurred. Moreover, only if we know that certain physical laws are valid can an operational meaning be assigned to the state variables (sec. 3). Therefore, the type of physical laws which are to be set up is dependent upon the kind of state variables which are used.

48. Object and Subject in Wave Mechanics

Frequently we are told that in the new physics the role of the perceiving subject is greater than in Newtonian mechanics. The observing physicist has to be introduced explicitly into physics, while previously physics has dealt only with the observed object.

This assertion must be taken with a grain of salt. We have seen that in wave mechanics the state of a mass point can never be described without including the instrument of measurement which is used. We must state explicitly whether we use a narrow slit or a wide slit in a diaphragm in order to describe the position of a mass point. If we call the mass point the physical object which we wish to describe, we must include into the description the instrument of measurement. This instrument is a medium-sized mass and can be described by using the language of Newton's mechanics. But there is

no sense in calling this instrument of measurement the "observing subject."

If we use the expression "observing subject" as it is used in the psychology of everyday life, we have to say that the observing subject (the living physicist) observes the instrument of measurement in the same way as he observes any object of Newtonian physics. Therefore, by "observed objects" we have to understand in the new mechanics, as well as in Newton's mechanics, the medium-sized mass, the instrument of measurement, the scale or balance which one observes immediately. The electron which passes a diaphragm must not be called the "observed object" if we want to avoid ambiguities. The "electron" is a set of physical quantities which we introduce to state a system of principles from which we can logically derive the pointer readings on the instruments of measurement. If we call these instruments "physical objects", the "electron" or "photon" can be called so only by a certain shift in the sense of the term 'physical object'. But we must never forget that there is no meaning in the question of what is "really" a "physical object". The only problem is to agree about an unambiguous use of this word. The loose way of speaking which is so customary in the borderland between physics and psychology has even succeeded in confusing the instrument of measurement with the "perceiving subject".

To say that the observer destroys by the very act of observation some property of the object which he wants to observe is a misleading formulation too. The observer has nothing to do at all in this matter. His role is exactly the same as in Newtonian physics. The "momentum of a particle" cannot be destroyed by observing the position because this momentum has never existed except in so far as we have a setup which allows the definition of a "momentum." All this sensational "destruction by observation" is an oversimplification. We found a similar situation in the theory of relativity. The introduction of the observing subject was misleading too. The correct thing to say is that, by setting up an experimental arrangement which would allow us to define 'the position of a particle', we do not "destroy the momentum of this particle". What is "destroyed" is only the possibility of setting up the "complementary" arrangement, which would allow us to define the term 'momentum'. The role of the living observer is also in "Relativity" only to observe "clocks" or "yardsticks" which can be described in the language of everyday life.

49. Metaphysical Interpretations of Wave Mechanics

Quite a few authors have maintained that by the new mechanics an "irrational element is introduced into physics" or that "physics is now supporting an idealistic word picture" or that "physics is now in agreement with the doctrine of free will" or even that "by the wave mechanics for the first time in the history of human thought the conflict between religion and science has been settled." If we pursue precisely the analysis of the logical structure of the new mechanics, we will understand that there is no foundation for all these philosophical interpretations of the new mechanics.

The basis of these interpretations has been provided by some metaphysical formulations of the new mechanics which have been given by a great many philosophers and, for that matter, by quite a few physicists and mathematicians. Such a formulation is, e.g., the introduction of the observing subject as a mental, immaterial "entity" into physics itself. This way of speaking, as we have already seen, is inadequate. By others the new mechanics is described as introducing a physical object which is both particle and wave. Some authors speak of it as "particle and/or wave" and some authors have even coined new words like "wavicle" to denote this hybrid object. As a matter of fact, such an object resembling a centaur, who was half-man and half-horse, does not exist in the new mechanics. There are experimental setups which can be described by using the term 'position of a particle' and others which can be described by using the terms 'momentum' or 'wave-lengths'. All the confusion is produced by speaking of an object instead of the way in which some words are used. We have again an example of what Carnap calls the fallacy of using the "material mode" of speaking instead of the "formal mode". In a great many presentations of the new mechanics it is not even pointed out clearly that the so-called "dualism between particle and wave" is only another expression for the complementary use of the words 'position' and 'momentum of a particle'. As in a great many other cases, an ambiguous way of speaking is introduced by the predilection of quite a few authors for an "ontological" way of speaking. Some scientists who are very competent in their fields succumb to the temptation to make statements on ontology. This is quite natural since ontology is nothing but the use of our everyday language in a domain where it loses its meaning.

The mental or idealistic character of the new mechanics is occasionally demonstrated by calling the De Broglie waves "waves of probability". Since "probability" is often used as a psychological concept, we describe in wave mechanics the physical world by using terms of psychology. This interpretation is certainly a misleading one. The new mechanics describes the percentage of electrons which strike on the average a certain region of the screen. There is nothing psychological involved, or at least there is no more psychology than in any branch of physics. For even in Newtonian physics every statement is accompanied by its operational definition. And this definition contains the description of physical instruments. This description again makes use of words which express sense data like 'blue', 'warm', 'soft', etc. These words are the only

psychological element in modern as well as in traditional physics.

Another way in which metaphysical misinterpretation enters into physics is the use which is made of the word 'real' in contrast to the word 'fictitious'. Quite a few authors would say that the De Broglie waves are "not real waves" but only a "fiction" which serves to describe the path of particles. This way of speaking is also misleading and is not in agreement with a coherent presentation of physics. We have learned that the movement of photons is determined by the electromagnetic waves exactly in the same way as a movement of small particles (electrons) is determined by the De Broglie waves. The reality of the electromagnetic waves can be demonstrated only by checking the effect of the photons on the bodies with which the waves come in contact. In the same way the reality of the De Broglie waves can be demonstrated too.

Quite a few philosophers and scientists have claimed that the "Indeterminacy Relation" gives a support to the philosophical doctrine of "Free Will". However, no statement can be supported by physical theories which cannot be formulated in terms of physical operations (sec. 2). The statement "the will is free" has, certainly, no operational meaning. It is a purely metaphysical statement and cannot be supported by any physical theory (secs. 2 and 4). Therefore, all the talk about the intrusion of mental and psychological elements into physics has its source only in an inadequate presentation of the recent parts of physics.

VIII. Structure of Matter

50. Continuity or Discontinuity of Matter

It has been an age-old dispute whether matter tightly fills the world space or whether matter consists of small indivisible particles, "atoms", between which there is empty space. The existence of these atoms has been suggested by many facts, in particular the law of constant proportions in the chemical compounds, the elastic properties of gases, etc. However, it has been held out against the old atomistic theories that nothing can be won by assuming that a big piece of iron consists of very small pieces of iron, called atoms. For then the question has to be raised: What is the structure of the small piece of iron called atom? If it is iron, it must have the same structure as the big piece, and so on.

The physics of the twentieth century has come to the firm conviction that the mass density inside solid bodies shows conspicuous maxima and minima which have specific distances from each other. These distances are characteristics of the physical and chemical constitution of the body. They are usually interpreted as the intervals between the centers of atoms. But it would not be correct to say that in these intervals between the atoms the space is "empty" and that the density of matter has the exact value zero. As the "position of a particle" is not an element of "physical reality", statements like "This vol-

ume of space is empty" cannot describe a physical reality either. Since (according to sec. 46) there is in any volume element of space a chance that a point event may happen, we can say, vaguely, that "matter is nowhere and everywhere". Certainly, we cannot say that atoms are small pieces of matter.

51. Operational Meaning of 'Matter'

A great part of this trouble comes from a misuse of the word 'matter'. In our everyday language we know very well what the operational meaning of 'matter' is. We know that iron or wood or the human body consist of "matter". But if we consider physical quantities which enter into the hypothetical setup of our physical science, we encounter words of which it is hard to say whether they mean "matter" or not. We use words like 'electricity', 'magnetism', 'ether', 'human mind', etc. The operational meaning of 'matter' is very clear if we speak of a piece of iron or wood or meat as being matter. Such a piece is identified as being "matter" by giving us the experience of resistance against penetration, of temperature, of color, of observable motion, etc. But if we ask whether we should call an electric charge or the ether "matter", we begin to doubt. For only a part of the operations which allow us to identify a piece of iron as matter are applicable in the case of an electric charge or of the ether. It is arbitrary to take just some particular operations as a characteristic of matter. The situation has been

483

dramatized by the fight which quite a few philosophers have put up in order to keep in use the word 'matter' without any regard for its operational meaning. It has been suggested that we have to distinguish between the physical concept of matter and the philosophical one. While in physics "structure of matter" means the way our observable bodies are built up from protons, electrons, etc., we are told that the philosophical concept of matter should mean everything which exists objectively independent of our subjective sense experience.

In order to avoid any ambiguity and to keep strictly to the operational meaning, it seems that the most reasonable thing to do is to use the word 'matter' in the sense it is used in our everyday language. This means to call a table a piece of matter and our brain a piece of matter, but not to refer by the word 'matter' to concepts like electrons or photons, let alone the "ether" or the "mind".

52. Electromagnetic Mass

At the end of the nineteenth century it became clear that every electrically charged particle has a mass, m, which can be calculated from the charge and the size of the particle. The mass m is (sec. 11) defined by $f = ma$. If we have a spherical particle with the charge e (in electrostatic units) and the radius a, the mass is approximately e^2/ac^2, where c is the speed of light. This follows from the well-known law of the electromagnetic field, according to which an accelerated particle is

equivalent to an electric current of increasing intensity. According to the law of self-induction, every current has a certain inertia against increase of its intensity. Therefore, an electrically charged particle is "recalcitrant" against an attempt to accelerate it. Therefore, it possesses an "inert mass" according to the operational definition of this term.

Occasionally one has distinguished between a "real mass" in the Newtonian sense of this word and an "apparent mass" which is feigned by an electric charge. However, these names can be misleading. They may suggest the metaphysical idea that an "apparent mass" is nothing "material" but something more subtle, something immaterial—electricity. It has been suggested on good grounds that every mass may be an apparent one and have its origin in an electric charge. For a neutral particle could consist of a positive and a negative charge each of which would have an apparent mass. They would not reveal their charges, which are neutralized. This electromagnetic theory of matter has been described by the slogan "matter has disappeared". It has been used largely in a crusade against materialism and on behalf of idealism. It is noteworthy that Lenin made this crusade the starting-point of his principal philosophical book. Actually the word 'matter' does not mean anything beyond the realm of the operational meaning mentioned above. The "hard fact" is that every particle carrying a charge e has, according to the operational definition of inert mass, a mass e^2/ac^2. Whether one has to introduce also a "mass" which is not covered by this formula is a question of the adjustment of our physical symbols to our experiences. It certainly has nothing to do with the question of the "reality" of matter.

53. The Number of Molecules in a Gas

To examine the structure of matter, it is obviously convenient to consider matter in the gaseous state. For, in this state, matter can be expanded arbitrarily and we can find out whether by this expansion we obtain the same type of matter in an infinitely diluted state or whether we obtain some small solid masses floating in a largely emptied space. If a gas is inclosed in a container, the phenomena taking place can be interpreted conveniently by the assumption that the gas consists of small particles which fly around until they hit the walls. The average kinetic energy is, according to the kinetic theory (sec. 19), proportional to the absolute temperature. These particles are called "molecules". They are certainly not pieces of "matter" in the sense in which we suggested in section 51 that this word be used. For such a particle has no temperature, no density, no color, but it has at every instant of time a momentum and a kinetic energy. It has certainly no surface as a piece of iron has, but it has a certain symmetry which may be spherical, cylindrical, etc. The salient point is whether the number of these molecules in one cubic centimeter can be figured out. What is the operational

meaning of this number? If there are several independent operations by which it can be figured out, we would say, with P. W. Bridgman, that these molecules are a "physical reality". This number can indeed be figured out. It can be done, e.g., by the observation of the kinetic energy of particles of microscopic visibility floating around in the gas and pushed around by the molecules. From the statistical hypothesis it can be derived that such a particle has the same kinetic energy as a gas molecule. Since we know the total energy of all molecules in the container from the observation of the pressure exerted upon the walls, we can divide this energy by the energy of a single molecule and obtain the number we are looking for. One obtains $L =$ approximately 6.8×10^{23} molecules in one mole (Avogadro's number).

Today there is even a method by which this number can be obtained in a much more direct way. In some cases one can spread out a substance in a layer which has the thickness of only one molecule. If we know the original volume of the substance and the area of the monomolecular layer, we can calculate the number of molecules. We obtain again the same number L. This number is the clue to the structure of matter. If we condense a gas, the molecules will touch each other approximately. We can obtain from the volume of the condensed state and the number L of the molecules the size of a molecule. We find that such a molecule is approximately

10^{-8} cm. across. This length—10^{-8} cm. —is a characteristic distance in the realm of molecules. It is called one "Ångstroem unit" (1 Å).

In a solid body, e.g., in a crystal, the molecules are almost in touch with each other. Therefore, the distance between two molecules in a solid body is also approximately equal to one Ångstroem. If the substance considered is not a chemical element but a compound, the molecule consists of a number of smaller particles called atoms. The table-salt molecule (sodium chloride, NaCl) consists of one atom of sodium and one atom of chlorine. The size of the atom is of the same order of magnitude as the size of a molecule. The molecule of a chemical element may consist of one atom, like mercury, or of several equal atoms, like hydrogen (H_2). The operational meaning of the number of molecules is very clear in a gas. In a container of the volume V at the temperature T, each molecule exerts the pressure KT/V on the wall, where K is a universal constant. In a solid body it is not unambiguous which atoms must be regarded as belonging to one and the same molecule. If we consider, e.g., a sodium chloride crystal, the sodium and chlorine atoms form a lattice within which there are no subdivisions. It is very arbitrary to pick just one sodium atom and one of the neighboring chlorine atoms and to call them a molecule. The statement that "one particular Na atom with one particular Cl atom forms a molecule" has no operational meaning.

54. The Problem of Chemical Binding

It has been an old problem to explain how the atoms in the chemical compound are kept in their position. A certain equilibrium between attraction and repulsion is needed. Is it possible to account for these actions on the basis of Newton's laws of motion and the laws of electromagnetic forces? The forces binding the atoms in the molecule are referred to as "valence forces". In the sodium chloride molecule the atoms are ionized. There is a positive and a negative electric charge. The attractive valence forces are electrostatic attractions between unlike charges. The repulsive forces are exerted by the shells of the atoms which are all negatively charged (secs. 58 and 59). But the matter is different if we have to derive the binding between the two hydrogen atoms in the hydrogen molecule. Between these neutral atoms there can be no attractive force in the Newtonian sense. The phenomenon of binding can be explained only on the basis of wave mechanics in which the concept of the path of a particle, and, in particular, of a particle at rest, has no operational meaning. This point will become clear when we discuss the structure of the atom (sec. 57).

55. The Structure of the Hydrogen Atom

We shall discuss first the simplest atom—the hydrogen atom—and describe later how the atoms of other elements differ from it. As already mentioned, the hydrogen atom, which has a size of approximately 10^{-8} cm. = 1Å, consists of two smaller particles, the proton and the electron. It has been accepted as a fundamental hypothesis in physics that every electrical charge is an integer multiple of the elementary charge $e = 4.8 \times 10^{-10}$ e.s.u. The proton and electron are each charged with one elementary charge. But the electron has a negative, and the proton a positive, charge. If we know the charge, we can determine the mass of a particle by simple experiments on the basis of Newton's laws of motion. If we denote the acceleration by a, the mass by m, the intensity of the electrostatic field in which the particle is moving by E, Newton's law of motion says that $ma = eE$ or $a = e/mE$. By observing a and E, we can determine e/m, and, from knowing e, we can calculate m. Since we have good reason to identify the cathode rays with flying electrons, the canal rays with flying protons, we find, by measuring the acceleration of these rays in electric and magnetic fields, that the mass of the electron is 9×10^{-28} gm., while the mass M of the proton is 1,837 times greater. These subatomic particles are, of course, of much smaller size than the atom.

We can estimate their size in different ways. If we assume, e.g., that the mass m of the electron can be derived entirely from its charge, we have $e^2/r = mc^2$, where r is a radius of the electron (sec. 52). If we substitute into this formula for e the charge and for m the mass of the electron, we find for the radius the value $r = $ approx. 10^{-13} cm. We call the proton the "nucleus" of the hydrogen atom because its mass M is much greater (about 1,800 times greater) than the mass of the electron. According to the original presentation of atomic physics, due to Rutherford, the electron traverses an orbit around the nucleus. The nucleus is of a size similar to that of the electron. We shall discuss later how the nuclear size can be figured out. This presentation kept strictly to the language of Newton's mechanics. In this pattern of description the space between electron and nucleus is empty. There is no "matter" within the atom, and one could use this argument in favor of the philosophy of "dynamism", according to which there is no matter at all in the world but only centers of forces. The nucleus attracts the electron according to Coulomb's law, $e'e/r^2$, when e' is the charge of the nucleus. If we calculate the smallest orbit of the electron around the nucleus, we find again that this radius is approximately $r = 10^{-8}$ cm. (sec. 39). We found the same value for the size of the atom starting from the number L of molecules (sec. 53).

If we remember our presentation of wave mechanics, we must understand that it is a very perfunctory way of describing the hydrogen atom to say that the hydrogen atom consists of two particles which are very small compared with the diameter of the atoms, the space between them being empty.

As a matter of fact, these statements have operational meaning only within the range of Newton's mechanics; since the radius of curvature of the "orbit" of an electron would be comparable with the De Broglie wavelength (sec. 40), Newton's concept of the "path of a particle" has no operational meaning for an electronic orbit near the nucleus. The real description of the hydrogen atom is given by a function which describes to us the chance of finding the electron at a certain point in space if we make an experiment which could reveal the presence of an electron at this point. Or, speaking more exactly: If we bring into the vicinity of the nucleus a measuring instrument which is able to register point events (e.g., scintillations on a screen), we find that along the circle with the radius $r = 1$ Å such an event happens much more frequently than at other points. However, the event may happen at any point in space which is near to the nucleus. To say, therefore, that the space within the atom is "empty" is to use the word 'empty' in a sense which has no operational meaning. For the distinction between the part of space which is penetrated by a certain orbit and the part of space which is empty is a distinction which has an operational meaning only within the realm of Newtonian mechanics. Even the space "outside" the atom cannot be called "empty" in the original sense of this word. For at any point in space there is a slight chance that "the electron may be caught"; speaking exactly, that a "point event" can happen (sec. 46).

56. Electron, Proton, and Neutron

The mass of the hydrogen atom is equal to the sum of the mass of the proton plus the mass of the electron. This sum is not very different from the mass of the proton itself. The chemical reactions of hydrogen are determined by the fact that its nucleus has a charge $+e$ and its shell consists of one electron with the charge $-e$. It was discovered recently that there is an element which has the same chemical properties as hydrogen but has a nucleus with a double mass. This means that the new element, "heavy hydrogen", has the same nuclear charge $+e$ and the same shell as ordinary hydrogen. But its nucleus contains besides the proton also a second particle which has nearly the same mass as the proton but does not carry any electric charge. This particle which has played a great role in recent nuclear physics is called "neutron". Ordinary hydrogen and "heavy hydrogen" are the simplest examples of a couple of "isotopes", i.e., atoms which have one and the same nuclear charge but different nuclear masses. Speaking accurately, the mass of the neutron is a little larger than the mass of the proton. Perhaps the neutron itself consists of one proton and of one electron, for the charges of these particles would neutralize each other. If this were so, the neutron would be, in a certain way, a hydrogen atom on a smaller scale. The hydrogen atom consists of a proton and an electron at a distance of 10^{-8} *cm.* = 1 Å, but the neutron would consist of the same two particles but at a distance of only 10^{-12} cm. = 10^{-4} Å from each other (nuclear size). As we have seen, the smallest orbit of an electron around the proton has a distance of 10^{-8} cm. according to wave mechanics as derived from Coulomb's law of attraction and Bohr's theory of the atom.

The nucleus of heavy hydrogen is also called "deuteron". Its existence and stability give clear evidence that orbits of such a small size (10^{-12} cm.) cannot be derived from Newton's and Coulomb's laws. For the constituents of a deuteron, the positively charged proton and the electrically neutral neutron, do not exert any force upon each other according to the laws of mechanics and electrostatics. The fact that they can form a coherent particle, the deuteron, can be derived from the new mechanics of small particles. If we assume that the neutron consists of a proton and an electron, the deuteron consists of two protons and one electron. If we examine the field of force produced by both protons according to traditional physics, the electron could be at rest only in the symmetry plane of both positive charges. However, according to wave mechanics, there is for any space element a certain fraction of the time of observation during which the electron is staying in this region. Therefore, it can also be near to either of the positive charges. As a matter of fact, it can easily be shown that the electron changes periodically its position between the two

positive charges. 'Position of a particle' is to be interpreted in the complementary language of wave mechanics. Then each part of the system becomes alternately positive and negative. In this sense, the deuteron consists at any instant of time of two parts with different signs of charge. The sign of the charge on each side changes periodically. We call this way of attraction "exchange force", for this attraction is produced, vaguely speaking, from the fact that the electron changes its position among the two protons. These "exchange forces" are, of course, not "forces" in the strictly Newtonian sense. They determine no acceleration but only the frequency of a certain constellation of particles (speaking more exactly, of certain point events).

57. The Forces of Chemical Binding

The existence of the neutron and the deuteron throws some light upon the dark spots in the problem of valence forces in chemistry. As a neutron is on the nuclear scale of 10^{-12} cm. a reproduction of the hydrogen atom, the deuteron nucleus is a reproduction of another well-known particle—the ionized hydrogen molecule. The hydrogen molecule consists of two hydrogen atoms, i.e., of two protons and two electrons. If we remove one electron, we obtain a particle with one elementary charge $+e$, called a positive "hydrogen molecule ion". It consists obviously of two protons and one electron exactly as does the deuteron nucleus. But, while this nucleus has a size of 10^{-12} cm., the ion is of atomic size. The distance of the electron from the protons is approximately 10^{-8} cm. However, the existence of such a compound of two positive charges can be explained exactly as in the case of the neutron-proton compound by exchange forces. One can understand that the introduction of wave mechanics has contributed much toward the understanding of valence forces, the saturation of valences, and other problems of chemical binding (sec. 54).

58. The Structure of the Atoms of Different Chemical Elements

Two proton-neutron pairs form a new particle which has a charge $+2e$ and a mass of four proton masses. It is called a helium nucleus and is of nuclear size. If two electrons move around this nucleus at a distance of approximately 10^{-8} cm., we obtain a helium atom. The helium nucleus has been known for a long time as the alpha particle emitted by radioactive substances.

On the basis of this hypothesis about the structure of the simplest atoms, we can attack the ancient question about the structure of matter: Are the different chemical elements, iron, gold, sulphur, etc., really different substances, or are they only different configurations of one and the same substance? In physics this question has the following operational meaning: Is it possible to derive logically the chemical and physical properties of different elements like hydrogen, oxygen, iron, gold, etc., from the hypothesis that an atom of gold is built up from exactly the same particles as an atom of oxygen, only in a different configuration? Previously the question was whether the atoms of gold, mercury, etc., are configurations of the simplest atom—the hydrogen atom. This seemed to be at least plausible. For the atomic mass of all elements is approximately an integer multiple of the atomic mass of hydrogen. Oxygen has an atomic mass which is approximately 16, nitrogen 14, and sulphur nearly 32 times the mass of the hydrogen atom. However, there are some serious exceptions. Chlorine, e.g., has the atomic mass 35.457. But exact measurements have shown that even the apparently integer numbers are only approximately integers. Nitrogen, e.g., is not exactly 14, but 14.008. Since we know that the atom itself is a compound of subatomic particles, at least of protons and neutrons, we understand that it was an oversimplification to expect all atoms to be just different configurations of hydrogen atoms.

We can expect only that all nuclei are compounds of neutrons and protons. Then, of course, the manifold of possible configurations is much greater, and we must not expect that the atomic masses are all integer multiples of the hydrogen mass. The generally accepted hypothesis is now that every atom has a nucleus which consists of N neutrons and P protons. They are packed so tightly together that they form a particle of nuclear size 10^{-12} cm. or 10^{-4} Å. The electric charge of such a nucleus is obviously

$+Pe$ and the mass approximately $P + N$ proton masses. This nucleus is surrounded at a distance of about 10^{-8} cm. by P negatively charged electrons. The total charge of the atom is zero. It is a neutral particle. Each nucleus is characterized by two integer numbers: the "atomic number" P and the "mass number" $A = P + N$. For the "ordinary hydrogen" $P = 1, A = 1$, for the "heavy hydrogen" $P = 1, A = 2$, for helium $P = 2, A = 4$, for "ordinary nitrogen" $P = 7, A = 14$, etc. As there is besides the "ordinary hydrogen" the isotope "heavy hydrogen" with the same P and a different A, there are besides the "ordinary nitrogen" several "isotopes of nitrogen" which have all the number $P = 7$ protons in common but have "atomic masses" A different from 14.

59. Chemical Properties and Atomic Weights

The chemical properties of an element, e.g., oxygen, depend upon the nuclear charge $+Pe$, which gives us also the number P of the electrons in the shell. Speaking exactly, the electrons are distributed among several shells. But the distance of the nearest shell from the nucleus is of the order of magnitude of 10^{-8} cm. Without going into details, we understand that chemical reactions take place in the following way: If two atoms (e.g., one sodium and one chlorine atom) approach each other, one electron of the outer shell of chlorine leaves its place and enters the outer shell of sodium. In this way these two atoms become of unlike electric charge (are ionized) and attract each other.

The chemical reactions of atoms are not affected by the number N of neutrons in the nucleus, provided that the number P of protons in the nucleus and electrons in the shell is not altered. We know already, as an example, the heavy hydrogen as distinct from the ordinary hydrogen. If atoms have equal nuclear charges Pe (i.e., equal atomic numbers P) but different numbers N of neutrons in the nucleus, they are "isotopes" and have different "mass numbers" A. Besides the ordinary oxygen $P = 8, A = 16$, there are isotopes with $P = 8$, but $A = 15$ and $A = 17$.

Two isotopes cannot be separated by chemical reactions. But as the masses of the nuclei are different, they can be separated by physical methods which make use of this difference, e.g., diffusion and centrifuging. If we have a piece of iron or a container filled with chlorine, we have obtained these samples by purifying them by chemical methods. But it can happen that our sample which is chemically pure contains not only atoms with one and the same number of neutrons but perhaps a mixture of isotopes which contain all the same number of electrons P which is characteristic of this chemical element. But the number of neutrons is different, and therefore the mass numbers are different too. If we have, e.g., a mixture of x atoms of ordinary hydrogen and y atoms of heavy hydrogen, the mass of the mixture would be $x + 2y = x(1 + 2y/x)$

proton masses. If y/x is, e.g., $1/10$, the mass of the mixture would be $1.2x$ proton masses. The number $1.2x$ is called the "atomic mass" of the mixture. It is the same number which is known from elementary chemistry as the "atomic weight" of an element like hydrogen. Actually it is the mass of a mixture of several isotopes of hydrogen. This "atomic mass" is certainly not an integer multiple of x. The departure of the atomic mass of elements from the integer multiples of hydrogen can be understood if we assume that, e.g., the gas which we call chlorine is actually a mixture of several isotopes of the ordinary chlorine. The departure from the integer value depends upon the composition y/x of this mixture.

However, even if we consider a chemical substance which is not a mixture of isotopes, the atomic mass cannot be expected to be an exact integer multiple of the proton mass (sec. 30). The "atomic mass" is not exactly equal to the "mass number". For example, in the case of helium ($A = 4$) we know that the nucleus consists of exactly four particles. If they were separated, their "atomic mass" would be "four". But by the formation of the helium nucleus potential energy decreases, kinetic energy is produced, and the rest mass must decrease (sec. 30). We know from the theory of relativity that the tight packing of the protons and neutrons in the nucleus must be responsible for a mass defect. The mass of the nucleus must be smaller than the sum of the masses of the protons and neutrons which constitute the nucleus. This fact accounts for the small departure of the atomic masses of chemical elements from the exact multiples of the proton mass, while the mixture of isotopes accounts for the "great" departures.

60. The Nuclear Forces

If we bombard atoms (e.g., a foil of gold) with alpha particles (helium nuclei with a charge $2e$), these particles are deflected by the repulsion of the positively charged gold nucleus. From the examination of the hyperbolic orbits of the particles under the influence of the deflecting forces we notice that this deflection is in some cases so strong that we have to assume that the alpha particles get very close to the center of the gold nucleus. Since the repulsive force is the Coulomb force (Pe^2/r^2), we can find from an easy calculation that the particle must approach the center of the gold nucleus to a distance which is smaller than 10^{-12} cm. One can also find from this calculation that for a distance greater than 10^{-12} cm. the Coulomb inverse-square law accounts well for the observed orbit. But it is equally certain that at a distance of 10^{-13} cm. from the center the inverse-square law is no longer valid. At a distance of 10^{-13} cm. a new type of force comes into play—the "nuclear force".

We have to do here not with a "force" in the Newtonian sense but rather with a kind of "exchange force". These "forces" are negligible at a distance which is considerably more than 10^{-13} cm. from the center. The radius of the electron can be calculated (sec. 52) as $a = e^2/mc^2 = 10^{-13}$ cm., if we assume that the mass of the electron has its origin only in its electric charge. If we assumed the same thing for the proton, its radius would be e^2/Mc^2, where M is the mass of the proton: $M = 1,837$ m. Therefore, the size of the proton would be only about 10^{-16} cm. The size of the nuclei would be much smaller than the size of the electrons. However, the examination of the deviation of alpha particles by the nuclei of different substances shows that the size of the nucleus cannot be so small. It must be, as a matter of fact, of the order of magnitude of 10^{-13} cm. This means it must be approximately the same as the size of the electron. This consideration shows that the proton mass cannot be completely derived from the electric charge; there must be forces other than the electromagnetic ones. These nuclear forces, which are analogous to the chemical valence forces, distinguish the phenomena in the nucleus from the phenomena at a great distance.

The chemical reactions and the emission of spectral lines (of visible light and X-rays) are phenomena in the shell or, exactly speaking, phenomena which can be predicted from what happens in the shell. However, there are other phenomena which have their foundations in the happenings within the nucleus.

61. The Phenomena of Radioactivity

The spontaneous emission of alpha particles by some atoms, in particular the radium atom, was one of the earliest discoveries which foreshadowed the twentieth-century physics. This emission is the first example of an atomic phenomenon which originates in the nucleus and cannot be understood on the basis of Newton's mechanics. According to section 46, it can be derived from wave mechanics. From this theory we can predict the percentage of alpha particles which leave the nucleus per second. But we cannot predict which specific particle will succeed in leaving the nucleus. It has been known for a long time that no physical law predicts which alpha particle will leave the nucleus but only what percentage of all particles present in the nucleus are emitted per second. This percentage accounts for the rate of decay which is characteristic for every radioactive substance.

The radioactive phenomena which were originally regarded as peculiar exceptions to the well-known laws of physics have now become a typical example of the application of wave mechanics. Originally one knew only an emission of alpha particles to be a spontaneous reaction within the nucleus of radioactive atoms which occurs without any interference from without. But later it was discovered that disintegration of nuclei can be produced artificially if we bombard nuclei with protons, neutrons, or alpha particles. The "stable" nuclei can be disintegrated by bombardment, while the unstable (radioactive) ones disintegrate spontaneously. It was also discovered that by bombardment a stable nucleus can be converted into an unstable one (artificial radioactivity).

62. Production of Positrons, Electrons, and Photons by Nuclear Reactions

As a result of bombardment experiments not only the particles which have been known before can emerge but a new particle was discovered, the "positron", which has the positive electronic charge $+e$ and a mass which is as small as the electron mass.

Whereas protons and neutrons are believed to be present in the nuclei of all chemical elements, the electrons, positrons, and photons which are emitted by the nucleus during a great many reactions have not been parts of the nucleus before but are produced by the reaction. We know that the atom emits light (photons) when its energy drops, and nonetheless these photons have not been present in the atom before. In the same way the nucleus emits photons which are produced by the transition from one state of the nucleus into another state with a lower energy. The photons emitted by the nucleus are, however, of a much higher frequency than the light emitted by the atoms. The "nuclear photons" are the "gamma rays" which are known from radioactive radiation.

These considerations make it very clear that the nuclear reactions can-

not be described by using the language of the mechanics of everyday life. We cannot simply say that the beta particles or the alpha particles are emitted from the nucleus as bullets are shot from a gun. Even the removal of a particle by bombardment cannot be described in this language of the mechanics of medium-sized bodies. According to Niels Bohr, a particle (e.g., a neutron) which strikes the nucleus sticks first in the nucleus like a bullet fired into a box of sand. Then a reaction takes place between the intruding particles and the nucleus. By this reaction new particles are produced. The intruding particle does not simply throw out other particles as a ruffian intruding in a crowded streetcar would throw out other passengers. The emission of a beta particle (electron) from the nucleus follows laws which are more involved. If the nucleus experiences a certain energy drop ΔE, a photon (gamma ray) of a definite frequency ν is emitted. ν is determined by $h\nu = \Delta E$ (sec. 39). The beta particles which are emitted have a continuum of kinetic energies.

63. The Structure of the Nucleus

If we try to make a picture of the structure of the nucleus, we can do it at least in a perfunctory way by using the model of a "drop of fluid" suggested by Bohr. The forces by which the nucleus is kept together do not belong to the electromagnetic type but to the exchange forces which account for a great part of chemical binding between the atoms in a molecule and for the cohesion between the atoms and molecules in a piece of matter.

These nuclear forces produce an effect which can be compared to the surface tension which is responsible for the cohesion of a drop of fluid. On the other hand, the positively charged protons exert repulsive forces upon each other. A stable nucleus can exist only if there is equilibrium between this "surface tension" which is proportional to the area of the surface and the repulsive forces of the protons.

If we consider a nucleus with few protons and neutrons, we notice that the number of protons and neutrons are almost equal. For light nuclei we find approximately $P = N$ or $A = 2P$. For oxygen, e.g., we know that $P = 8$. The atomic mass A is for "ordinary oxygen" exactly $A = 16$. There are, however, isotopes with $A = 17$ and $A = 18$, where $A = 2P$ is only approximately fulfilled. We can conclude, therefore, that for these light nuclei surface tension and electrostatic repulsion are in equilibrium if the number of neutrons equals approximately the number of protons. But, if we consider heavy nuclei, the area of their surface by unit of volume becomes smaller. Therefore, the number of neutrons which are needed to produce a sufficient surface tension becomes greater. In the case of the heaviest nucleus ($P = 92$) there are 146 neutrons in the "ordinary uranium", while for a rare isotope we find $N = 143$. This means that the mass numbers are $A = 238$ and $A = 235$,

which are both considerably greater than $2P = 184$.

It is obvious, therefore, that, by splitting a heavy nucleus, such as uranium, into two lighter nuclei, some of the neutrons are liberated, as they are not needed to secure the stability of the lighter nuclei. There is no stable nucleus with $P = 90$ or greater.

As we know from the theory of relativity, the "binding energy" which keeps the particles together is connected with the "mass defect" of the nucleus (sec. 30). If we compare the sum of the masses of the particles (protons and neutrons) with the measured atomic weight of a chemical element, we find a certain difference which is due to the energy which has been used in the formation of the nucleus. If we assume that we have to do not with a mixture of isotopes but with a sample which consists of identical nuclei of the atomic number P and the mass number A, the atomic weight W which is measured would be $W = A = P + N$, if we computed the atomic weight by adding up the masses of the particles from which the nucleus is built (neglecting the small difference between proton mass M and neutron mass). As a matter of fact, the measured atomic weight W is a little smaller than the mass number A. If this mass defect is $(A - W)M$, the "binding energy" of the nucleus B is $B = c^2 M (A - W)$, which has to be supplied in order to break up the nucleus into its elementary particles. This energy is also liberated by the formation of the nucleus.

It is particularly important to know the binding energy of a nucleus divided by the number of particles of which this nucleus is built. This number B/A has, in first approximation, one and the same value for all nuclei. This fact checks well with the picture of the nucleus as a drop of fluid. For in a fluid the energy is proportional to the volume, and the volume of a nucleus is proportional to the number A of particles.

If we examine the values of B/A more closely, we notice a clear trend. The "binding energy per particle" is increasing from the lightest nuclei to the value $A = 60$ (nickel) and is again decreasing toward the heavy nuclei. The nuclei in the middle of the series are the most strongly bound.

If we speak of a nucleus being comparable to a drop of fluid, we do not mean, of course, that the behavior of a nucleus can be predicted exactly by the equations of fluid motions as treated in ordinary hydrodynamics. For this would mean that the particles in the nucleus obey Newton's laws, which is outruled by the experiments on the behavior of small bodies (Part VII). The similarity to a fluid is restricted to the validity of some algebraical and dimensional relations which have the same form as the relations between surface tension and repulsive forces in a fluid. It is noteworthy, however, that the behavior of these extremely small particles can be predicted, as far as some general features are concerned, by this picture. It would, however, be a great mistake

if anyone should take this picture literally and draw conclusions from Newton's laws in an attempt to show that the motions of fluid particles contradict the behavior of smaller particles as predicted by wave mechanics. One must never forget that the term 'particle' in the description of the nucleus has not the same operational meaning as this term has in ordinary mechanics. 'Particle' in nuclear physics is only an abbreviation which refers to "point events" which can be statistically predicted.

64. Atomic Power[28]

We learned from section 63 that the formation or the breaking-up of a nucleus can become a source of energy (atomic power). We consider not only the formation of a nucleus from the "elementary particles", protons and neutrons, but from any smaller nuclei. If the binding energy of the resulting nucleus is smaller than the sum of the binding energies of the "building stones", the formation produces energy and rest mass disappears. If, however, the binding energy of the resulting nucleus is greater than the energy sum of the constituents, energy is liberated in breaking up this nucleus, and the fragments together have a smaller mass than the original nucleus.

If A increases, the binding energy per particle, B/A, decreases among the lighter nuclei and increases among the heavier nuclei. By this fact two ways are suggested for the production of energy from nuclear reactions. One could either use the formation of a light nucleus or the disintegration of a heavy nucleus.

The first way was used by nature in our sun, the second one by human inventors in the "atomic bomb". According to a widely accepted astrophysical theory, the heat of our sun is reproduced by a cycle of nuclear reactions (Bethe's Carbon Cycle) which boils down eventually to the formation of helium nuclei from protons.

In the first atomic bomb an isotope of uranium ($P = 92$, $A = 235$) was used. This heavy nucleus can be broken up by neutron bombardment. Every uranium nucleus of 235 particles is split into two nuclei of nearly equal size and some free neutrons. As the binding energy of all the fragments together is much smaller than the energy of the original uranium nucleus, there is a considerable production of kinetic and radiant energy accompanied by a disappearance of rest mass. The liberated neutrons hit other uranium ($A = 235$) nuclei and produce a further disintegration. In this way a "chain reaction" is started, which produces in a short time a considerable amount of energy in a way similar to that in which a lighted match can start a forest fire, by a "self-sustained" reaction.

This kind of splitting into two particles of similar size has been known under the name of "uranium fission". In order to start this chain reaction, one does not need to produce neutron projectiles first. There are always some neutrons in the air, originating from cosmic rays or radioactive substances. According to wave mechanics (sec. 46), there is even a certain chance that fission starts spontaneously. By the disintegration of 1 pound of uranium (235) an energy of nearly eleven million kilowatt hours is produced.

The process which takes place in the atomic bomb is a good example to offer in showing that disintegration of nuclei is a very common phenomenon which is not restricted to the radioactive processes in the older sense (emission of alpha and beta particles).

Besides the uranium bomb, a second type has been used which provides a good example of fission combined with artificial radioactivity. If ordinary uranium ($P = 92$, $A = 238$) is bombarded with neutrons, sometimes a neutron is captured by a nucleus. When this happens, a radioactive emission (of beta particles) takes place. This can be interpreted by assuming that one neutron in the nucleus emits an electron and is converted into a proton (sec. 62). The number of protons P now becomes 93, and we have a new element "neptunium". By again capturing a neutron, and again emitting an electron, the neptunium becomes "plutonium", with $P = 94$. This new element is the material of the second atomic bomb. It suffers fission like uranium ($P = 235$) and can start a chain reaction.

The explosion of the bomb has proved that Einstein's law $E = m_0c^2$ (sec. 30) plays a role not only in subtle laboratory experiments but also in power engineering. The introduction of the speed of light c into the equation of mechanics, which has been thought of as a far-fetched sophistication, becomes now conspicuous by the enormous energy production which can be understood only by the enormous value of c^2, the square of light velocity.

The explosion of the atomic bomb provided a very conspicuous example of the possibility that "mass can disappear". As a matter of fact, the disintegration of one pound of uranium ($P = 235$) left only 0.999 of a pound of fragments, while 0.001 of a pound was "converted into energy" (radiant and kinetic). We must not forget, however, that it would be misleading to describe this fact by saying that "matter has disappeared". We can say this if we mean "rest mass" by the word 'matter'. But if we gave to the sum of rest mass and E/c^2 the name 'matter', we could say again that "matter was conserved" even through the atomic explosion. If we use the term 'matter' beyond its common-sense meaning (sec. 51), we must give it an arbitrary operational definition. As in the case of 'time' or 'length', some operations which are to measure the "quantity of matter" yield identical results, if

we use the word 'matter' in the traditional sense and assume that the laws of traditional physics are valid. But if the laws of physics were different, it might happen that these operations would render different results, and it becomes ambiguous which of them is the "real quantity of matter". An example of such operations is provided by the measuring of masses before and after an atomic explosion. If the relativistic mechanics is true, the "masses" before and after the explosions are not equal. Therefore, "matter" has disappeared, if we identify "matter" with "rest mass". But if we choose a different operational definition of matter, we could not say that "matter has disappeared".

This statement has only an unambiguous operational meaning if we stick to traditional physics and its terminology. But in the new physics we cannot say whether "matter"does disappear or not, because the word 'matter' can no longer be used unambiguously, without introducing a new operational definition. As we learned in section 2, the operational meaning of an expression can only be defined unambiguously if we assume the validity of certain physical laws.

IX. Conclusion

It is hardly possible to decide accurately whether a certain presentation of physics is fit to become a part of a unified presentation of the sciences. However, it is easy to ascertain that quite a few presentations of physics have decidedly produced confusion when one has tried to fit them into the structure of other fields of knowledge.

In biology, for instance, the physics of the twentieth century has been used to bring about a decision in the age-old conflict between the "mechanistic" and the "vitalistic or organismic" conception. We have been told that the "new physics" decides in favor of spontaneity and emergent evolution and against rigid mechanical explanations.

In medicine the "new physics" has been quoted as favoring the case of all sorts of "practitioners" in their old fight against "orthodox school medicine". For, by the principle of uncertainty, a certain flexible element has been brought into science which, according to these people, eases the rigidity of scientific rules and provides a place for methods which are based more on intuition than on systematic knowledge.

In sociology and economics the new physics has been employed to bolster the "organismic view" of human society sponsored by some religious and political groups against the "mechanistic view of society" which has been allegedly the view of liberal and Marxist economists.

In theology the new physics has been put into the service of the fight of religion against materialism by pointing out that now "freedom of will" has won a certain place in the formerly rigidly deterministic universe.

Even the "occult sciences" have seen a green light in the physics of the twentieth century. The theory of relativity introduced the conversion of matter into energy and has supported the belief in "dematerialization".

As we learned in Part II, every physical theory consists of three kinds of statement: equations between physical quantities (relations between symbols), logical rules, and semantical rules (operational definitions).

The sensational results of the application of the "new physics" to biology, sociology, medicine, etc., have been very frequently achieved by the following method: The symbols of physics (e.g., waves and particles) have been inserted into the sciences of biology, sociology, medicine, etc., but the operational definitions of the symbols have been omitted. This means that the physical theories have been applied in a crippled condition. As symbols without operational meaning do not lead to any palpable result, some kind of operational definition had to be introduced. Actually, those symbols have been interpreted as meaning what they traditionally mean in biology or sociology. This meaning is, of course, very different from their operational meaning in physics.

We see that the application of the "new physics" to other fields of knowledge has not always been made by fitting the physical theories in their scientific form into the structure of other sciences. No wonder that nothing but confusion could be the outcome of such an attempt to "integrate human knowledge".

Examples are obvious. If we say in physics that in the realm of the subatomic phenomena the concepts of mechanics cannot be applied, we mean to say that the symbols of classical mechanics, like 'position of a particle', have no operational meaning in the realm of the subatomic phenomena. However, in biology there has been an ancient conflict between the mechanistic and the vitalistic conception. To say that the "conceptions of mechanics cannot be applied to the phenomena of life" has always meant that life is something autonomous, spontaneous, emergent, etc. The symbols of the new subatomic physics, if understood in connection with their original operational meaning, allow a description of the subatomic phenomena which has nothing to do with spontaneity, emergent evolution, or purposiveness. This "new physics" is, in the language of biology, not less "mechanistic" than classical physics. Therefore, if we understand all statements of physics in their operational meaning, we cannot draw any conclusion in favor of a vitalistic biology.

If the statement "length is relative" is understood in its operational meaning, it is a statement about the fact that certain procedures of measurement yield different results, whereas it had formerly been believed that they yield identical results. But if we transplant the statement "length is relative" without its operational meaning into psychology or sociology or medicine, the word 'relative' is interpreted, of course, in the way in which this word has been traditionally used in these fields of knowledge. It has meant there that all knowledge is subjective or his-

torically and ethnically conditioned. If we carried the original operational meaning of "relativity" from physics into psychology or medicine, we could never arrive at a statement about subjectivity or vagueness, let alone "agnosticism".

What we can learn from these examples is simply the fact that a presentation of physics is fit to be a part of the unified sciences only if the operational meaning of every statement is explicitly formulated and carefully carried along when the statement is applied to other sciences.

While a physicist works within his own domain, the operational meaning of his statements is "understood". No explicit logical analysis is needed, and such an analysis would even be tiresome and pedantic. When, however, a statement of physics is transplanted into a foreign soil, the operational meaning is no longer "understood", and a careful logical analysis is indispensable. Otherwise, we have to approve statements like: "The first explosion of an atomic bomb exploded once forever the doctrine of the materialists" or "was a crucial experiment in favor of dematerialisation" or "spiritualism is confirmed by science".

While one lives within a limited group of people, it is "understood" what we mean by calling a man a "nice fellow"; but if one comes in contact with an unknown group, there is not even a guess as to whether by calling a man a "nice fellow" we hint that he goes all-out for prohibition or that he likes a good drink.

In a presentation of the "new physics" we must, therefore, distinguish very carefully between statements which ascribe to words or symbols a new operational meaning and statements which describe the results of new experiments.

But we must not fail, on the other hand, to understand the close interconnection between these two kinds of statement. As we learned in Part II, and confirmed by the examples of Parts V (relativity) and VII (wave mechanics), symbols and expressions are introduced into physics only if they are of some help in formulating physical laws—the laws predicting observable phenomena. This means that physical operations can serve as definitions of physical quantities only if a great many operations lead to an identical numerical result. "In the discovery of what operations may be usefully employed in describing nature is buried almost all physical experience. . . . . The discovery that the number obtained by counting the number of times a stick may be applied to an object can be simply used in describing natural phenomena was one of the most important discoveries ever made by man".[29]

As we learned in the Introduction, quite a few physicists have suggested that the misuse of physics in other fields of knowledge could be prevented by isolating physics from any contact with people who want to interpret its meaning for human thought in general. By such an ostrich policy physics would be converted into a collection of rules which might be useful in technology. But physical science would lose its traditional role as the vanguard of progressive thought.

A danger of this kind has arisen again and again in the history of thought. In the year 1907 the French philosopher and historian of science Abel Rey wrote: "If physical science which has had an essentially emancipating effect in history goes down in a crisis which leaves it only with the significance of a technically useful collection but robs it of every value in connection with the cognition of nature, this must bring about a complete revolution in the logical art. The emancipation of the mind, as we owe it to physics, is a most erroneous idea. . . . . One must restore to a mystical sense of reality everything that we believed had been taken away from it" [30]

This danger cannot be met by isolationism in physics. The only thing to do is to make "physics safe for its role as a part of the unified sciences". This means again to give careful thought to the logical analysis of physics and to be always aware that only the combination of relations between symbols, logical rules, and operational definitions constitute the science of physics. There are quite a few philosophers who have felt sorry that "behaviorism" and "logical empiricism" have banned words like 'soul' and 'mind' from scientific psychology. They may find comfort in the fact that the same schools of thought have banned words like 'matter' from scientific physics. A "soulless" psychology and a "matterless" physics have been established as parts of "Unified Science". Words like 'matter' and 'mind' are left to the language of every-day life where they have their legitimate place and are understood by the famous "man in the street" unambiguously.[31]

Notes

1. "I have not attempted to introduce a new philosophy into science, but rather to remove an obsolete philosophy which has occasionally survived in the writings of scientists for a longer time than in the writings of the philosophers themselves" (Ernst Mach, *Erkenntnis und Irrtum* [1905]). "To neglect philosophy when engaged in the re-formation of ideas is to assume the absolute correctness of the chance philosophic prejudices imbibed from a nurse or a schoolmaster or current modes of expression" (A. N. Whitehead, *The Principle of Relativity* [1922]).

2. Philipp Frank, "Why Do Scientists and Philosophers So Often Disagree?" *Review of Modern Physics*, Vol. XIII (1941), and "The Philosophical Meaning of the Copernican Revolution," *Proceedings of the American Philosophical Society*, Vol. CLXXXVII (1944).

3. W. A. Wick, *Metaphysics and the New Logic* (1942); A. C. Benjamin, "The Unholy Alliance between Operationalism and Positivism," *Journal of Philosophy*, Vol. XXXIX (1942).

4. This *Encyclopedia*, Vol. I, No. 3, § 23 (Carnap).

5. *Ibid.*, § 8.

6. *Ibid.*, § 4.

7. *Ibid.*, § 24.

8. P. W. Bridgman, *Logic of Modern Physics* (1927).

9. It has been suggested by some authors that some rules of traditional logic be altered in order to fit modern quantum mechanics into a formalized system: M. Strauss, in *Erkenntnis*, Vol. VI (1936); L. Rougier, "La Relativité de la logique," *Journal of Unified Science*, Vol. VIII (1936); and G. Birkhoff and J. v. Neumann, "The Logic of Quantum Mechanics," *Annals of Mathematics*, Vol. XXXVII (1936).

10. Bertrand Russell, *The Scientific Outlook* (1931).

11. P. W. Bridgman, *The Nature of Thermodynamics* (1941).

12. Conventionalism has been the starting-point of a great many attempts to justify metaphysical creeds by arguing that "science" cannot tell us anything about physical reality and that therefore a vacuum is produced within which extrascientific methods may operate at will (see E. Le Roy, *Rev. de met. et de mor.*, Vols. XVII and XVIII [1899, 1900]; Pierre Duhem, "Physique de Croyant," *Annales de phil. chrét.* [1905/6]). The scientific angle is stressed by E. Nagel, "Nature and Convention," *Journal of Philosophy*, Vol. XXVI ((1929).

13. Henri Poincaré, *Science et hypothèse* (1903); P. Duhem, *La Theorie physique, son objet et sa structure* (1926); Philipp Frank, *Le Principe de causalité et ses limites* (1937), p. 184; and P. W. Bridgman, *The Nature of Thermodynamics*.

14. P. Frank, *Le Principe de causalité et ses limites*, p. 73.

15. P. Frank, *La Fin de la physique mécaniste* (1936).

16. P. Frank, "The Mechanical vs. the Mathematical Conception of Nature," *Philosophy of Science*, Vol. IV (1937).

17. P. Frank, "Modern Physics and Common Sense," *Scripta mathematica* (1939).

18. P. Frank, "Relativity and Its Astronomical Applications," *Sky and Telescope*, 1942, p. 9.

19. R. v. Mises, "Die Krise der Mechanik," *Proc. Congr. f. Appl. Mech.* (Delft), 1924.

20. *Logic of Modern Physics*, p. 59.

21. "It is the recoverability of the original situation that is important, not the detailed reversal of the steps which led to the original departure from the initial situation" (Bridgman, *Nature of Thermodynamics*, p. 122). In a great many writings this distinction has not been made and the word 'reversibility' is used for 'recoverability'.

22. Lord Kelvin (1852) cautiously assumed the validity of the Second Law for "inanimate" systems only. For a discussion of the apparent conflict between organic evolution and increase of entropy see Lecomte du Nouy, *Biological Time* (1936), and Bridgman, *Nature of Thermodynamics*, pp. 208 ff.

23. Bridgman, *Nature of Thermodynamics*, pp. 148 ff.

24. Quite a few philosophers suggested the introduction of a "philosophical time concept", without a clear-cut operational meaning, in addition to the time concept based on physical operations as used in the theory of relativity: H. Bergson, *Durée et simultanéité* (1922); A. O. Lovejoy, "The Time-retarding Journey," *Philosophical Review* (1931). A time concept based on biological operations was introduced by Lecomte du Nouy, *op. cit.*

25. "Can Quantum-Mechanical Description of Physical Reality Be Considered Complete?" *Physical Review*, Vol. XLVIII (1935).

26. In Newtonian mechanics the operational meaning of 'force' is based on the measurement of the acceleration of a particle. In the mechanics of subatomic particles the operational meaning of 'force' can be reduced to the statistical distribution of point-events or impulse-events in space at a certain time.

27. Albert Einstein, "Physics and Reality,' *Franklin Institute Journal*, Vol. CCXXI (1936).

28. H. D. Smyth, *Atomic Energy for Military Purposes* (Princeton, 1945).

29. P. W. Bridgman, *The Logic of Modern Physics*, p. 27.

30. *La théorie de la physique chez les physiciens contemporains* (Paris, 1907).

31. "Philosophy deals with positive truth, yet contents itself with observations such as come within the range of every man's normal experience" (C. S. Pierce, *Collected Works*, Vol. I, sec. 241).

Selected Bibliography

I. RECENT WORKS ON THE FOUNDATIONS OF PHYSICS

BERGMANN, G. "Outline of an Empiricist Philosophy of Physics," *American Journal of Physics*, XI (1943), 248–58, 335–42.

BOHR, NIELS. *Atomic Theory and the Description of Nature*. New York, 1934.

BORN, M. *The Restless Universe*. London & Glasgow, 1935.

BRIDGMAN, P. W. *The Logic of Modern Physics*. New York, 1929.

———. *The Nature of Physical Theory*. Princeton, 1936.

———. *The Nature of Thermodynamics*. Cambridge, Mass., 1941.

BROGLIE, L. DE. *Matter and Light*. New York, 1939.

DARWIN, C. G. *The New Concept of Matter*. London, 1931.

DAVIS, H. T. *Philosophy and Modern Science*. Bloomington, Ind., 1931.

EDDINGTON, A. S. *Space, Time, and Gravitation*. Cambridge, 1929.

EINSTEIN, A., and INFELD, L. *The Evolution of Physics*. New York, 1938.

ELDRIDGE, J. A. *The Physical Basis of Things*. New York, 1934.

FRANK, PHILIPP. *Le Principe de causalité a ses limites*. Paris, 1937.

———. *Between Physics and Philosophy*. Cambridge, Mass., 1941.

LENZEN, V. F. *The Nature of Physical Theory*. New York, 1931.

LINDSAY, R. B., and MARGENAU, H. *Foundations of Physics*. New York, 1936.

MISES, R. VON. *Kleines Lehrbuch des Positivismus*. The Hague, 1939.

REICHENBACH, HANS. *Philosophical Foundations of Quantum Mechanics*. Berkeley, 1944.

RUSSELL, BERTRAND. *Mysticism and Logic*. New York, 1929.

———. *The Scientific Outlook*. New York, 1931.

STEBBING, SUSAN. *Philosophy and the Physicists*. London, 1937.

SWANN, W. F. G. *The Architecture of the Universe*. New York, 1934.

WATSON, W. H. *On Understanding Physics*. Cambridge, 1938.

Selected Bibliography

II. CLASSICS ON THE FOUNDATIONS OF PHYSICS

DUHEM, PIERRE. *La Théorie physique: son objet et sa structure*. Paris, 1906.

ENRIQUES, F. *Problems of Science*. Chicago and London, 1914.

MACH, ERNST. *Science of Mechanics*. Chicago and London, 1919.

POINCARÉ, HENRI. *The Foundations of Science: Science and Hypothesis, The Value of Science, and Science and Method*. New York, 1919.

III. PHILOSOPHICAL INTERPRETATIONS OF PHYSICS

BRAHMA, N. K. *Causality and Science*. London, 1939.

CASSIRER, E. *Einstein's Theory of Relativity*. Chicago and London, 1923.

———. *Determinismus und Indeterminismus in der modernen Physik*. Goteborg, 1936.

DINGLE, H. *Through Science to Philosophy*. London, 1937.

EDDINGTON, A. *The Nature of the Physical World*. New York, 1928.

———. *Philosophy of Physical Science*. New York, 1939.

HALDANE, J. B. S. *Marxist Philosophy and Science*. London, 1938.

JEANS, J. *The Mysterious Universe*. New York, 1931.

———. *Physics and Philosophy*. New York, 1943.

MACKAYE, J. *The Dynamic Universe*. New York, 1931.

MARITAIN, J. *La Philosophie de-la nature*. Paris, 1936.

———. *The Degrees of Knowledge*. New York, 1938.

MONTAGUE, W. P. *The Ways of Things*. New York, 1940.

NORTHROP, F. S. C. *Science and First Principles*. New York, 1931.

REISER, O. *Philosophy and the Concepts of Modern Science*. New York, 1935.

WHITEHEAD, A. N. *The Concept of Nature*. Cambridge, Mass., 1920.

———. *Science and the Modern World*. New York, 1926.

Cosmology

E. Finlay-Freundlich

Cosmology

Contents:

PAGE

1. INTRODUCTION 507

2. EARLIER DISCUSSIONS OF THE COSMOLOGICAL PROBLEM ON THE
   BASIS OF NEWTON'S LAW OF GRAVITATION 509
   2,1. The Observational Background 509
   2,2. Olbers' and Seeliger's Objection to Such a Universe . . . 511

3. THE OBSERVATIONAL BACKGROUND FOR MODERN COSMOLOGICAL
   CONSIDERATIONS 515
   3,1. The Various Definitions of Distance 515
   3,2. Remarks on the General Red-Shift 520
   3,3. The Apparent Distribution of Matter in Space 522

4. THE COSMOLOGICAL POSTULATE 523
   4,1. The Nature of the Postulate 523
   4,2. The Theory of an Expanding Universe on Classical Grounds . 524
   4,3. Summary 531
   4,4. Supplement 533

5. THE RELATIVISTIC TREATMENT OF THE COSMOLOGICAL PROBLEM . 539
   5,1. General Introduction 539
   5,2. Supplement 549
   5,3. The Cosmological Constant 554
   5,4. General Remarks concerning Singularities in the Cosmological
        Problem 555
   5,5. The Importance of the Conception of a Finite World . . . 557
   5,6. The Observational Test of the Relativistic Theory 559

6. CONCLUDING REMARKS ON THE TIME SCALE OF THE UNIVERSE . 562

NOTES 564

BIBLIOGRAPHY 565

506

Cosmology

E. Finlay-Freundlich

1. Introduction

1,1. In cosmology an attempt is made to answer several questions. How is the universe built up as a whole? Do the laws of nature, which we derive from experience gained in our "neighborhood"—be this the earth or the solar system or the galaxy—remain applicable if we imagine this "neighborhood" extended until it comprises the whole universe? Is an infinite world compatible with the laws of nature or must we restrict the world to a finite size if we wish to avoid insurmountable difficulties of principle?

Cosmology has to face the problem of infinity not as a mathematical abstraction but as a physical reality. This confers on all its problems a particular speculative character and, at the same time, a particular attraction; for many of us are anxious to obtain an answer to the question: Can we form a picture of the whole universe? Naturally, to such a question a definite and final answer cannot be expected in the near future, and it is not surprising that all attempts so far have come to an early standstill on account of our much too limited and scanty knowledge. Despite that, attempts to frame a unified picture of the whole world have not ceased, and they have been of great scientific value because they have considerably widened and deepened our understanding of the fundamental laws of nature.

1,2. Remarks on the History of the Problem

From 1917 on, cosmology has again been in the foreground of interest in connection with the development of the general theory of relativity. It seemed as if this theory could bring us a decisive step nearer to a solution of the cosmological problem. The actual situation, however, is not so simple. It is true that

507

Introduction

cosmology has taken a decisive step forward during the past few years, but this advance was predominantly due to the amazing advance of astronomical knowledge beyond the realm of our galaxy into remote depths of the universe. Our "neighborhood," until then confined to small parts of the galaxy, was extended, by a rapid advance into extragalactic space, to distances which possibly may no longer be considered negligibly small as compared with the dimensions of the whole universe. The cosmological problem took full advantage of this enrichment of our knowledge. The theory of general relativity at the same time opened an incomparably deeper insight into its foundations. As far as the representation of observational facts is concerned, classical mechanics, no less than the theory of relativity, offers explanations for practically all observations. In this respect neither of them may claim to be definitely superior to the other. But the theory of relativity discloses the full variety of possible solutions and promises in the end a very much deeper understanding of the structure of the universe.

The first idea, which later proved to be really essential for a solution of the cosmological problem, may be called Lambert's idea of a hierarchic structure of the universe (see 2,2). This idea has lost, in the present situation, its finality. But at an earlier stage it removed, when applied in the right way, the main obstacles which seemed to bar the first efforts to solve the cosmological problem within the framework of Newton's mechanics. These first efforts were based, as we now know, on empirical foundations much too narrow and weak for such a complex problem. Thus the possible solutions that were offered had a rather academic character and never fully satisfied the imaginations of natural scientists. As a result, cosmological considerations remained in the background until the great breakthrough into the world of spiral systems with its new discoveries gave a tremendous impetus to the whole problem. When that occurred, the guidance was taken over by the theory of relativity, which, no less than Newton's theory of gravitation, had to stand the test of its cosmological consequences and which offered more efficient tools to attack the whole problem.

2. Earlier Discussions of the Cosmological Problem on the Basis of Newton's Law of Gravitation

2,1. The Observational Background

Every investigation of the structure of the universe has to start from a knowledge of, or from assumptions about, the distribution of matter in space. And since obviously all experience will always be restricted to a part, probably to a small part, of the universe, no exact knowledge of the distribution of matter over the whole universe can ever be claimed. We therefore attempt to infer from the observed distribution in our "neighborhood" how matter may be distributed beyond the surveyed part of space. This is the basic assumption in all cosmological considerations.

As long as astronomy had not penetrated into extragalactic space beyond the boundaries of the star-system to which the sun belongs, stars were believed to be the elements, the "bricks," so to speak, of which the universe was built. We believed ourselves to be imbedded in an ocean of stars. From star distances derived by the astronomers in various ways, the law according to which the stars are distributed in space was inferred.

The most simple initial assumption was that of uniform distribution. If the stars were uniformly distributed in space and, in addition, were all of equal luminosity, i.e., of equal absolute magnitude,[1] or, alternatively, if the relative frequency of stars of different luminosities for every element of space were the same, their numbers—counted in order of increasing magnitudes m—should increase according to the formula:

$$\frac{N_{m+1}}{N_m} = 3.982 ,$$

or

$$\log N_m = 0.6 \cdot m + \text{constant} .$$

(2,1.1)

In this formula N_m measures the number of stars as bright as, or brighter than, the magnitude class m. The formula assumes, moreover, that space is transparent, so that the brightness of any light source decreases inversely with the square of its distance. The derivation of the formula can be briefly outlined as follows:

Suppose the star-systems to be uniformly distributed in space and all of equal luminosity; the number N_m of systems down to the limiting magnitude m is obviously proportional to the volume of the sphere on the surface of which a system reaches the limiting brightness m. Hence $N_m \sim r_m^3$, where r_m denotes the limiting distance. On the other hand, since the inverse-square law for the decrease of apparent brightness i_m with distance is assumed to hold, we have

$$\frac{i_m}{i_{m+1}} = \frac{r_{m+1}^2}{r_m^2} = 2.5 ;$$

for, by definition, the ratio of intensity of two light sources differing by 1 magnitude in apparent brightness is equal to 2.5. Consequently,

$$\frac{N_{m+1}}{N_m} = \frac{r_{m+1}^3}{r_m^3} = (2.5)^{3/2} = 3.982$$

(or roughly $= 4$) and this, since

$$N_m = 3.982^m N_0 ,$$

yields (N_0 being a constant)

$$\log N_m = 0.6 + \text{const} .$$

It was soon realized that in the star-system surrounding us, i.e., the galaxy, the assumption of a uniform distribution does not hold good; and the correct conclusion was drawn that the galactic system is of finite size. Despite that, the preceding formula still dominates all cosmological considerations. However, it is no longer applied to the distribution of stars in the galaxy but to that of star-systems in the extragalactic space, now that it has been realized that star-systems are the elements from which the universe is built up. All statistical data concerning the distribution of matter in space in the following cosmological considerations refer to the distribution of star-systems in the universe and will be based on formula (2,1.1).

For the moment only the most simple case of uniform distribution of matter has been considered. If space is infinite and the density of matter everywhere finite, i.e., not zero, an infinite world, filled with an infinite amount of gravitating matter, would result. Is such a universe conceivable? Are we permitted to perform this transition to infinity without encountering difficulties of principle?

2,2. Olbers' and Seeliger's Objection to Such a Universe

In a world in which the mean density ρ of matter is finite, even if its value is as small as we please, the limit of the gravitational potential

$$\phi = \int_V \rho \, \frac{dV}{r} \qquad (2,2.1)$$

has no definite value when V, the volume, tends to infinity.[2] Also the expression for the gravitational stress becomes indefinite. Seeliger's objection against a universe filled with an infinite amount of matter is based on this fact.

If the world contained infinitely many bright stars, the night sky would shine with a brightness corresponding to the average surface brightness of these stars; this is Olbers' objection.

Both objections can be removed by applying the idea of a hierarchic structure of the world which prescribes a special distribution of the infinite amount of matter over the infinite Euclidean space. The singularities mentioned above may thus be avoided. By "a hierarchic structure" the following is meant: matter is distributed in space so that stars combine to form greater systems, star-systems or galaxies; galaxies again combine to form still greater systems, supergalaxies; and so on. From each rank on the hierarchic ladder we can step to a rank of higher order consisting of elements of the preceding rank, and so forth to infinity. If on such a hierarchic structure we impose certain additional restrictions, the singularities of the above-mentioned character need not appear. In the resulting world the mean density of matter becomes zero.

Let G_0^0 denote a system of lowest rank (the upper index indicates one particular system of this rank); it will be a single star, for instance, our sun.

Let G_1^0 denote the system of the next higher rank, a star-system or a spiral system like the galaxy, containing N_1 stars, among these, in particular, also the element G_0^0, the sun.

Let G_2^0 denote a system of next higher rank, consisting of N_2 spirals, again containing, in particular, the element G_1^0, our galaxy, and so on.

In addition, let us simplify the picture by attributing to all systems a spherical shape, of radius R_0, R_1, R_2, . . . , respectively, each system having the total mass M_0, M_1, M_2, . . . , respectively, and let us finally assume that, in particular, G_0^0 lies on the surface of G_1^0; G_1^0 on the surface of G_2^0. . . . Then G_0^0, the sun, will be subject to a gravitational force derivable from a potential which, on account of the spherical symmetry of every system of any rank i, is equal to the sum of terms M_i/R_i, each term representing the value of the gravitational potential at the surface of a spherical mass of radius R_i and mass M_i. The gravitational potential at G_0^0 is thus given by the infinite series:

$$\frac{M_1}{R_1} + \frac{M_2}{R_2} + \ldots \frac{M_i}{R_i} + \ldots \qquad (2,2.2)$$

This sum converges if

$$\frac{M_i/R_i}{M_{i-1}/R_{i-1}} = \gamma < 1 , \qquad (2,2.3)$$

i.e., if

$$\frac{R_i}{R_{i-1}} > \frac{M_i}{M_{i-1}} \frac{1}{\gamma}.$$

Now, G_1^0 consists of N_1 stars of mass M_0; hence $M_1 = N_1 M_0$. Similarly,

$$M_2 = N_2 M_1 = N_1 N_2 M_0 ,$$

$$M_3 = N_3 M_2 = N_1 N_2 N_3 M_0 .$$

Consequently,

$$\frac{M_i}{M_{i-1}} = N_i , \qquad (2,2.4)$$

and the potential from which the gravitational force acting upon G_0^0 is derived and which results from a superposition of infinitely many systems in the hierarchic order will be finite if the radii of the various systems satisfy the inequality,

$$\frac{R_i}{R_{i-1}} \geqq \frac{1}{\gamma} N_i \qquad \text{where} \qquad \gamma < 1 . \qquad (2,2.5)$$

The special assumptions made here with regard to the spherical shape of each system and its surface position in the next higher system are without influence upon the general conclusions concerning the convergence of the infinite series for the gravitational force.

Olbers' objection is similarly removed by satisfying the same inequality, as is easily shown. This objection is not of the same importance in principle as is Seeliger's objection, for it can also be removed simply by assuming that the universe contains sufficient obscuring dark matter to reduce the integrated light of the sky to the observed value. Thus, in an infinite world consisting of infinitely many bright stars, the total gravitational attraction upon any star (our sun plus the earth, for instance) could remain regular, i.e., have in every point a defined finite value, if the matter were distributed according to the rule outlined above; and also the brightness of the night sky could be kept sufficiently low. In such a universe the mean density, $\bar{\rho}$, of the matter, taken over a volume increasing toward the infinite Euclidean space, converges to zero, although the total amount of matter in the universe would increase beyond all limits when the volume increases toward infinity.

It may be worth mentioning that so far as our present experience goes—it actually covers only the first two steps—a distribution of the stars similar to such a hierarchic structure seems really to be indicated in the universe. The stars of rank 0 combine to form star-systems, spiral nebulae, or star-systems of different structure, of rank 1, and these again tend to agglomerate into systems of the next higher order. The Coma Virgo cluster and other clusters of spiral systems represent such systems of rank 2. This structural feature of the universe has, however, not yet found its place in modern cosmological theories. All models of the world still confine themselves to the most simple case of uniform distribution of matter in space and disregard the special features of the actual distribution. It has therefore not yet been carefully investigated how far the observed hierarchic structure in the universe actually would satisfy the special conditions which should insure finite values of the gravitational forces in each volume element of an infinite space.

Another difficulty which was encountered in the first attempts to solve the cosmological problem was the expectation of unduly high velocities in an infinite universe containing an infinite amount of matter. This difficulty is also eliminated in the hierarchic structure. However, at the time when the idea of a hierarchic world order was developed, the high and systematic velocities of distant spiral systems were still unknown and the necessity had not then been realized of conceiving the cosmological problem as a dynamical, and not a static, problem. The universe would be called "static" if—despite local motions of celestial bodies, as, for instance, the orbital motion of the planets around the sun—no large-scale changes in the distribution of matter in the universe were going on. In such a "static" world the mean density in the distribution of matter would remain constant when referred to sufficiently large volumes of space and sufficiently long intervals of time.

When our knowledge extended far beyond the limits of our galaxy and high velocities of the spiral systems were discovered[3] which no longer could be considered small even when compared with the velocity of light, these velocities were of a character very different from the high velocities expected on account of increasingly large potential differences. This discovery, together with the new methods of determining distances in stellar space, extending our "neighborhood" far beyond the limits reached until then, changed the whole outlook on the cosmological problem to such an extent that all previous results had to be completely revised.

At practically the same time the general theory of relativity took up the study of the cosmological problem from a very different point of view, for it had to test its new laws on the cosmological problem.

However, also within the framework of Newton's mechanics, the new empirical facts—in particular, the newly discovered expansion of the spiral systems—demanded a satisfactory explanation. All empirical facts will therefore be rediscussed from both the classical and the relativistic points of view.

3. The Observational Background for Modern Cosmological Considerations

3,1. The Various Definitions of Distance

The original method of determining the distance of a celestial body is based on the measurement of its trigonometric parallax. Using as base line the diameter of the earth's orbit, periodic

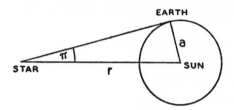

changes in a star's apparent position due to the orbital motion of the earth in the course of a year, when measurable, enable us to derive its distance. The mean radius of the earth's orbit is $a = 1.495 \cdot 10^{13}$ cm. The parallax, π, of a star, defined as the angle subtended by the radius of the earth's orbit as seen from the star (see Fig. 1), is thus connected with the distance r by the equation

$$\sin \pi = \frac{a}{r}. \tag{3,1.1}$$

Since π is always a very small angle, smaller than $1''$, the preceding formula may be written as follows:

$$\pi'' \sin 1'' = \frac{a}{r},$$

when π'' denotes the value of the parallax in seconds of arc; hence

$$\pi'' = \frac{a}{r} \frac{1}{\sin 1''} = 206265 \times \frac{a}{r}. \tag{3,1.2}$$

The distance r corresponding to a value of $\pi = 1''$ has been

515

chosen as a convenient unit of distance and is called 1 parsec; its numerical value is $3.084 \cdot 10^{18}$ cm $= 3.26$ light years.

Reliable measurements of trigonometric parallaxes are not possible for values of π far below $0.01''$; thus distances beyond 100 parsecs are not within the range of measurable trigonometric parallaxes. It is chiefly due to this fact that all cosmological considerations before, let us say, 1920, to which the preceding section refers, have been of a rather academic character. It is now obvious that only when our knowledge extends to distances of the order of millions of parsecs can we hope to draw conclusions concerning the structure of the universe at large, for instance, conclusions for or against a finite or infinite world.

Thus cosmology experienced its revival in modern times when the great break-through to star-systems far beyond the limits of the galaxy succeeded. This advance from distances of the order of 10^3 parsecs to distances of the order of 10^8 parsecs became possible when methods of determining distances in interstellar space were developed which were independent of a direct reference to a base line. For the limitation of the method of measuring trigonometric parallaxes is due to the extreme smallness of the radius a of the earth's orbit as compared with the average distances of stars in our galaxy, not to mention its smallness relative to distances of extragalactic star-systems.

The emancipation from a direct reference to a base line succeeded when a more detailed astrophysical knowledge of various types of stars was obtained. This knowledge made it possible to reverse in special cases the equation,

$$m - M = 5 \log r - 5 , \qquad (3,1.3)$$

which connects the absolute magnitude M of a star, i.e., its luminosity (see note 1), with its apparent magnitude, m, and its distance, r, that is, not to use this equation to determine M but to derive from it the value of r. Originally, this equation was formulated to define the luminosity of a star in terms of magnitude classes. The apparent brightness, m, of a star depends on two factors, namely, the luminosity of the star, i.e., the radiation emitted per second from its surface, and its distance. A very bright star may appear very faint, i.e. its m is large (the value for the magnitude class increases with decreasing brightness, a star of first magnitude being very bright), if it is situated at a sufficiently great distance from the earth. Consequently, M can be derived only when the distance r of the star, i.e., its parallax, has been measured. Then, if we define M as the star's apparent magnitude at the standard distance of 10 parsecs and suppose space to be absolutely transparent, M can be determined from equation (3,1.3) if r is known. But the process can be reversed, and the value of $(m - M)$ can be derived, if independent means become available to determine the value of M, for a star or for a whole star-system. Then the distance r can be calculated from our equation; for this reason $(m - M)$ is called the "distance-modulus."

Distances obtained in this way may be called "photometric" distances; for they depend on the knowledge of the absolute magnitude M, i.e., of the luminosity of the celestial body under consideration. No special base line limits the possibility of deriving distances from the distance-modulus, as in the case of trigonometric parallaxes.

The radius of the earth's orbit, the fundamental base line for measuring star parallaxes, is, of course, implicitly involved; for any independent statement on the absolute magnitude M of a star from purely astrophysical data is possible only with the help of stars of which the parallaxes are known.

The base line a enters thus into the calibration of the rule that discloses the star's luminosities. When this rule has been established, a star needs only to be identified as a member of the special class to which the rule refers, and its distance results from equation (3,1.3).

With the general widening of our astrophysical knowledge, various rules were discovered yielding the calculation of luminosities of special stars or whole star-systems from special astrophysical characteristics. As far as individual stars are concerned, such information refers to the absolute magnitude and the periods of Cepheid variables, or to the upper limit of luminosity reached by novae, or to the upper limit of luminosity observed for normal stars.

For Cepheid variables a relation between the period of their variability and their luminosity was discovered in 1912 by a study of the numerous Cepheids found in the Magellanic star clouds. This period-luminosity relation has been carefully calibrated by the use of Cepheids of measured distances within our galaxy. After that was done, it was only necessary to measure the period of light change for any Cepheid, observed anywhere in space, then to read the corresponding value of M from a graph giving the relation between period and luminosity and finally to find r from equation (3,1.3).

As to novae, observations have disclosed that in the phase of greatest brightness the absolute magnitude scatters only moderately around a mean value near to $M = -5.7$. So, whenever in a distant star-system the flashing-up of a normal nova (not a supernova, which belongs to a different class of objects) is observed near its maximum of brightness and when the apparent brightness m in this phase has been measured, the value of $(m - M)$ may be estimated with reasonable certainty, and thus also the distance r. Here again the "base line," i.e., the radius of the earth's orbit, enters by means of the calibration from which the approximately constant value of the luminosity in the maximal phase for various novae had been derived.

Third, statistical investigations of the absolute magnitudes of stars of known parallaxes in the galaxy revealed the fact that for normal stars a well-defined upper limit of M exists. The mean absolute magnitude of, say, the 25 brightest stars observed in a distant star-cluster may therefore be supposed to fall close to this upper limit; this again provides an estimated value of $(m - M)$, and so also of r.

By applying such methods, the distances of globular star-clusters and various other star-systems were determined, and the range of measurable distances was extended from a few hundred parsecs to millions of parsecs. It was recognized that the globular clusters, although members of our galaxy, are situated outside its main body at distances ranging from 6 to

50 kiloparsecs (1 kiloparsec = 10^3 parsecs). A still deeper penetration into space by similar methods brought the world of extragalactic star-systems within reach at about 300 kiloparsecs, and valuable knowledge was obtained concerning the next higher class of systems in the hierarchy of world structure. Only very few of these extragalactic systems are near enough to allow observations of individual stars belonging to them. A further extension of our knowledge to still greater distances had to be based on methods specially developed from a study of their apparent sizes, from the internal structure and integrated luminosity of extragalactic systems, etc. It was found that the absolute magnitudes, M, of whole star-systems similar in structure—spiral systems or globular clusters—scatter only moderately around a mean value. This mean value of M can therefore be used to estimate with sufficient accuracy the distances of spiral systems or globular clusters from the distance-modulus, in which m and M now denote the apparent and absolute magnitudes of a whole star-system. It is true, distances derived in this way locate a star-system only with a certain statistical probability as at a certain distance; they are not necessarily individually accurate. Similarly, the apparent diameters may be used to derive what are called "spatial distances," based on the fact that the linear dimensions of systems of similar structure appear to scatter only moderately around a mean value.

The next step was to correlate distances of star-systems with the observed values of red-shifts of the lines in the spectra of distant star-systems.

A general red-shift of the lines in a spectrum may result from a recession of the light source, star or star-system, from which the spectrum is produced. In this case the speed of recession dr/dt, r denoting the distance of the light source, is related to the red-shift, i.e., the observed increase of the wave length λ of the spectral lines, by the equation

$$\frac{d\,r}{dt} = c\,\frac{d\lambda}{\lambda}, \qquad (3,1.4)$$

where c is the velocity of light.

The correlation between the observed red-shifts of spiral systems and their distances yielded the equation

$$\frac{d\,r}{dt} = \kappa\,r\,, \qquad\qquad (3,1.5)$$

where κ is a constant; this is the law of the general expansion of the universe.

The formula (3,1.5) calibrated for the nearer systems with known distances, furnished a further possibility of estimating distances when only observed red-shifts are available. In this way our knowledge concerning the distribution of matter in space was widened step by step through a great variety of methods of measuring or estimating distances.

When the problem of cosmology was revived in connection with the theory of relativity, the surveyed part of the world extended to distances of nearly 10^8 parsecs, containing in its volume about one hundred million star-systems, each consisting of some thousand million stars. This was an empirical background very different from that on which astronomy had had to rely when it first attempted to tackle this complex problem. The space surveyed is no longer so insignificantly small as to make all considerations purely speculative. In particular, the newly discovered red-shifts, interpreted as an expansion of the universe, demanded a solution of a character very different from the static world (see p. 8) which formerly was taken for granted.

3,2. Remarks on the General Red-Shift

The correct interpretation of the red-shift is the key to a full understanding of the cosmological problem. Strong doubts about the right of interpreting the observed red-shifts of extra-galactic star-systems as Doppler effects have been expressed. These doubts originate chiefly in the desire to keep the door open for a static solution, according to which the world as a whole may be understood as being at rest, in equilibrium, and not itself involved in a gigantic evolution.

Are there any sound reasons to justify such doubts? During

the past century the measurement of shifts of spectral lines and their interpretation as resulting from velocities in the line of sight has proved to be one of the most powerful and never failing tools of astrophysics. Neither within the range of globular star-clusters, that is, up to distances of 10^4 to 10^5 parsecs, nor within the range of the nearest spiral systems, from 3.10^5 to 5.10^5 parsecs, do the observed displacements of the spectral lines in their spectra give any reason for suspecting that these shifts of spectral lines may not be interpreted as Doppler effects. Positive and negative values are measured. Beyond a certain distance, it is true, red-shifts, i.e., positive values for $d\lambda/\lambda$, prevail, indicating a general recession of the more distant spiral systems, systems which definitely no longer belong to the cluster of star-systems of which our galaxy appears to be a member. A general expansion of the world is indicated. On the other hand, as we shall see, theoretical considerations show that no solution of the cosmological problem is to be expected yielding a static world in which gravitating matter fills the universe with a finite value of the density. A world at rest, filled with gravitating matter, is not possible. So observational facts indicating systematic motions of the world matter and changes in its density distribution had to be expected. The only surprising fact might perhaps be that the observations disclose such a simple change, which in the first approximation might be described as a universal, regular, isotropic expansion of the universe. Nothing yet justifies the assumption that in the red-shift of the spectral lines of extragalactic systems we are dealing with a fundamentally new and hitherto unknown phenomenon of light emission.

It is naturally possible that in the future we may have to change this view under the weight of new discoveries. However, until these come forward, we should follow the route that so far has not brought us to a wrong turn and interpret the red-shifts as Doppler effects.

The numerical value of the constant factor κ in the formula (3,1.5) which expresses the expansion of the world, is near to 500 km/sec, when r is measured in units of 10^6 parsecs.

Background for Modern Cosmological Considerations

Spiral systems at about 1,000,000 parsecs distance recede with a speed of about 500 km/sec; at 10^8 parsecs for the most distant spiral systems, so far observed, the speed of recession is already of the order of one-sixth of the velocity of light. Such unexpectedly large values for the velocities, hitherto never reached in nature unless for electrons or particles of atomic size, have naturally intensified the doubts as to whether we are entitled to interpret the red-shifts of star-systems as Doppler effects at all.

3,3. The Apparent Distribution of Matter in Space

So far the survey of star-systems indicates, if the facts in favor of a hierarchic structure are disregarded, a uniform distribution of matter in space in accordance with formula (2,1.1), i.e. $\log N_m = 0.6 \cdot m + \text{const.}$, where N_m is now the number of nebulae brighter than the limiting magnitude m. If the unit of volume is taken sufficiently large to smooth out local irregularities in the distribution, the resulting number of star-systems per unit volume is roughly constant at all distances and in all directions. Spreading the matter, concentrated in the stars, evenly over the whole space, a constant value ρ for the density of matter is obtained. Accordingly, the constant value for ρ will be assumed in all following considerations.

This result, however, is not independent of the interpretation of the observed red-shifts $d\lambda/\lambda$; for the values of m in formula (2,1.1) are systematically influenced by the increase of the values of $d\lambda$ with increasing distances. The apparent brightness, m, of a nebula is in two respects affected by a red-shift. The quantity m measures the energy, i.e., the number of light quanta, received from the light source in question within a certain range of wave lengths per unit of time and unit of receiving surface. The energy E of a light quantum is given by $E = h\nu$, where ν denotes the frequency and h is Planck's constant. Since $c = \lambda\nu$, and the light velocity is constant, $E\lambda = hc$ is also constant. Therefore, if E^* measures the energy of the quantum affected by a change of wave length equal to $\Delta\lambda$, the product $E^*(\lambda + \Delta\lambda)$ must be equal to $E\lambda$; hence

$$E = E^* \left(1 + \frac{\Delta\lambda}{\lambda} \right). \qquad (3,3.1)$$

Thus, if $\Delta\lambda$ is positive, the energy of the quantum is reduced by the factor $[1 + (\Delta\lambda/\lambda)]$. This reduction of the energy, and thus also of m, must be considered, whatever physical phenomenon produces the observed shift, as long as the equation $E = h\nu$ is maintained.

The total amount of this correction, when all further selective influences, like that of the earth's atmosphere, light losses in the telescope, etc., on the bolometric luminosity of the star-systems are included, is about $3\Delta\lambda/\lambda$. The Bootes group of spirals, for instance, for which $\Delta\lambda/\lambda = 0.131$, thus calls for a correction of about 0.39 magnitude, by no means an insignificant correction.

When this correction is applied to the observed m of the star-systems, we obtain an apparently uniform distribution of matter in space, in agreement with the formula: $\log N_m = 0.6 \cdot m$ + constant.[4]

When, however, the red-shift is interpreted as a Doppler effect, an additional correction, amounting to an additional term $\Delta\lambda/\lambda$, has to be considered. For now one has to account for the additional fact that the brightness of a receding light source is reduced because, owing to its speed of recession, less quanta reach the observer's eye or the photographic plate, per unit of time, than would reach it if the light source remained at the same distance. The application of this recession correction changes the relation between m and $\log N_m$. There is no longer an agreement with the assumption of a uniform distribution of matter in space. The significance of this fact will be discussed further on.

4. The Cosmological Postulate

4,1. The Nature of the Postulate

Although, as has just been pointed out, observations do not unconditionally agree with the hypothesis of an expanding universe filled with matter of a constant mean density ρ, the cosmological models of the world have hitherto been developed on this assumption. Otherwise the problem would still defy all our efforts to find a solution. Do the laws of nature give a satisfactory description of such an expanding world, and are the available data sufficient to decide whether the world is finite or infinite? In what respects does the relativistic approach to the problem yield more satisfactory results? These are the questions to which we wish to give an answer.

We shall approach these questions by making the postulate that the observed phenomena are representative of the whole universe; only then will the results represent a cosmological theory. This cosmological postulate will be:

Every observer, wherever placed in the universe, describes the observed phenomena identically, saying that he is placed in an isotropically expanding universe in which matter is evenly distributed.

This postulate brings to conclusion a development of science which had its beginning with Copernicus and Kepler. In much the same way as they removed the earth, and thus man as well, from the central position which man in those days claimed for himself in his restricted picture of the world, the cosmological postulate deprives the picture of the whole universe of all features that tend to grant to man any privileged central position.

4,2. The Theory of an Expanding Universe on Classical Grounds

When, about thirty years ago, the attempts to solve the cosmological problem were resumed, the point of departure was very different from that of former attempts.

The space actually surveyed is no longer a merely infinitesimal part of the star-system to which the solar system belongs. It comprises many more star-systems than was the number of single stars included in the earlier surveys. We know that a stationary world in which matter may, in the first approximation, be considered as being at rest is, on principle, not possible —apart from a very special case (see p. 52)—and we know also that actually such a world is not indicated by observations.

All this rightly raised hopes that a resumption of the cosmological problem would bring us nearer to a solution than earlier attempts could have hoped to achieve. Will a decisive advance towards this solution come from the vast extention of our knowledge of the distribution of matter in space, including the phenomenon of expansion, or from the development in our knowledge of the principles of physics and geometry, stimulated by the theory of relativity? It is worth while to resume once more the cosmological considerations on classical, nonrelativistic grounds, in order to answer this question.

We consider matter, condensed in star-systems, as uniformly distributed in space and treat it as a continuous substratum filling space like the molecules of a fluid. Neglecting, first, the gravitational forces, we ask: Which form of motion of this matter in space is compatible with the cosmological postulate? Is, in particular, an isotropic expansion of the spiral systems, derived from the general red-shifts as represented by the formula $dr/dt = r \cdot$ constant, compatible with the postulate that every observer, wherever placed in the universe, should form the same picture of the world? A more exact mathematical discussion will be given in the supplement to this section; here only the general trend of the ideas and results will be outlined.

Every observer considers himself at rest while the outside world is moving relative to his position in a very special way, namely, receding in all directions symmetrically, with speeds increasing proportionately to the increase of distance. No rotation of the world relative to any position is observed. If an observer 1 attributes to the universal substratum at an arbitrary place P_1, characterized by space coordinates, x_1, x_2, x_3, a certain density $\rho(x_i, t)$ at a moment t, any other observer 2, receding from 1 but considering himself again at rest in his own world, must attribute to the substratum at the corresponding place, i.e., the place located relative to his position, as P_1 was located relative to observer 1—a position characterized now by coordinates x_1', x_2', x_3'—the same density $\rho(x_i', t)$ at the same moment. This must be true for any position whatsoever. Obviously, the density ρ of the substratum cannot be a function of the spatial coordinates x_i at all, but only of the time t.

The substratum must consequently fill space with uniform

density ρ; the value of ρ may, however, change with time, provided that at every moment this change is the same at every place. A change of just this character is produced by an isotropic expansion, during which the amount of matter, supposed to be invariant within every finite volume of space, continues filling with uniform density the steadily increasing volume of space allotted to it. Thus we find, from kinematical considerations only, that the conception of a uniformly expanding (or also contracting) universe, filled with matter of constant density (the density being a function of time only), introduces, in principle, no new difficulties into the cosmological problem.

The difficulties which were encountered in earlier attempts to solve this problem originated in the gravitational field set up by the world matter, if space was supposed to be of infinite volume and filled in its entirety with matter. It was found impossible to define the gravitational potential in a satisfactory way when a static solution of an infinite world was envisaged.

Now the situation is changed for two reasons; the general expansion of the universe demands the solution of a very definite dynamical problem, namely, the dynamical explanation of the expansion from universal gravitation; the cosmological postulate, on the other hand, reduces strongly the possible variety of solutions. For the general expansion only the gravitational field produced by the world substratum can be made responsible. This makes it necessary to define clearly the potential arising from this field at every point, and in doing this some of the uncertainties which formerly hampered the progress of the whole problem are avoided. The general field originating from the evenly distributed expanding world substratum will be called the "inertial field," to indicate that this field—in accordance with the principle of equivalence in the theory of relativity— may be made responsible for the appearance of the inert mass of bodies but not for local orbital motions as they are observed, for instance, in the neighborhood of the sun (see also p. 52).

Owing to the existence of this inertial field, no observer in the world should experience any accelerations which would make it impossible for him to consider himself at rest in his world. Therefore, if ϕ denotes the value of the potential at an arbitrary point in a system S, characterized by the coordinates x_1, x_2, x_3, relative to the observer's position in the center, and $\partial\phi/\partial x_1$, $\partial\phi/\partial x_2$, $\partial\phi/\partial x_3$, the components of the force derived from this potential, these components must vanish whenever $x_1 = x_2 = x_3 = 0$, that means, at the observer's place; and this must be true for every imaginable system S connected with an observer. The mathematical expression for ϕ must consequently have terms of second order as terms of lowest order in the space coordinates, so that, when $x_i = 0$, $i = 1, 2, 3$, all three components $\partial\phi/\partial x_i$ vanish.

In view of the symmetrical character of the distribution of matter, it is natural to postulate for ϕ an expression of the special form $\phi = A(x_1^2 + x_2^2 + x_3^2)$, where A may be a function of the time t. If, for instance, we take A equal to $\frac{2}{3}\pi\kappa_0\rho(t)$, we would obtain

$$\frac{\partial\phi}{\partial r} = \frac{\kappa M(r)}{r^2}, \qquad (4,2.1)$$

where $M(r)$ denotes the mass inclosed in a sphere of radius r. With such an expression for ϕ, the difference of the value of the potential at any two points P_1 and P_2 depends on their relative distance only. Therefore, taking P_1 as the center of a system S and describing a sphere around P_1, passing through P_2, the value of ϕ will be the same at all points on this sphere, and every point may be considered as screened off from the world outside this sphere. A similar statement can be made if P_2 is taken as the observer's place, i.e., as the center of the system, and P_1 is located on the sphere around P_2. The expansion of the universe in the field defined by this potential ϕ becomes a radial, rectilinear motion of P_2 relative to P_1 (P_2 being quite an arbitrary point), describable by the equation of motion

$$\frac{d^2 r}{dt^2} = \frac{\partial\phi}{\partial r},$$

which, if $M(r)$ again denotes the mass inclosed in the sphere of radius r and k_0 the gravitational constant, takes the form:

$$\frac{d^2 r}{dt^2} = \frac{- \kappa_0 M \ (r)}{r^2}. \tag{4,2.2}$$

To r we may give the special form: $r = r_0 R(t)$, thus attributing the isotropic expansion to the change in time of a general scale factor R, which is a function of t, while r_0 denotes the arbitrary distance of P_2 from P_1 at an arbitrarily chosen moment $t = 0$. The last equation then becomes independent of the value r_0, i.e., independent of the special choice of P_1 and P_2, and reads:

$$\frac{d^2 R}{dt^2} = \frac{- \kappa_0 M}{R^2}, \tag{4,2.3}$$

where M is a constant, stating that in an expanding volume the inclosed mass remains constant.

The dynamical problem inherent in the expansion of the universe is thus formally reduced to a very simple problem of celestial mechanics, namely, that of the expansion of a spherical system of bodies in which every body moves radially and the value of the density ρ remains constant throughout any volume, the numerical value of ρ naturally decreasing with increasing volume.

The dynamical problem to which we have reduced the expansion is easily solved. The character of the solution depends on the value and sign of the constant of integration h, which enters when a first integration is performed. This yields

$$\tfrac{1}{2}\dot{R}^2 = \frac{\kappa_0 M}{R} + h, \tag{4,2.4}$$

where $\dot{R} = dR/dt$ measures the speed of expansion. We obtain expanding universes of a periodic character when the constant h is negative. In this case the expansion should come, in a finite time, to a standstill and reverse into a contraction, which, after a finite time, should lead to a singularity when for $R = 0$ the density ρ becomes infinitely large, since $M = \tfrac{4}{3}\pi\rho R^3$ is a constant.

This case, $h < 0$, is called in celestial mechanics the "elliptic" case. The parabolic and hyperbolic cases correspond to $h = 0$ and $h > 0$, respectively. They would not predict any catastrophic singularity of the universe in the future, for in both cases R will steadily increase to infinity. But the expansion must have started from a singular initial state, if we follow the expansion backward until R becomes zero. The dynamical problem of the universal expansion introduces in every case a catastrophic beginning or ending of the world.

In this model of an expanding universe the transition to infinity, i.e., to a space of infinite volume, has to be done by exhausting the space with an infinite sequence of concentric spherical shells, all centered around the observer's place. The value of the potential, being proportional to the square of the radius of each shell, increases then in a definite way beyond all limits if the mean density of matter, taken over the finally infinite volume of space, is supposed to remain finite. Granted that, owing to the assumed absolute symmetry in the distribution of the substratum, no observer suffers an acceleration, the gravitational pull or "tear" would nevertheless become undetermined when an infinite world is envisaged.

This difficulty could be evaded by changing the assumption from which we started and considering the world as having a hierarchic structure as discussed in section 2. A finite regular value of the potential in the limit of an infinite volume of space can then be guaranteed. The hierarchic model which we had discussed formerly would have to be complemented only by attributing to it a general expansion.

The condition for regularity had been (2,2.5)

$$\frac{R_i}{R_{i-1}} \geqq \frac{1}{\gamma} N_i.$$

This inequality must remain valid for all times.

It is easily seen that, for instance, a speed of expansion proportional to the radius R_i of the system of rank i would satisfy this condition.

In such an expanding universe of hierarchic structure the limiting value of the potential would remain finite, and there would be no reason for giving up Newton's law of gravitation. But an expanding and contracting world introduces new singularities of a different character. They occur when the density tends to infinite values; and such singularities have to be envisaged also in an expanding world of hierarchic structure when Newton's law of gravitation is retained.

Various changes in the formulations of this law were considered when the cosmological considerations were still aiming at a static solution of the problem. One had considered, for instance, the possibility of a faint absorption of gravitation in space which would extinguish the gravitational effect of the infinitely distant masses. Such a correction of Newton's law would not remove the kind of singularity that appears in an expanding world. Another way to remove these singularities would be to supplement the expression for the potential $\phi = k[M(r)/r]$ in Poisson's equations by a term λr^2, which means introducing a repulsive force in addition to gravitation. The constant value of λ (λ has nothing to do with the wave length λ in the preceding section) would have to be adjusted to the actual conditions in the universe, so that for small parts of the universe (which, however, could still cover practically all the world so far explored) the field of the repulsive force would be very weak compared to the gravitational field of the world substratum.

This special correction will be discussed in more detail because it removes the singularities and because in the later relativistic treatment of the cosmological problem quite similar singularities are met by introducing a similar correcting term.

When the effect of such a cosmic repulsive force is considered, new solutions of the dynamical problem result. A contracting world may be brought to a standstill before R reaches the value zero and before ρ becomes infinite. For this purpose the constant h must be chosen <0, while λ has to be >0, and the value of M must be chosen appropriately.

If the actual universe corresponded to such a solution, we would have to assume that the expansion, which we observe at present, was preceded by a contraction which came to a standstill at regular values of R and ρ, respectively. When this happened, the tendency to contract was counterbalanced by the tendency to expand due to the repulsive force. After that the universe began to expand again, and we are experiencing this phase at present.

Such a solution of the cosmological problem avoids the occurrence of catastrophic singularities in the life of our universe. It must, however, be realized that this is achieved only by the *ad hoc* hypothesis of a cosmic repulsive force, for which, so far, no other proof or indication exists.

The classical treatment does not lead to a deeper understanding of the cosmological problem; in particular, no insight into the question is gained as to whether the universe might be finite or infinite. In the relativistic treatment, as we shall see, the equivalent correcting term, called the "cosmological constant," opens quite novel aspects of this fundamental question.

The uncertainty with regard to the possible finiteness of space could be eluded simply by attributing to space the metric properties of a finite, closed, non-Euclidean space. However, since the classical theory does not establish any link between geometry and gravitation, this would represent a further *ad hoc* hypothesis, which could be neither proved nor disproved, unless our empirical knowledge of the structure of the universe could be extended to distances not small compared with the radius of curvature of space. However, a priori no possibility exists of predicting the value of this radius of curvature. And even should experience prove the correctness of such a hypothesis, our understanding of the origin of the metric properties of space would not have been advanced.

4,3. Summary

Therefore, in summarizing the results so far obtained, we can state:

The tremendous extension of our knowledge of the distribution of matter in the universe has, as far as an advancement of the cosmological problem within the framework of classical physics is concerned, not brought us much nearer to a solution of the problem. The general expansion of the universe which has now to be considered and which demands a dynamical solution of the cosmological problem does not produce new complications in principle. On the contrary, it leads to an unambiguous definition of the potential function to satisfy the cosmological postulate and to describe dynamically the expansion; thus uncertainties which formerly remained are avoided. The expansion can be easily incorporated and explained on the ground of known problems in celestial mechanics. As a new feature, singularities appear which make it impossible to apply the dynamical solution for the expansion to the whole lifetime of the universe (see later, sec. 6,1).

The one difficulty of the cosmological problem, i.e., the fact that Newton's law of gravitation combined with the conception of an infinite Euclidean space, filled in its whole extent with matter of a finite average density, leads to infinite values of the gravitational potential, can be removed by assuming a hierarchic structure of the universe.

Indications of such a hierarchic structure actually exist. However, the survey of spiral systems does not contradict, so far, the assumption of a uniform distribution of matter, if the scale is chosen sufficiently large. This most simple assumption of uniform distribution has, therefore, until now been preferred for the model of the universe.

The hierarchic structure can be chosen so that the potential at every place remains regular when the transition to one finite space is performed, and Newton's law of gravitation may be retained without restriction for the whole universe. Also the phenomenon of expansion could easily be incorporated into the picture of a universe of hierarchic structure. But singularities, characteristic of the dynamical problem which describes the expansion, necessarily appear; and they can be removed only by a change of Poisson's fundamental equation for the potential. This means in every case an abandonment of Newton's law. Various amendments of Newton's law offer themselves. One is to allow for a repulsive force apart from gravitation, this repulsive force becoming effective only in dimensions large compared with the space so far explored, so that all local orbital motions of the whole galaxy are mainly ruled by gravitation. This amendment also permits the removal of the singularities arising from a catastrophic crash (the initial explosion of the universe) and offers in addition a solution in which matter is, on the average, uniformly distributed through space. No special hierarchic structure need be further envisaged.

Such amendments would in every case be based on an *ad hoc* hypothesis, made only for the special purpose of offering an acceptable solution of the cosmological problem free from singularities within the framework of Newton's mechanics. They do not essentially enrich our understanding of the whole problem. In the relativistic treatment which follows, similar amendments have also to be made *ad hoc;* but, by opening new aspects of the greatest importance, they obtain special significance.

The inability of Newton's theory to relate the forces in nature to the metric properties of space deprives it of the possibility of bringing the fundamental question of cosmology: "Is the universe finite or infinite?" within the reach of cosmological considerations.

4,4. Supplement

A. The infinite Euclidean space is filled uniformly by a "substratum" of spiral nebulae, which will be treated like the molecules of a gas distributed uniformly through space.

Let x_1, x_2, x_3, be the three Cartesian coordinates of a point referred to the observer's position in system S, and let x'_1, x'_2, x'_3, be its coordinates with reference to another system S', supposed to be in motion relative to S. The equations connecting the coordinates in S and S' are $x_i = a_i + x'_i$, where the a_i, functions of time, define the relative position of the observers in the systems S and S', respectively. The streaming of matter at a point (x_i) is described by velocity components v_i:

$$v_i(x, t) = \frac{da_i}{dt} + v'_i(x', t), \qquad i = 1, 2, 3 . \quad (a)$$

Every observer, placed at the zero point of his system of reference, S or S', considers himself at rest with regard to his immediate neighborhood. Consequently, $v_i(x, t) \equiv 0$ when all three coordinates $x_i = 0$, and equally $v'_i(x', t) \equiv 0$ when all $x' = 0$.

According to the cosmological postulate, each observer arrives at the same description of the world surrounding him. The density ρ and pressure p ascribed to the substratum at any point P_1 by an observer in S must be equal to the corresponding values ρ' and p' ascribed at the same moment by an observer in S' to the substratum at a point P_1' having, relative to the observer in S', the same coordinates as P relative to the observer in S. That means $\rho(x_i, t) \equiv \rho'(x_i', t)$ when $x_i = x_i'$. Similar equations hold for p and v_i. Since the values of the coordinates x_i may be arbitrarily chosen, it follows that $\rho = \rho(t)$ and $p = p(t)$, i.e., ρ and p are functions of the time t only. The velocity components v_i may still depend on the space coordinates x_i; for they are defined only relative to an arbitrarily chosen point and do not have the character of local properties of space.

It can easily be shown that the v_i must be linear functions of the space coordinates.

If in equation (a), we put $x_i' = 0$, we find

$$v_i(a_i, t) = \frac{d a_i}{d t};$$

and, since $x_i' = x_i - a_i$, this leads to the functional equations

$$v_i(x_i - a_i, t) = v_i(x_i, t) - v_i(a_i, t),$$

which must be satisfied for any value of a_i. This implies

$$v_i = a_{i1}x_1 + a_{i2}x_2 + a_{i3}x_3, \qquad i = 1, 2, 3 \quad (b)$$

where the nine functions a_{ik} are functions of time only.

A world expanding (or contracting) symmetrically in all directions corresponds to the special case $a_{11} = a_{22} = a_{33} = f(t)$, $a_{ik} = 0$, when $i \neq k$. If we set $x_i = x_i^0 R(t)$, where $R(t)$ is a function of t only and represents a universal scale factor, while the x_i^0 are arbitrary initial values of the coordinates of any point in a system S, the expansion (or contraction) is described by the equations:

$$v_i = \frac{d x_i}{d t} = x_i^0 \frac{dR}{dt} = x_i \frac{\dot{R}}{R}, \qquad \text{since } x_i^0 = \frac{x_i}{R(t)}; \qquad \left(\dot{R} = \frac{dR}{dt} \right).$$

We thus obtain a law of expansion corresponding to that derived from the observed red-shift of the spiral systems.

So far the considerations have been of a purely kinematical

nature. They have not taken account of the fact that all matter in the world is gravitating and, when moving, transports momentum. Conditions therefore have to be added which make sure that (1) all matter is preserved, i.e., no matter is lost, unless transformed into radiation; this last possibility will be, however, disregarded; thus the law of continuity must be satisfied; (2) the preservation of momentum is assured as well; (3) finally, an assumption about the law of gravitation has to be made; it will be here that matter gravitates according to Newton's law.

The following equations have thus to be satisfied:

$$\frac{\partial \rho}{\partial t} + \frac{\partial}{\partial x_i}(\rho\, v_i) = 0\,, \qquad \text{i.e., equation of continuity;} \quad (4,4.1)$$

$$\frac{\partial}{\partial t}(\rho\, v_i) = -\frac{\partial}{\partial x_k}(\rho\, v_i v_k) - \rho\frac{\partial \phi}{\partial x_i}\,, \qquad \begin{array}{c}\text{conservation of}\\ \text{momentum;}\end{array} \quad (4,4.2)$$

$$\frac{\partial^2 \phi}{\partial x_1^2} + \frac{\partial^2 \phi}{\partial x_2^2} + \frac{\partial^2 \phi}{\partial x_3^2} = 4\pi\,\kappa_0 \rho\,, \qquad \begin{array}{c}\text{Poisson's equation}\\ (\,\kappa_0 = \text{gravitational constant})\,.\end{array} \quad (4,4.3)$$

In these equations summation from 1 to 3 has to be performed for every index appearing twice in the same term. In equation (4,4.2) no term depending on a hydrostatic pressure p enters, for, since p has been assumed to be independent of the space coordinates, the term $\partial p / \partial x_i$ which otherwise would appear drops out. The first term on the right-hand side of equation (4,4.2) takes into account a possible influx of momentum due to streaming. About ϕ, the gravitational potential, special assumptions will be made shortly.

If we restrict the considerations to the case of an isotropic expansion, only the diagonal terms of the matrix (a_{ik}) are $\neq 0$; so $v_i = a_{ii}x_i$, and equation (4,4.1) yields

$$\rho + \rho\,(a_{11} + a_{22} + a_{33}) = 0\,, \qquad (4,4.4)$$

while equation (4,4.2), which, owing to formula (4,4.1), may be written in the form

$$\frac{\partial v_i}{\partial t} + v_k\frac{\partial v_i}{\partial x_k} = -\frac{\partial \phi}{\partial x_i}\,,$$

yields

$$\dot{a}_{ik}x_k + a_{ik}a_{ks}x_s = \frac{\partial \phi}{\partial x_i}\,, \qquad (4,4.5)$$

where, on the left-hand side, only the terms enter for which $i = k$. Hence

$$-\left[\frac{\partial^2\phi}{\partial x_1^2} + \frac{\partial^2\phi}{\partial x_2^2} + \frac{\partial^2\phi}{\partial x_3^2}\right] = \dot{a}_{11} + \dot{a}_{22} + \dot{a}_{33} + \Sigma i a_{ii}^2 = -4\pi\kappa_0\rho \, ;$$

and this, introduced into the differential equation of continuity,

$$\frac{d}{dt}\left(\frac{\dot{\rho}}{\rho}\right) = -(\dot{a}_{11} + \dot{a}_{22} + \dot{a}_{33}) \, ,$$

gives

$$\frac{d}{dt}\left(\frac{\dot{\rho}}{\rho}\right) = 4\pi\kappa_0\rho + \Sigma i a_{ii}^2 \, , \qquad \text{which is} \quad \geq 0 \, . \qquad (4,4.6)$$

This equation is of great and general significance. It is incompatible with the assumption of a static world with $\rho = $ constant > 0; for $\dot{\rho} = 0$ and $a_{ii} = 0$ entails $\rho \equiv 0$. It is thus impossible to imagine a world in which gravitating matter remains at rest; it is also impossible to let matter stream in such a way that the density remains constant.

B. The Gravitational Potential

According to the cosmological postulate, every observer placed at the origin of an admissible system S considers himself at rest relative to his immediate neighborhood. Hence the field characterized by the potential ϕ must not give rise to any forces tending to displace the observer. It must be realized that we are only considering the "inertial" field arising from the uniformly distributed world substratum which is not responsible for the movements of any free particles. The study of local orbital motions at any special place—for instance, in the immediate neighborhood of the sun or within a star-system—owing to a special distribution of the gravitating matter is not considered. The potential ϕ must, in consequence, have such a form that its derivatives, $\partial\phi/\partial x_i$, $(i = 1, 2, 3)$, vanish for $x_i = 0$, thus producing no resultant force; i.e., the representation of ϕ by power series in x_i must begin with terms of at least the second order. We shall choose ϕ as a quadratic form of the coordinates—in particular, in view of the isotropic character of the field, as

$$\phi(x_i) = A(t)(x_1^2 + x_2^2 + x_3^2) \, . \qquad (4,4.7)$$

Then the potential difference between two points P_1 and P_2 will depend only on their relative distance. Equation (4,4.3) yields $A = \frac{2}{3}\pi\kappa_0\rho(t)$, if only the field arising from gravitating influence of the matter inside a sphere around P_1 reaching to P_2 is considered. However, it must be understood that, in general, a further term λr^2 with an arbitrary factor λ may be added. The field arising from such a term will be considered later. From the special formulation of the potential it follows that its value at an arbitrary point P_2 depends only on its distance from the origin P_1 of the system. Therefore, if we produce a sphere around P_1 reaching to P_2, each point on the sphere may be regarded as screened off from all gravitational influence of the masses lying outside the sphere.

C. The Differential Equation for the Scale Factor R(t)

It is obvious that our description of the phenomenon of the universal expansion (or contraction) must be independent of the choice of the distance $r = r_0R(t)$ of P_2. We must be able to describe it in terms of the changes of the scale factor $R(t)$ only. While the volume of a sphere contracts or expands, the density ρ of matter contained in it changes continuously but has the same value at all points inside.

Introducing the value $r = r_0R(t)$, into the differential equation

$$\frac{d^2 r}{dt^2} = \frac{-\kappa_0 M(r)}{r^2} \qquad (4,4.8)$$

for a point at the surface of a sphere containing the mass $M(r)$, the differential equation for R follows:

$$\frac{d^2R}{dt^2} = \frac{-\kappa_0 M}{R^2}, \qquad (4.4.9)$$

where $M = \frac{4}{3}\pi\rho R^3$ is a constant; it is the mass contained in a volume of unit radius $r_0 = 1$. The mass remains unchanged within any expanding closed surface.

The differential equation immediately yields the integral

$$\tfrac{1}{2}\dot{R}^2 = \frac{\kappa_0 M}{R} + h, \qquad (4,4.10)$$

where the constant of integration h may be negative, positive, or zero. Accordingly, we obtain various types of expansions which may be called "elliptic," "hyperbolic," and "parabolic," since the character of the solutions closely resembles solutions obtained in the theory of rectilinear motion of two gravitating bodies.

In the preceding general discussion of the problem it has been shown that the dynamical problem of an expanding or contracting universe always gives rise to singularities when, for $R = 0$, ρ becomes infinite. The singularities can be removed by amending Poisson's equation for the gravitational potential by an additional term.

D. The Cosmological Constant

If ϕ is supplemented by a term with the new constant λ,

$$\phi = \frac{\kappa_0 M\,(r)}{r} + \frac{\lambda}{6}\,r^2\,, \qquad (4,4.11)$$

the differential equation for the scale factor $R(t)$, all other assumptions left as before, becomes

$$\frac{d^2R}{dt^2} = \frac{\kappa_0 M}{R^2} + \frac{\lambda}{3}\,R \qquad (4,4.12)$$

and yields the integral

$$\tfrac{1}{2}\dot{R}^2 = \frac{\kappa_0 M}{R} + \frac{\lambda}{6}\,R^2 + h\,. \qquad (4,4.13)$$

The singularity $\rho \to \infty$, which formerly occurred for $R = 0$, does not appear again, if λ is chosen > 0, $h < 0$, and M is appropriately small.

In this case the equation

$$\frac{\kappa_0 M}{R} + \frac{\lambda}{6}\,R^2 + h = 0\,, \qquad (4,4.14)$$

corresponding to the case $\dot{R} = 0$, yields two solutions, for, say, $R = R_1$ and $R = R_2$, where $R_1 < R_2$. For $R > R_2$, real solutions exist, for which R, coming from infinity—corresponding thus to a contracting universe—approaches the limit $R = R_2$, where $\dot{R}$ becomes zero. The contraction comes to a standstill for a regular finite value of ρ and changes into an expansion; at this turning point the repulsive force just balances the attraction due to the general gravitational field of the substratum.

According as λ and h are chosen $\gtrless 0$ and the appropriate values for M are introduced, a great variety of solutions results. But only the one class, just mentioned, corresponding to $\lambda > 0$, $h < 0$, and M—that means also ρ—small, yields solutions which are free from singularities and hence do not necessitate the assumption that the universe has to pass through a critical phase. But it must be emphasized that such regular solutions are obtainable only by introducing the *ad hoc* hypothesis of a new and otherwise not traceable repulsive force in the universe.

5. The Relativistic Treatment of the Cosmological Problem

5,1. General Introduction

The relativistic approach to the cosmological problem proceeds apparently on such absolutely different lines that to arrive at the same final equations which describe the expansion may appear rather surprising. It is, however, no more surprising than the fact that the field equations of the general theory of relativity, which are framed in a completely different set of conceptions, lead in the first approximation to the same laws of motion as Newton's theory.

The fundamentally new feature in the relativistic approach is the initial establishment of a correspondence between the phenomenon of gravitation in space and the geometry of physical space. The universal gravitational field is no longer placed in an empty space endowed from some mysterious source with the special laws of Euclidean geometry. Metric field and gravitational field become different expressions of one and the same fact, namely that in a space filled with matter and energy individual bodies are no longer freely movable, owing to the phenomenon of gravitation. In the theory of relativity they move according to purely geometric principles through a space of continuously changing metrical properties;

in the old interpretation, gravitational forces, acting upon them in a space of presupposed metrical properties, moved them. In the classical theory there was thus a duality of gravitational field and geometrical properties of space, without any possibility of linking them explicitly together. For the geometry of space was not related to any physical phenomena, although the axioms of Euclidean geometry are actually based on a physical assumption, namely that all bodies are freely movable without change of size and shape. Now both fields are merged into one field.

The metric field, determined by the distribution of all gravitating matter in the universe, will be generally of a non-Euclidean character, and it is just this deviation from Euclidean geometry which we interpret as gravitation.

Since the gravitational field changes continuously when we proceed in space from one place to another, it must be characterizable by the differential expression of the line element which, according to Riemann's foundation of geometry, defines the metrics of space. Thus the problem of gravitation—in our case specifically the cosmological problem—is reduced to the question: What is the expression for the line element at an arbitrary point of an expanding universe filled in its whole extent with matter, when gravitation is geometrized along the lines followed by the general theory of relativity? In particular, since a closed finite space is characterized by a special expression for the line element, do we get a line element which indicates that our universe is finite?

We see that the geometrization of gravitation throws light immediately upon this fundamental question of cosmology. For this reason the relativistic treatment of the cosmological problem proves much more fruitful, even though, as we shall see, it does not yet give us a conclusive answer to this fundamental question.

Before entering into the specific cosmological problems, we shall investigate briefly the empirical background which justifies such a geometrization of gravitation.

That we are accustomed to use rectangular Cartesian coordinates in which to formulate the laws of motion is merely a matter of convenience. Instead of fixing the position of a mass point in space by its three coordinates x, y, z, with reference to three straight, orthogonal axes, we can equally well use more general Gaussian co-ordinates x_i, $i = 1, 2, 3$. Expressed in such general coordinates, the distance, ds, between two infinitely near points takes the form:

$$d s^2 = \sum_1^3 g_{\mu\nu} d x^\mu d x^\nu , \qquad (1a)$$

where the coefficients $g_{\mu\nu}$ are functions of the coordinates x_i. In special cases this expression may take the simple form

$$d s^2 = d x_1^2 + d x_2^2 + d x_3^2 . \qquad (1b)$$

If, in particular, Euclidean geometry is assumed to prevail in space, i.e., if the law of Pythagoras holds good for finite dimensions, then the finite distance s of any point of space from the observer, assumed to be at $x_1 = x_2 = x_3 = 0$, is given by

$$s^2 = x_1^2 + x_2^2 + x_3^2 . \qquad (1c)$$

To have based the laws of motion on the more general but also much more complicated expression $(1a)$ of the line element would have meant a useless and unproductive complication of all formulae of physics unless the deviations from the Euclidean form $(1b)$ found a natural physical interpretation. That could not be achieved as long as the metrical properties of space were considered to be of purely mathematical character. However, by superimposing a gravitational field upon a space of presupposed Euclidean geometry, as Newton's theory had to do, certain features were impressed upon the theory of motions which, from the outset, were felt to be extremely unsatisfactory in principle. For instance, the fact that all observations refer truly to bodies moving *relatively* to each other could not be incorporated into the theory, except in the special case of uniform rectilinear motions. Furthermore, empty space had to be made responsible for the appearance of forces like the centrifugal forces, although they are indistinguishable from gravitational forces on account of the apparently absolute equality of inert and heavy mass for every body.

It is the great discovery of Einstein's general theory of relativity that such fundamental shortcomings of the classical theory of motion can be eliminated by a geometrization of the gravitational field, i.e., by reducing the physical phenomenon of gravitation to the metrical properties of space.

Let us, for instance, consider an infinitely small volume of space near the earth's surface. Owing to the equality of inert and heavy mass, every falling body experiences the same acceleration in the gravitational field of the earth. In this small volume the effect of gravitation can be "transformed away" by a suitable transformation of the coordinates which cancels the acceleration; thus in this restricted area of space Euclidean geometry would be restored for an observer falling with all other bodies.

For a finite volume of space no such transformation is generally available. Owing to the change of the gravitational acceleration with increasing distance from the earth and owing to the gravitation of sun, moon, and planets, distant stars, etc., the gravitational acceleration cannot be simply transformed away. In geometrical terms this means that for an infinitely small volume of space the reduction of the general expression of the line element (1a) to the simpler Euclidean form (1b) is always possible—for finite dimensions, however, only if we presuppose the validity of Euclidean geometry in space.

Thus the expression for the line element,

$$d s^2 = \Sigma\, g_{\mu\nu} d\, x^\mu d\, x^\nu\,,$$

can be used as an indicator of the existence of a gravitational field; the gravitational and metrical fields of space are then merged into one. The relativity of all motions may be postulated; the equality of inert and heavy mass, so far only empirically ascertained, becomes incorporated in the definition of mass; the unnatural duality in the definition of mass is avoided. Centrifugal phenomena, practically indistinguishable from gravitational phenomena, become special manifestations of gravitation.

These very general remarks on the principles of general relativity need some essential amendments. First, all considerations have to be based on the four-dimensional line element

$$d s^2 = \sum_1^4 g_{\mu\nu} d x^\mu d x^\nu,$$ (5,1.1)

including a fourth parameter, x^4, which may have the character of a pure time variable.

Second, the metric field of space is determined not only by the distribution of matter in space but also by the distribution and density of energy in its various forms, such as radiation, momentum, etc. Already the special theory of relativity has taught that energy is endowed with inertia, and the identity of the inert and heavy mass of every body which now becomes a cornerstone of the general theory of relativity thus attributes gravitation to all forms of energy.

In order to establish *de facto* the relationship between geometry and gravitation, the line element ds in its general form defined by ten functions $g_{\mu\nu}(\mu, \nu = 1, 2, 3, 4)$, $g_{\mu\nu} = g_{\nu\mu}$, has to be connected with the ten components of the mass-energy tensor which describes the distribution of all quantities on which the gravitational field depends. The gravitational field is no longer described by one function of scalar nature, the potential ϕ. It is described now by a tensor with ten components; the ten functions $g_{\mu\nu}$ replace the one potential ϕ. This makes the relativistic treatment of all dynamical problems, to which our cosmological problem also belongs, extremely intricate.

The cosmological problem approached within this new frame of ideas asks: What is the expression for the line element in a universe which is uniformly filled with matter while expanding symmetrically in all directions? Can we draw from the form of the line element general conclusions with regard to the structure of the world, in particular with regard to the alternative: finite or infinite world?

To obtain an answer to these questions, we must start from

The Relativistic Treatment of the Cosmological Problem

the most general expression for the line element, i.e. (5,1.1), and apply, step by step, the various special assumptions which we assume to hold good in the universe.

First, since all gravitating matter is supposed to be uniformly distributed over the available space and since, moreover, the appearance of the universe is to be the same for every observer, the problem will be solved when the expression for ds is known for one and only one observer in an arbitrary position x^1, x^2, x^3, x^4, of the universe. Second, we shall assume that every observer measures his proper time t with a clock at rest in his position; his own coordinates, in a system of reference fixed to his position as origin, at any arbitrary moment will be: $x^1 = x^2 = x^3 = 0$, $x^4 = t$. Consequently, without any sacrifice in generality, we may start from the more simple expression:

$$d s^2 = dt^2 + \sum_1^3 g_{\mu\nu} d x^\mu d x^\nu . \qquad (5,1.2)$$

The coefficients $g_{\mu\nu}$ in the last equation might still be functions of all four variables x^1, x^2, x^3, t. Third, the observed expansion is the same in all directions; it will consequently affect the scale but not the metrical properties of space in the course of time. Changes of the $g_{\mu\nu}$ in time must therefore be describable in terms of a changing scale factor $R(t)$ acting upon all three coordinates in the same sense. We may consequently split the scale factor off and put

and
$$g_{\mu\nu} = -R^2 (t) \gamma_{\mu\nu}$$
$$d s^2 = dt^2 - R^2 (t) d\sigma^2 ,$$
where
$$d\sigma^2 = \sum_1^3 \gamma_{\mu\nu} d x^\mu d x^\nu \qquad (5,1.3)$$

depends on the three space coordinates only.

So far we have made use only of the cosmological postulate, of special assumptions concerning the distribution of matter in space, and of the character of the general expansion; these assumptions reduce the expression for the line element to the most simple possible form, given above. The field equations of the

544

theory of relativity, i.e., the equations which actually link the metrical functions $g_{\mu\nu}$ to the gravitational field, have not yet been used. These field equations, which will be considered more closely in the supplement to this section, can be considerably simplified. Of the ten components of the mass-energy tensor which defines the gravitational (inertial) field, we can disregard all but one component, namely, the one which enters also into the classical theory. It is the density ρ of the cosmic matter in space. This is justified because the gravitational field arising from the radiation filling the interstellar space is negligibly faint compared with the field produced by solidified matter. The same is true of the field originating from the energy manifested in the irregular random motions of the cosmic matter, i.e., the spiral systems. Thus the only quantity which is made responsible for the observed (inertial) field is the density of matter which we find condensed to stars in the star-systems. With these restrictions, the field equations take a particularly simple form, from which very general and, for the cosmological problem, essential conclusions can be drawn. One equation assumes the simple form

$$3\frac{\ddot{R}}{R} = \tfrac{1}{2}x\rho\,,$$

where $\ddot{R}$ is written for d^2R/dt^2. This equation does not contain the metrical functions $\gamma_{\mu\nu}$ at all; and, since the left-hand side depends on the time t only, we may conclude immediately that the right-hand side, i.e., the density ρ, is also a function of the time t only.

We arrive thus, as in the classical case, at the result that the matter filling the universe must be uniformly distributed, the density being a function of t only. In the remaining field equations, which contain the metrical functions $\gamma_{\mu\nu}$ and their derivatives, the latter appear cloaked in a tensor $G^*_{\mu\nu}$ in such a way that the conclusion $G^*_{\mu\nu}/\gamma_{\mu\nu} = $ const. may be drawn immediately. This equation, however, is known as the equation to be satisfied in a space of constant curvature. Whether the curvature is positive and space consequently closed and finite, or negative, or zero (in which latter cases space would be of infinite volume) remains still an open question. The answer would depend on a final solution of the field equations, which, however, is not to be expected at present. For, apart from the expansion term $\dot{R} = dR/dt$, the equations contain also terms containing $\ddot{R} = d^2R/dt^2$, i.e., terms depending on the acceleration (or deceleration) of the expansion, about which so far nothing is known.

But the relativistic treatment promises to give us a more direct insight into the one alternative so fundamental in the cosmological problem, namely, that of the finiteness or infinity of space. Apart from these general conclusions (see supplement), one integral can be derived from the field equations which, in appropriately chosen constants, takes the same form as the integral obtained for the scale factor R in the classical case; it reads:

$$\tfrac{1}{2}\dot{R}^2 = \frac{\kappa_0 M}{R} + \text{const.}$$

The only difference is that in the classical treatment the constant on the right-hand side appeared as, apart from numerical factors, an unspecified constant of integration, h, the energy constant. In the relativistic treatment the most interesting quantity is a factor ϵ, which can assume any of the three values

$$\epsilon = \left\{ \begin{array}{c} +1 \\ 0 \\ -1 \end{array} \right. ,$$

which decide whether space is finite or infinite.

The fact that in both treatments we obtain the same equation for R is an interesting, but by no means surprising, result. The expansion of the universe is in both cases explained according to the rules which govern regular motions; in the classical case it is reduced to a known problem in celestial mechanics, in the relativistic treatment the field equations must lead in the first approximation to laws of motion similar to those of classical mechanics. Thus both treatments could be expected to yield similar results in the first approximation.

From this point on the discussion follows, therefore, the same line as in the classical case. We have to distinguish again between solutions of the elliptic, parabolic, and hyperbolic types, the first class being of a periodic character. Also the same singularities appear when R, tending to zero, causes ρ to increase beyond all finite limits.

If we summarize the results obtained so far, we see that the relativistic treatment does not produce, as far as the representation of observable facts is concerned, results which extend beyond those obtainable also by classical methods. However, there can be no doubt that the relativistic approach promises to bring us much nearer to the roots of the whole problem, as soon as the empirical data are complete enough to make possible an exhaustive treatment of the field equations. For the equations lead immediately toward an answer to the fundamental question: Is space finite or infinite? This gain is due to the geometrization of gravitation, which automatically includes a geometrization of the cosmological problem, too.

The singularities which enter as a consequence of the dynamical character of the solution may be treated in the relativistic case in a manner very similar to the classical case. In the latter they were removed by adding to Poisson's equations for the gravitational potential a term with a new arbitrary constant λ, now called the "cosmological constant." Such a term means the introduction of a repulsive force which has to counterbalance the effect of gravitation.

Although the removal of the singularities proceeds in the relativistic treatment along similar lines, there are distinguishing features of a very essential character.

Both the amendment of Poisson's equation and that of the field equations in the relativistic theory were originally introduced not to remove singularities which appear in the *dynamical* problem of an expanding universe but to remove the difficulties which were encountered when the aim was to establish the existence of a static, infinite world. Einstein's first solution of the cosmological problem in 1917 still aimed at a static solution. The general red-shift of the spectral lines of distant star-systems, i.e., the expansion of the universe, was not yet known. In both the classical and the relativistic case the static solution encounters difficulties which make it necessary to introduce additional terms into the equations connecting the potential ϕ and the metrical functions $g_{\mu\nu}$. Now a static world is no longer asked for. The amendment has to remove the singularities entering the dynamical solution. A λ-term, supplementing Poisson's equation of the Newtonian potential, added to the left-hand side of the relativistic field equations serves this purpose. The introduction of such a term is in both cases an *ad hoc* hypothesis. In the relativistic treatment, however, the new term can be incorporated in a more natural way.

Poisson's equation, which states that the divergence of the gravitational force through a closed surface is proportional to the density of the enclosed gravitating matter, loses, by the addition of the λ-term, the possibility of such an interpretation. The field equations of general relativity, on the other hand, in which the components of the matter-energy tensor are set proportional to a differential expression of the second order in the metrical functions $g_{\mu\nu}$, $(\mu, \nu = 1, 2, 3, 4)$, and their derivatives, permit, without restricting their applicability, the addition of a term $\lambda g_{\mu\nu}$; this drops out when the divergence is calculated. Thus a cosmological constant λ may be introduced into the relativistic treatment without any fundamental influence upon the theory. In addition, the term has a purely metrical meaning, apart from representing a repulsive force counteracting in large dimensions the action of the universal gravitational field. From the metrical point of view the λ-term enables us to calibrate all measurements of lengths in a finite universe, when both λ and ϵ are chosen as > 0. The constant radius of curvature in this model of the universe defines the scale to which the length of the meter as the unit of length may be referred. And, since the field equations in their most general form definitely require the addition of a λ-term, its introduction into the theory of relativity is claimed by some—for instance, by Eddington—to be an essential and indispensable feature of the theory. On the other hand, it cannot be denied that, owing to the fact that no further evidence supports the introduction of such a term, the beauty of the relativistic theory is to a certain extent disturbed by such a term. Without the urge of the cosmological considerations, the relativistic theory could rightly boast of being able to account for all observed facts without introducing any constants apart from the known fundamental constants of nature, i.e., the gravitational constant κ and the value of the light velocity c. The λ-term upsets this claim. This and the fact that the inclusion of the λ-term actually does not yet give a final solution of the cosmological problem have induced Einstein and others to drop this term completely. They thus sacrifice, perhaps, very attractive solutions of the cosmological problem, but they free the theory from not sufficiently well-founded assumptions.

When, however, the λ-term is accepted, the integral derived from the field equations (see below) takes a new form; in appropriate units it reads,

$$\tfrac{1}{2}\dot{R}^2 = \frac{\kappa_0 M}{R} + \frac{\lambda}{6} R^2 c^2 - \tfrac{1}{2}\epsilon c^2, \qquad (5,1.4)$$

where c is the velocity of light and $\epsilon = \begin{cases} +1 \\ 0 \\ -1 \end{cases}$ is the above-mentioned factor the value of which decides whether the expression for the line element at any position in the universe corresponds to that of a closed finite space of positive curvature ($\epsilon = +1$), or of a Euclidean flat space ($\epsilon = 0$), or of a hyperbolic space ($\epsilon = -1$). The new form of the integral, which was also obtained in the classical case, after amending Poisson's equations by a λ-term, offers solutions which are free from the disturbing singularities encountered without the λ-term. When λ is chosen >0, $\epsilon = +1$, and when also the value of M is chosen appropriately, solutions result which no longer lead to a catastrophic collapse of the universe for $R = 0$. In addition, solutions are obtained, when $\epsilon = +1$, which yield a closed and finite universe.

5,2. Supplement

In the general theory of relativity the scalar potential ϕ of the gravitational field is replaced by the ten functions $g_{\mu\nu}$ in the expression of the line element

$$d s^2 = \sum_{1}^{4} g_{\mu\nu} d x^\mu d x^\nu .$$

The density ρ of the matter filling space has to be replaced by the ten components of the energy-momentum tensor

$$T_{\mu\nu} = \rho \sum_{\alpha\beta} g_{\mu\alpha} g_{\nu\beta} \frac{d x^\alpha}{d s} \frac{d x^\beta}{d s} , \qquad (5,2.1)$$

for not only matter but all forms of energy contribute to establishing the universal inertial field.

Poisson's equations for the potential ϕ are replaced by the field equations of the theory of relativity, which postulate that

$$G^{\mu\nu} - \tfrac{1}{2} g^{\mu\nu} G = - \kappa T^{\mu\nu} .$$

On the left-hand side the metric field is represented by the components of a tensor, containing the $g_{\mu\nu}$ and their first and second derivatives,

$$G^{\mu\nu} - \tfrac{1}{2} g^{\mu\nu} G ,$$

which satisfy the equation

$$\text{div } (G^{\mu\nu} - \tfrac{1}{2} g^{\mu\nu} G) = 0 .$$

The $G^{\mu\nu}$ are the components of a tensor called the Riemann-Christoffel tensor; G is the invariant derived from it: κ is a constant connected with the constant of gravitation κ_0 by $\kappa = (8\pi/c^2) \kappa_0$; c is the velocity of light. The field equations may also be written in the form:

$$G_{\mu\nu} = - \kappa [T_{\mu\nu} - \tfrac{1}{2} g_{\mu\nu} T] .$$

These equations lead in the first approximation to Newton's laws of motion in a gravitational field. Terms of higher order predict the correct amount of the advance of the perihelion of Mercury which could not be explained in the classical theory.

Into these field equations we have to introduce the empirical facts furnished by astronomical observations and have to decide which values must be given to the components of the energy-momentum tensor in a world over which the matter is spread at every moment with a constant mean density ρ. It is beyond doubt that in the first approximation the gravitational field arising from the radiation filling stellar and intergalactic space may be completely disregarded; also the field arising from the momentum of the streaming matter need not be considered. Thus, apart from the component $T^{44} = \rho$, none of the ten components of the energy-momentum tensor

$$T^{\mu\nu} = \rho u^\mu u^\nu \qquad \mu, \nu = 1, 2, 3 ,$$

where

$$u^i = \frac{d x^i}{d s} ,$$

needs to be taken into account. Only $T^{44} = \rho$, the density of the matter, remains, as in the classical case.

But the density ρ may still be a function of the time $x_4 = t$ and of the spatial coordinates; i.e., $\rho = \rho(x^i, t)$.

With regard to the line element, (5,1.1), considerable simplifications are also possible, without reducing the generality of the following considerations. We may treat the world as a "3 + 1 dimensional" manifold, splitting off the "time-term," and write

$$d s^2 = d \tau^2 + \sum_1^3 g_{\mu\nu} d x^\mu d x^\nu , \qquad (5,2.2)$$

where in common units

$$d \tau = c \, d t .$$

The $g_{\mu\nu}$ may still be functions of all four coordinates: x_1, x_2, x_3, t. This line element allows a further reduction to a more simple form when we consider that the observed expansion is the same in all directions. Every spatial direction is consequently absolutely equivalent to any other; in other words, space is *isotropic*. This reduces the $g_{\mu\nu}$ to the special form:

$$g_{\mu\nu} = -R^2 (\tau) \, \gamma_{\mu\nu} ,$$

and the line element takes the final form:

$$d s^2 = d \tau^2 - R^2 (\tau) - d \sigma^2 , \qquad (5,2.3)$$

where

$$d \sigma^2 = \sum_1^3 \gamma_{\mu\nu} d x^\mu d x^\nu ,$$

and R, representing a universal scale factor, is still an unknown function of the time τ. Since every observer is expected to experience a completely identical picture of the universe, the knowledge of ds at only one arbitrary point of the universe is sufficient for a complete solution of the cosmological problem.

The differential equation for R, which describes the expansion, is obtained by introducing the special values of $T^{\mu\nu}$ and $g_{\mu\nu}$, just derived, into the field equations, after having calculated the values of the $G^{\mu\nu}$. Two differential equations of second order in R result; they read

$$G^*_{\mu\nu} - 2\dot{R}^2\gamma_{\mu\nu} - R\ddot{R}\gamma_{\mu\nu} = -\tfrac{1}{2}\,\kappa R^2\rho\gamma_{\mu\nu} \qquad (5,2.4)$$

and, for the term for which $\mu = \nu = 4$,

$$3\,\frac{\ddot{R}}{R} = -\tfrac{1}{2}\,\kappa\rho\,, \qquad (5,2.5)$$

where $\dot{R}$ and $\ddot{R}$ denote the first and second derivatives of R with regard to the time; $G^*_{\mu\nu}$ denotes a tensor resulting from contraction of the Riemann-Christoffel tensor with regard to the three spatial coordinates only.

From these equations two essential conclusions can immediately be drawn. Since R is a function of τ only, it follows from equation (5,2.5) that the right-hand side also must be a function of τ only; hence $\rho = \rho(\tau)$, the density of the matter, is at any fixed moment the same at every point of space. This is the same result as that which in the classical treatment in section 4,2 followed from Poisson's equations for the potential ϕ.

Moreover, if we divide equation (5,2.4) by $\gamma_{\mu\nu}$, we obtain

$$\frac{G^*_{\mu\nu}}{\gamma_{\mu\nu}} = 2\dot{R}^2 + R\ddot{R} - \tfrac{1}{2}\,\kappa R^2\rho\,. \qquad (5,2.6)$$

Here the left-hand side is a function of the space coordinates x^i ($i = 1, 2, 3$) only, the right-hand side of the time t only; from that we conclude

$$\frac{G^*_{\mu\nu}}{\gamma_{\mu\nu}} = \text{const.}$$

This equation, written in its general form,

$$G^*_{\mu\nu} + (n-1)\,C\,g_{\mu\nu} = 0 \qquad (5,2.7)$$

is the necessary condition for a constant curvature of space; C is a constant; n, the number of dimensions, is in our case equal to 3.

Therefore, under the special assumptions concerning the distribution of matter and the cosmological postulate, the field equations yield the preliminary result that the three-dimensional space must be one of constant curvature. We can write the last equation in the form

$$G^*_{\mu\nu} + 2\epsilon\, C'\, g_{\mu\nu} = 0 \ , \qquad (5,2.8)$$

C' representing a positive constant; then $\epsilon = +1$ corresponds to a spherical or elliptic space of finite content, $\epsilon = 0$ to infinite Euclidean space, and $\epsilon = -1$ to infinite hyperbolic space. No indication is so far given as to which of these three cases our universe actually belongs to.

The law of expansion, giving R as a function of τ, follows by the elimination of the quantities depending on the spatial coordinates from (5,2.4–8), that is, by eliminating $G^*_{\mu\nu}$ and $\gamma_{\mu\nu}$. Introducing into equation (5,2.4) the value of $G^*_{\mu\nu}/\gamma_{\mu\nu}$ from equation (5,2.8) and replacing from equation (5,2.5) the term $R\ddot{R}$ by $-\frac{1}{6}kR^2\rho$, we obtain the equation

$$\dot{R}^2 = \tfrac{1}{3} k\,\rho R^2 - \epsilon - C' \qquad (5,2.9)$$

Differentiating this equation and replacing the terms containing $\ddot{R}$ again from equation (5,2.2) produces

$$\frac{\dot{\rho}}{\rho} + 3\frac{\dot{R}}{R} = 0 \ .$$

This last equation can be written

$$d \log (\rho R^3) = 0$$

and yields the integral

$$\tfrac{4}{3}\,\pi \rho R^3 = M = \text{const.} \qquad (5,2.10)$$

The arbitrary numerical factor $4\pi/3$ has been attached to make the constant M equal to the corresponding constant M which resulted in the classical case from the corresponding integration of the differential equation of the scale factor R (see p. 22).

The integral expresses the conservation of matter and energy. If we introduce this value into equation (5,2.4), we obtain

$$\dot{R}^2 = \frac{\kappa M}{4\pi R} - \epsilon\, C' \,.$$

To reduce this equation to the common units used in the differential equation in the classical case (sec. 4,2), $d\tau$ has to be replaced by cdt and κ by $(8\pi/c^2)\kappa_0$. The resulting equation,

$$\tfrac{1}{2}\dot{R}^2 = \frac{\kappa_0 M}{R} - \tfrac{1}{2}\epsilon c^2 C' \,, \tag{5,2.11}$$

is the same as the equations obtained in the discussion based on Newton's law of gravitation. The further discussion proceeds, therefore, on parallel lines to those of art. 5,1.

5,3. The Cosmological Constant

If we definitely wish to free the model of our world from any possible catastrophe, the singularity which appears in the solutions of the last question can be removed in a similar way by a λ-term. The tensor $G^{\mu\nu} - \tfrac{1}{2}g^{\mu\nu}G$, which enters into the field equations, may be complemented by an additional term of the form $\lambda g^{\mu\nu}$.

λ is the "cosmological constant" originally introduced by Einstein into the problem and later abandoned. The field equations now read

$$G^{\mu\nu} - \tfrac{1}{2}g^{\mu\nu}G + \lambda g^{\mu\nu} = -\kappa T^{\mu\nu}, \tag{5,3.1}$$

and equations (5,2.4) and (5,2.5) have to be replaced by

$$G^*_{\mu\nu} - 2\dot{R}^2\gamma_{\mu\nu} - R\ddot{R}\gamma_{\mu\nu} + \lambda R^2\gamma_{\mu\nu} = -\tfrac{1}{2}\kappa\rho R^2\gamma_{\mu\nu} \,, \tag{5,3.2}$$

$$3\frac{\ddot{R}}{R} - \lambda = -\tfrac{1}{2}\kappa\rho \,. \tag{5,3.3}$$

The new differential equation for R reads:

$$\dot{R}^2 = \tfrac{1}{3}\kappa\rho R^2 + \frac{\lambda}{3}R^2 - \epsilon\, C \,, \tag{5,3.4}$$

which, after again introducing cdt for $d\tau$ and the gravitational constant, κ_0, and after dropping superfluous constants, gives us

$$\tfrac{1}{2}\dot{R}^2 = \frac{\kappa_0 M}{R} + \frac{\lambda}{6}R^2 c^2 - \tfrac{1}{2}\epsilon c^2 \,, \text{ i.e. (5,1.4)} \,.$$

While formerly $\frac{4}{3}\pi\rho R^3 = M = $ const. inevitably involved a singularity of $\rho \to \infty$ for $R = 0$, the solutions of the new equation no longer necessarily have a singularity of this kind, either in the past or in the future. The amended integral offers nine separate classes of solutions corresponding to the possible combinations of $\lambda \gtrless 0$ and $\epsilon \gtrless 0$; among them are, for $\lambda > 0$, $\epsilon = +1$, and M appropriately chosen, solutions free of the disturbing singularities.

5,4. General Remarks concerning Singularities in the Cosmological Problem

As far as the representation of observable facts is concerned, neither the classical nor the relativistic treatment can claim superiority over the other for the cosmological problem. Both lead to the same main results: that a static world filled with gravitating matter is not possible and that an isotropic expansion is compatible with the cosmological postulate and describable in terms of their laws of motion; both treatments lead in the first approximation to the same equation describing the expansion.

The only disturbing factor is the appearance of singularities which make it impossible to apply the solution to the whole lifetime of the universe. In the classical case, such singularities can be avoided only by a departure from Newton's law of gravitation—a very serious step; in the relativistic case the theory is flexible enough to allow the necessary amendments. However, they introduce into the laws of nature a new universal constant, the cosmological constant λ, for which there appears otherwise no necessity.

We must ask therefore, since the singularities arise when we get excessively large values of the density ρ, whether our fear of excessive densities, the opposite to the *horror vacui*, is justified and not perhaps the last reminder of a subconscious yearning for a harmonious universe. If all observable facts in our present universe were intelligible, we would not have any reason to suspect that the universe has gone through conditions essentially different in the past from those we observe at present, conditions which we would call "singular." But the universe definitely reveals features which we are so far unable to explain within the frame of the actually accepted laws of nature.

The matter in space from which the value of the density ρ in all preceding considerations has been derived appears condensed in radiating stars and in clouds of a widely dispersed dark matter. In the stars the matter is highly condensed at a high temperature; in the clouds both temperature and density are very low. But both forms of matter contain atoms of large atomic weight, i.e., atoms heavier than hydrogen and helium.

According to atomic physics, it seems securely established that the atoms of all chemical elements are built up from protons, neutrons, and electrons and that in the building-up of the various elements large amounts of atomic energy are freed. The latter fact offers a solution to the problem of the source of energy of stellar radiation. In particular, the transformation of hydrogen into helium proves to be a source which, at the temperature prevailing in the central cores of many stars, yields sufficient energy to supply their radiation. Hydrogen becomes the prevalent element in the early stages of stellar life.

In accordance with this fact, the theory of the internal structure of stars was forced to assume a high percentage of hydrogen in stellar matter to remove discrepancies which otherwise would have arisen between astrophysical observations and atomic physics. However, the creation of helium from hydrogen goes on only in the presence of sufficient amounts of nitrogen and carbon. No difficulties in principle would arise if, apart from the building-up of helium, the building-up of heavier atoms like nitrogen and carbon also appeared possible in the stars. This is, however, not the case. Temperatures higher than $2 \cdot 10^7$ degrees in the center of a star of solar mass are not to be expected, and this value is far too low to account for the building-up of atoms heavier than helium. Thus the origin of heavy atoms as they are found in the earth's crust, the sun, and all stars in roughly the same relative abundance remains inexplicable in the present state in which we find the stellar matter.

Temperatures of the order of 10^{10} degrees would be needed. We have reason, therefore, to assume that the universe passed through a stage when the heavy elements were created. Temperatures and densities must have been much higher than the values which we experience in the present universe.

Also the fact that all estimates of the age of the universe (see the paragraph on p. 56) point toward a surprisingly short time scale of the order of only 10^9–10^{10} years leads to the conjecture that the present universe is in a temporary state which may be separated from a preceding state by a critical phase of transition. Our efforts to remove singularities from the solution of the cosmological problem may therefore prove to be vain and unnecessary attempts.

5,5. The Importance of the Conception of a Finite World

The one fundamentally new feature introduced by the relativistic theory is the existence of solutions of the field equations yielding a closed finite world. It is a problem of the greatest importance for future investigations to make sure whether the distribution of matter in space indicates a line element corresponding to such a spherical or elliptic space. In the next section we shall discuss this question with regard to the results derivable from present observations.

The first solution of the cosmological problem in the general theory of relativity by Einstein in 1917, which still aimed at a static world—the general expansion of the universe was then still unknown—already offered such a solution. It brought for the first time the concept of a finite world within the reach of exact scientific research. Although this special solution is no longer of practical interest, because it does not account for the expansion, besides corresponding to an unstable world, it marks a milestone in the advance of the cosmological problem. The inclusion of the general expansion does not deprive the problem of such solutions; they correspond now to expanding or contracting closed spaces.

If we knew that our universe was finite and thus contained only a finite amount of gravitating matter, the delicate problem of infinity would be definitely removed from the realm of physical science. In addition, such a model of the world would open the way to a deeper understanding of the intrinsic properties of mass, gravitation, and inertia.

Mach had already realized that the property of inertia, i.e., of inert mass, manifested by every body and strictly proportional to its gravitating mass, should be reducible to the universal gravitational field produced by all matter filling the world. One cannot single out two bodies and attribute to them certain physical properties, as, for instance, inert mass, without keeping in mind that these properties may be brought into being by the very existence of all the other masses in the world. Mach says, without explicit reference to the cosmological problem, in chapter ii of his *Mechanik:* ". . . one has to realize that even in the most simple case, when apparently only the action between two bodies is considered, it is impossible to disregard the rest of the universe. Nature does not start from elements as we are forced to begin with. . . ." So we have to expect from a fully satisfactory solution of the cosmological problem a final answer also as to the question of the origin of inertia. Obviously, however, only in a finite world can we hope to relate the finite inertia of every body to the field induced by the finite total amount of gravitating matter in the universe.

Indeed, from the preceding equations (5,3.2), (5,3.3), and (5,3.4), in the special case: $\dot{R} = \ddot{R} = 0$ and $\epsilon = +1$, we obtain $2/R^2 = \kappa\rho$, and $\lambda = 1/R^2$. In this case the finite radius of curvature of the world is determined by the mean density of the matter which produces the universal field of inertia. No boundary conditions for infinity are needed because the finite density of matter in our model produces a finite world and abolishes the necessity of defining boundary conditions at infinity for the gravitational "potentials" $g_{\mu\nu}$.

The gravitating matter of the universe *produces*, so to speak, the space of the universe, and all manifestations of matter, like inertia, are defined only with regard to the gravitational field set up in this space.

In the static solution to which Einstein was led, the mean

density ρ alone defines the value of the radius of curvature. For nonstatic solutions—and obviously they alone apply to the actual world—$\dot{R}$ and $\ddot{R}$ are no longer zero and must be known. This complication explains why the observational test of the relativistic theory of an expanding world is still incomplete.

5,6. The Observational Test of the Relativistic Theory

The necessity of admitting a new parameter λ, without being able to fix its sign and without, in addition, knowing the sign of ϵ in the expression for the line element, increases the multitude of possible solutions to such an extent that at present we are still unable to select out of this multitude the solution referring to the actual universe. Could we disregard the λ-term, i.e., put $\lambda = 0$, then we could immediately conclude from equation (5,3.3) that

$$\dot{R}^2 = \tfrac{1}{3} k R^2 \rho - \epsilon$$

which gives

$$\tfrac{1}{3} k \rho - \left(\frac{\dot{R}}{R}\right)^2 = \frac{\epsilon}{R^2}. \tag{5,6.1}$$

The geometry of a finite and closed space, $\epsilon = +1$, obviously entails

$$\rho > \frac{3}{k} \left(\frac{\dot{R}}{R}\right)^2,$$

while

$$\rho < \frac{3}{k} \left(\frac{\dot{R}}{R}\right)^2$$

corresponds to a metric of the hyperbolic type. The knowledge of the mean density ρ and of the expansion $\dot{R}$ would suffice to decide whether the world may be supposed to be finite or not. The observed rate of expansion yields

$$\frac{\dot{R}}{R} = \frac{1}{c} \frac{525 \text{ km sec}^{-1}}{10^6 \text{ parsecs}} = 5.68 \cdot 10^{-28} \text{ cm}^{-1}.$$

If we introduce this value into the preceding inequalities, we find

$$\rho \gtrless 5.2 \cdot 10^{-28} \text{ gm cm}^{-3}.$$

The Relativistic Treatment of the Cosmological Problem

The observational data, on the other hand, yield a value for ρ, when we spread all matter, condensed in spiral systems, evenly over space, only of the order of 10^{-31} gm cm^{-3}. For the survey of spiral systems gives about 10^8 systems in a sphere of about $3 \cdot 10^8$ parsecs diameter; each system contains about 10^9 stars of solar mass; hence each system has a total mass of about $2 \cdot 10^{42}$ gm. This gives a mean density,

$$\rho = \frac{2 \cdot 10^{50}}{4.4 \cdot 10^{80}} \cong 5 \cdot 10^{-31} \text{ gm cm}^{-3}.$$

This value would indicate an infinite world with a hyperbolic metric. But, in fact, the observational data are, as yet, far too uncertain to permit any such far-reaching conclusions.

Moreover, when the statistics of nebulae reach distances from which the light needs more than, say, 10^8 years to travel to the observer, the expansion of the universe may have undergone, during that time, an acceleration or deceleration which might not be negligible; thus terms containing $\ddot{R}/R$ may have to be considered. The formulae which represent the observed facts become very complicated; they have to be developed into series, and a careful estimate of the contribution of each term is needed. Various discussions on these lines have been carried out, using all available data. They show that the observational data are not yet sufficiently reliable to justify such an ambitious proceeding. The results of different discussions, based on practically the same material, are inconsistent with one another. An undetected, systematic error in the photometric scale of only 0.1 magnitude, for instance, for magnitudes larger than 19, could falsify the resulting photometric distances sufficiently to annul an otherwise clearly indicated positive curvature of space. Thus no final conclusion concerning the sign of the curvature of space or the sign of the cosmological constant can yet be drawn.

The main conclusion which so far may be drawn from a comparison of theory and observations is as follows:

The relativistic treatment of the cosmological problem promises to give in the future a definite answer to the one question which appears to be the highest prize of all efforts, namely, the question: Is the universe closed and finite? The possibility that the universe may be closed and finite is not excluded by the observational facts but neither, so far, is it supported by them. Solutions of the cosmological problem which lead to a finite world demand values for the mean density ρ of matter about a thousand times higher than observations indicate. Moreover, the value of the constant radius of curvature of space resulting from these solutions is not much larger than the greatest distances which have, so far, been actually reached, i.e., distances of the order of 10^8–10^9 parsecs. If the universe were really so small, the present survey would already cover about one-fourth of its whole extent. This appears extremely unlikely. We would, moreover, be forced to assume that the matter which we observe concentrated in the star-systems is only a small fraction of the matter that is responsible for the universal gravitational field and thus for the curvature of space. A mean density ρ of the order of 10^{-31} gm cm^{-3}, in accordance with the observational facts, does not justify the conclusion that the metric of space is of a spherical character. If the λ-term is completely dropped, we obtain a solution proposed by Einstein and De Sitter, corresponding to an expanding world with Euclidean geometry.

The distribution of matter in our neighborhood up to 10^8 parsecs seems to agree perfectly with a uniform distribution of matter in space only if we do not interpret the red-shift as a Doppler effect. The corrections which, according to section 3,3, must be applied if the red-shift is interpreted as a Doppler effect change the apparent distribution of matter so significantly that the assumption of a uniform distribution of matter in space no longer seems to hold good. An agreement with the assumption of uniformly distributed matter can be restored by introducing a positive curvature of space which just cancels the "disturbing" effect of the red-shift correction, if the red-shift is interpreted as a Doppler effect. The introduction of such a positive curvature of space offers a solution which yields a closed finite world. But this picture of the world seems hardly acceptable on account of its small size and the high density of matter, a density about a thousand times higher than that which observations seem to allow. Therefore, if we accept the present observed value of ρ, we seem to be forced to sacrifice the most precious prize, that of the finiteness of the universe. But it must be emphasized once more: the observational data are still quite incapable to carry the responsibility for such a heavy burden of theoretical conclusions.

All considerations so far have been of a preliminary character, simplifying to the utmost all assumptions concerning the distribution of matter in space and concerning its motion. It may turn out that, just as our knowledge was insufficient when we tried to solve the cosmological problem with a scanty knowledge of the distribution of stars in the galaxy, so now our present knowledge of the distribution of galaxies in the universe is not yet extensive enough to be adequate.

This, however, must not obscure the fact that the relativistic treatment of the cosmological problem is very much superior to the classical treatment. It opens up a fascinating insight into the complex relations which exist between the geometry of physical space and the forces of the material world. On this new basis of conceptions we may hope to obtain in the future a unified picture of the universe in which the mass, size, and geometry of the universe follow from a firmly established theory of matter and energy.

6. Concluding Remarks on the Time Scale of the Universe

The discovery of the expansion of the universe has given a very unexpected impetus to the problem of the time scale of the universe.

If we date the age of the universe from the beginning of its present expansion, it would apparently only slightly surpass the lower limit of $1.5 \cdot 10^9$ years, which is the age of the earth's crust. This assumes that the expansion proceeded steadily from zero to infinity. The result would not be substantially changed if our world belonged to the class of oscillating universes. A possible exception is offered by a model world expanding asymptotically from the nonstable static solution of Einstein. No hierarchic structure in the ages of the various systems, earth—star—star-system—universe, seems so far to exist. This evidently means that in the not so very distant past the present state of our universe started from a very special event which we may have to call a "singularity." This event forms the limiting frame of our picture of the world. Other facts known and already mentioned make it equally probable that our present picture of the universe does not cover its whole lifetime.

The final solution of the cosmological problem depends on further extension of our empirical knowledge of the structure of the universe. Observational astronomy will have to furnish this knowledge. It is in itself, however, an advance of fundamental importance that the whole cosmological problem, including the question "Is space finite?" has been brought into the sphere of the exact reasoning of natural science.

Notes

1. The brightness of a star is given in magnitude classes; the intensity of light received from a star of magnitude m is 2.5 times that received from a star of magnitude $(m + 1)$. The zero point of measurement is arbitrarily fixed. The absolute magnitude of a star is the magnitude it would have if placed at the standard distance of 10 parsecs. 1 parsec = $3.08 \cdot 10^{13}$ km.

2. When space is considered uniformly filled with gravitating matter, the mass contained in a small element dV of space will be equal to ρdV, where ρ measures the density of the distribution of matter; ρ may be in the most general case a function of the three space coordinates and time. The gravitational force exerted by any such element of mass upon an arbitrarily chosen point P at a distance r from the element is, according to Newton's law, inversely proportional to the square of r, having in each direction of coordinates a component X, Y, Z. The function

$$\phi = \int_V \rho \, \frac{dV}{r}$$

is called the "potential of all attracting particles." The components X, Y, Z, in terms of ϕ take the simple form:

$$X = \frac{\partial \phi}{\partial x}; \qquad Y = \frac{\partial \phi}{\partial y}; \qquad Z = \frac{\partial \phi}{\partial z}.$$

For a point P located deep inside the gravitating matter, the gravitational pull in any direction may be equal to that in the opposite direction. In this case a body placed at P will not suffer any acceleration, but it will be subject to a stress arising from the forces pulling it in all directions.

3. See, for details, sec. 3,2.

4. See E. Hubble, *The Observational Approach to Cosmology* (1937), Fig. 3, p. 57.

Bibliography

CHARLIER, C. V. L. *Arkiv för Matematik Astronomi och Fysik*, Vol. IV, No. 24 (1908), and Vol., XVI, No. 22 (1922).

EDDINGTON, A. S. "The Expanding Universe," *Proceedings of the Physical Society*, Vol. XLIV, Part I, No. 241 (1932).

EINSTEIN, A. "Kosmologische Betrachtungen zur allgemeinen Relativitätstheorie," *Sitzungsberichte der Preussischen Akademie der Wissenschaften* (1917), Part I.

HECKMANN, O. *Theorien der Kosmologie*. Berlin: Springer Verlag, 1942.

HUBBLE, E. *Observational Approach to Cosmology*. Oxford: Clarendon Press, 1937.

———. *The Realm of the Nebulae*. London: Oxford University Press, 1936.

McCREA, W. H., and MILNE, E. A. *Quarterly Journal of Mathematics*, Vol. V (1934).

McVITTIE, G. C. *Cosmological Theory*. ("Methuen's Monographs on Physical Subjects.") London: Methuen & Co., Ltd., 1937.

MILNE, E. A. *Quarterly Journal of Mathematics*, Vol. V, No. 17 (1934).

PAULI, W. "Relativitätstheorie," *Encyclopädie der mathematischen Wissenschaften*. Leipzig and Berlin: B. G. Teubner, 1921.

ROBERTSON, H. P. "Relativistic Cosmology," *Review of Modern Physics*, Vol. V, No. 1 (1933).

SELETY, F. *Annalen der Physik*, Vols. LXVIII, LXXII, and LXXIII.

DE SITTER, W. *The Astronomical Aspect of the Theory of Relativity*. ("University of California Publications in Mathematics.") Berkeley, 1933.

———. "On the Expanding Universe," *Proceedings Koninklyke Akademie van Wetenschappen te Amsterdam*, Vol. XXXV, No. 5 (1932).

———. "On the Expanding Universe and the Time Scale," *Monthly Notices of Royal Astronomical Society*, XCIII (1933), 628.

TOLMAN, R. C. *Relativity, Thermodynamics, and Cosmology*. Oxford: Clarendon Press, 1934.

Foundations of Biology

Felix Mainx

Foundations of Biology

Contents:

PAGE

I. INTRODUCTION 569

II. WAYS OF WORK IN BIOLOGY 570
  A. General Foundations 570
  B. The Elementary Points of View 575
    1. The Morphological Point of View 576
    2. The Physiological Point of View 583
    3. The Genetical Point of View 588
  C. The Complex Points of View 592
    1. The Organism as an Open System . . . 592
    2. Growth, Development, Reproduction: The
      Historical Character of the Organism . . 598
    3. Organic Diversity and Its Structure . . . 602
    4. The Population as the Natural Form of Exist-
      ence of Living Beings 608
    5. The History of Organisms 610

III. THE SIGNIFICANCE OF SPECULATION IN BIOLOGY . 621
  A. The Psychological Function of Speculation in
    Scientific Biology 623
  B. Parabiology: The Misuse of Speculation . . . 626
    1. Mechanism and Vitalism 628
    2. Subjectivity, Activity, Purposiveness, and
      Conformity to Plan in the Organic World . 634
    3. Wholeness and Causality 638
    4. Autonomy or Heteronomy of Living Things:
      Ontological Questions 643
  C. World Picture and Philosophy of Life . . . 649

IV. CONCLUSION 652

SELECTED BIBLIOGRAPHY 653

Foundations of Biology

Felix Mainx

I. Introduction[1]

The word 'biology' is here to be understood in the sense of the fundamental science of living natural objects and thus as denoting all the disciplines of zoölogy, botany, physical anthropology, and the parts of neighboring sciences which are relevant to this field. Applied biology in its various branches, medicine among others, will be excluded. Human psychology, which, on account of the peculiarity of its special methods, requires a separate treatment, is also excluded. The notion of biology is not to be set up in opposition to the sciences of zoölogy, botany, etc., nor are these sciences to be subordinated to it, as sometimes happens (see Chap. III, B). Understood in this sense, biology is a branch of natural science, and it will be shown in the following pages that in its structure, in the kind of statements it contains, and in its mode of work, it represents an empirical science like the other sciences of nature.

The delimitation of biology according to its subject matter is a purely practical question. The concept of life was originally derived from the subjective awareness of human beings and then applied later to a greater and greater range of natural objects, when it was thought that these had properties in common with human behavior. In the present state of knowledge there is in practice no difficulty in distinguishing living objects from dead ones. Where there may be some doubt, as in the case of the viruses, it is purely a matter of convention whether, for scientific purposes, we classify such things under biology or not, just as drawing a boundary between zoölogy and botany is purely a matter of convention. Some authors attach great importance to

1. I am greatly indebted to Professor Joseph H. Woodger for his translation of this monograph from the German manuscript and for his criticism, and to Professor Victor Kraft for his reading of the manuscript and his valuable advice.

establishing the boundary between the living and the dead by definition and, from a metaphysical need, wish to treat this question as an ontological one (see Chap. III, B, 4). Even if we find that drawing a sharp boundary is not empirically possible, the independence of biology as a science will not disappear. Its independence rests on the observed peculiarity of its object and on the development of its own methods of research and points of view, which are required by that object. The division of biology into its various branches is also purely practical and for that reason depends, at a given epoch, on the prevailing direction of interest among the various possible points of view. Also for that reason, it is not meaningful to ascribe a fundamental importance to a "system" of biology. The vague separation into zoölogy and botany, according to the kind of object studied, is traditional and therefore still largely provides the basis for the organization of teaching and research. The subdivision into morphology, physiology, genetics, ecology, etc., according to point of view or direction of approach, cuts across both realms of living things. The branches of biophysics, biochemistry, paleontology, biogeography, etc., form bridges to the related sciences, so that a sharp delimitation of biology from other sciences from the point of view of method is not possible, either in the practical work of research or in scientific organization. Efforts to achieve a synthesis, particularly on the border lines of biology, are very active and successful at the present day.

II. Ways of Work in Biology

A. General Foundations

Corresponding to the character of biology as an empirical science, its statements have the general features of empirical statements. In the face of the great variety with which we are confronted by the world of living things and by the problems which this variety presents, the part played by pure description among biological statements is very great. The elementary constituent of a descriptive statement is the report of an act of observation—for example, the report of a color, of the measurement of a magnitude, of counting, of weighing, etc. It must, however, be remembered that even in these elementary descriptive statements the beginning of hypothesis construction must be recognized. The description chiefly serves for the characterization of a state of affairs in a way which will facilitate the recognition of other states of affairs of like kind. The observer will therefore emphasize, in the description of a fact, those sense impressions which seem to him essential and will neglect those which seem inessential. The manner in which a descriptive statement comes about is thus in a certain degree dependent upon the scientific attitude of the observer. This feature shows itself in an enhanced degree when descriptive reports are generalized and a fact is described not as unique but as "typical." Then the statement which is regarded as purely descriptive also takes on the character of a hypothetical statement. It is clear that this holds for all descriptions of processes which are characterized as following a typical course after repeated observations. Statements which express a functional connection between particular concepts—e.g., the statement "Cell respiration stands in a definite, regular connection with temperature"—in themselves exhibit the complete type of a hypothesis belonging to the empirical sciences. The concepts employed in them are defined by means of rules of operation, in which it is unambiguously stated how the practical observations are to be correlated with the concepts. In this way the concepts become constituents of the language of science, and their verbal meaning deviates more or less from that of everyday language. For that reason the words are often replaced by symbols which are unambiguously defined by the rules of operation. There is no sharp logical boundary between descriptive and functional statements. In many statements of biology descriptive and functional elements are mixed.

The general logical foundations of the formation of biological concepts and theories need not be discussed here in detail, since they hold for all the natural sciences and are therefore expounded in detail in other parts of this *Encyclopedia* (Lenzen, 1938; Woodger, 1939; Hempel, 1952).[2] For the same reason the characters which an empirical statement must possess if it is to be given a legitimate place in an empirical science will be only briefly enumerated. Such a statement connects a definite number of concepts in a manner which is free from objection from the formal logical standpoint. These concepts must be defined in a manner which makes their correlation with observable elements of experience possible. If this is done, the statement itself is testable by experience. By the making of an observation or by an experiment the statement can be verified or falsified. By 'verification,' here and in what follows, only confirmation by experience is to be understood, and by 'falsification' the absence of such confirmation. Nevertheless, by the use of these terms the question of the "truth" or truth-content of the empirical sciences is not touched upon. This question has no significance in biology other than that which it has in all other empirical science, and it is treated in other parts of this *Encyclopedia*. The uses of a statement of empirical science can also be stated in the following way: On the basis of such a statement, predictions about what is observable can be made. The occurrence of the predicted experience then means the verification of the statement, its absence the falsification of it. A correctly formulated hypothesis therefore has heuristic value in the development of science. On the basis of such a hypothesis new observations can be made, new experiments set up, which lead to the confirmation or falsification of the hypothesis. Hypothetical formulations with which observable processes, in the course of time, can be correlated and according to which such processes can therefore be predicted are called "rules" or "laws," e.g., Mendel's rules of inheritance. This terminology is also often chosen for formulations in which a relation between simultaneously existing observable elements is given, e.g., the law of the constancy of the chromosome number in species. The use of the word 'law' or 'rule' has often led to misunderstandings with which we shall be occupied later. This applies especially to the expression 'nat-

2. See Selected Bibliography at end for full data.

ural law,' now rarely used in biology. In principle we are free to construct hypotheses as we like—i.e., every formation and combination of concepts which satisfies the requirements sketched above is admissible. Psychologically speaking, the imagination of the investigator plays a great part in the conceiving of hypotheses and is frequently called "intuition." In practice, the conceiving of hypotheses chiefly takes place against a background of the state of the science already reached, having regard to the complex of statements which have already become more or less fixed constituents of this science.

Statements about biological facts which according to their structure could be testable by experience but which for practical reasons cannot at the time, or perhaps ever, be tested are admissible as hypotheses even though no judgment about their confirmation is possible. Statements which according to their structure are not testable by experience and which, on account of the absence of an unambiguous correlation between the symbols and the observable facts, can never in principle be verified or falsified have no function in the system of an empirical science. They are neither false nor correct in the sense of empirical science but cannot be used in science and have no legitimate place there. Whether they are, in general, scientifically meaningless is a question which will not be discussed here. But there is still another type of statement whose confirmation is in principle untestable. These are statements which, although externally they have the form of a hypothesis of the empirical sciences, are in fact of such a nature that the rules of correlation between the concepts used in the statement and what is observable constitute the whole content of the statement itself. These statements are occupied exclusively with repeating the rules of operation. These are tautologous statements. If in the statement "The positive phototactic reaction of a *Euglena* is proportional to its light-requirement" the concept "light-requirement" is only testably defined by means of the establishment of the behavior under the stimulus of light, this is a tautologous statement of the above kind. Such tautologies have no direct value for science. Nevertheless, as will be discussed later, tautologous statements frequently occur in scientific expositions.

If a hypothesis is falsified by a reliable observation, it is abandoned. If it is confirmed by the observation, it is used further in science. A hypothesis which is frequently verified comes to be regarded more and more as a stable constituent of the science and often becomes a habit of thought. If a hypothesis can be tested in various ways in accordance with the rules of correlation and if all these tests verify it, then the hypothesis is regarded as an especially well-confirmed one. If the incorporation of a new hypothesis in the permanent structure of the science takes place without contradiction, this is regarded as a proof of the "correctness" of the hypothesis or of the "truth" of its content. If the incorporation does not succeed without contradiction, then, by a thorough logical analysis of the contradictory statements and their elements, we must investigate whether the contradiction is not merely apparent and whether it cannot be removed by a logical change. In other cases the contradiction can be bridged over by means of accessory hypotheses which restrict or extend the validity of the hypothesis and which, in turn, must satisfy the requirements of testability by experience. Naturally, they must not be introduced only *ad hoc*, i.e., they must not be merely tautological or formulated without any connection with the rest of the system of the relevant science, because in that case they would in principle be removed from any testability. On account of their heuristic value they can often give birth to a new development in science. If a far-reaching contradiction persists between old and new hypotheses, this leads to a "crisis" in the empirical science concerned. This, in turn, leads to a revision of the system of statements hitherto in use. It may then prove to be the case that certain theories that have hitherto been accepted must be abandoned because they no longer correspond to the augmented experience or that the old theory represents a special case of the more comprehensive, newer theory. Growing experience leads to an enrichment of our stock of scientific possessions and necessitates a perpetual revision of the complex of statements of the science, and in this lies its progress. The contradiction is never in the experience but always only in our formulations. Empirical science knows no "aporia" in the philosophical sense.

If, by the observation of a phenomenon, an existing hypothesis is verified, we say that the phenomenon is "explained" by the hypothesis, especially when we have to do with a frequently confirmed hypothesis. The "explanatory value" of a hypothesis seems to us to be especially great if many different experiences can be correlated with it and if it can be added to the whole edifice of science without contradiction. Some statements are formulated or interpreted with the intention of making a process or a complex of relations "intuitable" or "picturable" to human beings. Many formulations of natural science, to which a high explanatory value is ascribed, are so abstract that this feature of intuitability is entirely absent. The difficulties and misunderstandings which can result in science from the human need for intuitability and from the confusion of "explanation," in the sense here defined, with "intuitability" will be discussed later with examples.

In the following pages an attempt will be made to show that the statements of biology in its various subdivisions satisfy the requirements described above and that biology can be regarded as a system of such statements. According to the way of considering and of investigating biological data, we shall here carry out a twofold division into elementary and complex points of view. This subdivision is purely practical and is intended to make our methodological exposition easier.

B. *The Elementary Points of View*

In considering an organism, we can direct our attention, according to our inclination and aim, either to its visible peculiarities or to its behavior and performances or to its significance as a member of a reproductive chain, and in accordance with these possibilities it is customary to distinguish the subdivisions of morphology, physiology, and genetics. The individual investigator will, according to his gifts, his training, or his particular problem, direct himself exclusively or predominantly to the one or the other point of view. But, in spite of the necessarily high specialization of the methods involved, it is naturally impossible for him to devote himself exclusively to the one or the other point of view, so that here also the boundaries are not sharp. In dealing with complex problems, the synthesis of the various points of view is even unconditionally necessary.

1. The Morphological Point of View

The enormous multiplicity exhibited by living things necessitates at the outset a description of the visible "characters" of particular organisms—their forms, proportions, colors, and measurable magnitudes. This morphological description relates not only to the external, but also, in the form of anatomy, to the internal, structures, the construction and mutual spatial relations of the organs. It continues right down to a morphology of the tissues, of the cells and their parts, in the branches of histology and descriptive cytology. An essential feature of descriptive morphology, in the widest sense of the word, is the discovery that with its help a classification of organisms is possible, that their recognition on the basis of a sufficient description is possible. The process of classifying in this way involves at the same time the setting-up of a hypothesis—namely, the conception of the "type" of the organism concerned, by means of which we can undertake, in the particular case, to correlate the objects found in nature with the appropriate type in a manner sufficient for practical purposes.

To begin with, the expression 'type' is to be understood in the widest sense of the word. The process of setting up types occurs in every morphological as well as physiological classification, for the reason that the descriptive elements used for the purpose never merely represent the description of an observed result but are constructed from a series of such results by abstraction of the features which are common and which seem essential to the investigator. Where the process of classification into types is carried out in a deliberately critical spirit by the investigator— say, in connection with genetical or biometrical problems—certain conventional rules are set up for carrying it out. For example, a classificatory measure of size is given as the mean value of a definite number of measurements, with its mean error and the width of variation of the character given by the standard deviation. A pattern of markings is arranged according to a series of standardized categories, and so on. But such a strict interpretation of the method of classifying into types is usually avoided, and we content ourselves with specifying such characters as will satisfy practical requirements. The most diverse categories which are descriptively specifiable can be the object of the typifying process. It may be a race, a species, a higher classification unit, a group of organisms brought together from some standpoint other than the classificatory (e.g., all water plants), which is described and so made an object of the typifying process. The species concept of taxonomy is thus only a special case for the delimitation of which, as will be discussed later, other than classificatory notions are required. But an organ, a tissue, a cell form, a metabolic process, a mode of instinctive behavior, etc., can all be objects of the typifying process and in this way undergo a classification. Not only do we work in this way in the morphological branches of biology, but the description of physiological processes on a comparative basis leads to a classification by a like process.

The establishment of the "type" of a form, of a mode of behavior, or of a process involves the formation of a hypothesis, in so far as an arbitrary selection of results of observation is compared and emphasized as essential for classification, and thereby the hypothesis of a regular, common repetition of these characters suitable for classification is set up. In this way a law of coexistence of characters, which is empirically testable, is asserted. To the comparison of results of observation there is, of course, added, in this procedure, an idealization, brought about by the fact that the choice of what is described is to some extent arbitrary. This idealization appears in increased measure in the working method of comparative morphology when this proceeds to the setting-up of higher types of higher order. By the comparison of types of similarly organized living beings, we reach in comparative morphology the highly abstract concept of the "constructional plan." In this comparison similarities and differences are established, and essential characters are distinguished from inessential ones; and in this way a "fundamental" constructional plan, common to several organic types or a large group of organisms, is constructed. By further comparisons more and more comprehensive, more general plans of construction can be set up. The bodily parts, organs, and parts of organs which correspond to one another in the constructional plan are called "homologous" structures.

By comparing the structural plan, say, of a vertebrate, with the concrete natural objects, e.g., with a horse, a whale, and a bird, the procedure of "homologizing" is carried out. The anterior pair of legs of the horse is homologous with the anterior fins of the whale and with the wings of the bird. Homologous organs are thus such as correspond to one another in their place in the structural plan, those which can take the place of one another in the structural plan and which coincide in the plan of higher order. Homologous organs exhibit a construction out of fundamental elements which correspond to one another—e.g., homologous parts of the skeleton, of the musculature, of the innervation.

The formulation of the concept of homology has the typical form of a hypothesis of empirical science. It represents a system of schemata, the structural plan with the spatial relations of characters and organs which are characteristic of it, and it can be empirically tested in so far as the presence of the homologous organs can be verified in every concrete organism by a morphological investigation. It permits predictions of the kind which assert that, if an organism shows itself to be a vertebrate by one characteristic feature, then we may also expect the other homologous structures of this group. Since the domain of validity of the hypothesis is restricted to a particular group of organisms, we should not, in the case of a falsification, place the organism under investigation in this group, i.e., the Vertebrata. It will "correspond" to the structural plan of another group.

In spite of the high degree of idealization which is involved in the setting-up of structural plans of high order and in an extended use of the concept of homology, the statements of comparative anatomy are nevertheless empirical statements. The elements for the construction of structural plans are derived from experience, and the concepts used for this purpose remain, through their definitions, empirically testable. In this way the extreme abstractions of morphology, in so far as they remain scientifically useful, are distinguished from the entities of geometry, the construction of which takes place on the basis of a purely intellectual conception and for which the requirement of testability by experience is meaningless.

As an example of a morphological statement of wide scope, we may mention the law of the constancy of the chromosome number for any species. It holds for all organisms (with the exception of microörganisms which have no nucleus) and states that in all mitoses of a member of a group defined as a species (we shall return to this definition later) the same number of chromosomes is always to be observed. Instances of the falsification of this rule require the setting-up of testable accessory hypotheses, e.g., the hypotheses of chromosome races, of polyploid races, and the like.

Statements like the above are reached by a process which M. Hartmann (1948) calls "generalizing induction." The verification of a statement by an experiment he calls the method of "exact induction." These expressions do not seem to me to be well chosen, since an essential methodological distinction between the two procedures cannot be maintained on closer examination. It would, perhaps, be better to contrast the two procedures as "comparative" and "experimental." It is by no means the case that obtaining and testing statements by comparison is especially characteristic of the morphological point of view, let alone the only one open to it. A large part of morphology is called "experimental" morphology because the method of obtaining and testing statements is that of experiment, and thus "exact induction." The statement "A certain plant forms finely divided leaves under water but on the surface has undivided floating leaves" can be tested not only by observations on cases occurring in nature but also by experiment, as by artificially setting up the conditions mentioned and seeing whether the predicted connection is realized. This statement connects concepts descriptive of the environment with morphological concepts.

There is a tendency to assign to experimentally testable statements a greater "demonstrative power," or at least a greater picturability, than is assigned to statements which are testable by comparison. It is customary to contrast them, as "causal-analytical" statements, with "systematic" or "order-analytical" statements. The expressions 'causal-analytical' and 'order-analytical' are not very fortunately chosen. In the first place, it is a question not of analytical statements in the logical sense but of synthetic ones. Moreover, neither a causal connection nor an ordering is "analyzed." It would be more correct to say that these statements set up rules or laws according to which in the first case a succession of states in time, in the other case a co-existence of characteristics or a correlation between properties which is empirically testable, is predicted. We could—with certain reservations—also speak of "the search for a causality of process" or of "a search for an order." If we wish to restrict the concept of "order" to the establishment of a law of coexistence of properties, then only those statements are to be called "order-analytical" which are formed on the basis of repeated observation of such a coexistence and are formulated so as to be testable on further observations of the same kind: for example, statements about anatomical structures or about the results of biochemical analyses. If a statement is formulated on the basis of repeated observations of processes in which one particular order passes regularly into another, then this statement is also obtained by comparison and is testable by further comparative observations. Nevertheless, in such cases it is customary to speak of "causality of process," and we should have to regard these statements as causal-analytical ones, although they have been obtained by comparison. Experiment is then a special case of this procedure in which we deliberately bring about a particular initial situation in order to test the succession of changes of state demanded by the hypothesis. By arbitrarily varying the initial situation, experiment opens up new possibilities for the setting-up and testing of hypotheses and in that way leads to the enrichment of science. Since the expressions 'causal-analytical' and 'order-analytical' are much used in the literature, they will be retained in the following pages, in spite of the reservations just mentioned.

Both types of statements are genuine statements of empirical science if they satisfy the requirements sketched in Chapter I. The changes of state in dead, and especially in living, nature to which our experiences relate bring it about that we are often placed under the necessity of transforming order-analytical statements into causal-analytical ones. In this way one established order is traced back to another, and a rule is set up for the transformation of one arrangement into another which is empirically testable. It is a question of great importance from the point of view of the theory of knowledge whether, in principle— if only under certain, perhaps not always given, assumptions— *every* order-analytical statement can be transformed into a causal-analytical formulation. The question then arises about the "final order" to which everything can be referred back. Since this and similar fundamental epistemological questions are common to all empirical sciences, and biology occupies no special position in relation to them, they will not be discussed here. In any case, it very often happens in biology that order-analytical statements are transformed into causal-analytical ones and that attempting to do this is of great heuristic value.

In the field of morphology this is the case when questions of form and structural plan are expressed as questions of formation and development, in the ontogenetic or phylogenetic sense. M. Hartmann (1948) has expressed the opinion that the "exact inductive" method is superior to the "generalizing inductive" method and, by way of example, compares the following statements: (1) "All carnivores have a relatively short, all herbivores a relatively long, gut" (generalizing inductive); (2) "On an animal diet tadpoles acquire a relatively short gut; on a vegetable diet they acquire a relatively long one" (exact inductive). The second statement, which can at any time be verified by experiment, carries more conviction, according to Hartmann, than the first. W. Zimmermann (1948) on the other hand, objects that the example is not well chosen, because, for testing the efficiency of the two methods, two very nonequivalent facts are compared. In the case of the tadpole we are dealing with an ontogenetic developmental process which is dependent upon diet; the example from comparative anatomy has a scientific meaning only if it is formulated as a phylogenetic question. The assumption must then be made that, in adaptation to dietetic conditions, among the various animal species the different relative lengths of gut have in the course of phylogeny become species-specific, genetically fixed characters. I agree with Zimmermann that in the examples chosen by Hartmann we have to do with very different facts, but I am not of his opinion that the first statement, which belongs to comparative anatomy, must be denied all scientific value, inasmuch as it is not formulated as a phylogenetic statement. As a pure order-analytical statement, it satisfies the general requirements of an empirical statement and at the same time calls attention to an existing regularity. If we wish to transform it into a phylogenetic statement, then we assume that similar statements can be formulated about all processes occurring in phylogeny and that these would be—at least in principle—empirically testable. But then the statement is transformed from an order-analytical one into a series of causal-analytical statements.

A greater measure of picturability attaches to the concepts and formulations of morphology than is the case for the statements of other branches of biology. In using these concepts, people are often tempted, in consequence of this picturability, to choose a linguistic expression which, although very picturable, nevertheless does not, strictly speaking, agree at all with the logical foundations of empirical science. This is so, for example, when it is said that in the development of an organism a structural plan is "realized," that the development "deviates from the structural plan," or that it "follows the structural plan" or "is governed by it." The dubiousness of this mode of expression becomes still clearer when we speak in comparative morphology of the "metamorphosis of homologous organs," of a "transformation of shoots into thorns," or of a "transformation of the various biting mouth parts of one insect into the corresponding differently formed sucking mouth parts of another." Everyone who is scientifically educated naturally knows that such linguistic expressions are intended only in a metaphorical sense. The "structural plan" exists, indeed, only as an idea in our consciousness, and the "metamorphosis" of homologous organs is also only an imaginary process. The case is somewhat different when we speak of "ontogenetic" or "phylogenetic" developmental processes, but then other rules hold concerning the correlation with the facts of experience. In the picturability of the notions often used in the discussion of morphological problems there lies a danger of falling into a realism of ideas which, as a legacy of the traditional philosophy of nature in the form of idealistic morphology, has played a great part in the historical development of biology. Some modern biologists do not seem to be quite consistent in avoiding this danger, and some profess quite openly a more or less concealed realism of ideas (see Chap. III, B, 2–3).

There is no sharp delimitation between the morphological and the other points of view in biology. The simplest descriptive data often point to the function of the organ or other physiological relations. Every anatomy, especially every histology and cytology, is in some measure functional anatomy and so leads into physiology or forms its foundation. The morphological point of view has special relations to the problems belonging to systematics and evolution which we shall discuss among the complex points of view.

2. The Physiological Point of View

Physiology is commonly defined as the theory of the functions of the organism, of its organs and tissues, or, better, as the theory of the processes which take place in the organism and between the organism and the environment. Sometimes we are told that only physiology deals with the material and energetic properties of organisms, morphology dealing with their forms.

A statement about physical forms, however, has a meaning for empirical science only if it is testable by experience and hence on material structures. Everything independent of "material" carries with it the danger of a realism of concepts. If a form or structure concept in a biological statement is defined in a way which is completely independent of the "substance," the whole statement loses its testability by experience and with that its usefulness to empirical science. Other assertions of idealistic morphologists teach that causal-analytical thinking is characteristic of physiology, while for morphology we have "prototype" thinking. Although the notion of "prototype," conceived as a kind of intuitive picture, is not defined so as to be testable in experience, yet idealistic morphologists try to set up a kind of causal relation between the prototype and the empirically testable forms and structures of organisms. We shall return later to these differences of view regarding the concept of cause (Chap. III, B, 3).

It is also incorrect to say that physiology is the domain of "exact induction," while morphology is the field of "generalizing induction." Moreover, many statements of physiology are constructed of purely descriptive elements and are obtained by comparison and thus order-analytically. Again, physiology provides many examples of the possibility of transforming order-analytical statements into causal-analytical ones. Physiology often works with purely physical or chemical methods, as in demonstrating the transformations of materials in metabolism or in the analysis of the processes in semipermeable membranes. It frequently expresses its statements in formulas of a high degree of abstraction in which quantitatively expressible functional relations between two or more variable magnitudes are asserted to hold. On account of this structure such formulations are regarded as especially "exact," as strictly scientific or demonstrative. They are, however, not more and not less demonstrative than every other empirical statement which conforms to the requirements of the logic of science.

The following is chosen as an example of the properties of physiological statements described above: "The respiratory quotient CO_2/O_2 is approximately 1 in herbivores, somewhat less in omnivores, and still less in carnivores." The descriptive elements of the kind of diet of various organisms and their gas exchange, used in the statement, can be defined by specifying the particular diet and by giving methods for the exact determination of the gas exchange. In this way unambiguous correlation of the statement with the testing facts is made possible. The stipulation about the gas exchange can be formulated quantitatively, as well as the physicochemical composition of the food, if necessary. The whole statement has the character of one obtained by the generalizing inductive method and thus of an order-analytical statement. But, from it, single statements can easily be derived which have the character of experimentally testable causal-analytical statements. This becomes especially clear if we set up more restricted statements about the utilization of nutrient materials in metabolism—for example, about the utilization of carbohydrates, fat, and protein as respiratory material—and in this way connect a whole series of testable causal-analytical statements, which make the rule expressed in the wider statement easily understood. In this way the statement shows itself to be a "blanket statement"[3] about an exceedingly complex occurrence which we can—in this case—analyze quantitatively, and to a great extent, into its partial processes. In consequence of this we regard the process as especially well "explained," since we can, from the more comprehensive statement, derive a diversity of statements which are testable by various methods and which can be connected with one another and with other experiences without contradiction. In this way single statements of a particular kind, e.g., those about the utilization of proteins in metabolism, can be related to more comprehensive statements, e.g., about the respiratory processes.

Let us take another example from the physiology of irritability. The phototactic behavior of *Pandorina*, a free-swimming

3. [The term 'blanket statement' is a translation of *Pauschalaussage*. In correspondence the author explains this term as follows: "By 'Pauschalaussage' I understand a simple statement, the background of which is a complex occurrence, and which can therefore be expressed—under certain circumstances—as a complex of other statements."—EDITORS.]

colony-forming green alga, is characterized by a lower threshold, i.e., the light-intensity at which the first positive phototaxis is observed and below which no reaction occurs; the turning point, i.e., a stronger light-intensity, at which the positive phototaxis turns into the negative; and the upper threshold, i.e., a still stronger light-intensity, at which the negative reaction is also absent. The position of these cardinal points of phototactic behavior is dependent in a definite way upon the hydrogen-ion concentration of the solution in which the colonies live. Here the concepts which are connected in the statement are, on one side, a physical property of the environment (the light of a particular intensity coming from one direction), on the other, a mode of biological behavior (the unilaterally directed movement of the alga colony, which is produced by a particular sequence of strokes of its flagella). The whole statement is thus undoubtedly a blanket statement about a very complex occurrence within the organism, extending from perception of the stimulus up to the response. This blanket statement can also be analyzed into a series of testable single statements, e.g., a statement about the place of perception of the stimulus and its structure, one about the conduction of the stimulus, one about the structure of the flagella and their function, one about the changes of state of the whole reacting system which are connected with the changes of the hydrogen-ion concentration, and so forth. The light is then called the "stimulus" (or, better, "cause of the stimulus") and the behavior of the organism its "answer" (or response) to the stimulus. The dependence of the position of the cardinal points on the hydrogen-ion concentration is described as a *Stimmungsphänomen* ("phenomenon of mood") and the shifting of the cardinal points according to experimental changes of the hydrogen-ion concentration as *Umstimmung* ("change of mood") of the organism. This simple example from the physiology of irritability illustrates the point that, in formulating statements about such processes, linguistic expressions are frequently used which are taken from the psychology of human experience. In this way statements become more "picturable," without a special "explanatory value" being given to these statements by the use of such linguistic expressions. The possibility of empirically testing them remains instead just the same as for every other physiological, i.e., empirical scientific, statement. If we wish to define the concept of "mood" (*Stimmung*) in a scientifically useful way, we can do this only by correlation with definite experimental observation, and the same holds for the concepts "stimulus," "stimulus-perception," etc.

The use of anthropomorphic terms from human psychology is widespread, especially in the whole of the physiology of the nervous system and special senses of the higher animal organisms, including man. With a consistent application of fundamental scientific criticism, no difficulties result from the use of these linguistic expressions. Even the complex behavior of higher animals can be treated throughout as a problem in purely empirical science. But that a great danger lies in these anthropomorphic modes of expression, which can lead to a conscious or unconscious psychologizing of biological facts, will be shown later (Chap. III, B, 2).

Only in the case of man does a really new problem arise— from the fact that to him is given a new kind of experience, the inner experience, in the form of his own feelings and sensations. One and the same state of affairs can be described in the form of neurological statements and thus in accordance with empirical science and the requirement of experimental testability, or it can be described as sensation in the expressions of psychology. In this way arises the complex of problems of the correlation of physiological with psychological experiences, the psychophysical problem. Since this cannot be discussed without a fundamental critical analysis of psychology, we shall not deal further with it here. It need only be pointed out that an excellent and unambiguous clarification of the boundaries between the physiological and the psychological points of view from the biological side has been carried out repeatedly in recent times (e.g., Bünning, 1949).

A special role is played by the physiological standpoint in the complex problems of developmental physiology, in ecology, and in the investigation of behavior.

3. The Genetical Point of View

In opposition to the usual scheme, the fundamental point of view of genetics will here be given a separate treatment, since it shows, in practice, a certain independence in biology and plays a part in all complex problems of biology, on account of the great importance of modern genetics. The geneticist does not consider, as does the morphologist, the structure of an organism or those of several distinct organisms comparatively; neither does he, in the first instance, investigate the processes which occur in organisms, as does the physiologist. The geneticist deals comparatively with the individuals of a population which form a continuing reproductive chain.

The word 'heredity' (*Vererbung, l'hérédité*), which is used in genetics in order to denote the reappearance of characters of the ancestors among the descendants, conceals within itself the danger of misunderstandings which even today still often lead to senseless debates. It is taken from the language of everyday life, in which it signifies a process in which an external possession passes from ancestors to descendants. Here we have actual possessors, e.g., father and son, and the inherited object. In a derivative sense we speak in genetics of "inheritance" and of "inheritable properties." The word 'property' is used here as though there were a "bearer," independent of the properties, by which these properties are inherited. Nevertheless, no scientific statement can be formulated in which the notion of the bearer is so defined that it could be tested in experience according to definite rules of correlation, independently of all properties. On the contrary, it is always the properties themselves, and these only, which, in accordance with definitions, can be empirically tested. The organism itself is definable only by means of the properties which serve to characterize it. The idea of a bearer independent of the properties is an analogy with the idea of self among human beings, and its transference to things and organisms is an anthropomorphism, or conceptual realism. In the scientific language of biology, therefore, 'inheritance' and 'inheritable property' mean only that, in the comparison between ancestors and descendants, like or deviating characters can be established for the classification of the individuals. The comparative method leads to the setting-up of statements of the order-analytical type. To see in the connections so expressed a process of inheritance implies a further hypothesis, through which the order-analytical statements are transformed into causal-analytical ones.

The best-known general statements of genetics are the Mendelian rules or laws of inheritance. In the statement "After crossing a bean with red flowers with one with white flowers, we find (after self-fertilization in the F_1 generation) in the F_2 generation red-flowered and white-flowered plants approximately in the proportions 3:1," there is expressed, in the first place, on the basis of the comparison of the results of repeated experiments, a definite correlation between the characters of the parents and those of the succeeding generations. Owing to the fact that the organisms compared belong to a reproductive chain, the causal-analytical treatment of this relation is appropriate. This was formulated by Mendel in the form of the hypothesis of the hereditary factors. In its later form this is composed of the following statements: "The factor R 'produces' the red, the factor r the white, flower color"; "Every plant possesses a pair of hereditary factors, either RR or rr or [the F_1 hybrid in the above example] Rr"; "Whenever the sex cells are formed, the members of these pairs are separated, and each sex cell possesses only an R or an r"; "Through random combination of the gametes in fertilization, the ratios of the red- and white-flowered plants in the F_2 generation are explained." The assumption that a factor R "produces" the red flower color is conceived after the pattern of the "forces" of physics and, to begin with, means no more than that, in this way, a correlation is proposed between the hypothetical concept "factor" and something observable. If the statement were restricted to this correlation, i.e., had no other content than this rule of operation, then it would be a tautology and therefore not testable by experience. But, owing to the fact that the concept "factor" is invested with still other defining characters which are empirically testable, such as its behavior during gamete formation and fertilization, the whole hypothesis becomes scientifically useful. It asserts something about hypothetical units which are distributed and about a hypothetical mechanism which distributes them. The heuristic value which the hypothesis possesses is shown by the various experimental arrangements by which it can be tested, e.g., the backcrossing of the F_1 hybrids with the dominant or the recessive parent, the analysis of the F_2 generation by crossing with the recessive grandparent, the testing of the assumption of the purity of the gametes in the hereditary processes in the case of haploid parthenogenesis and haploid organisms, the testing of the mechanism of distribution by tetrad analysis, and so on. Beyond this, the heuristic value of the Mendelian hypothesis has proved itself precisely in those cases which seemed to falsify the hypothesis. The setting-up of testable accessory hypotheses has empirically disclosed new connections and, just through these cases, has led to a significant extension of knowledge. We feel that observations about inheritance are "explained" (for instance, the relative frequencies in the F_2 generation) when they can be derived from the general hypothesis. By combining the methods of hybrid analysis with the cytological and embryological methods of modern genetics, such observations can also be derived from yet other hypotheses, such as those about the minute structure of the chromosomes. Through the consistent combining of all these observations, the explanatory value of the hypotheses involved seems to us to be especially satisfactory.

The Mendelian rules are statements about the relative frequency of the various hereditary types in the generations following a hybridization; they are thus statements about the structure of a collective, i.e., statistical statements. The Mendelian hypothesis of the hereditary factors thus characterizes the distributive processes, assumed in gamete formation and in fertilization, only in a statistical way. Naturally, this does not mean that these processes in the individual case, and consequently the individual hereditary fate of a member of the F_2 generation, are indeterminate—say, in the sense of microphysical uncertainty. All these processes which decide the fate of individual descendants after hybridization are, in fact, definite macrophysical processes and could be described in a determinate way by other formulations for the individual case, although this would be very difficult in practice. Statistical statements also occur in other ways in genetics, as in the domain of the statistics of variation. All statistical statements require for their empirical testing a treatment of the observational material by the theory of errors. This is a general requirement of scientific method, has nothing to do with the object of the investigation, and therefore does not mean, as some biologists believe, an encroachment of methods not belonging to biology. The frequent use of statistical statements has, especially among some theoretical authors, given rise to remarkable and erroneous judgments about genetics. One observer sees, for example, in the Mendelian rule, because it appears in mathematical form, the ideal example of a biological "law," while another, for the same reason, would deny it every biological character. The lack of reflection from the point of view of the logic of science is especially noticeable in such discussions.

The concept "hereditary factor," which governed the early days of genetics after the rediscovery of the Mendelian rules, has, in the course of the development of this science, merged into the wider concept of the "gene" or "allele." The notion of the gene, as it is used today, is a good example of biological theory and concept formation. Today the gene is defined by so many points of determination that a great number of different testable formulations can be derived from statements in which this concept occurs. In the experiment of hybrid analysis the gene appears as a segregating unit, in cytogenetic experiments as a localization unit, in mutation experiments as a unit of mutation, in embryological experiments as a unit of action. Nevertheless, the concept of the gene is by no means dogmatically frozen, as outsiders sometimes believe, but is in a state of constant change, and a series of heuristically valuable tendencies in modern genetics points the way to a revaluation, perhaps even to a fundamental revision, of this concept.

During recent years the genetical point of view has occupied an increasingly important place in biological research and continues to gain ground. It seems to be destined to form a connecting link between the other aspects of biology and at the same time to promote a synthesis of biological thinking. It therefore plays a large part in all complex questions of biology, especially in the domain of developmental physiology and, in the form of the genetics of populations, in the questions of systematics and evolution.

C. The Complex Points of View

In this section we shall try to show by means of some examples how statements about complex states of affairs are formulated in biology. Every simple biological experience can be represented in various ways on the basis of the elementary points of view described in the previous section. In the case of more complex situations a synthesis of the various points of view is unavoidably necessary. The subdivision of this section in accordance with some fundamental problems of biology is quite arbitrary and in no way exhaustive.

1. The Organism as an Open System

The attempt to reach more general statements which will be true of organisms on the basis of the various points of view leads, among other things, to the definition of the organism as a system. A system is a structure in which all processes are connected functionally in a more or less complicated way. The rules which assert something about *single* processes occurring in a system then only hold conditionally, with reservations, since in such cases the mutual relations of the processes considered to all other processes are neglected or deliberately simplified. Almost all biological statements struggle with this difficulty. An exhaustive total representation of the mutual relations prevailing in a complex system by means of a single statement is not possible. The more general and comprehensive such a statement is, the more indefinite are the concepts used in it and the less testable are they in experience. That every organism is a system is a very general statement, which must first be supplemented, in every single case, by a whole series of special statements if it is to be testable in experience.

There are systems in nonliving nature also, so that the system character of the organism does not provide the means for a strict delimitation of the living from the nonliving. Naturally, the relations in living systems are for the most part of a much higher degree of complication. Moreover, the nonliving systems are not, in principle, closed systems, although most of them can, in practice, be regarded as such. This is not possible with organisms. On account of the persistent mutual exchange of material and energy with the environment, organisms are open systems to a great extent. All possible changes in the environment can have far-reaching effects in the system. The attempt to treat the living organism experimentally as a closed system leads to its destruction, to death. In consequence of the open-system character it is still more difficult in the case of organisms than it is in nonliving systems to draw a boundary between the system and the environment. Every biological statement, strictly speaking, not only relates to the organism but takes in a part of the environment as well. The "thinghood" of a living being is only a psychological experience of the human observer; in a strictly scientific statement the organism, or one of its parts, occurs never as a completely closed definable "thing" but only as a conventionally marked-off part of what is accessible to empirical science. From the latter point of view it is therefore not meaningful to wish to draw a boundary between organism and environment or to distinguish in principle between living and nonliving parts or substances in the organism. The materials and energies which enter the organism from the environment take part more or less intensively in the processes occurring in it, and the same may be said of those passing out of it. It cannot be meaningfully stated when these chemically or physically definable things are "alive" or when they are not. Naturally, for practical purposes, we can often make a delimitation between organism and environment which is exact enough for the purpose in hand. Similarly, we can often distinguish between living and dead parts of an organism in a way which suffices in practice.

The processes occurring in organisms are so connected with one another that, in spite of the perpetual change of the materials and energies composing the system and in spite of the most diverse disturbances and variations of state in the environment, the actual individuality of the system in question is, within certain limits, preserved or is restored after being disturbed. Only if the disturbances by the environment or the processes in the interior of the organism exceed certain limits is the system irreversibly destroyed. This ability of organisms has been called "self-regulation." The state in which a living organism finds itself is a state of equilibrium (variable within limits which are characteristic for each organism) in which, in contrast to the forms of equilibrium prevailing in the practically closed inorganic systems, a whole series of processes occurs in disequilibrium—a state which has therefore been called "dynamic equilibrium" (Hopkins) or, very appropriately, "flow equilibrium" (Bertalanffy). Many complexes of biological statements of fundamental importance contain more special statements about equilibrium states or differences of potential of the most diverse kinds—for example, statements about the function of the enzyme system in living cells, about the osmotic relations in cells, about changes of state in the conduction of nervous impulses in nerves, and so forth. Such statements need not always owe their origin to the physiological point of view; statements about states of equilibrium also play a part in the morphological point of view.

The most general statements about the state of living systems are especially concerned with the energy relations in the living world. The maintenance of the flow equilibrium or, in other words, the preservation of the potential differences necessarily present in the living organism, is connected with incessant performance of work; the organism raises in this way the entropy level of its environment; it feeds, so to speak, on "negative entropy" (Schrödinger). In connection with such considerations and similar ones, it is often customary to speak of the "activity" of the organism as one of its special characteristics. So long as this word is used only to denote the mutual energy relations between organism and environment, it is meaningful from the point of view of empirical science. Unfortunately, the use of such words and ones similar to them often leads more or less consciously to anthropomorphic ideas, like the ideas of active "forces" in physics, which have long been superseded.

In order to make the complicated system relations in organisms more easily intelligible, recourse is often had to so-called "models." In biology experimental arrangements or theoretical constructions are called "models" if they show, with deliberate simplification, one or more far-reaching analogies with the processes in the organic world. As examples, we need only mention the experimental permeability model of Osterhout, the theoretical derivation of the average cell size and of cell division from the data of metabolism by Rashevsky, and finally the molecule representation of the gene. Such models are not hypotheses in the usual sense, since they cannot be directly tested by biological observation in the individual case. Rather, they are complexes of statements which have already been verified by physical or chemical experience and are applied to biological data almost allegorically with conscious restriction and certain reservations. They share with the biological hypotheses the property of heuristic value. The model serves as a general pattern in accordance with which special hypotheses about processes in organisms are constructed which are then directly testable by biological observation. A misunderstanding of this function of models in biology may easily lead to the superficial opinion that through them a series of important properties of organisms is "explained."

Among the various mutual relations both in the structure and in the function of living systems, it is possible to distinguish more or less clearly those of greater from those of lesser importance: a kind of hierarchical order prevails within these mutual relations, which is also called "organization." This feature is also not entirely absent from nonliving systems, yet in living systems it shows a far higher degree of complication. In this general form the statement about the "order" prevailing in living systems has only a very vague scientific meaning. If we give it a purely physical meaning, then it is a statement about the uncommonly high degree of atomistic order which we find in the structure of an organism in contrast to its environment and which, in fact, occurs nowhere else in nature. If we wish to make use of the statement about the given order for the clarification of biological facts, then we must set up more special statements about the kind of this order, its rules, and their correlation with experience, in order to make it testable by observation. Failure to attend to this requirement often leads to the purely tautological use of general statements about order in the organic world.

It is not only to the whole organism that the notion of the open system can be applied; this notion is also applicable to the parts. This holds especially for the cell, which, even in the union of the many-celled organism, in both its structure and its functions clearly shows the properties of a complex open system and is therefore rightly called an "elementary organism." Since more has become known about the cell nucleus, especially about the highly complicated structure and function of the chromosome apparatus, these structures must also be given the character of complicated systems. The morphological and physiological points of view lead to a complex of statements about the system relations prevailing in the organism among its organs, cells and cell parts, and the processes taking place in them. The standpoint of genetics reveals in the genome a system of special order, which is a guiding center for all these processes and is therefore directive for the establishment of the properties and modes of reaction which can be demonstrated morphologically and physiologically in the organism. The rules for the causal connections between the system of the genome and that of the organism form further complexes of statements of a genetico-embryological point of view. With the high degree of complication which statements acquire in this way, it is understandable that every hasty generalization or simplification jeopardizes the scientific value of such statements and that only the strict requirement of empirical testability guarantees the wider utility of such formulations. For the most part, this requirement is satisfied in the domain of modern genetics in an exemplary manner.

The organism is connected with the environment by a series

of complex mutual actions. If we direct our attention to this, as we do, for example, in ecology or in the study of the behavior of the higher animal organisms, then we consider the organism and the environment, or a particular portion of the environment, as a system. In the biocoenoses which inhabit a particular environment, various organisms stand in regular reciprocal action with one another. In an increased degree this holds for closer relations between different organisms, as these are given in the various forms of parasitism and symbiosis. We can in this way extend the system notion to systems of higher order in which equilibrium states with certain capabilities of self-regulation can likewise be established. What has been concluded above from a consideration of the single organism as a system holds also for statements belonging to these domains of biology in all essential respects.

The high degree of complexity which belongs to the systems of higher or lower order considered in biology brings it about that many of the statements of biology are blanket statements. These establish a regular relation, for example, between the nutrient materials taken in and the end-products of metabolism given out or between the temperature and the formation of a pigment in the body covering or between the moisture of habitat and the occurrence of a particular plant species or between the development of a parasite and the conditions in its intermediate host, and so on. Such statements deliberately refrain from describing in detail all those processes which must be assumed as part-processes in the domain of facts under consideration. They are, nevertheless, typical statements of empirical science. The great frequency of such statements of quite a special kind distinguishes biology from other branches of natural science. Nevertheless, it is not correct to regard only statements of this kind as genuine "biological" statements and to deny this title to statements which deal with single processes in the organism and therefore are or have the type of physical or chemical statements. There are in biology no "more or less biological" statements but only statements belonging to empirical science.

2. Growth, Development, Reproduction: The Historical Character of the Organism

It is a peculiarity of living systems that many processes in them, including very essential ones, are irreversible, or run in only one direction. Every organism shows a chain of changes of state following one another in a manner which is characteristic for the organism. From a plant seed, if it does not die, a plant will grow which, if it does not die, will form seeds in its turn. An insect passes from the fertilized egg to the imago through an irreversibly directed metamorphosis, finally forming gametes itself. The fundamental processes of this event in organisms are called "growth and development," by which are understood the most complicated processes—not only the intake of substances but also the differentiation and formation of structures and functions. Moreover, processes are included in a regular manner which lead to the production of new systems of like kind and thus to the reproduction of organisms. These regular changes of state give to the organism a thoroughly historical character, i.e., its present state is essentially dependent upon its previous states and determines its future state. Although this does not distinguish organisms in principle from all other natural objects, yet this historical character is especially clearly seen in them. This rests, on the one hand, on the more or less strong dynamic of the persistent material and energetic exchange of the organism and, on the other hand, on its complex system character, which imposes an irreversible, unidirectional course upon most of the processes concerned. Corresponding to the properties of the organism as an open system, the environment plays a great part in all this as a complex of factors making possible and partly conditioning growth, development, and reproduction. Organisms are normally adapted to their environment in such a way that, in spite of the persistent threat to their existence, a maintenance of the organic forms is guaranteed.

A whole series of the most diverse biological statement complexes deals with the peculiarities of organisms here briefly sketched. What is most striking is the co-ordinated occurrence of temporally ordered processes which lead to the phenomena of growth and development. The statements in question attempt, especially whenever they go beyond pure description, to set up rules, which are empirically testable, according to which predictions can be made about the temporal succession of growth and developmental processes. Such statements can be set up and tested from a great many different standpoints. We frequently speak of the "problem of development" when we have in mind the developmental processes from the fertilized egg to the complete realization of the structural plan, or other growth processes, from the purely morphological point of view. It may here be once more emphasized that, even with such complicated problems, statements of scientific value are possible within the purely morphological point of view. As an example, the allometric growth equation of Huxley and Tessier may be mentioned. It permits predictions about the relative volume and surface changes of body parts or organs in a living system which changes its form through disproportionate growth of its parts. The validity of this law has been demonstrated for a great number of the most diverse transformations of form which an organism can experience during its ontogenesis. For exceptions, supplementary hypotheses were set up of basically the same structure. It is clear that statements like the allometry law are blanket statements under which a whole series of the most diverse and complicated processes is subsumed. The law implicitly assumes that among these processes such a degree of integration exists that a regular blanket statement is not only possible but actually holds good. This feature gives to such statements in biology their special value.

The germ-layer theory may be mentioned as a further example of a predominantly morphologically oriented statement about developmental processes. In contrast to the previous example, we have here a formulation which does not appear in mathematical clothing but works with descriptive concepts and yet, like the allometry statement, is order-analytical in origin. It gives a scheme for the course of the early development of the most diverse animal organisms and for the fundamental processes of organ formation. It also is a blanket statement about very complicated processes of growth and differentiation which are represented in it according to precepts from the morphological viewpoint.

Purely morphological statements about questions of growth and development carry with them the danger of losing their empirical scientific character by slipping into a realism of concepts. If we say that a growth process "obeys the allometric equation," we naturally do not mean that this equation is a "force" which pushes the growth in this direction, and that in this way the manner of growth is satisfactorily "explained." When we see the statements of the germ-layer theory confirmed in the ontogeny of an organism and observe the developmental processes by which the "structural plan of the organism is realized," this, again, does not mean that this structural plan and its various developmental phases are present, as it were, as pre-existent molds into which the "stuff" is poured like a plastic mass. Usually only thoughtless formulations of this kind or such as are chosen in an effort to reach better picturability can lead to serious errors in biological thinking.

Morphological concepts can be defined, in a statement which is useful in empirical science and is free from logical objections, only as assertions about structures which are testable in experience, and they cannot occur as independent concepts in abstraction from this testability. The morphological consideration of processes of growth and development which is applicable to the individual case embraces only visible forms of proportions and spatial relations. But these forms are constructed of living cells, cell aggregates, and their secretory products in the course of complicated processes of multiplication, growth, and differentiation of these living systems. All these processes are also accessible to the other elementary standpoints of biology, and only a synthesis of all possible points of view opens up the prospect of penetrating into these most complex connections. This path has been successfully followed, for example, by experimental embryology. Spemann's theory of organizers is a complex of statements which are empirically testable by experimental interference with embryogeny. They say something about the material actions which issue from certain embryonic regions and about the way in which other embryonic regions react to these actions by growth and differentiation processes. In addition to these, we have statements about the relations between the state of the environment and the developmental processes. The complex interplay of the single processes of embryogeny, governed by organizers of various orders and modifiable by the influence of environment, corresponds with the nature of the organism as a complex open system. The whole physiology of growth and reproduction, the doctrine of hormones and active substances, constitutes a statement complex of a similar kind. In all these statements it is customary to call the magnitudes, measured in experiments, which lie outside the practical surface limits of the organism against the environment the "outer factors," those established within these limits the "inner factors," of growth, of development, etc. But a separation in principle of "outer" and "inner" in such statements is not possible. A hormone which is produced by certain cells of the body is also an "outer" factor for other cells, and the same holds for the action of certain embryonic parts upon the others. Quite frequently, theoretical misunderstandings have arisen through the transformation of the purely conventional significance of the words 'outer' and 'inner' into one of principle.

In a much narrower sense of the word, we can speak of inner factors of development if, turning to the genetical standpoint, we make statements about the part played by the genome as a guiding center of growth and differentiation processes. In an embryological experiment, by operative interference with development, by altering the outer conditions and the like, the present parameters of state of the open system in process of development are changed, and the ensuing changes in the course of development are followed. In a geneticophysiological experiment, by the introduction of a mutated gene, of a genetic defect, and the like, a different state of the whole system is given from the beginning. The typical reaction of the whole system and its parts is altered. Nevertheless, under those changed basic conditions the physiological method of embryological experiment can still be applied. In this way the possibilities of causal-analytical exploration of the processes are significantly enlarged. Statements can be formulated, for example, in which a gene-conditioned active material is represented as an enzyme for the production of certain pigments. Or the moment in development can be demonstrated in which a gene-determined rhythm of certain cell divisions determines the later form of an organ. Or certain complicated transformations in the metabolism are shown to be a chain of particular gene-conditioned enzyme reactions. Statements of this kind connect the concept of the gene or of its alleles, which is empirically testable by means of hereditary analysis or cytogenetic investigations, with morphological and physiological concepts which are verifiable by observation of structures and by physiological and biochemical investigations. They connect structural and functional properties of the genome with structural and functional properties of the organism and are thus statements about the rules correlating two systems of different order with one another. In this borderland between genetics, physiology, and embryology, biological research reaches the highest degree of integration of its technical possibilities but also the highest degree of heuristic fruitfulness.

3. Organic Diversity and Its Structure

The organisms which live on the earth today appear to us in a very great multiplicity of distinguishable types or forms. This multiplicity is a discontinuous one; the types are not connected with one another by all conceivable transitions. This multiplicity is, in fact, very great, but it is finite; we can determine the number of distinguishable types. This work has been completed, at least for certain groups of the animal and plant kingdoms. The given multiplicity does not admit of an ordering by purely quantitative characters, as is the case in, say, Mendeleev's system of the chemical elements. In the domain of organisms, therefore, we could not predict with certainty the properties of a form at present unknown, as was the case prior to the discovery of still unknown elements. Nevertheless, the given multiplicity of organisms is not completely devoid of order, that is, free from any characteristic which permits us to set up a definite order in the multiplicity. On that account men have, from the earliest times, attempted to set up a system of the animal and of the plant kingdom. Attempts at systematic classification according to some one or a few arbitrarily chosen characters have led to the so-called "artificial" systems, e.g., the Linnean system of the flowering plants. We call our present-day system a "natural" system because, in setting it up, as many characters as possible which have proved to be the essential comparable characters, after a careful comparison of the material, are taken into account. The system is based on the concept of species, which will occupy us later and is, in addition, constructed in accordance with the more or less great morphological, possibly also physiological, similarity of the types. Similar species are united to form a genus, similar genera to form a family, and so on. In the establishment of these higher units the systematist proceeds in such a way that he uses for their characterization those properties which strike him, in a comparison of several like species, genera, etc., as special common properties of all members of the groups concerned and which are specially suited to their delimitation from other groups similarly formed. These properties are then distinguished as the generic, family, or order properties. The higher systematic unit becomes in this way an entity which is defined by a few general characters. These are indeed demonstrable—although often in modified form—among the representatives of all species which belong to the higher unit in question; yet the representatives also have the special characters of the species. It is agreed that the higher systematic units, in the first place, are only categories created by the ordering human understanding. Only the doctrine of evolution attempts to see in them natural units, namely, related groups of species with common ancestry. But, quite apart from this theory, it is very peculiar from the purely systematic standpoint that it is possible to unite species into higher systematic units by means of the same ordering principles in all the various types of living things, in spite of their great diversity of organization. In the great multiplicity of living things an order of a very peculiar structure seems to prevail, which is not comparable to the multiplicity of inorganic things.

Every classificatory approach is based on the concept of species. The differences of opinion about the definition and use of this concept in biology are indicative of the great difficulties of this group of problems. The concept of species is used in very different senses by different biologists, according to their point of view and the problem in hand. Nevertheless, most biologists are of the opinion that "the species," in contrast to the higher systematic units, is to be regarded as a *primary* natural unit. The contradiction which seems to occur here is in some degree understandable in view of the present state of biological knowledge but still often leads to misunderstandings. The pure systematist understands by 'species' a type, a "species picture," which is defined by the enumeration of a series of morphological, possibly also physiological and ecological, characters. Organisms found in nature can be recognized as "members of this species" for the most part with certainty by comparison with this type species. It is left to the tact of the specialist—schooled by experience—to regard certain deviatory types as varieties or races within a species or as another species or, in particular cases, as an aberration or anomaly. When we say that "within this genus so and so many species have so far been described," we are using the word 'species' in this purely systematic sense. That this use of the species notion has persisted for the practical purposes of determining individual organisms and assigning them a place in the given multiplicity is for us a proof of the finiteness and discontinuity of this multiplicity and of the constancy—at least relatively—of its ordering structure. In the view of the "species" as an elementary unit of just this order there lies a weighty argument for its estimation as a natural unit. Yet, even for this purely systematic construction of the concept of species, it is necessary in many cases to institute investigations in the field or the laboratory which go beyond the limits of the comparative method. The membership of the very different female forms among butterflies with unisexual polymorphism within one and the same species only becomes clear when their complete fertility with the uniform-appearing males is established. The recognition that drought and rainy-season forms in other butterflies, which look so different, are included in one species takes place only when the natural connection between these generations is ascertained. Here a factor plays a decisive part which is not directly derivable from the comparison of forms—that of the genetic connection.

The "members of a species" usually occupy a continuous habitat in which they form a reproductive community. The requirement of complete fertility within the species is for the most part silently presupposed. Wherever the region of distribution is discontinuous, the virtual fertility between members of a species is taken into account for the delimitation of the concept of species even from the standpoint of the systematist, although not in the first instance. But there are cases in which two groups of organisms, which are distinguishable morphologically either not at all or only with such difficulty that the systematist would at most speak of varieties, are accepted as two species because the two groups of organisms are not fertile with one another and thus form—in spite of a common habitat—two completely separate reproductive communities. On the other hand, there are cases in which species which are well distinguished morphologically, ecologically, and geographically can produce completely fertile hybrids. In yet other cases there are, among the members of a systematically unobjectionably defined species, various groups of individuals which are not fertile with one another within their own group but are fertile with the individuals of another group. This is the effect of genetical sterility factors which coexist in the population. The definition of the species as a reproductive community or at least as a potential reproductive community is thus not quite satisfactory, but it at least denotes a real biological unit.

With the inclusion of the criterion of the reproductive community in the species concept, but especially with the application of this concept in every point of view other than the taxonomic, the concept of species changes its meaning and is to be differently defined accordingly. In such statements, for example, as "This species occupies this or that region, it lives under these particular ecological conditions, it has a particular genic structure," we no longer mean the species type of the systematist but mean instead the totality of the individuals which we customarily include in this species type on the basis of the taxonomic rules and thus a collective of empirically given individual organisms. It is a source of frequent misunderstandings that the word 'species' is used both in the sense of the systematist for the abstraction "species type" and in the other sense for a collective of individuals, but without noticing the diversity of meaning. When we speak of the species as a sum of empirically given individuals, we must specify the point of view according to which this collective is to be marked out and the characters which specify it as a collective of a particular structure. The concept of "species" then passes over into the concept of a "population," which will be discussed in the next chapter. It is impossible within the limits of this monograph to enter into the whole complex of problems connected with the concept of species. This much only is established: that in every biological statement in which the notion of species occurs the definition of species must be so formulated that it is unambiguously explained how the content of the statement is testable in experience.

In the drawing-up of natural systems the systematist works chiefly with the methods of comparative morphology and hence according to order-analytical principles. Experience allows him to choose those statements as the most productive which enable the given multiplicity of forms to be ordered in a way which corresponds best with all biological requirements. The use of the concept of homology, its extension to the developmental stages of animals, the knowledge of the essential significance of certain characters (e.g., of the flower structure in plants) for the setting-up of homologous structural plans, and the like, will cause him to introduce a very diverse valuation of the characters of an organism and to build up the system in accordance with these valuations. The system obtained in this way makes use of the concept of the "relationship" of the various living forms. Two species are said to be nearest related when, within the same genus, they have the greatest number of essential characters in common. In the same sense this concept of relationship is extended to the higher taxonomic units. We are so accustomed today to thinking in evolutionary terms that we often forget that this concept of systematic relationship was originally not meant in the evolutionary sense at all and cannot be used in this sense in pure taxonomy. Without any evolutionary accessory meanings, taxonomy is a scientifically unexceptionable and essential part of biology. It leads to statements about the given organic multiplicity which reveal a series of noteworthy properties of this multiplicity. It is, in the first place, noteworthy that, generally speaking, in the most diverse domains of the organic world groups of species can, according to like ordering principles, be united to form genera and these again to form higher units and that in consequence similar "relationship relations" seem to govern the multiplicity. Further, it is very peculiar that the degree of multiplicity now prevailing in the various domains of the organic world can in no way be predicted according to a general principle or derived from more general laws. While some genera show a great number of species, others are represented by only a single species, and the same holds for the other higher taxonomic units. The potential multiplicity, which we can imagine to ourselves on the basis of what is given in nature as a purely intellectual construction, seems to be realized in the various groups of the world of organisms in very different densities, with very notable gaps or accumulations. Here, also, the multiplicity among living things differs essentially from that in nonliving nature, for the given multiplicity of the elements or of the crystal forms can be derived, more or less satisfactorily, from the more general statements of physics. The purely order-analytical method of taxonomy is not able to undertake such derivations for the multiplicity of living things. The theory of evolution attempts to do this in so far as it transforms the order-analytical statements of taxonomists into causal-analytical ones. In this way the concepts of systematics, for instance, the species concept or the concept of relationship, gain new definitions and new correlations with experience. These new problems will occupy us later (Chap. II, C, 5).

Ways of Work in Biology

4. The Population as the Natural Form of Existence of Living Beings

In nature there are no species in the systematist's sense but only individuals of various ages, in various stages of development, and in various physiological states, which we speak of as belonging to a particular species—in accordance with the rules of correlation chosen for the definition of the species concept. All together, they form the population, the real form of existence of the species in question at a particular moment of time. Various statements about an organism which are not to hold only for a particular individual and which, moreover, are to be empirically testable in many ways are therefore always statements about the population, about a natural collective. There are various considerations which compel us to formulate collective statements about a population. The mode of adjustment of an organism to its environment; its preservation and successful self-assertion in the environment; the whole complex of ecological, biogeographical, and evolutionary problems requires statements in which not individuals but the natural collective of the population is spoken of and within which the single individuals are distinguished by an average similar norm of reaction. In this way the population is regarded as a system of higher order within which definite relations prevail between the individuals and which, as an open system, stands in mutual relation to the environment. The inner system property of the population becomes especially clear in cases in which the individuals are united into partial systems, accompanied by far-reaching differentiation among themselves, as in the social insects. Also, a united herd as a community of action and a pair in process of reproduction furnish examples of partial systems in this sense. In these and similar cases the whole population is constructed from such more or less firmly united subsystems to form a single supersystem. The population of every organism has its particular structure as a system which can be described by means of a series of empirically testable statements.

The genetical point of view applied to a population likewise reveals its system character. The potential multiplicity of the genome, resulting from the number and spontaneous mutability of the genes, is realized and distributed in various degrees in the reality of a population. The genetics of populations formulates statements about this genetical multiplicity on the basis of experimental analysis. For the distribution of a mutated allele, for the origin of new allele combinations, for the composition of the population out of such combinations and similar processes, the way in which the reproduction and maintenance of the organism concerned are regulated is of the utmost importance. For organisms without sexual reproduction the rules are quite different from those for organisms with obligatory or facultative sexual reproduction. In the case of obligatory autogamous hermaphrodites (for example, many flowering plants) other presuppositions are given than is the case for organisms with separate sexes or even for those the population of which is composed of groups which are sexually isolated from one another. The genetics of populations regards the organism and its population as a genetical system, the peculiarity of which lies not only in its stock of virtual and actual genetic multiplicity but also in the individuality of the apparatus which serves for the distribution, enhancement, or diminution of this multiplicity.

The relations of the genetical system to the environment are conditioned, above all else, by the testing which the gene-conditioned type of reaction gives to the organisms in their struggle for existence. The concept of the selection value of the various gene combinations is defined, in statements belonging to the genetics of populations, not merely speculatively but in accordance with testability by observation and experiment. The study of the population with respect to its potential and actual fertility and its relations to the environment forms the foundation. Statements about the nature of the reproductive relations are therefore of the greatest importance for all population problems. We must take into account whether the reproduction in a population shows panmixia; to what degree the virtual panmixia is actually realized in the population; whether genetic, sexual, ecological, or geographical isolating factors break up the population into several or few separate reproductive communities; and the like.

Since statements about populations concern a collective, they very often have a statistical character. The mathematical methods of population statistics form the equipment for all biological statements about populations. Making more or less arbitrary assumptions about the initial magnitudes, population statistics formulates, in a purely mathematical way, the consequences which result from the peculiarity of the organism concerned as a reproductive and as a genetical system. These general statistical statements about the waxing and waning of a population, about the increase and decrease of its genetic multiplicity, about the formation of races, etc., are models which, by the substitution of special magnitudes, become transformed into empirically testable hypotheses in the special, individual case. The whole population of an organism is, in its inner system relations and in its relations to the environment, in an equilibrium, which can experience certain fluctuations and which, by analogy with the state of a single organism, can be called a "flow equilibrium" or a "dynamic equilibrium."

5. The History of Organisms

It is not only the consideration of the single organism which compels us to regard the organism as an open system with a historical character (see Chap. II, C, 2). The momentary state of a whole population is also dependent on its previous states and determinative for the future states of the population, subject to perpetual mutual interaction with the environment. If we extend this point of view to the whole organic world, we come necessarily to the total view that the populations of the various living things stand in mutual relation to one another and to the environment and thus, together with the total environment, form a complex system in which dynamic equilibrium prevails, subject to certain fluctuations and displacements. In such a total view the question of the earlier states of the organic world cannot be separated from the question of the history of the earth, which is regarded as a historical object by the remaining branches of natural science.

The question of the history of organisms would therefore have to be regarded as a genuine problem of empirical science even if we possessed no witnesses concerning earlier states of the world of organisms. But we do possess such witnesses, in the form of fossils which give us a picture, if only an extremely incomplete and defective one, of the diversity of living things which have formerly lived on this earth. The information which these remains give us is especially defective regarding the very long period of the earth's history prior to the Cambrian, during which living things must also have existed. In spite of this defect, paleontology shows us, in a convincing way, *at least* that the multiplicity of organic forms was formerly different from what it is today, that this multiplicity has undergone great changes during the history of the earth, that certain forms of life have died out without leaving any descendants behind, and that certain great groups of the plant and animal worlds, which we regard as the most highly organized, have not existed in early epochs. In this way the question of the processes which have taken place in the history of organisms and the investigation of such processes by the methods of empirical science have become essential problems of biology.

The totality of such processes is usually called the "evolution of organisms," and the part of biology concerned with it is called the "doctrine of descent" or "theory of evolution." No subdivision of biology is to such a degree choked up with unrestricted theorizing or fogged by fanciful speculation or made the battleground of extrascientific differences of opinion as this. Especially when the question of the descent of mankind is taken into account, this purely biological problem becomes involved in the arena of philosophical controversies. A clarification of this situation, which is so interesting from the cultural and logical points of view, will be attempted later (Chap. III, C). Here it must be emphatically pointed out that, in the domain of evolution also, only those questions have a legitimate place in scientific biology which are formulated as problems of empirical science. Moreover, the whole complex of evolutionary problems can be treated in biology only in accordance with the principles of empirical science and in this respect is in no way distinguished from other branches of biology. Statements about the history of the organic world have, of course, the peculiarity that they relate in part to epochs of the remote past from which no witnesses of the organisms, or only indirectly accessible ones, have come down to us. For that reason a large part of the statements about evolution will forever have the peculiarity of not being empirically testable in practice and will thus consist of hypotheses about the confirmation of which we cannot judge. If these hypotheses are to maintain their place in science, they must, of course, be so formulated as to be in *principle* empirically testable. They then often have the property that, by analogies with practically testable statements, something can be inferred with a certain probability about the degree of confirmation which these hypotheses might have in case there was a practical possibility of testing them. In no case are hypotheses admissible if their structure is such as to preclude the possibility of empirical test *in principle*.

It is customary to treat the whole complex of evolutionary problems in three groups of statements: (1) Statements which assert that evolutionary processes have, in fact, taken place. Such statements seem to be fully verifiable, in view of the paleontological findings. (2) Statements about the way in which the evolution of the existing multiplicity of living things has, in general or in particular, been brought about—thus the assumption, say, of the gradual transformation of one species into another or the assumption of a phylogenetic tree with dichotomous branching and so of a real derivation of several species from a common ancestral form. In a more comprehensive form, such statements set up a common phylogenetic tree for the whole organic world ("phylogeny"). (3) Statements about the processes in living beings, and between them and the environment, which have taken place during evolution and are going on still and confront us phenomenologically in their totality as the evolution of forms. This group of considerations is often referred to as the question of the "forces, causes, factors, or mechanisms" of evolution. Such formulations conceal within themselves the danger of conceptual realism leading to misinterpretation, as though there were certain scientifically accessible realities, in addition to the processes, in the organism or between this and the environment which "guide" the events of evolution, just as in an obsolete, primitive view of physics the "forces" push and pull the inert particles of matter. For formulating statements belonging to groups 2 and 3 and for their empirical testing, the following possible points of view especially are available: (1) the paleontological; (2) the biogeographical; (3) the comparative anatomical and comparative embryological; and (4) the genetical, that of the genetics of populations and related experimental methods.

Paleontology contains statements about hypothetical evolutionary series, which can be tested by fossil material. When the material lies in easily datable, uninterrupted strata, statements about limited parts of a phylogenetic tree can be regarded as empirically verifiable, although immediate proofs for the natural connection of the generations and the statistical properties of the populations in their full extent are not obtainable from fossil material. Statements about greater phyletic connections and about the whole phylogenetic tree of organisms can be verified or falsified by the material of paleontology only with greater or lesser probability. Nevertheless, paleontology can set up certain more general hypotheses about phylogeny which, being obtained by the method of generalizing induction, can be tested on fossil material. As examples, we may mention the theory of typogenesis and typostasis of Schindewolf and the so-called "Dollo's law." The paleontological point of view is thus quite capable of setting up empirically testable hypotheses about the presumed course of evolution without the help of other possible points of view. Regarding the processes which have taken place in the organisms and between them and the environment during evolution, paleontology naturally has little to contribute. It is misleading when some paleontologists make use of concepts borrowed from other branches of biology—such as the concept of "mutation"—which are scarcely testable on fossil material.

The point of view of plant and animal geography proceeds

from the distribution of the organic multiplicity prevailing at present in its relations to the inhabitable space, and from the displacements in this distribution which are still unambiguously determinable historically. It formulates hypotheses about phyletic connections which are testable by this distribution. The theory of *Rassenkreise* of Rensch and the theory of race and species formation by geographical isolation, as on islands, are examples of the way in which evolutionary hypotheses are formulated as statements which are testable by biogeographical observations. In its retrospective formulation of questions, this point of view finds its connection with paleontology; in its ecological aspect its affinities are with the statistics of populations and the genetic and other experimental methods. Statements obtained by the biogeographical point of view are naturally restricted to fragments of the total event of phylogeny— mostly to the youngest branchings of the phylogenetic tree— while questions concerning its fundamental structure are beyond the grasp of this method. On the other hand, the statements of biogeography and ecology are often connected with the question of the "causes" of evolution, since they deal with the observed distribution of species and races in their testable connection with the environment.

The comparative anatomical and embryological point of view is that from which the idea of evolution chiefly took its origin and from which even at the present day argumentation takes its starting point. The system obtained purely order-analytically is transformed into a causal-analytical one by the introduction of the idea of evolution. The ideal "relationship" according to greater or smaller similarity is reinterpreted as a narrower or broader real blood-relationship. For that reason the statements of systematics and comparative anatomy become more picturable for the thinking human being, and therein, chiefly, lies the charm of this idea. This is all the more intelligible when we consider that the given multiplicity of recent and fossil organic forms cannot otherwise, in its organization, be reduced to any other rules of empirical science and therefore would seem completely unintelligible without the idea of evolution, i.e., would be without evident connection with other data of natural science. Statements about the supposed phylogenetic tree or single developmental series, which we owe to the comparative morphological point of view, still clearly betray their order-analytical origin. The systematist distinguishes his species by means of certain characters and brings together in one genus all those species which have certain common characters, which are now called the "generic" characters. In a similar way the comparative anatomical point of view constructs the hypothetical phylogenetic tree by regarding the special characters of the race or species as the phylogenetically youngest, those of the genus as phylogenetically older, and so on. The hypothetical "common ancestral forms" and "missing links" of the doctr ne of descent arise from similar order-analytical thinking, and in this abstract form, in which they issue from the comparative method, they naturally cannot correspond in all points with a real living being. A phylogenetic tree constructed in this way consists at first of structural plans and developmental series of an abstract kind, the correlation of which with experience now must take place on the basis of the hypothesis of evolution, according to rules of correlation other than those used in the purely morphological point of view. For the constructed phylogenetic series the real genetic connection must be demonstrated or made probable in experience or must be refuted. This requirement is often overlooked, and statements about "relationship" or about developmental series in the comparative anatomical sense, obtained by purely order-analytical methods, are formulated as statements about real evolutionary processes; in such cases the idea of evolution enters as a kind of habit of thought which is taken for granted. This holds in a high degree for comparative embryological statements, which, in the form of the so-called "biogenetic basic law," were formerly often very uncritically treated. It must nevertheless be emphasized that all statements obtained by the comparative methods have their legitimate place in biology as hypotheses about the evolutionary process, if they are formulated as statements which are testable in principle, and thus even when they are not testable in practice, perhaps because the fossils can give no information about this part of the process, and the other points of view, such as the biogeographical and the genetical, also offer no point of attack for a test.

Comparative anatomical research chiefly formulates statements about the supposed course of evolution. Through a functional consideration of the anatomical structures, its theoretical procedure is, however, always connected with the question of the evolutionary factors. The concept of "adaptation" comes from the general experience that organisms in their structures and their functions are so well fitted to their specific environment and are accustomed to react to fluctuations in the state of the environment in such a way that, as a rule, their lives as individuals and as populations are to a sufficient extent guaranteed, in spite of the relatively great threat to their existence. The concept is thus ambiguous from the start: it denotes both a state, that of "being adapted," and also certain processes, that of "adapting one's self," here, in the first instance, in the sense of an individual type of reaction. Yet in the doctrine of descent the concept of adaptation is chiefly used in the sense that in the assumed changes of the species type or in the emergence of new species is seen a process of "adaptation" to changing, or new, complexes of environmental conditions; or this process of adaptation is called the "cause" of the evolutionary process. In such formulations we very frequently encounter, in more or less explicit form, errors which are traceable to conceptual realism, which define "adaptation" as an "active principle" and the change of forms as the "effects thereby brought about." Formulations are only free from objections from the scientific standpoint if regular connections are hypothetically expressed by them between the supposed changes of form and the states of being adapted to the environment, in such a way that they are in principle testable in experience. The attempt to make such formulations picturable very often misleads us into representing evolution as though it had led out of a state of not being adapted, or of being ill adapted, into one of being well adapted. From this there result paradoxical discussions such as that of how, in that case, such a complicated organ as the eye could come into existence by virtue of nothing but small steps of adaptation, when it can only perform its function properly in its completely developed state. But we have no right to assume that there have ever existed nonadapted or ill-adapted organisms, for these would not have been viable. Even the most simply organized living things are ideally adapted to *their* specific environment. Types in which the new adaptation to a changed environment has not been sufficiently successful have obviously died out. In a correct formulation we have to do with the assumption of processes in evolution through which particular adaptations have passed over into other adaptations and thus not properly with adaptation but only with changes in adaptation. The same care in the use of concepts is called for when we speak of "higher development," of "perfecting," or of "progress in evolution." Admittedly, the organization of the higher animals and plants exhibits a higher degree of complication than that of the Protista. Yet this does not mean that the former have acquired a higher degree of adaptation and, with that, of life-efficiency or that we are otherwise justified in introducing a value-judgment. The biological survival value of a species in its proper environment is the only scientifically usable measuring rod.

In dealing with the complex of theories constituting the doctrine of descent, an essential change has come about, since it has become possible, through the progress of genetics, of population genetics, and of related experimental tendencies, to make subdivisions of this field immediately testable either experimentally or by comparative methods. The fundamental assumption of the doctrine of descent presupposes that all those processes which have led during evolution to change in the organic multiplicity in principle also take place today and that the state of this multiplicity at present, as well as every past state at a particular moment of time, is a transition state from a previous to a succeeding one. It must therefore be possible in principle to demonstrate the elements of the evolutionary processes in the organic events of the present day. Those tendencies of research which deal with these questions on the borders of genetics and evolution formulate, for example, experimentally testable statements about the selective value of genetically conditioned differences, about the genetical differences between geographical races, about the genetical differences between species which form fertile hybrids, about the genetical basis of the sexual isolation of races and species, and so on. Comparative genetics formulates statements about the homologizing of the gene stock of related species and about the properties of organisms as genetical systems with regard to the evolution problem. The inference of the genome—of its mutability, its possibilities of new combinations, its structure as a system, and the possibilities of change in its structural basis—opens up undreamed-of new aspects for the treatment of evolutionary questions in the form of scientific, directly testable statements. Questions about the evolution of organisms are often transformed in this way into questions about the evolution of the genome. Admittedly, they do not thereby become simpler, but they become clearer and become accessible in many respects to experimental testing. But, especially, the evolutionary problems become in this way removed from a purely phenomenological study of the types of organisms and become related to elements of which we have some knowledge that they stand as guiding centers behind the phenomena of organic forms and their modes of reaction and that they possess quite definite, unambiguously definable properties. It is therefore not correct to see in the new synthesis of genetics and evolutionary research only a continuation and a more or less important supplementation of the former comparative morphological method. For evolutionary questions, it is far more important when it is demonstrated that in the genome of species there are homologous genes, the properties and function of which in the genome are very well definable and empirically testable by various methods, than when a homologizing of properties is carried out under the morphological point of view. Only those will fail to recognize this who are accustomed to treat everything that can be empirically established as "properties" of a "bearer" existing independently of these properties as a "subject." In this conceptual-realistic view, genes, chromosomal structures, and physiological properties are, just like the peculiarities of form, only "properties" of a species or race, which, from being an abstract entity, becomes the "bearer" of all these properties. Such "species," as well as their supposed phylogenetic changes, are naturally inaccessible to empirical scientific treatment. The individuals or populations whose properties are empirically inferred are scientifically defined only by the totality of these scientifically accessible properties.

Naturally, only questions of race and species formation, of genetic differences between races, species, and genera, are accessible to treatment by the experimental methods just sketched, while these are applicable only to a slight extent or not at all to the question of a genetic connection between the higher systematic groups of organisms. The genetico-experimental point of view restricts its statements, therefore, as a rule to the finer branches of the phylogenetic tree, and leaves open many questions about the total course of phylogeny. On the other hand, it offers the great advantage of contributing important statements about processes during evolution and of subjecting the generally applicable questions about the "factors" and "forces" of evolution to a treatment which is free from objections from the point of view of empirical science. From the point of view of method, we often distinguish today between the realm of "microevolution," which is accessible to genetical methods, and that of "macroevolution," which is less accessible to them, without at the same time wishing to attribute to this distinction too great an importance in principle.

It is well known that for a long time two theories under various forms have played a great part in evolutionary research: the theory of natural selection, sometimes rather inappropriately called "Darwinism," and the theory of direct adaptation, or Lamarckism. Most modern biologists, especially under the impress of the development of genetics, support some variant of the selection theory, often called "Neo-Darwinism." Lamarckism has few supporters today, especially on account of the repeated failure of attempts to verify special hypotheses derived from it. It may be emphasized that it is quite possible to formulate entirely different views about the origin of the given organic multiplicity as empirical scientific hypotheses, e.g., the assumption of an origin of new species from dead substances or from "nothing." Such hypotheses, however, have little prospect of verification by experience and present great difficulties to a consistent incorporation into existing knowledge. The question of the first origin of living things on the earth can likewise only be formulated as an empirical question within scientific biology. It has led to the setting-up of a series of more or less unobjectionably formulated hypotheses, the verification of which has, however, not yet been possible, partly for practical reasons contained in the hypotheses themselves. This question also has been excessively emphasized for extrascientific reasons and made a topic for polemics—often, unfortunately, in a very unobjective form.

Although the whole complex of problems thrown up by the theory of evolution must in many of its parts always remain in the stage of a hypothesis which is not testable in practice, yet it has done invaluable heuristic service in all branches of biology and has therefore become an indispensable part of the method of biology. In view of the multiplicity of points of view regarding evolutionary questions indicated above, it is not surprising that representatives of the various subdivisions of biology, such as systematists, morphologists, paleontologists, biogeographers, and geneticists often put the problems differently and give the hypotheses a different meaning from their several points of view or estimate their empirical confirmation differently. Differences of opinion which come to light in this way have often originated clarifying discussions and so led to fruitful new efforts. Unfortunately, such conflicts are often unpleasant and fruitless, owing to a lack of understanding of the logic of science. In recent years a clear and far-reaching approximation of the various viewpoints has taken place, and various books, as well as discussions at congresses and symposia, allow us to recognize clearly the development of a new synthesis of all possible points of view.

Moreover, in this most difficult branch of biological investigation a phenomenon has become clear which in many other branches of biology, and in all pure empirical sciences, can be regarded as a touchstone for the fundamental confirmation of the methodical path of these sciences: the spontaneous convergence of all lines of development in the science toward a closed, consistent picture of the world.

III. The Significance of Speculation in Biology

Poesis doctrinae tamquam somnium.—FRANCIS BACON

To follow the historical development of the methods and concepts of modern natural history from its prescientific sources and to point out the various phases of this development in its interaction with the contemporary spiritual and social situation would carry us beyond the limits of this monograph. Such an investigation would make much in the present state of biology more understandable than would a criticism purely from the methodological point of view. It would also show that the significance of speculative thinking has played a very different role in the development of this science at different times. Under the indefinite and vague blanket label of "speculation" there are brought together here all possible teachings, doctrines, systems, theories, and statements which deal with living things but which do not satisfy those prerequisites which we have learned to regard as fundamental properties of biological hypotheses and theories. Speculative theories are distinguished from the typical form of statements belonging to empirical science chiefly by the great generality of their sentences, by the imprecise definition of the concepts used, by the lack of unambiguous rules of operation for their use, and by various other features which make the empirical testing of these statements, or the derivation of special testable statements from these general theses, impossible. These general statements are often tautological. They frequently deceive on account of their high degree of "picturability" or by claiming a higher "explanatory value," or they win the reader by seeming to give a simple common solution to very heterogeneous problems of the most diverse fields of natural science.

The situation in biology at the present day is characterized by the fact that the scientific drive of its various disciplines is almost exclusively oriented in the empirical direction and scarcely takes note of the existence of speculative movements. In the textbooks and manuals, in the numerous periodicals well known to the specialist, we shall look almost in vain for essays of a speculative kind. At the great international conferences in which the latest advances in the biological disciplines are reported and in which, by lively exchange of ideas, the directions for further work are obtained, and in narrowly limited symposia serving to synthesize various research tendencies, speculative ideas are scarcely mentioned. And yet there is a large genus of literature of this kind which is distributed in the form of corpulent books, of brochures, of discussions in philosophical journals, of reviews of philosophical congresses, and of articles in general or popular scientific magazines. They are rarely read by experts and are absent from the specialized libraries, and yet many books having a speculative biological content reach higher sales than the best textbooks. Their readers are chiefly educated laymen, the representatives of neighboring sciences seeking a general orientation, and persons of general education. Through the reading of such works the erroneous opinion is often propagated that "biology" is a peculiar, predominantly speculative science which is antithetical to the empirical sciences of zoölogy, botany, etc., and even that these are subordinate to it. Among the authors of the speculative literature are some philosophical experts who have themselves never worked at biology and often have only a very imperfect and superficial understanding of the biological literature which they have read. But we also often find among them biological experts who have acquired great merit by research but have abandoned it in their later years and devoted themselves wholly to speculation. Another group consists of prominent representatives of biological disciplines, frequently of those belonging to applied biology—

doctors, technicians, and the like—who in their busy lives have found few opportunities for occupying themselves with fundamental scientific questions but occasionally feel themselves obliged to publish speculative effusions. To follow the psychological roots of these phenomena would be an attractive task but one which lies quite outside the limits of this work.

It is in any case a phenomenon which is so clearly impressed on no other branch of natural science and is especially characteristic of biology that alongside the system of expert science such a rich speculative literature should grow up, without there being at the present day a living connection worth mentioning between the two fields. There are a few leading biologists who have undertaken the troublesome task of a critical analysis of speculation from the empiricist standpoint: among them are M. Hartmann, A. Bünning, H. Winterstein, and W. Zimmermann. It is certainly no unjustified reproach when the theorizing authors complain of the complete lack of interest in speculative ideas on the part of most scientific biologists. Much leading-astray by worthless literature, much conscious or unconscious misuse of science, much baseless spiritual conflict, could have been avoided if more interest in general questions had been shown among empiricists. Many platitudes and crudities of the empirical specialists, many errors of the universalist striving after knowledge, could be done away with, if, in the course of a lively discussion based on considerations from the logic of science, it could be shown more clearly what the tasks and limits of empirical scientific thought and the prospects for a synthesis with other intellectual possibilities are, especially in the field of biology. In this positive sense the following attempt at a short survey should be judged, in spite of its partly negative criticism.

A. The Psychological Function of Speculation in Scientific Biology

It is not to be denied that speculative ideas of a general kind, which cannot be regarded as empirical hypotheses in the proper sense, can have a legitimate function in biology by promoting the development of this science. There are even some authors who see a danger to the future development of modern biology in its impoverishment of speculative inquiries. It is without doubt the case that in the field of biology, as well as in other fields, the present-day empirical structure of science has gradually developed from a state allied to the philosophy of nature of former times, whereby the speculative views of prescientific and early scientific times have formed the historical, spiritual background. After the European Middle Ages had for the last time offered a world picture of astonishing unity and completeness, we witnessed a gradual dissociation between empirical research and the philosophy of nature. After a period of empirical progress in the sixteenth and seventeenth centuries, there followed in the eighteenth century a time of stronger emphasis on ideas belonging to the philosophy of nature, which exerted their influence far into the nineteenth century and in which are to be sought the roots of many concepts of modern natural science. These concepts have, of course, strongly changed their meaning in the course of development, and many problems and alternatives of that time have become pointless through the progress of empirical scientific work. Nevertheless, it was these concepts that stimulated the research and first gave it direction.

An example of this is furnished by the opposed views of preformation and epigenesis in the ontogeny of organisms, which in various transformations are to be recognized throughout the history of biology. Both have their roots in primitive, prescientific ideas. While the philosophical views of antiquity and of the Aristotelian medieval Scholastics seem rather to have supported the ideas of preformation, the Christian philosophy of Augustinian origin preferred epigenesis. In the eighteenth and nineteenth centuries the two views stood opposed to each other in the form of fundamental biological theories. The present state of knowledge has shown this disjunction to be pointless. These discussions between the two tendencies, carried out chiefly in the field of the philosophy of nature, were often helpful for empirical research by giving it direction but were also sometimes cramping and misleading. In any case, the historical roots of the concepts of modern genetics are often traceable to the theory of preformation, while the concepts of developmental physiology often have their origin in epigenetic views. The general theory of sexuality of M. Hartmann provides an example from recent times. The author of this monograph has attempted to show that the general statements of this theory are tautologous and do not have the properties of unambiguously defined and heuristically useful hypotheses (1933). Nevertheless, the theory of sexuality of Hartmann has given a stimulus to a great number of valuable experimental investigations which represent an essential enrichment of our knowledge. The special hypothetical starting points of these investigations come, of course, in all cases from the complex of theories of stimulus physiology, developmental physiology, biochemistry, and genetics, and the results accordingly fit into these fields without either proving or disproving the general statements of the theory of sexuality. Their interpretation by means of the ideas of this theory seems only a superfluous decoration. The function of such speculative theories of a general kind obviously consists only in forming a conceptual background offering incentives of a general kind, revealing possibilities to thought, and in this way enlivening the investigation, while the setting-up of working hypotheses always arises from the current state of the science in the form described in the first chapters of this work. The importance of speculation seems to decrease more and more with the progressive increase in knowledge and the perfecting of empirical methods. The author does not venture an opinion on the question whether this process necessarily takes place according to a law of intellectual development.

In this connection the question arises whether there exists a "theoretical biology" as an independent science. Some authors, especially von Bertalanffy, have declared themselves firmly in favor of taking this branch into consideration in the organization of teaching and research and have drawn attention to the parallel case of theoretical physics. But I believe that this is not a correct comparison. In the case of physics the mastery of a specific apparatus of applied mathematics presupposes special gifts and special methods. Yet even here the boundaries between theoretical physics and experimental physics have become very much blurred in recent years and were in part only set up through the misunderstanding of certain physicists about the development of their science. In the field of biology the situation is quite different, inasmuch as here at present no special mathematical apparatus is necessary for the setting-up of a system of theories but, on the contrary, an enduring contact with experimental research. Even the biophysical ideas of Rashevsky, constructed with a great display of mathematics and abstraction, or the apparatus of biostatistics and the mathematical methods of genetics maintain their significance only in continuous connection with experimental research. A deliberate separation of a "theoretical biology" would today mean an intellectual decline or even the encouragement of speculative tendencies which would not promote the development of the science. A *purely* theoretical biology would be unable to make any scientific assertion which would say more than the statements of the special branches about living things or to which the latter would be subordinate. The concept of "general biology," as it is used in the present work, means nothing more than a synthesis of the single biological disciplines. In this sense it is certainly justified from the teaching point of view but not as a branch of research, since specialization is not then admissible. With the increase of knowledge the integration of the results of the special branches of biology becomes an ever greater, but at the same time a more and more necessary, task. It will best be solved by the working together of experimental biologists who possess the endowments necessary for the task. But the demand for the establishment of a special field of research of "general biology" with its own characteristic methods is not in the least justified.

B. Parabiology: The Misuse of Speculation

Under this heading are gathered together speculative tendencies, systems, and statements which in my view either do not exercise at all the legitimate function of speculation described in the previous section or do so only in part or are inclined to surpass the boundaries of this function. This designation is not intended to have a derogatory sense but only to indicate that these tendencies stand "alongside" scientific biology, without showing any scientifically unobjectionable connection with it. That is not to say that these tendencies cannot have other, perhaps very valuable, functions in the life and thought of mankind. For the progress of biological science, however, they seem to be useless, or even harmful in so far as they mislead nonbiologists about the real character of scientific biology and divert the expert from a correct formulation of his problems.

Many fundamental misunderstandings owe their origin to the ambiguity of the word 'life,' by which we denote not only the state of living natural objects as it can be dealt with in empirical science but also the subjective recognition of our existence—and thus introduce a problem of quite another kind, which has nothing to do with the scientific tasks of biology. Even for the expert it is not always easy to avoid these overtones of meaning of the word 'life' when he wishes to use it only in the sense of his science. Connected with this is also the fact that in many parabiological systems an open or concealed conceptual realism is involved in using the word 'life.' In popular expositions we often read: "Whatever new knowledge biological science may have brought to light, what life really is, what its nature may be, we know just as little as formerly; it is, moreover, an unsolved an insoluble problem." In such and similar contexts the concept "life" is personified into a natural object existing independently alongside the living organism. In this sense "biology" then becomes the proper science of this "life," separated from the biology of scientific experts, or having the latter subordinate to it. Some other concepts of biology are also involved in a similar conceptual realism, such as "form," "configuration" ("Gestalt"), "wholeness," "species," "type." That no scientifically useful beginning can be made with such ideas, that they must, on the contrary, always remain sterile speculations, without connection with scientific biology, is indeed self-evident. In what follows, a short survey will be given of various systems of a parabiological kind, without attempting a complete or exhaustive treatment.

The Significance of Speculation in Biology

1. Mechanism and Vitalism

The alleged antithesis between systems of mechanism and systems of vitalism is frequently regarded as a very important and fundamental problem of biology. These two views have appeared in various special forms and modifications and today still have their representatives here and there among scientific biologists. The majority of experts take, of course, a rather detached attitude toward them. On the contrary, these views receive lively discussion in circles interested in the philosophy of nature, in which, today, various kinds of neovitalism find supporters. Regarded from the empiricist standpoint, the mechanistic and vitalistic systems show a fundamental affinity, while the contrasts between them seem to lie more in the formulation. Common to both systems, in the first place, is the view of causality in the sense of an executive causality, while in the empirical science of the present day the causal connection is usually conceived only in the sense of a consecutive causality. Both systems seek for an "explanation" of the organic event by reducing it to a "principle" or a "category" which is "at work" in it, which "causes" it. The concepts of "principle," "category," etc., are reified into "working causes." It is of secondary importance whether the mechanist locates these causes in definite hypothetical "elementary vital units" or in the fundamental properties of "matter" or whether the vitalist seeks them in an "entelechy." There have, indeed, been remarkable combined systems, as, for example, Reinke's theory of "dominants" or the "biodynamic guiding fields" or Pflüger's theory of "life-stuffs" which veil themselves in a physicochemical garb but are essentially vitalistic. On the other hand, H. Driesch has called himself, not without justice, a "consistent mechanist" because his theory presupposes a machine model of the organism and an extreme preformationist view, in which the entelechy "is the engineer who sets this machine in motion."

A further common feature of mechanistic and vitalistic systems is the tautological character of their general statements. When the mechanist ascribes properties to his "biomolecules," when he assumes structures in the fertilized egg from which all vital and developmental processes necessarily follow, this amounts to no more than a pure tautology, in which what is to be explained is already put into the definition of the concepts serving for explanation. The same is done by the vitalist with the entelechy, which is always required to do what is expected of it for the explanation of the vital and developmental processes. In the statements in which the concept of entelechy or something similar occurs, these concepts are, for the most part, defined only by their connection with the processes which they are to cause. In this way such statements escape empirical testing either by experiment or by observation. It is for this reason that the formulations of mechanism and vitalism have been discussed for many decades on the purely speculative plane, without the great progress meanwhile reached in research having permitted a decision in this controversy. Every new result can be interpreted in the expressions of mechanism or in those of vitalism. There is no *experimentum crucis* which can decide between the opposed standpoints. Naturally, attempts have not been wanting which sought to formulate the statements of the great speculative systems in a manner which is empirically testable or to derive special formulations from the general statements which are empirically testable. The famous examples of the older vitalism—for instance, the experimental separation of cells during cleavage stages—were admittedly proofs against special hypotheses derived from preformationist ideas, such as Weismann's theory of the "genetically unequal" cleavage division, but are not proofs for or against the general theorems of vitalism. General assertions about the "entelechy" or the "forces immanent in the system," for example, avoid giving any information about the difference between the mosaic and the regulative type of cleavage; they cheerfully neglect the question of why the regulative ability has limits and where they are. If from these general statements we try to derive special ones which are testable in experience, then this can only be done by means of the transition to hypothesis formation on the basis of the existing state of empirical research, in accordance with the rules sketched in the earlier chapters of this work. The researches of

Spemann, Hörstadius, Plough, and others have shown, that special, more or less narrow limits are set to the equipotential system and that these limits can be defined by causal analysis of the conditions given in the system or in the environment which are empirically testable. Development takes an atypical course just as consistently and in a manner just as much open to experimental analysis as it does a typical one.

It is characteristic of the newer vitalism that its supporters present a special, perhaps intentional, lack of understanding of the results of genetics and their connection with the achievements of developmental physiology—for example, Woltereck (1931) and Driesch (1944). The classical formulations of the great mechanism-versus-vitalism antithesis are adapted to the state of biology which prevailed at the time of its origin. Like all speculative systems which strive after "final ontological" statements, they cling to the current state of empirical science, which they attempt to petrify in its main features. The results of genetics are therefore as far as possible represented as in principle unimportant or are reinterpreted into "problems of higher order." An idea of genetical developmental physiology—the temporal co-ordination of gene or enzyme action—is emphasized as the problem of "insertion," and this is ascribed to the intervention of the entelechy. Such an assumption could only hold as a true scientific hypothesis if the laws of the relations between the entelechy and its effects were empirically testable by an appropriate definition of the concepts used, under varying experimental conditions. But as soon as we attempt such a formulation, the vitalistic notion of entelechy disappears and gives way to the empirical determinations of the magnitudes of the system under consideration. Driesch, the most consistent thinker among the vitalists, therefore tried in his later works to remove the concept of entelechy more and more from the grasp of empirical scientific methods. While some vitalists would see in the entelechy something "in principle analogous to the forces of inorganic nature," Driesch regarded it as "not energy, not force, not intensity, and not constant" but as something "immaterial, nonspatial." It is not so much this vague definition which makes the central concept of vitalism scientifically unusable as, rather, the complete lack of hypothesis construction which could contribute to the enlargement of the science by empirical testing. The same is true of the materialistically formulated general basic concepts of mechanistic systems.

A feature common to many parabiological systems is the more or less anthropomorphic character of their formulations, on which, in part, their seemingly explanatory value and their effect on the public depend. As one of many examples, we may mention the considerations which Ballauff (1940) connects with the establishment of structures or processes of a complex kind in the organic world. He holds that in such cases we always speak of "order," but we must always ask where the "orderer" is. A point of view which for every order requires an orderer and for every "law" a lawgiver—some neovitalists say that the role of the entelechy is not "executive" but "legislative"—is typically anthropomorphic, and attempts have been made to describe the vital processes and the behavior of all organisms, right down to the unicellulars, in the language of human psychology. Such an attempt would be perfectly admissible methodologically if only we had the possibility of making statements about the psychic life of organisms which are testable by inner experience in the same way in which statements about the conscious processes of human beings are testable. For that reason extreme psycho-vitalism is the only form of vitalism which has been thought out thoroughly. That such a procedure is for many reasons not open to biology is clear to everyone, and therefore also the concealed forms of the anthropomorphic point of view must be rejected as inadmissible.

Biologists of the empiricist persuasion are usually called "mechanists" by vitalists. It is objected against them that they attempt to reduce all organic events to the simple physico-chemical laws of inorganic nature, which is held to be impossible. What we are to understand by "reduce" and "the simple physicochemical laws" of inorganic nature is, of course, quite inadequately defined. The scientific biologist is far from beginning work with the preconceived opinion of such a reducibility.

What matters to him is that he should use those exact empirical methods which have proved themselves to be successful in his science. *In principle* these are, of course, the same as those used in the inorganic natural sciences. No biologist will assert that the results of the physics and chemistry of inorganic nature will alone suffice to elucidate scientifically the relations of living things. In organisms we find materials united in the most complex metabolic relations with one another and with the environment in the form of highly complex open systems of peculiar hierarchically ordered structure, such as we do not find in nonliving nature. Without the knowledge of living things we should never have had examples of such structures, and an essential part of what is given in nature would have remained unknown to us. The laborious scientific exploration of this field has therefore also given rise to sciences with independent special methods within the boundaries of natural science as a whole. Nevertheless, no man doubts that even within organisms the same general physicochemical laws hold as in inorganic nature. The elements show the same reactive properties; the same general kinetic rules hold, and the same thermodynamic relations. Moreover, the often-cited apparent contradiction to the second law of thermodynamics holds only if we falsely regard organisms as closed systems. But it vanishes as soon as we consider the organism with its environment and thus a larger section of nature.

An argument often brought forward from the side of a philosophy of nature is the seeming "improbability" of organisms. We cannot here discuss the ambiguous use of the notion of probability by many authors. In so far as this objection relates to the structure of living things and their energy relations, the matter has been made so clear in recent years from the physiological side (Bünning, 1949) and from the physical side (Schrödinger, 1944) that a general discussion of it in vague terms should be superfluous. Structures which are improbable when considered from the point of view of inorganic models are formed by the special biochemical and biophysical situation in the organism and are thus very probable. The peculiarity of organic events does not lie in peculiar "forces" which do not occur in inorganic nature but in the exceedingly complicated systematic connection of the materials in their interactions. In view of this organization, the energy changes in the organism are no less "probable" than every other physical occurrence; on the contrary, they are the only probable ones. Peculiar relations prevail in the parts of nature represented by organisms just in so far as the high degree of order (or the thermodynamic potential) becomes greater by an increasing disorder in the rest of nature. Such relations would scarcely be conceivable or predictable from a knowledge of inorganic nature alone. Nevertheless, the experience of inorganic and that of organic science join together to give a unitary and consistent picture of the world, derived from the fundamental unity of method. This characterization of the present-day situation in biology does not signify mechanism in the above sense, still less the acknowledgment of a philosophical doctrine in the sense of materialism.

The attitude taken by biologists toward the great speculative systems is often characterized by some uncertainty and too much aloofness (e.g., Spemann, 1936). There is no point in contesting the doctrines of vitalists and mechanists by factual arguments, as M. Hartmann and L. von Bertalanffy do in so praiseworthy a manner, if the essential point is not emphasized, namely, the methodological defects of these systems. For that reason it is a mistake to adopt the standpoint that, although in the present state of investigation a decision between the alternatives presented by parabiological ideas is not possible, yet such a decision is to be expected of some future state of biology. Empirical science will never make possible a decision about questions which are formulated in a way which excludes treatment by empirical methods. From the point of view of method it is also very precarious to use the concepts of vitalism in order to denote the boundary of the unexplored or unexplorable in an indeterminate manner, without trying to seek a purely epistemological clarification of the question of such a boundary. Again, other biologists are of the opinion that the mechanism-vitalism antithesis is to be overcome by a synthesis to which both terms are subordinate; yet such attempts easily lead to new formulations of a parabiological kind. Finally, there is a series of speculative theories which, although not openly basing themselves on the systems mentioned, nevertheless have essential features in common with them and, in spite of their often cautious and careful formulations, have a parabiological character. Some of these will be discussed in the following sections.

2. Subjectivity, Activity, Purposiveness, and Conformity to Plan in the Organic World

Some authors believe that living beings are essentially distinguished from inorganic things by their "subjectivity." The notion of "individual" is often linked with this idea. It is clear that the roots of this view lie in the personal experience of human beings, who feel themselves to be indivisible subjects. Closely connected with these ideas is the assertion that organisms are distinguished from the rest of nature by a special "activity." The difficulty of using the notion of "individual" empirically in an unobjectionable way in a physiological view of organisms has already been discussed. According to the level of organization and other circumstances, an "individual" is divisible by the formation of complete new "individuals," or it can regenerate essential parts after loss or support such loss without regeneration. The notion of "individual" is not used in biology in the original sense of the word but only conventionally and must be specially defined for the special case. If this is done, then the notion of "subjectivity" also disappears from physiological analysis, unless we wish to withhold it, as a vitalistic fiction, from such analysis. The same holds for the "activity" of the organism, in so far as more is meant than a statement complex about the energy exchanges between organism and environment. That the study of behavior and other branches of biology must often use these and similar expressions in its statements in order to make them picturable is self-evident. It will not be difficult for the methodologically educated to estimate the borrowed expressions correctly, serving as these do the interests of picturability, without attaching an ontological significance to the concepts used in them. This holds especially for the distinction made in various connections between "active" or "living" and "passive" or "dead" parts in organisms or cells. Here, also, such a distinction serves only to illustrate certain physiological facts but has no ontological significance. Analysis always leads only to complicated reciprocal actions of the equally indispensable constituents all contributing to the construction of a highly organized system, but without a special degree of "vitality" or "activity" being ascribed to any one of these materials or structures. It would be a mistake to seek in the organism for special "living" structures, for "centers of activity" or an "active principle" of any kind which makes use of the remaining materials and structures as only the passive "substrate" of its action, as is done in principle in all mechanistic and vitalistic systems. The same holds for such expressions—very often used in biology—as 'potency,' 'primordium,' 'capacity,' 'factor,' 'tendency,' etc. In genuine scientific statements these words never denote independent powers concealed in organisms from which some special activity proceeds but are only illustrative circumlocutions for the assertion of a law or for the definition of a set of conditions which are testable in experience. Although we are again and again compelled by the high degree of complication of organic events deliberately to simplify by considering single causal connections outside their relation to the system, and although it is advisable to use language illustratively in the exposition of such connections, still we must always be aware of the danger of misunderstanding which this involves. Such commonly used notions of scientific biology as "gene," "enzyme," "hormone," "stimulus," "excitation," and the like, are also never quite safe from such misinterpretations and are often used, especially in popular expositions, in a parabiological sense without the author intending, or even being aware of, the fact. It is perhaps more difficult in biology than in the other natural sciences to be popular without being misleading.

The frequently used expressions 'purposiveness' and 'conformity to plan' are also dangerously ambiguous. When we describe the structures of an organism and their functions within the system and in reciprocal action with the environment, we are naturally at liberty to call all these arrangements and functions "purposeful" in so far as the ability to live and the peculiarity of the system depend on just these arrangements. A system having unpurposeful arrangements is simply not possible under the given circumstances. We can with equal justification ascribe "purposefulness" to every inorganic thing if we regard being as it is as its purpose. It is clear that such a highly complicated open system as a living organism, whose existence is so strongly threatened, must possess a great number of arrangements in order to preserve itself. The fact that in spite of these arrangements a very great number of living things always perish for external reasons, that in the case of certain types of organization every organism perishes at a certain age for internal reasons, shows how difficult it is to preserve these complicated open systems in nature. Moreover, many species of animals and plants can be shown to have died out in the course of the earth's history. The organizations of organisms thus prove to be successful only under particular constellations of external and internal conditions and are purposeful in this sense for the maintenance of the individual and the species. The significance and origin of a purposeful arrangement, such as the elaborately constructed mouth parts of an insect, can be depicted scientifically only from the following three points of view and thereby "explained": (1) its structure and its function and their significance for the preservation of the life of the animal are described; (2) the genetical and embryological conditions of its origin in ontogeny are investigated; (3) as far as this is possible, the processes in the phylogenetical origin of organisms with such arrangements are investigated. The employment of the concept of purposefulness in biology in this sense is perfectly correct. When the question of the function of an organ and of its significance for the survival of the individual or the species is called the question of its "purpose," then this kind of teleological point of view is quite common in biology and has high heuristic value. But it does not *fundamentally* distinguish the procedure of biology from that of other sciences of nature. For even in nonliving systems it is possible to ask about the significance of the single constituents for the preservation of the system. That it plays such a significantly greater part in biology and is so uncommonly successful heuristically rests only on the fact that living systems exhibit such a high degree of complication. It is part of the definition of a highly complex system that each of its structures and functions—allowing for variations within limits—must be necessary for the maintenance of the system in its particularity and is in this sense purposeful. But in many speculations of a teleological tendency the concept of purposefulness is used, more or less disguised, in that other sense which it is given in everyday language. Here arrangements or actions are called "purposeful" if their aim is imagined by human beings as a goal which is striven after. Extreme finalists seek to "explain" all organic events, even processes of phylogenetic and ontogenetic development, by a "striving for a goal," which they see as a special "principle" which is "at work" in organic things. As in other vitalistic concepts, here also a consistent psychologizing, according to the pattern of human conscious processes, provides the only possibility of making such formulations testable in experience. Later we shall discuss the noteworthy attitude taken by finalists toward the notion of causation.

The situation is similar in the case of certain general statements about the "conformity to plan" or "planned order" in organic beings. A general statement about a persistent order easily becomes tautologous if special testable statements about the rules of this order are not derivable from it. Also, the order-analytical statements of biology, obtained by purely comparative means, possess scientific value only through their quite special content which is testable in experience. Statements about the degree or the height of an order, also, are only useful when they provide a means for measuring this degree. The same holds for the concepts "plan," "constructional plan," "functional pattern," and the like. It is characteristic of many parabiological theories that they turn such concepts into things to which they attribute an action on the "substrate" of organic events. In this way some philosophers of nature speak of an "antecedence" of the structural plan in the developmental processes in ontogeny. The apparent explanatory value of such ideas lies wholly in the purely anthropomorphic conception which is behind them.

It is not to be denied that the teleological mode of expression makes biological topics very clear and is therefore almost indispensable, especially in teaching biological subjects. Nevertheless, it is doubtful whether the statement "So long as an organic system has not yet reached its greatest possible development, it strives toward it" is to be regarded as a "general principle" of organic development. If such a sentence is not intended in an extreme psychovitalistic sense, it says nothing at all.

3. Wholeness and Causality

In many recent philosophical systems the concept of "wholeness" (*Ganzheit*) plays a central part. Its fundamental assertion is that "the whole is more than the sum of its parts." By many authors the peculiarity of being a whole is regarded as especially characteristic of living things, where types of wholeness of still higher degree are distinguished, e.g., species, biocoenoses, peoples, races, etc. Other authors allow nonliving objects such as atoms also to have this property of wholeness. The principal difficulty for a critical evaluation of such statements lies chiefly in the fact that in them the concepts "the whole," "the parts," and "the sum" are not unambiguously enough defined to be useful in special testable hypotheses. In some of these statements "the whole" signifies no more than our idea of a living thing which we consciously experience as a closed unity, and "the parts" are those ideas of parts into which we divide this whole conceptually according to taste. In such statements the question of the relation between the whole and its parts is a logical or psychological, but not a biological, one. In so far as it is a question of empirically comprehensible parts, or part-processes, in an organism, it is already laid down, in the more or less satisfactory definition of these parts, in what regular relations they stand to other parts of the whole. The word 'sum' in a purely mathematical sense is then never sufficient for a characterization of these relations. Even where we speak, for example, of a summative action of individual processes, a derivative and deliberately simplified mode of expression is being used. Physiological analysis leads us again and again to the complicated mutual actions in which all the constituent parts of a living being are systematically involved with one another and with the environment. It would be a mistake to expect that the processes in an organism could be understood from the sum of all those reactions which its parts would perform if they were not in the union of the living system, for then they would be under quite different sets of conditions and would therefore react quite differently. If the above-stated principle says no more than that in organisms such complicated systems of conditions always prevail, then it is quite justified. Then the notion of "wholeness" is only a paraphrase of the concept of system and is also applicable to systems of higher order, such as species and biocoenoses, as well as to inorganic systems. The same holds for another frequently repeated formulation, namely, that "all processes in an organism are related to the whole." There is nothing objectionable about this if it is only an attempt to characterize the type of a complicated system. But it is necessary, if error is to be avoided, also to take into account the relationship to the environment. In a more special sense the "purposefulness" of the partial processes in an organism is also intended by this statement, i.e., the partial processes run their course in accordance with the maintenance of the whole system (see Chap. II).

Many philosophers of nature wish, of course, to give to the concept of wholeness an explanatory value extending far beyond these limits. This is expressed, for example, in the formulation "The whole determines the parts, not the parts the whole" or "The whole is the cause of the particular course taken by the single processes." Such assertions presuppose that there is another path to the immediate comprehension and definition of the "whole," as well as the analysis of all its structures and partial processes, than the methods of empirical science available to biologists. In making use of the Gestalt experience which we have in considering a living thing, the morphologist is especially prone to fall into a realism of ideas in the sense of an "intuition of wholeness" or an "experience of a formed totality." It is an age-old need of human beings in considering nature to raise the deep emotional impression which nature and her creatures make upon us into an explanatory principle. Thus we often try to see, behind the forms of living things, especially of the human body, laws which derive their meaning from purely mathematical considerations, aesthetic sensations, or ethical feelings and thus from regions of the inner experience of mankind. The theory of proportions of Albrecht Dürer is one of many examples. But such hypotheses are useful in empirical science only when their elements are testable in outer experience. The concept of "Gestalt" as an immediately comprehended totality has its legitimate place, therefore, only in statements belonging to the study of behavior, to the physiology of the senses, or to experimental psychology, in which the reactions of living things are dealt with. For the more highly organized living things react to all the stimulations in their environment *as if* this were a "Gestalt." Thus, in the statements mentioned, something is expressed about the relations between a particular environment and a particular mode of reaction of the organism which can be tested empirically, with the help of the hypothetical concept "Gestalt." This has nothing to do, however, with the question of in what manner the forms and structures of living things are scientifically accessible in external experience. An idealistic morphology, which tries to adopt toward biology somewhat the same relation that geometry has to physics, will always exhaust itself in tautological assertions. Zimmermann (1938) has called the lack of subject/object separation the fundamental error of this way of thinking and has discussed the arguments against a fundamentally different procedure in the treatment of morphological and physiological questions.

In connection with the discussions of this and of the previous chapter, reference should be made to the various views about the notion of cause which prevail in speculative systems (see also P. Frank, 1932). The teleologists distinguish between "efficient causes" and "final causes," in which they regard "final causality" (in the sense of aim or purpose causality) as an essential characteristic of fundamental biological processes. A statement formulated according to empirical science—"The state *B* follows regularly on the state *A*"—is customarily interpreted in such a way in ordinary language that *A* is called the "cause" of *B*. Nevertheless, nothing at all is altered in the statement and its empirical testability if we call *B* the "cause" of *A*. Only in the case of human actions is it otherwise, in so far as the state *B*, intended by the acting human being, is already consciously present as an aimed-at goal; and it is the knowledge of this connection which permits us so to transform the above statement that we include this fact in the definition of the state *A*. Thus the attempt to make a distinction between efficient and final causes seems meaningless without an anthropomorphic psychologizing of natural processes. Idealistic morphology also seeks to introduce, in its idea of the "archetype" and its relations to development in ontogeny and phylogeny, a kind of final causality into the morphological point of view (cf. Troll, 1948). If rules which are to be testable in experience are set up for the occurrence of an ontogenetic or phylogenetic developmental process, it is again quite irrelevant whether we wish to call the initial state the "cause" of the final state or, vice versa, to call the "form" reached in development the "cause" of the developmental processes. Only in the case of an inadmissible reification of ideas into natural objects and an anthropomorphic reinterpretation of the goal or aim concept does the so-called "archetype" thinking of the morphologist seem to be different from the so-called "causal" thinking of the physiologist. But then the formulation at once loses its usefulness for empirical science. In close connection with the mode of procedure of physiology some authors have believed that they could recognize a special form of causality in "impulse causality" or "catalytical causality" which ostensibly occurs only in certain processes in organisms, for example, processes in stimulus physiology or processes governed by enzymes (cf. Mittasch, 1938). But the peculiarity of the causal process in such occurrences persists only so long as we isolate partial causes from the event and do not take fully into account, or do not fully know, the constellations of conditions given in the system itself. It is here also, as with the processes supposedly to be explained only by "wholeness," that empirical science makes use of just such effects, which seem inexplicable by the causal connections hitherto known, in order to discover new parts and new mutual relations in organisms. The use of the methods which have proved themselves in science always leads to a consistent synthesis with the existing knowledge and to an enlargement of the system.

Some developments of speculative biology have concerned themselves especially with the question of the indeterminateness of the processes in organisms. It is clear that many biological laws can be only statistical laws in the sense of classical statistics. These are all statements about collectives and collective processes, like the Mendelian rules of inheritance. Almost all biological processes take place in the macrophysical domain, so that in their case the indeterminateness of the microphysical elementary processes plays no part. This also holds for enzymatic, hormonal, and stimulus-physiological processes, for which an origin from the microphysical domain has often been incorrectly asserted on the basis of superficial estimates. The single exception is the very important process of mutation. The role of the mutation frequency within a population or of the somatic mutative occurrences within a body is therefore only statistically describable. But the action of a mutation when once it has appeared in the life of a cell or of an organism and in their descendants again lies entirely in the macrophysical domain. The idea of P. Jordan of an amplifying mechanism which could allow the processes in the microphysical domain to influence the macrophysical organic event is quite conceivable. But we know no such phenomenon in the behavior of organisms, and it would be difficult to imagine how such effects within an organism, which are only statistically predictable, can be reconciled with its existence as a highly complicated system. The conceptually quite mistaken relation into which Jordan has brought his "acausal amplifier theory" with the question of the human freedom of will has created for the whole theory an undeserved popularity. Meanwhile, clarification from various directions (e.g., Hartmann, 1948, and Bünning, 1935) renders a discussion of it unnecessary. On the clear conceptual distinction between the fundamental quantum-theoretical indeterminacy (e.g., of a single event of mutation) and the classical statistical indeterminacy (e.g., of the processes of selection) and on the correct evaluation of the great importance of such events for processes on the phylogenetic plane, especially clear ideas have been expressed by Möglich, Rompe, and Timoféeff-Ressovsky (1944).

4. Autonomy or Heteronomy of Living Things: Ontological Questions

Both from the side of the philosophy of nature and from that of empirical biology, well-intended attempts have repeatedly been made in recent years to bridge the chasm which exists between the two ways of thinking. In these efforts attention is often deliberately restricted to the question whether an unambiguous "characterization" of living things is possible without the assumption, at the same time, of ontological statements in the metaphysical sense. Or the question is raised how far a special type of conformity to law could be found in the organic world which would allow a special position to the methods of research in this domain of nature. Moreover, the question has often been regarded as decisive whether these special laws of the living are reducible to the laws which hold in the inorganic world. In this way the question arises which stands at the center of many discussions—the question of the "autonomy or heteronomy" of living things. In many of these discussions there is a lack of unambiguous definitions—for instance, concerning the demands which are to be made of a general "characterization" of living things or of the fundamental "peculiarity" of a law.

Von Bertalanffy, especially, has tried, in numerous weighty writings, beginning on the basis of an extensive biological knowledge, to bridge by means of an "organismic view" the antitheses between the various speculative systems in biology. Supported by the guiding principle of assuming "a special form of law peculiar to living things," he tries to avoid all metaphysics and is therefore rejected by most vitalists as a mechanist. In his earlier works we find such formulations as "alongside chemical differentiation, still another factor, a special form factor, plays a part." But such an analysis of an organic event into separate processes, divided in accordance with the morphological and physicochemical points of view but without a metaphysical conceptual realism, is meaningless and for that reason can never lead to heuristically useful investigations. The same holds, for example, for the following alternatives set up by von Bertalanffy (1930/31): organizer action is *either* "chemical," conditioned by materials having a formative action, *or* "dynamic," in so far as in the organizer "stronger formative powers are localized" and "an energy gradient exists between it and the environment," in which case it "forces its dynamic" upon its environment, and so "material differentiations are created by the formative movements." If from such statements all vitalistic, executive-causal, and anthropomorphic elements are removed, then these formulations cease to assert anything. Like von Bertalanffy, Ungerer (e.g., 1942) does not wish to be classed as a vitalist but nevertheless uses vitalistically colored statements as guiding principles of a "total or organismic outlook." A noteworthy part is played in the writings of these and other authors by the distinction between "qualitative" and "quantitative" processes, these being opposed to one another as alternatives in the description of formative processes. A distinction is also often made between qualitative and quantitative "causal laws" in biology. But the concepts "quality" and "quantity" have a distinguishing value only for our perceptions of things and not for the scientific reproduction of experiences. It is true that to aid the imagination in predominantly descriptive discussions, especially of a morphological kind, we are accustomed to use the terminology of our "qualitative" ideas; yet we could also reproduce all these facts in a language which, although more prolix, uses only "quantitative" expressions. For a principal distinction or characterization of particular processes or laws, the alternative of "qualitative" or "quantitative" is in any case not useful.

It is interesting that in the course of the development of his organismic view von Bertalanffy has gradually laid aside the vitalistic elements of his theory, so that his characterization of living things today coincides with the view of the organism as a complex open system, which is customary elsewhere in biology. von Bertalanffy has made valuable contributions to the theory of open systems (e.g., 1949, 1950). Ungerer (1942), Alverdes (1939), and some other theoreticians often define the concept of "wholeness," to which they give a central place in their systems, in such a way that it quite, or almost quite, coincides with the biological concept of system. Their demand for a point of view which considers the whole organism is perhaps due to the fact that the unfortunately unavoidable and far-reaching specialization of modern biologists has often led to simplifications and one-sidedness in the representation of biological facts which do not do justice to the system character of living things. This demand is certainly justified. But, unfortunately, behind its residue or kernel there lurks a parabiological view, especially when a general characterization of living things is attempted from a "wholeness" point of view. Living things can certainly be distinguished by a series of peculiarities from nonliving things. This characterization becomes all the clearer the more special the statements become. For a completely satisfactory characterization the whole complex of biological statements must, of course, be invoked. The more we try to make this characterization general, the more the statements become lost in indefinite formulas. The parabiological character of such formulations becomes especially clear when it is sought to distinguish living things from nonliving ones by the "introduction" of a single special "force," of a special "principle," or of a special "category." What scientific biology is able to contribute to the characterization of living things will never be able to satisfy the wish of the metaphysician for an "ontological" definition of the living. The assertions of an empirical science are never "last words" or "reality statements" in the sense of ontology. The empirical scientist must decline to reinterpret the theorems of his particular science as ontological ones and thereby to rob them of their proper meaning, as, unfortunately, often happens from the side of the philosophy of nature (cf. Wenzl, 1938). The same holds for the sharp delimitation between living and nonliving or even between animals and plants, again and again demanded by metaphysics, as well as for the special importance which is attributed to the question of the origin of life on the earth. When Troll (1951) devoted much labor and sagacity to deciding whether, "in the view of ontology," the true viruses are living or not, this really only showed that suggestions toward further developments in the empirical sciences were hardly to be obtained from such considerations. Yet these and other boundaries erected in the name of ontology might prove to be unfortunate for this development if we wished to treat them as untouchable obstacles to the methods of research.

The question of a special kind of law for living things, of "laws special to biology," is particularly dear to the heart of many theoreticians. Some satisfy themselves with the general requirement of such laws or with the general reference to the "character of wholeness" of such laws. We have pointed out in our discussions about the various aspects of biology that it is not possible, from the standpoint of methodological criticism, to call some of its statements more and others less "biological." Nevertheless, some biologists are inclined to consider morphological and ecological statements or statements belonging to the study of behavior "biological" statements, but purely physiological or biochemical statements they regard as "nonbiological." Behind such distinctions is concealed the above rejected assumption of "living" and "nonliving" parts of organisms, the assumption of a special specific "activity" in organisms or a mechanistically or vitalistically oriented idea of wholeness. Where concrete examples for "specifically biological laws" are brought forward, we find mentioned, for example, the following: the allometry law, certain laws relating to metabolism, Child's gradient theory, Mendel's rules, certain phylogenetic laws which can be derived from the evaluation of series of fossil forms, and so forth. In all these cases we are dealing with formulations having the type of genuine empirical scientific statements. What distinguishes them from other biological statements, although of course not sharply, is the fact that they are blanket statements, statements about a very complex occurrence, the initial and final states of which they connect causally by giving a rule. It is clear to everyone that such statements represent the subsumption of a great number of single processes which take place within the system and in reciprocal action with the environment and are causally connected in the most complicated manner with one another and with the environmental constellations. Many of these laws are of the classical statistical type, in which the individual processes, about which they say something collectively, certainly take place on the macrophysical level and according to classical causality. Such blanket statements are not, however, at all confined to biology but occur in every branch of natural science, as in the description of geological processes. The fact that they seem to be especially common in biology depends, indeed, only on the high degree of complication of the processes about which something is to be said. These assertions only preserve their empirical scientific character if we define the concepts occurring in them in a manner which is empirically testable. Statements about "Gestalten" thus only do so if physical structures are intended. But as soon as "form" is viewed in the conceptually realistic sense of idealistic morphology, the statements lose this character. But the same holds also for "form" or "Gestalt" in the statements of inorganic natural sciences and is not therefore the special peculiarity of "specific biological laws."

It is a further question how far blanket statements of this kind can be analyzed into statements about all single processes which take place in the total event, whether they can be reduced to these simpler statements without remainder, or, as is often said, how far the total event can be satisfactorily "explained" in terms of the single processes. It is certainly the case that for none of the examples of biological laws mentioned above has this analysis been carried out completely, nor does the prospect of such an analysis exist. Even the wish for such a *complete* analysis scarcely exists, for it would not be likely to yield any noteworthy results. Moreover, it would occur to no one to demand of a statement expressing laws of geological processes information about the individual fate of every grain of sand or molecule contributing to the composition of the rocks. It seems much more essential that some important individual processes within such a biological total event should be subject to analysis by the most diverse methods and that the laws thus obtained should not stand in contradiction to the blanket statement about the total event. Where such contradictions have seemed to result, they have always led to the discovery of new, hitherto unknown connections or constellations of conditions. But contradictions *in principle* between blanket statements in biology regarded as "specifically biological" and the statements about the single processes felt perhaps to be "less biological" have nowhere resulted. On this circumstance the scientific picture of the world, which is, on the whole, free from gaps and from contradiction, rests. Naturally, this is not to say that a direct reduction of biological blanket statements to some simple laws or "principles" of inorganic nature can be carried out. Such a requirement fails to recognize the high degree of complication of living nature and arises chiefly from the metaphysical requirement of certain philosophical tendencies. The requirement of an answer in principle to the alternative "autonomy or heteronomy of organic life" is thoroughly metaphysical and cannot be made within empirical science.

It is interesting to note what a philosopher who has a sufficient book knowledge of biology, for example, Ballauff (1943), says about this question. It is striking to an empiricist that in the discussion of biological matters (e.g., the "system problem" or the "Gestalt problem") the most diverse theories are brought forward as co-ordinate, equivalent possibilities for thought or are set up in opposition to one another as alternatives, although some of them are formulations that are free from objection from the standpoint of empirical science, while others are pure metaphysical speculations. In this lack of critical examination there arises a chaos of obscurities, contradictions, and absurdities which has very little to do with scientific biology. Ballauff himself feels this and, from the philosophical standpoint, applies the sharpest criticisms against the systems which we call "parabiological," especially against their realism of ideas and concepts and the anthropomorphism of their formulations. He proclaims a "gnoseological crisis" in the present-day philosophy of nature. Even biologists like Ungerer (1942), for example, who wish deliberately to exclude metaphysics from their view, are insufficiently critical in drawing the boundary between methodologically unobjectionable theories and parabiological speculation. Ungerer sympathizes, like some other biologists—for example, M. Hartmann—with the "stratigraphical systems" of the philosophy of nature, as they have been developed as a "theory of levels of being," especially by Nicolai Hartmann. Although the careful formulations of this theory approximate closely the form of empirical scientific theories by the utmost avoidance of metaphysical speculations, yet to the empiricist they seem, as general statements, to be suitable only to serve as tautological statements, not as sources for the derivation of special statements which are empirically testable. It seems to be the case that every attempt to build, on the basis of the progress of the special sciences, by way of an "inductive metaphysics," systems which have more to say than the special sciences is condemned to remain a play of ideas which is not able to help the empirical sciences in their work but may only hinder them.

C. World Picture and Philosophy of Life

The restricted and very doubtful function which speculative theories today exercise in the real working of biology, and in the whole structure of the empirical natural sciences, would scarcely justify the efforts which so many thinkers devote to this field and the many interested readers which publications of this sort find. Between the speculative systems of biology and philosophical, literary, religious, and political systems of thought, we find all kinds of transitions. Among these are found systems of high aesthetic charm and of deep ethical content. Philosophical speculation about nature obviously exercises psychological and social functions which lie within the domain of philosophies of life.

The Significance of Speculation in Biology

Many theoretical biologists, for example, Ungerer, Driesch, etc., especially emphasize the importance of biological theory for a world view and from this derive the demand for a stronger emphasis on speculation in biology. The true basis for the growth of speculation thus lies in a need for a philosophy of life.

By 'world picture' (*Weltbild*) and 'philosophy of life' (*Weltanschauung*) we shall here understand well-defined domains of human spiritual life. By 'world picture' will be meant the total representation of the world which we can form on the basis of all statements of empirical science. It seeks to embrace everything which can be accessible to us on the basis of outer experience. The world pictures which various peoples at various times have formed have been very diverse, according to the way of handling empirical methods and the intellectual elaboration of the material of experience. Today, also, the world pictures of different contemporaries are very diverse in content and extent according to the degree of education in empirical science and the spiritual capacity of the individual. A world picture can, in any case, be objectified in principle in a transtemporal and transindividual manner, in so far as it has outer experience as its basis, as a common gift to all men. The world picture is liable to errors and incompletenesses which can be corrected by outer experience. In their practical behavior for the attainment of external goals men guide themselves according to the current world picture. From this practical application of the world picture there result the applied sciences, technology, medicine, agriculture, and the like.

On the other hand, by 'philosophy of life' is to be understood the totality of all the emotionally tinged strivings and ideas of a man, the totality of his value-judgments, of his ethical and aesthetic maxims. Ideas about the meaning and the tasks of human life, about behavior toward his neighbors, about the meaning of human societies, and therewith about the whole field of ethical, social, religious, and political activities of a man are rooted in his philosophy of life. While a man thus reaches decisions regarding inner and outer goals of his personal and suprapersonal life on the basis of his philosophy of life, he obtains the means for the attainment of outer goals from his knowledge of the world.

It is the conviction of the author of this monograph that *every* philosophy of life rests on a faith, on a decision to trust, which can only be reached from an inner human experience. But on the basis of the world picture of empirical science a philosophy of life can never be built. From the statements of empirical science not a single decision in matters concerning a philosophy of life or valuation can be reached. Biology as an empirical science can therefore never give an answer to those "great questions of life" which move men from within. Naturally, this does not mean that occupying one's self with biology or with another empirical science or the mere contemplation of the empirical world picture is not an experience of high ethical and aesthetic content. Yet the sources of valuation or of emotion flow from the domain of the philosophy of life. The synthesis between philosophy of life and world picture, as a spiritual task of every time and of every man, is therefore never only a matter of empirical science but always a matter of faith. The author of this monograph, who counts himself fortunate in having, as a Catholic Christian, a positive belief, has always felt this consistent synthesis to be the greatest fulfilment of the spiritual life. Nevertheless, he believes that the strong emphasis on speculation in the empirical science, and the various philosophical systems of nature found within modern European spiritual history are symptoms of the weakness of faith and the uncertainty and disunity of our times. Many of these systems clearly bear the stamp of religious substitutes. They seek to fill the substance of faith, hollowed out by the Enlightenment and by liberalism, with pseudo-scientific content. In this way elements foreign to science are imposed upon the empirical sciences, and their methodological foundations are threatened.

This becomes especially clear in the great totalitarian ideologies of our time, which have taken over this tendency as a spiritual inheritance of the nineteenth century. In their efforts to subordinate all the religious and philosophical convictions and feelings of mankind to their program, at the same time denying the latter's foundation in belief and pseudo-religion, they attempt a scientific justification of this program on a basis of philosophy and empirical science. German National Socialism attached great importance to the parabiological "wholeness" theories, because it wished to justify pseudo-scientifically on this basis the totalitarian demands of *Volk* and *Rasse*. Marxism transfers the same mode of argument to "class." Of the biological subdisciplines, genetics, especially, is repeatedly drawn into the conflict of opinions about philosophies of life. While the representatives of exact genetics were called "materialists" by the vitalists, they were branded as "idealists" by the biologists of the school of Lysenko. The philosophical and political ideologues would also like to make their doctrines the foundation and measuring rod for the method of natural science. The author of the present work heard in a programmatic speech of a leading German biologist in the year 1938 these words: "The connection [between "race" and culture] is intuitively grasped by the *Führer*, and it is the one and only task of German science to buttress this intuition scientifically." Similarly, in the condemnation of exact genetics by Lysenko the highest authority was the pronouncement of the party leader, not scientific observation.

IV. Conclusion

In this monograph an attempt has been made to show that biology, in the structure of its statements and in its methods of work, is a purely empirical science. Corresponding to the peculiarity of its object, it has developed special methods and special subdivisions, but these are nevertheless not different in principle from those of other natural sciences. Its assertions, in so far as they exceed the boundaries of pure descriptions, have the form of empirical scientific hypotheses or theories. The explanatory value of its statements rests on foundations similar to those of other natural sciences. The derivative modes of expression or tautological formulations of certain statements which aid picturability are, corresponding to the peculiarity of its object, perhaps more frequent in biology than in other natural sciences but do not detract from its empirical character.

As an unavoidable consequence of its rich development, biology has experienced an especially marked subdivision into special branches, and this carries with it a certain danger of one-sidedness. The synthesis of the results of biology nevertheless goes on throughout consistently and fruitfully and leads to a constant development of the science. There is in biology no "crisis," as has sometimes unjustly been stated. The synthesis of the results of biology with those of the remaining natural sciences has been fruitfully established in many borderline regions and leads to an empirical world picture which is on the whole consistent and unified, if incomplete.

A critical study of foundations and methods which could only be hinted at in this monograph would certainly be very useful in biology. But here biology occupies no special position, because such problems are common to all of the empirical sciences.

Selected Bibliography

ALVERDES, F. 1939. "Biologische Ganzheitsbetrachtung," *Zeitschrift für die gesamte Naturwissenschaft*, Vol. V.

BALLAUFF, T. 1940. "Über das Problem der autonomen Entwicklung im organischen Seinsbereich," *Blätter für deutsche Philosophie*, Vol. XIV.

———. 1943. "Die gegenwärtige Lage der Problematik des organischen Seins," *ibid.*, Vol. XVII.

———. 1949. *Das Problem des Lebendigen.* Bonn: Humboldt-Verlag.

BERTALANFFY, L. VON. 1932, 1942. *Theoretische Biologie*,Vols. I and II. Berlin: Verlag Borntraeger.

———. 1949. *Das biologische Weltbild.* Bern: A. Francke A.G. (English trans., 1952, *Problems of Life* [New York: John Wiley & Sons; London: Watts & Co.]).

———. 1950. "An Outline of General System Theory," *British Journal for the Philosophy of Science*, Vol. I.

BÜNNING, E. 1935. "Sind die Organismen mikrophysikalische Systeme?" *Erkenntnis*, Vol. V.

———. 1949. *Theoretische Grundfragen der Physiologie.* 2d ed. Stuttgart: Piscator-Verlag.

Selected Bibliography

DRIESCH, H. 1928. *Philosophie des Organischen.* 4th ed. Leipzig: Quelle & Meyer.

———. 1944. *Biologische Probleme höherer Ordnung.* 2d ed. Leipzig: Barth.

FRANK, P. 1932. *Das Kausalgesetz und seine Grenzen.* Vienna: Springer.

———. 1950. *Modern Science and Its Philosophy.* Cambridge, Mass.: Harvard University Press.

HARTMANN, M. 1948. *Die philosophischen Grundlagen der Naturwissenschaften.* Jena: Fischer.

HEMPEL, C. G. 1952. *Fundamentals of Concept Formation in Empirical Science,* in *International Encyclopedia of Unified Science,* Vol. II, No. 7. Chicago: University of Chicago Press.

LENZEN, V. F. 1938. *Procedures of Empirical Science,* in *International Encyclopedia of Unified Science,* Vol. I, No. 5. Chicago: University of Chicago Press.

MAINX, F. 1932. *Die Sexualität als Problem der Genetik.* Jena: Fischer.

MITTASCH, A. 1938. *Katalyse und Determinismus.* Berlin: Springer.

MÖGLICH, F.; ROMPE, R.; and TIMOFÉEFF-RESSOVSKY, N. W. 1944. Über die Indeterminiertheit und die Verstärkererscheinungen in der Biologie," *Naturwissenschaften,* Vol. XXXII.

SCHRÖDINGER, E. 1944. *What Is Life?* New York: Cambridge University Press.

SPEMANN, H. 1936. *Experimentelle Beiträge zu einer Theorie der Entwicklung.* Berlin: Springer.

TROLL, W. 1948. "Urbild und Ursache in der Biologie," *Sitzungsberichte der Heidelberger Akademie der Wissenschaften,* Vol. VI.

———. 1951. *Das Virusproblem in ontologischer Sicht.* Wiesbaden: Steiner.

UNGERER, E. 1942. "Die Erkenntnisgrundlagen der Biologie," in *Handbuch der Biologie,* Vol. I. Potsdam: Athenaion.

WENZL, A. 1938. *Metaphysik der Biologie von heute.* Leipzig: Meiner.

WINTERSTEIN, H. 1928. *Kausalität und Vitalismus vom Standpunkt der Denkökonomie.* 2d ed. Berlin: Springer.

———. 1938. "Der mikrophysikalische Vitalismus," *Erkenntnis,* Vol. VII.

WOLTERECK, R. 1931. "Vererbung und Erbänderung," in H. DRIESCH and R. WOLTERECK (eds.), *Das Lebensproblem.* Leipzig: Quelle & Meyer.

WOODGER, J. H. 1939. *The Technique of Theory Construction,* in *International Encyclopedia of Unified Science,* Vol. II, No. 5. Chicago: University of Chicago Press.

ZIMMERMANN, W. 1938. "Strenge Objekt/Subjekt-Scheidung als Voraussetzung wissenschaftlicher Biologie," *Erkenntnis,* Vol. VII.

———. 1948. *Grundfragen der Evolution.* Frankfurt an der Oder: Klostermann.

The Conceptual Framework
of Psychology

Egon Brunswik

The Conceptual Framework
of Psychology

Contents:

PAGE

I. EXPERIENCE AND THE EMERGENCE OF THE OBJECTIVE AP-
PROACH 659

   1. The Primacy of Mind in Philosophical Dualism. Sensationism 659

   2. Privacy and Limited Communicability in Phenomenological Introspection 661

   3. The "World of Things" and Its Residue of Ambiguity . . 663

   4. Objectivity, Methodological Physicalism, Operational vs.
Experiential Positivism 668

   5. The Futile Search for "Criteria of Consciousness." Verbalization 672

II. THE FUNCTIONAL UNIT OF BEHAVIOR AND THE LEVEL OF COM-
PLEXITY OF PSYCHOLOGICAL RESEARCH 674

   6. Stabilized Achievement and Vicarious Mediation. The
Lens Model 674

   7. Organismic Adjustment as a Probability Function . . . 679

   8. The Molar Approach: Focal and Macro-mediational Reference 683

   9. The Molar Approach: Probability Laws and Representative
Experimental Design 686

   10. Descriptive and Reductive Theories. Law, Inference, Explanation 691

III. MISCONCEPTIONS OF EXACTITUDE IN PSYCHOLOGY 694

   11. Thematic Physicalism 694

   12. Fear of Preliminaries 702

   13. Hostility to Theory and to Central Inference 705

Contents

IV. TRADITIONAL APPROACH AND CONSTRUCTIVE CRISIS IN PSYCHOL-
OGY 708

14. Historical Schema and Interpretative Hypotheses . . . 708
15. Sensory Psychology 710
16. Intentionalism, Early American Functionalism, Psychoanalysis 711
17. The Decline of Methodological and Nomological Dualism 718
18. Antithetic Divergence: Gestaltpsychology and Classical
Behaviorism 720

V. CONVERGENCE TOWARD AN OBJECTIVE FUNCTIONAL APPROACH 723

19. General Trends toward Realization of Norms 723
20. Distal-Central Reference in Molar Behaviorism, Probabilistic Functionalism, Factor Analysis 725
21. Dynamic Personality Theory 733
22. Nomothetic Encapsulation in Topological Psychology,
Postulational Behavioristics, Mathematical Biophysics . . 735
23. Brain Models and Statistical Extrapolations in Cybernetics
and Communication Theory 743

BIBLIOGRAPHICAL NOTES 751

The Conceptual Framework of Psychology

Egon Brunswik

I. Experience and the Emergence of the Objective Approach

On its way to becoming a science, psychology had to face certain requirements of procedural policy or general methodology.[1] The issues involved fall into two major groups. One deals with the rigor of fact-finding, inference, and communication. In this respect, there must be methodological unity of psychology with the other sciences, especially physics. In contemporary psychological discussion this requirement is often expressed by saying that we must make psychology an "objective" or an "operational" discipline in the general manner attempted by behaviorism (Chap. I). A second set of problems arises in connection with efforts not to lose sight of the specific tasks of psychology in the process of objectifying it but to establish exact study on an adequate level of complexity, sometimes called "molar" or "functional." The thematic identity of psychology can be, and can only be, established by the recognition and programmatic employment of specific research "designs" and aims relatively uncustomary in the other natural sciences. Such diversity is not only compatible with, but necessary within, the basic unity of the sciences (Chap. II).

1. The Primacy of Mind in Philosophical Dualism. Sensationism

Perhaps in no other science is the problem of objectivity as troublesome and as urgent as in psychology. This is primarily due to the fact that psychology uniquely lends itself to entanglement with dualistic metaphysics.

The rationalist answer to the quest for certainty. Introvert meditation.—There are two ways in which the "quest for certainty"[1] has been handled in the history of philosophy. One of them, basically subjectivistic, mentalistic, or "introspectionistic" in character, is associated with the rationalistic and idealistic outlook prevalent on the European Continent since Descartes. The other, related to objectivism, is affiliated with the empiricism and positivism of England and America.

The mentalistic concept of certainty is implied in Descartes's *cogito ergo sum* in which our awareness of thought is raised to the ultimate criterion of existence; in the same vein, subjective clarity and distinctness was regarded as the criterion of truth. This was in line with the rationalist tendency to consider reason, and especially mathematical reasoning with its feelings of intuitive self-evidence, as a source of knowledge superior to empirical observation, and of identifying knowledge in general with mathematical knowledge.[2]

The study of conscious experience and the method of introspection thus acquired the stature of an enterprise as worthy and distinguished as the study of matter in physics, if not more so. There existed a challenge to achieve for mind what physics had so successfully achieved for matter. In time, psychology became firmly associated with this task. The subjectivistic conception of psychology remained unchallenged up to the onset of the behavioristic redefinition of psychology at the beginning of the present century. It lent a certain internal coherence to psychology, temporarily lost after the demise of introspectionism as the unchallenged ideology, but about to be re-established in terms of objective features of research design.

Descartes's philosophical method was one of contemplative "meditation," as had been that of Augustine and the mystics. This was introspectionism in the narrower, introvert sense. Aside from the higher forms of cognition, its favorite topics were the "passions" and other complex emotional states. It may be schematically depicted as a relatively static, self-contained form of central encapsulation (Fig. 2, presented later for comparison with subsequent developments).

Sensationism and experiential positivism.—It remained for the English empiricists to shift the introspectionist's attitude from the so-called higher mental functions to sensation. The peripheral kind of extraversion developed in this process (represented in Fig. 3 by dislocating the reference point of introspection to a position out of center in the direction of the sensory or the motor periphery) became more elaborately cultivated during the first major period of experimental psychology in the second half of the nineteenth century (sec. 15). Upon analysis in "hard introspective labor," conscious data are said to reveal themselves as a bundle of punctiform sensations of color, sound, and the like, including sensory feelings and images; the latter are considered, after the fashion of Hume, as nothing but "faint copies" of sensations. As Titchener has put it: "All the genuine findings of psychology must consist of sensations." For its static, piecemeal character this point of view has been characterized as "structuralism."

The same kind of elementaristic introspection underlies a variety of "impressionistic" manifestations in the late nineteenth century. The most remarkable of these is Mach's conception of sensation as the common starting point for psychology and physics. Its introspective origin is best revealed in Mach's report, in his *Analysis of Sensations,* of a primordial philosophical experience in which "the world suddenly appeared as . . . [a] mass of sensations." According to present gestaltpsychological and psychopathological views, such a position lifts to undeserved prominence a relatively insignificant and uncommon, artificial, distorted, and indirect type of experience.

Mach's philosophical sensationism defines an idealistic position in the tradition of Berkeley's tenet that "to be is to be perceived" (*esse est percipi*). It may be taken as the climactic core of what Boring[50] has recently labeled "experiential positivism," in contradistinction to the "operational positivism" which stresses the objective methods of physical measurement and thus leads to behaviorism rather than to sensationism (sec. 4).

2. Privacy and Limited Communicability in Phenomenological Introspection

From sensationism, two opposite lines of development may be distinguished. One represents a swinging-back of introspection to more natural and global forms (sec. 2). The other constitutes a direct bridge to the empiricist development of the concept of objectivity in which self-evident "clarity" gives way to cooperational "rigor" (secs. 3 and 4).

Constancy hypothesis and gestalt qualities.—The first of these developments takes place within the modern psychology of form-perception (see sec. 18) and of thinking. The classical notion of sensations as the ultimate building-bricks of perceptual experience is ascribed by Koffka to a "confusion of the phenomenal with the functional." Our knowledge of the anatomy of the sensory surfaces with their mosaic of receptor cells and the bundles of nerve fibers issuing from them, much emphasized in nineteenth-century physiological psychology, is said to have polluted introspection and to have exerted a suggestive influence toward a sensationist interpretation of perception and of consciousness in general. An unrecognized "constancy hypothesis," tacitly taking for granted a one-to-one correspondence of sensory stimulus elements and conscious sensations (represented in Fig. 3 by the solid circle at the periphery and the bracketed broken double arc issuing from it toward the introspective data), is said to underlie sensationism.

Descriptive phenomenology. The "nonpictorial" contents of "imageless thought."—The basic deficiencies of introspection as a scientific tool become evident, more drastically than in perception, in the efforts to establish a descriptive "phenomenology" of thinking. One of the reasons for this is that the lack of a common external reference situation underscores what is known as the "private" rather than "public" character of introspective observation. Related to this is the so-called "noncommunicability" of introspection. There is hardly an instance in introspective psychology in which the helpless solitude of the observer is as clear as in the studies of the so-called Würzburg school.[50] Thinking is in the main described by such negative attributes as 'unanschaulich' (which has been variously translated by '*im*-palpable,' '*in*-tangible,' '*non*-visualizable,' '*non*-sensory,' '*non*-pictorial,' or 'image-*less*').

The "World of Things" and Its Residue of Ambiguity

An illustration of the artistic character of the language that alone might be able to convey some of the flavor of what was to be communicated is Bühler's term ' "aha"-experience,' used to describe sudden flashes of insight; a phenomenologist outside the Würzburg group, William James, spoke of "feelings of 'but' " in characterizing the frequent spontaneous self-criticism of subjects in thought experiments. Picturesque language of this kind merely appeals to the "empathy" of a richly experiencing subject, to a kind of introspection by proxy as elicited—but by no means precisely transmitted—by the report.

There were elementarist counterclaims such as Titchener's recourse to sensations of kinesthesia as the true vehicles of thought. Woodworth, for a while sympathetic to the concept of imageless thought, has more recently suggested that the problem be "shelved as permanently debatable and insoluble."[3] Actually, the Würzburg type of phenomenology has gone out of fashion after a life-span of barely more than two decades. The history of introspectionism is thus not unlike that of the tragic hero finding himself doomed by having become true to himself.

3. The "World of Things" and Its Residue of Ambiguity

Borrowed relative stability in sensationism.—The emphasis on sensation predominant in Engish empiricism had lent introspection a certain measure of stability. Since sensory experiences are usually supported by the presence of an external stimulus, they are sturdy and persistent in comparison with emotions, imagery, and thought. They withstand observation better, and there is greater conformity of report from subject to subject. In this manner sensation not only shared in, but for a while carried the full load of, a slowly developing methodological ideal of inter- and intra-subjective univocality of observation. It must be remembered, however, that the cognitive security that goes with an emphasis on the boundary between organism and environment comes not so much from the subjective certainty of intra-organismic actuality as it does from the temporal continuity and the social communality of the *measured* "geographic" surroundings. The emergence, pointed out by Bentley, of the peripheral region of contact, the "human skin," as "philosophy's last

663

line of defense,"[4] is thus the outgrowth of a basic fallacy. So far as psychology proper is concerned, there resulted the paradox that an introspectionism sidetracked from its course by a confusion of issues with the flourishing sensory anatomy and physiology of the period could, by virtue of the very type of distorted introspection practiced by it, appear to be more of an exact science than an unadulterated phenomenological introspection ever could be capable of becoming.

Thing-language and approximate perceptual thing-constancy. —The development of the psychology of perception from strict sensationism to gestaltpsychology has found its continuation and climax in a phenomenology of the "worlds" of color, touch, etc., with their unique types of global perceptual attributes (such as apparent "surface-texture,"[76] or "solidity"). This third type of approach restores the thing-language of daily life to its rightful place in the inventory of immediately given experience.

This development has found a rough parallel in a series of shifts in experiential positivism. Mach's punctiform sensations give way, in Russell,[5] to an emphasis on organized "perspectives" and the "aspects" of the things that may be constituted from them; a further step leads to Carnap's and Neurath's "thing-language" and "language of daily life" (this *Encyclopedia*, Vol. I, No. 1, pp. 52 ff. and 19).

In the same seemingly paradoxical manner in which sensationism was said to support the ideal of objectivity, the thing-language gains a spurious foothold in the realm of *measured* physical predicates through the surprisingly invariant yet not entirely foolproof correlations of the experienced "things" with these measured properties, known in psychology as "approximate perceptual thing constancy" (see sec. 20). Since the correlations of phenomenal thing-attributes are predominantly with the more permanent measured properties of remote solid bodies —sometimes called "distal" stimulus variables—rather than with the comparatively shifty direct—or "proximal"—stimuli impinging on the sensory surfaces, the semblance of stability is even greater than for sensation. As Lenzen (this *Encyclopedia*, Vol. I, No. 5, pp. 29–30) has phrased it, the mechanism of thing-constancy in a sense displaces the "partition between object and observer" into the distant environment. The phenomenon of perceptual thing-constancy, significant in its relation to certain notions in the mathematical theory of sets, has time and again drawn stimulus and response together in the minds of philosophers somewhat in the manner of naïve realism, and has helped to ban their separation as an "unnecessary duplication" of entities, to be eliminated in accordance with the parsimony requirement known as "Occam's razor." It is obviously for the same reason that the thing-language has sometimes been considered equivalent to the objective or "physicalistic" language. The inherent difficulties can in the end be resolved only by replacing the traditional dualism of "ontologically" contrasting realms or metaphysical substances, such as subject and object, or mind and matter, by a simple duality or distinction, within one and the same physicalistically conceived universe of discourse, between organism and environment, or between stimulus and response.

One of the most cogent reasons for the conceptual separation of stimulus and response in the psychology of perception is the fact, mentioned above, that the cognitive mechanism is far from perfect or foolproof or, more generally, that there is inter-regional ambiguity. Early gestaltpsychology had received its perhaps strongest impetus from the realization, notably by Benussi, that there is a marked "gestalt-ambiguity" in the perception of identical stimulus configurations. A picture first seen as two profiles facing each other may suddenly shift in perception to give the radically different impression of a goblet formed by the same outlines. Both the terms "profile" and "goblet" designate "thing-predicates" in the naïve-realistic language of perceptual units. Although figure-ground reversals like the one described may play a minor role when it comes to the description of physical apparatus in the laboratory, they are nonetheless significant enough in principle to upset the unique consistency required of a truly physicalistic language. No sharp line can be drawn between the ambiguities of ordinary thing-perception and those of "physiognomic" responses to human expression or even those of aesthetic appreciation. Ambiguity is the mainstay of the so-called "projective techniques" (such as the inkblot test by Rorschach) in which gestalt-ambiguity is purposely provoked to throw light on the organization tendencies and personality of the responding observers.

It is at this point that psychology must insist on greater precision and scrutiny than even the physicist may find necessary in his routine description of experimental situations or manipulations. For the psychologist, perceptual thing-impressions or their physiological counterparts—or the ensuing overt verbal or behavioral manifestations—are definitely "responses," to be contrasted with stimuli. The distinction which Carnap[6] draws between "predicates of the thing-language" and "perception terms" can be maintained only if the former are restricted to data of measurement, which often seems tacitly implied. Unless this is done, the two sets quoted must both be treated as part of the response world. Their separation is a case of undue duplication of experience similar to the juxtaposition of "content" and "immanent object" in classical intentionalism (sec. 16). It is here where we must apply Occam's razor. Wherever description of stimuli in the "thing-language" is taken as a practical substitute or temporary expedient, there must be ultimate recourse to the strict forms of measurement in all cases of disagreement or doubt.

Historical steps in the separation of subject and object. "Copernican revolutions."—In historical perspective the present discussion appears as part of the struggle for a necessary differentiation within the originally unstructured field of naïve personal experience. The fact that this struggle took place primarily within the framework of the rationalism-empiricism controversy makes this latter philosophical issue one of the most important for psychology (for its general discussion see Santillana and Zilsel, this *Encyclopedia*, Vol. II, No. 8; see further Northrop[7]).

The development of this differentiation begins with Democritus' skepticism of sensory qualities, occasioned by observations of the ambiguity of stimulus-response relationships in the field of smell and taste. His views are developed in Locke's doctrine of "secondary qualities." That the latter's "primary qualities," such as size, shape, and motion, were likewise not free of ambiguity and deception had not been made sufficiently pivotal prior to the psychological emphasis on gestalt-ambiguity, mentioned above, and on the related subject of gestalt-"inadequacy" (or "geometrical illusions"), toward the close of the last century. As has been pointed out especially by Metzger,[69] the gestaltpsychological emphasis on the contribution of the individual to the perception of form leans heavily upon Kant in the way of general background attitude of a rationalistic nature, especially upon the doctrine of the subjectivity of space. (Concerning Kant as a rationalist see also Reichenbach.[2])

Crucial turns in the history of ideas, such as that just mentioned, are sometimes described as "Copernican revolutions." They define a succession of increasingly threatening blows to the pride of the ego; in psychoanalytic terms, the history of science is one of "retreating narcissism," or disentanglement of the objective from the subjective and wishful. Copernicus himself dethroned man's planet as the center of a faraway universe; Darwin dethroned the human species as the absolute master of the animal kingdom; Freud went still further and dethroned the conscious ego as the true representative of our own motivational dynamics. Kant and gestaltpsychology complete the picture by showing the subjectivity of the thing-language. Discovery of an ambiguous rather than univocal relationship between distant regions or variables seems to be at the root of most of such revolutions; the new "schools" protest the respective "constancy hypotheses" (see secs. 2, 16, 21).

Since gestaltpsychology is based on the recognition of a previously neglected fundamental ambiguity, it is not astonishing that it has in some quarters—quite expressly so in the school of E. R. Jaensch which dominated the German psychology of the late 1930's—met with an emotional resistance quite comparable to that opposing psychoanalysis. The importance of a more or less generalized "intolerance of ambiguity" for the cognitive and emotional outlook of personality as a whole and for social attitudes has been pointed out by Frenkel-Brunswik.[8]

It may be noted that the brief but significant history of psychology by Hulin[9]—the only book on the subject written under the aspects of a single "leitmotif"—sees the development of psychology as a gradual recognition of the differentiation between the subjective and the objective, although the overcoming of animism and anthropomorphism are stressed to the comparative neglect of the perceptual or other developments set forth here.

That there are remnants of stimulus-response confusion even in the field of the sensory qualities proper may be exemplified by the fact that the response term 'pitch' is sometimes used as a synonym for the stimulus term 'sound frequency.' Such recent studies in "multidimensional psychophysics" as Stevens' stimulus-response analysis of tonal attributes,[23] showing an influence of sound intensity upon the pitch response, demonstrates the equivocal relation of the two vocabularies and thus renders their separation mandatory.

4. Objectivity, Methodological Physicalism, Operational vs. Experiential Positivism

Measurement as observation of point-coincidences in space-time.—We now turn to a more positive delimitation of physical measurement. It forms an important part, although only a small one, of thing-perception. Its selection is based on the challenge, inherent in the ideal of "objectivity," of finding a class of observations which, although not necessarily of the highest possible subjective convincingness, display the highest attainable degree of consistency in the sense of agreement among different observers and, for the same observer, from one observation to another. Observations satisfying this requirement of inter- and intrasubjective univocality may then be used to set up a network of knowledge maximally free of contradictions. As has been pointed out primarily by Schlick and by Eddington,[10] of all observations those of the coincidences of points in space-time come closest to fulfilling this requirement. They are, therefore, used as the protocol operations in objective, physical "measurement" (see also Lenzen, this *Encyclopedia*, Vol. I, No. 5, p. 90); the term 'introspection' in the technical sense has come to refer to all the vast remainder of givennesses, including most of thing-perception.

A further advantage of drawing the line between the subjective and the objective in the manner described is given by the fact that, even from a purely introspectionistic point of view, point coincidences can in all likelihood much more easily be demarcated from the rest of the "experiences" or "givennesses" than sensations or any of the other types of cognitive content can be from one another. The physicist's universe, and with it objective psychology, thus emerges as an offshoot of a particularly salient kind of experience.

"Observational reliability" as a statistical measure of objectivity.—In the sense of section 3, one may in general expect a gradual decrease of stability and agreement of response as one proceeds from the observation of point-coincidences to thing-perceptions, images, thoughts, valuations, and subjective feelings or moods. This may be used to lend an element of continuity to the distinction between introspection and objective observation. The traditional dichotomy between the subjective and the objective could thus be moved, as Comte would say, from the metaphysical to the scientific level or, as Lewin[89] would say, from the Aristotelian to the Galileian mode of approach.

We may relate the concept of objectivity to what is known to the psychological statistician as test reliability. In the establishment of the reliability of a psychological test there is repeated application of the test to a sample of individuals; in replacing "test" by "type of observation," and in applying this observation to a sample of environmental situations or situational elements, we may define objectivity as observational reliability.[30] Either only one individual (subject) is involved, repeating the observations at least once (intra-individual observational reliability), or each of at least two individuals has to do one observation per situational element (intersubjective observational reliability). "Objective," then, is a class of responses yielding maximum reliability coefficients within or between individuals facing a common geographic situation or situational element.

In taking a strictly empirical point of view, one may be satisfied to ascertain by statistical procedure that point-coincidences happen to be unsurpassed in this respect, i.e., that pointer-reading types of observation are the most reliable methods in testing the surrounding world. It will be only such comparative oddities as, say, hallucinations which will spoil the ideal correlation, underscoring the inescapable relativity of the concept of objectivity, and thus the need to express its degree quantitatively. An alternative, deductive defense of point-coincidence as an epistemologically unique kind of givenness may take its start from the observation, made below, that it is the only case of observational equality in *all* respects.

Coordinate language in communication and theory construction.—The requirement of inter- and intrasubjective univocality extends from observation to communication within and between individuals, as well as to the construction of hypotheses, theories, inferences, and predictions. This is fulfilled by the use of a "coordinate language" which permits us to locate observations and inferences in a system of mathematical coordinates in terms of topological or metric relationships, avoiding reference to "qualitative" terms.

Basic observational protocols dealing with point-coincidences use the relational terms 'equal' and 'unequal.' While in ordinary subjective equivalences as studied in psychophysics the surroundings or other characteristics are in general different for the two items to be compared, a point coincidence is an equivalence in *all* respects at the same time; this reduces to a minimum the sources of disturbance and renders unnecessary the "naming" of the judgment in terms of any sensory or other introspective "qualities" which subjectively designate the "attitude" taken in the process (see secs. 3 and 20). The greater persistency of impressions of inequality in the face of simultaneous impressions of equality has induced Dubislav[13] to consider "immediate judgments of simultaneous inequality"—rather than those of equality—to be "the most certain facts of perceptual observation." Stevens' emphasis on "discrimination" as the operational successor of sensory qualities[23] reflects a similar point of view.

Cooperational rigor vs. subjective clarity.—The shift of emphasis, in the search for certainty and truth, from intuitive clarity and subjective self-evidence as stressed by such rationalists or neo-Scholastics as Descartes (sec. 1) or Brentano (sec. 17), to rigor in the sense of inter- and intrasubjective consistency, univocality, and coherence as stressed by the empirical sciences, implies a fundamental expansion of scope from the one to the many, from the self-contained to the relational, the introvert to the extravert, the private and encapsulated to the public and socialized. Experiential units no longer stand in isolation but are viewed in connection with one another. The important question becomes that of stability within and among observers; there develops a certain self-imposed, democratic restraint with respect to all those types of givennesses which are found to lead to disagreement, regardless of their impact in consciousness. Ness's[11] term 'cooperational univocality' points to the social coordination of knowledge thus achieved. Or, to use a term by Morris, any sign used in science should be a "comsign";[12] it should have the same signification to the organism which produces it that it has to other organisms stimulated by it.

In further support of the suggested shift of emphasis from intuitive clarity to cooperational rigor it may be mentioned that some relatively unnoticed studies within the Würzburg school (sec. 2) have exposed, by means of objective methods of checking, the frequent fallaciousness of introspections regarding object-reference, certainty, and clarity.[48]

Methodological positivism and operationism.—Since criteria of objectivity were first developed in physics, the striving for objective standards may also be designated as "methodological physicalism" (see Carnap and Neurath, this *Encyclopedia*, Vol. I, No. 1). Concepts become scientific when they are anchored in physical observation, directly or by mathematized construction procedures. Anchoring may be either by definitive "definitions" after logical analysis or by tentative "reduction" sentences which will allow to put all statements concerned to the test of verification or, less stringently, to that of confirmation. The view that scientific statements have meaning only by virtue of the concrete operations that enter into the definitions of the concepts employed is known as "operationism."[1] Care must be taken that the concept of "operation" be sufficiently specified in the sense of methodological physicalism lest relatively casual testing procedures including introspection be placed on an equal footing with objective methods. Furthermore, equivalence of certain operations which are different from a purely manipulatory point of view—such as direct measurement vs. triangulation or other inferential ascertainment of length—must be properly stressed (see sec. 6). Equivalence of operations may have to be established by empirical correlation of results rather than on deductive grounds.

Methodological physicalism and operationism are part of a general "positivistic" tradition. Their psychological counterpart is the school of behaviorism (secs. 18 and 20) which attempts to base all its statements on the movement or physiology of organisms taken as physical bodies. Methodological or "operational" positivism thus must be distinguished from the epistemological, "experiential" positivism of the Mach tradition in which the immediately given of introspection is uncritically identified with sensation (see sec. 1). A distinction must further be made between methodological and thematic physicalism, the latter being given by the uncritical emulation of the real or alleged particular aims and problem content of physics by other disciplines (see sec. 11).

5. The Futile Search for "Criteria of Consciousness." Verbalization

Docility and other functional criteria.—As was pointed out by Bekhterev[70] and by others, among them Boring,[50] there is no objective indication of the presence of consciousness that could withstand scrutiny, and thus little possibility of successfully "reducing" consciousness to physical observation. This fact must be stressed in the face of numerous historical attempts to link "mind"—primarily in infants or animals—to physical criteria.

Some of these suggested criteria are "morphological," such as the presence of a nervous system or of neural specialization (Yerkes). Others are "functional," such as "discrimination" (selective response) or "choice" (Bergson), modifiability of reaction through individual experience in learning or "associative memory"—also called "docility"—(Romanes, Lloyd Morgan, Loeb), and variability of reaction—sometimes called "initiative"—(Yerkes, Margaret Washburn). Other authors have suggested such behavioral criteria as use of tools, specified response to facial expression, use of representational language or of other symbols, ability to play or to respond to fictitious situations, or apparent purposiveness as given by adaptation to an end, especially if the situation or the response is novel.

In surveying some of these suggestions, Washburn[14] comes to the conclusion that they all elude precise definition. Some of them, such as learning or docility, can readily be traced to unicellular organisms or even to the simplest inanimate processes. She further points out that Yerkes has rendered the problem a matter of degree by emphasizing the "rapidity" of learning as a criterion. Since she admits neural specialization as a second, likewise quantitative, criterion, the problem of "evidence of mind" is made to appear at best as a matter of multiple-criterion and probabilistic, rather than of one-criterion and absolute type of reduction. Carr[14] deplores our "lack of knowledge of any decisive and unambiguous criterion of ideas in man" which in itself precludes any satisfactory approach to the problem as applied to animals.

Introspectionistic and behavioristic use of verbalization.— Special attention should at this point be given to an issue all too often left in confusion, that is, the relationship of verbal report and introspection. Behaviorists have often felt that there was an unholy alliance, if not an intrinsic association, of verbal report and introspection. Watson somewhat naïvely decreed 'unverbalized' to be the behaviorist's translation of 'unconscious' and somewhat arbitrarily identified thought with "laryngeal habits"; there was a marked decline in the use of verbal report in favor of "overt behavior" during the heydays of classical behaviorism (see sec. 18).

It should be noted that the presence or absence of overt verbalization plays a relatively minor role in the historical list of suggested criteria of mind given at the beginning of this section. Introspectionism is possible without the use of verbal report, such as in all attempts at interpreting facial or pantomimic expression in an intuitive process of empathy or in an explicit inferential search into another person's mind by analogy with one's self. On the other hand, it is possible to utilize verbal report without falling back upon introspectionism. This is the case whenever words are not taken in their common-sense or dictionary meaning but anchored, by statistical correlation, in overt behavior or its results.

For example, placing a check mark in the space for "No" opposite the statement "I would like to be an actor"—a verbal response whose similarity to point-coincidence is obvious—has been found to have the pragmatic "meaning" that the subject has potentialities to become a successful engineer.[15]

Such semantic analyses as the anchoring of the traditionally "philosophical" concept of "truth" in the language system of daily use, attempted by Ness,[16] also belong in this context.

To be distinguished from the overt behavioral is the dynamic, or motivational, anchoring of verbal report which likewise often discards common-sense meaning in favor of objective methods of ascertainment (see sec. 16).

II. The Functional Unit of Behavior and the Level of Complexity of Psychological Research

While differing in important points of elaboration of their systems, behaviorists have been rather unanimous in their definitions of the basic structural characteristics of behavior. These definitions contain all the necessary conceptual elements for making explicit the general characteristics of an approach necessary to establish objective psychology on an adequate level of complexity. But they have left to the general methodologist the actual drawing of conclusions concerning the design of a type of research that would match the pattern of behavior itself.

6. Stabilized Achievement and Vicarious Mediation. The Lens Model

Definition of the subject matter of psychology seemed obvious, or else a matter to be left to the philosopher, so long as the fundamental conception of psychology was introspectionistic. Thus the chief exponents of early experimental psychology, Wundt and Titchener, acted in conformity with philosophers of the period when they declared, respectively, that psychology deals with "immediate"—as contrasted to "mediate"—experience, or with experience as "depending on"—as contrasted to "independent of"—an experiencing person.

Vicarious functioning and focusing on "ends." Purposive behavior.—All this changed with the shift toward a behavioristic conception of psychology at the beginning of the present century. The subject matter of psychology now was physical occurrences; but psychology could obviously not cover all the physical occurrences, not even all physical or physiological occurrences within or around organisms. Was the incidental stumbling of a person over an obstacle to be considered behavior or nonbehavior? The theorist of classical behaviorism, Albert P. Weiss,[17] provided the crucial model to deal with doubtful cases of this kind in his "raindrop analogy" of purposive behavior. He pointed at the convergence of originally diverse occurrences toward a common characteristic end stage in behavior and compared it with the way raindrops originally scattered over a wide area are eventually carried to a common point in the sea.

An effort had thus been made to link the most prominent descriptive feature of behavior, characterized as "striving toward a goal" or as "equifinality," to simple physical processes. The latter were called upon originally to demonstrate the possibility of a mechanistic reduction or "explanation." But by the same token the existence of the pattern itself, as a fundamental descriptive characteristic of behavior, was acknowledged.

Two features are involved in this pattern. One is what may be called "stabilization" of the end stage, to be labeled "terminal focus" henceforth. The second is the diversity of preceding stages. Another prominent representative of classical behaviorism, Hunter,[18] concentrates on the implications of this second feature upon the flexibility and exchangeability of pathways relative to an end when he elevates "vicarious functioning" to the role of the defining criterion of the subject matter of psychology. The same holds for certain ideas by Holt, while Hobhouse has conceived of the possibility of remote focusing.[19] In the case of Meyer's "psychology of the other-one"[19] the emphasis is not so much on vicariousness proper than it is on the related aspect of the "concerted" character of behavior as exemplified by the cooperation of several limbs in action; this is comparable to the "multiplicity" of mediation which is important in perception (see below). In listing the essential postulates for a human "robot," Boring[35] notes, under "vicarious response," that "there is little that will make our robot seem more human than this ability to choose one means after another until the goal is reached."

All concepts and considerations of this kind stress the equivalence and mutual intersubstitutability of certain activities, habits, sense departments, or bodily organs for one another in behavior. They form "hierarchies" either in the sense that some of the alternatives may be more useful than others (e.g., by possessing a greater probability of resulting in the characteristic end stage) and/or in the sense that some alternatives are better established in the organism than others (e.g., by more effective learning). It is primarily the latter feature which Hull had in mind when he coined the phrase 'habit-family-hierarchy'[90] to describe and to explain patterns of vicarious functioning in overt behavior.

The terminal reference which is always at least a tacit constituent of such models is made part of the label itself when Tolman[74] operationally redefines the "purposiveness" inherent in all behavior as "persistence through trial-and-error, and docility, relative to some end." There is a variety of "means" to each end, and this variety is changing, both variety and change being forms of vicarious functioning. Rats are found to persist in this manner until arrival at the goal and its consummation is followed by quiescence. And the stumbling of a person over an obstacle is, according to this definition, to be considered behavior if, and only if, it can be shown to be linked with a group of events (or habits) running off in series until a certain result is brought about, say, damage to the individual ("accident-proneness," which may be part of a broader "suicidal" purpose).

Backward extrapolation in perception. Variability and multiplicity of mediation.—In overt behavior an additional initial focus is frequently given by a hunger stimulus or by a "central" motivational state. The initial focus forms a counterpart to the external terminal focal state of arriving at the food. The earliest behaviorists have usually seen the main problem in the convergence of causal chains toward the terminal focus. However, in the field of cognitive processes, the initial focus is not an event within the organism as is motivation; rather, it is an external variable. In the case of perceptual thing-constancy, the measured sizes of physical objects in space—constituting a "distal" stimulus variable (see sec. 3)—turn out to be, in a figurative sense, backward extrapolated by the terminal perceptual size response irrespective of their distance from the observer. The latter feature implies a variability of "proximal" retinal projections; together with "cues" or "criteria" for a further distal variable, depth, in a process of "multiple" mediation,[77] these projections form patterns of vicarious functioning in establishing the final size response. External initial foci of this kind possess a status symmetrical to that of the external terminal foci encountered by the student of overt behavior; in both cases a mechanism of vicarious mediation is needed to correlate external to internal events with a sufficient degree of likelihood.

Since there is no perceptual cue which would be available under all circumstances or is completely trustworthy (see below), the perceptual system of higher organisms must for most types of perceptual attainment develop what the present writer has suggested calling an "or-collective" or an "or-assemblage" (*Oder-Verbindung*) of mutually interchangeable cues vicariously mediating distance or other situational circumstances to the organism (see sec. 20).[77] Since cues form a hierarchy just as do means, we may also speak of a "cue-family-hierarchy," in conformity with Hull's "habit-family-hierarchy." It is characteristic of the formal convergence of lines of psychological research differing in content (see sec. 19) that the two concepts were introduced independently in the same year.

The lens analogy of the unit of achievement.—A stabilized or relatively stabilized connection between focal variables taken as classes of events rather than as individual occurrences, as established through vicarious functioning either in perception or in overt behavior, may be characterized as an "accomplishment" or "achievement" (*Leistung*)[77] (for further specification see sec. 7; for discussion see also Carnap, this *Encyclopedia*, Vol. I, No. 1, p. 48). The term 'function' will likewise in the main be used to characterize wide-arched dependencies of this kind, in contradistinction to the more microscopic aspects of the "functioning" of physiological processes. Focal variables outside the organism may be called "functionally attained" (*intentional erreicht*). From a "molar" point of view, comprehensive patterns of this kind may be taken as the dynamically integrated and effective units of behavior. Organisms may in this sense be characterized as "stabilizers" of events or of relationships.

The total pattern involved, when viewed as a composite picture of numerous cases of individual mediation from initial to terminal focus, bears resemblance to a bundle of rays scattering from a light-source and brought back to convergence in a distant second point by a convex lens. A generalized "lens model" for stabilized functional units is shown in Figure 1.[20] While correlation between the focal variables is assumed to be relatively high although in general not perfect (see sec. 7), those of each

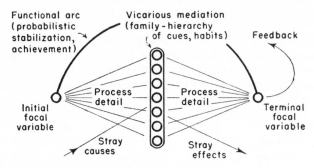

Fig. I. The lens model: Composite picture of the functional unit of behavior

focus with the single elements or chains of mediation may be low. A semicircular arrow is appended in the figure to the terminal focus to indicate that lens patterns do not stand in isolation but are apt to reflect back upon the organism in a future state in what is now sometimes called a "feedback" loop (see sec. 23), such as when arriving at the food is followed by satiation and reinforcement of the preceding behavior ("law of effect"; see sec. 18). Further lens patterns may be involved in this process.

Considering the relative chaos in the regions intervening between focal variables, focal connections may also be called "interrupted." Although they do not require action-at-distance, they are relations-at-distance. Further formal characterization of such connections may be given in terms of the mathematical classification of relations by Russell.[21] Under this aspect they appear as asymmetrical (directed) and as transitive (carrying over from one focal pattern to the next).[77] This analysis may also be used for an operational definition of "meaning."

The broadest context into which behavioral units may ultimately be fitted is "(probable) survival" as a descriptive fact of life. The concept of survival, established by Darwin, Tyler, and other biological and sociological functionalists of the nineteenth century, is extensively used by Hull in his recent formalization of classical behaviorist theory (sec. 22). For examples of artificial stabilization mechanisms see section 23. It may be added that even lenses in the literal sense of the word are, when taken as a class and as imbedded within a stabilization mechanism, to be found only within, or as products of, higher organisms.

The principle of vicarious functioning has also been used in semiotic by Morris,[12] in his concept of "sign-family." The necessity of establishing, in the "research behavior" of psychologists, equivalence among various "member-operations" with reference to the final term to be defined, as connected with the problem of avoiding unmanageable multiplicity of concepts in "operationism" (see sec. 4), has been discussed by Ness,[11] by Stevens,[1] and by Israel.[1]

Central-distal vs. peripheral focusing of achievement.—Recent psychology has shown that variables located in certain "areas," "layers," or "regions" of the environment or of the organism seem more often to be focal than those in others. Some of the most crucial changes of emphasis in contemporary psychology are based on the recognition of the relatively nonfocal, vicarious, "generalized" role of the sensory as well as of the motor periphery, coupled with the comparatively focal character of the central as well as the distal regions, both situational and historical, in the case of higher animals at least. This recognition defines the shifts from classical to molar behaviorism, from sensationism through gestaltpsychology to the perceptual thing-constancies, while the inhomogeneity of the developmental time continum—with a focal area in early childhood not anticipated in the academic psychology of learning—is a discovery of psychoanalysis (see Chaps. I, IV, and V; a survey of areas is given with Figs. 2–7).

7. Organismic Adjustment as a Probability Function

Limited "ecological validity" of cues and means. Partial vs. total cause-and-effect relationships.—As was pointed out by the writer in greater detail elsewhere,[77, 1] any organism has to cope with an environment full of uncertainties. Forced to react quickly or within reasonable limits of time, it must respond before direct contact with the relevant remote conditions in the environment, such as foodstuffs or traps, friends or enemies, can be established. The probability character of intra-environmental relationships, their limited "ecological validity,"[30] becomes of concern in two regional contexts: on the reception or stimulus side as the equivocality of relationships between distal physical or social objects and proximal sensory stimuli or cues, and on the effection or reaction side as the equivocality of relationships between proximal outgoing behavioral responses, or means, and their more remote distal results and effects.[78] In one or the other of these ways behavioral responses are, to apply a saying by Thurstone, of necessity based upon "insufficient evidence."

Another way of stating this state of affairs is by pointing toward the fact that the relationship between the mediated event and its mediators is one of probable partial causes and partial effects (see also Reichenbach[22]). In the case of the depth-cue of geometrical "perspective," a trapezium-shaped image on the retina may have any one of several possible lines of causal ancestry. It may be historically caused either by a rectangle actually tilted into the third dimension—the valid instance of the cue—or by an actual trapezium in frontal position —an invalid, misleading instance of the cue. This is due to the fact that neither an actual rectangle nor an actual trapezium would constitute the total cause of the proximal image, among the remaining conditions being the rotation of the object in space.

This picture must be set up to counterbalance the notion of univocality based on the tacit limitation of causal considerations to relationships between total cause and total effect as frequently taken for granted by the traditional logician or experimental physicist with his more fully developed, explicit access to the relevant elements of a situation and his rational means of analyzing them. Such relationships have been described by the "strict" laws of the classical natural sciences and have been idealized in such universal conceptions as Laplace's world formula.

Impossibility of foolproof distal achievement. "Functional validity." Quasi-rationality.—In line with the inherent probability character of object-cue and of means-end relationships, gross organismic coming-to-terms with the environment can thus never become foolproof, especially so far as the more vital remote distal variables are concerned. It is in this sense that, as William James has phrased it, perception is "of probable things." In the terminology of Reichenbach's probabilistic empiricism, behavior and the inferences implicit in it must retain a certain "wager-" or "posit"-character.[22] Perceptual and behavioral functioning is spoiled much in the manner in which stray rays (Fig. 1) are apt to interfere with perfect focusing. Imperfections of achievement may in part be ascribable to the "lens" itself, that is, to the organism as an imperfect machine. More essentially, however, they arise by virtue of the intrinsic undependability of the intra-environmental object-cue and means-end relationships that must be utilized by the organism; these are comparable uncontrolled lateral light-sources.

It is for this reason that the concept of achievement or of cognitive "correctness" must be defined in psychology in the generic terms of over-all statistical correlation between variables as classes rather than in terms of single hits or misses of judgment or of action. "Achievement," in the sense of the probability for an initial focal event (say, a measured stimulus) to be followed by its terminal counterpart (say, the correct perceptual estimate), may, then, be defined as "functional validity" and measured by a correlation coefficient. Here it is the environment which is being tested by the reacting organism, rather than vice versa as in "test validity" (compare with "observational reliability" *versus* "test reliability" as developed in sec. 4; for further details see Brunswik[30]).

All imperfections notwithstanding, the use of cues and the taking into account of several variables at the same time injects an element of reasoning into stabilization mechanisms.

681

Implicit reasoning of this kind may readily be made explicit by modern "cybernetics," or by "mathematical biophysics" (see secs. 22 and 23). In the field of the perceptual constancies, it has first been recognized in Helmholtz' doctrine of "unconscious inference" (*unbewusster Schluss*).

The present writer's emphasis on the cognitive achievement of the perceptual constancies, brought about by the utilization of cues, is, as Boring[23] has phrased it, a "modern equivalent" of the doctrine of unconscious inference. The reason why it is not merely a revival of this doctrine may be expanded as follows. It sheds the introspectionistic entanglements of "unconscious inference," such as of Helmholtz' untenable assumption that we are dealing with a mechanized form of originally conscious processes. Nor does it accept the notion that the perceptual constancies function exactly like rational inferences save for the fact that they are not conscious. Rather, the crucial change is a functional-objective one, based on the experimental demonstration of perception as a subsystem of the total personality that is relatively autonomous of reasoning proper[77] and distinguishable from the latter in terms of both achievement and mediation. Cues of perception proper are found to be sluggishly established as probabilistic stereotypes; once established, they act with a quick efficiency full of peculiar pitfalls. More recently, an attempt was made directly to compare perception and thinking in terms of differences in the statistical distribution of error.[30] All evidence may best be summarized by designating perception as a "quasi-rational" rather than a rational system. Perception is what Werner[24] has labeled an "analogous function (or process)" to reasoning, more primitive in its organization but vested with the same purpose (in the behavioristic sense of this term).

In an attempt at rational reconstruction of the ways of the quasi-rational, with its reliance on vicarious cues each of which is of limited validity, one may best refer to a remark of Thorndike comparing the impressionistic or intuitive judge of men to a device capable of performing what is known to statisticians as multiple correlation. This is a device, related to what cyberneticists have called redundant communication (sec. 23), by which the probability of individual correctness may be increased but not perfected to the point of certainty. By contrast, man-made gun or tank stabilizers and the related "thinking machines" may, within the limits of their technical purpose, perfom in a practically foolproof manner. This is due to the fact that they can usually be built with a concentration on a few cues of maximal trustworthiness and thus dispense with the services of cues of limited validity. In the case of distance or rotation from a frontal plane, to quote just one example, triangulation may be used in preference to the untrustworthy cues so often employed by perception proper, especially in monocular vision.

8. The Molar Approach: Focal and Macro-mediational Reference

The methodological postulate of behavior-research isomorphism.—The suggested definition of the functional unit of behavior in terms of the lens model is a case of "persuasive definition" in the sense of Stevenson.[25] It is well founded by the fact that—so far as is known to the present writer—no one attempting an objective definition of behavior in terms of a structured pattern was able to think of anything but vicariously mediated periodic equifinality (sec. 6). The fruitfulness of the definition is demonstrated by the fact, to be pointed out in Chapter V, that most contemporary movements in psychology explicitly or implicitly use this scheme or at least some of its basic features.

In applying the lens model to psychological methodology, one may metaphorically paraphrase Spinoza's parallelistic credo and demand that the "order," or pattern, of research "ideas," or design, should be the same as the pattern of the "things" studied, which in our case is behavior. Research may be said to have reached an adequate, "functional," or "molar" level of complexity only if it parallels, and is thus capable of representing, behavior in all its essential features. We may call this the methodological postulate of behavior-research isomorphism.

Focal alertness.—In particular there should, first, be no restrictions upon the scope of the search for focal relationships. Research should focus wherever behavior focuses or may focus. Since foci are defined by intercorrelation with other foci, a great variety of inter-variable and inter-area relationships must be studied, including some that arch out very far in space or time. One may adopt Boring's admonition to "tap" the organism at the right places and augment it by pointing out that the environment—both spatial and temporal—must also be tapped at the right places. In this process, central-distal emphasis seems to emerge as a particularly promising pattern of regional reference (see sec. 6 and Chap. V). The final breaking-away from traditional peripheralism was preceded by chains of painful disappointments in experimental research[3] which can be traced to a naïve oversight of the nonfocality of certain variables. Such nonfocality could have been readily anticipated if psychologists had been more effectively imbued with the principle of vicarious functioning.

Since possible reference variables and the possible intercombinations between them reach an unmanageable total, "blind" gathering of fact must seem increasingly hopeless in psychology. Like a prospector digging for oil, psychologists become increasingly reluctant to treat the field to be explored as homogeneous or nondescript. Tentative generalizations, hunches, and expectations regarding possible singularities of behavior furnish a promising basis for an imaginative theorizing-before-the-fact as to the probable "fruitfulness" of an approach.

Passive control in the macro-mediational study of vicarious functioning.—While focal correlation furnishes an over-all frame of reference, the gross degree of flexibility of vicarious functioning, that is, the range of mutually intersubstitutable processes in a family hierarchy of cognitive cues or of overt behavioral habits (means) mediating between two focal variables remains a second integral aspect of psychological inquiry. In terms of the lens model, this range is comparable to the aperture of the pupil of the eye. Tolman[74] has recently spoken of the "width" of the "cognitive map." This range or width characterizes the degree of "generality," and thus in turn the degree of functional efficiency and foolproofness, of a behavioral unit. Its study "from above," that is, under the aspect of the superordinate behavioral units, may be characterized as macro-mediational.

In controlling vicarious mediation, care must be exercised not to interfere with naturally established mediation patterns. These aspects of mediation must therefore be controlled "passively," that is, be studied in a permissive laissez faire manner with respect to their free dynamic flow; there must be deliberate neglect of "active" control at least up to a certain point, despite the fact that the conditions involved either are definitely known to be relevant or are at least potential mediators bridging the gap from one focus to another. In particular, mediation must not be "channeled"[30] by allowing, say, only one of the many perceptual distance cues to function, or by providing only one path to the goal, as was the case in earlier phases of experimental psychology. Channeling of mediation leaves no room for vicarious functioning; in consequence the entire relief of focal versus nonfocal variables or regions is obscured.

As the specific merits of a bombsight or a gun stabilizer can only be fully appreciated when the conditions to be "eliminated" by these mechanisms are kept variable rather than constant, organismic achievement can unfold in its full scope and specific focalization only when there is room for expansion of the vicarious mechanisms which funnel the divergent causal chains back to convergence in a terminal focus. Any machine to be tested must be exposed to the full challenge of conditions for which it is designed, and so must any organism.

Impairment of vicarious functioning implies the breakdown of higher functions. Mental disturbance or retardation in natural development has often been described in terms of "rigidity," "fixation," "concretism," etc. An example is Krech's finding that rats with brain lesions tend to move along more stereotyped paths to the goal than do normal rats, and even will try to avoid the challenge of choice points altogether whenever possible.[26] (See also Frenkel-Brunswik.[8])

Micro-mediationism and the atomistic approach to the functional arc.—Neglect of any of the methodological requirements listed in this chapter constitutes psychological atomism of some sort. In the traditional micro-mediational tracing of sensory-neural transmission (see secs. 10 and 15) one may easily lose sight of focal relationships. On the other hand, concentration on relationships between pairs of variables far removed from each other in space or time, to the exclusion of all mediational considerations, as sometimes found in mental testing or in the purely correlational treatment of the inheritance of traits (see sec. 20), is atomistic in the opposite sense; it stresses the focal arc at the expense of checking on the scope and intricacy of vicarious functioning through which (relative) stabilization is alone possible.

685

9. The Molar Approach: Probability Laws and Representative Experimental Design

Statistical approach in functional and differential psychology.—Since behavior as an adaptive function is inherently bound to have its imperfections and occasional failures, the traditional nomothetic search for strict laws becomes an insoluble task when applied to it (sec. 7). The degree of univocality of results cannot exceed the degree of univocality of behavioral achievements themselves, so long as the latter are studied as class-relationships between focal variables connected with each other in the manner of part-causes and part-effects. All we can hope for with respect to molar behavior is what Reichenbach and others have called "probability laws."[22] Among these are correlation coefficients and other statistical measures of concomitant variation (see also Nagel, this *Encyclopedia*, Vol. I, No. 6, p. 25). It must be stressed once more that the probabilistic character of behavioral laws is not primarily due to limitations in the researcher and his means of approach but rather to imperfections inherent in the potentialities of adjustment on the part of the behaving organism living in a semi-chaotic environmental medium. In this sense even an omniscient infinite intellect, when turning psychologist, would have to adopt a probabilistic approach (see further the discussion with Hull referred to in sec. 11).

As was pointed out for physics by Mises,[27] only differential equations which encapsulate within infinitesimally small space-time segments can be conceived of as laws in the strict, absolute sense. Their integration into the customary macroscopic form implies incorporation of further variables which inject a probabilistic element into the law. There is an inherent tie between nomothetic approach and confinement of scope which goes beyond the routine type of analytic effort involved in any finding of regularities. (For discussion of seemingly opposite further aspects of the difference between micro- vs. macro-approach see below in this section, and Philipp Frank.[28])

Under the pressure of conditions which are in important respects similar to those just referred to, correlation statistics and representative sampling as a deliberate lump treatment of inter-variable relationships was first developed in a field close to psychology, anthropometrics, soon to spread to the study of "individual differences" in psychology proper (see sec. 19). Imperfect correlation between variables, such as between the intelligence of parents and offspring, was from the beginning recognized as due to partial causes other than those included in the operational data.

The neglect of these remaining conditions, in spite of the fact that they obviously contributed to the results (see sec. 8), was, and still is, looked upon by most psychologists as a blemish on scientific exactitude, to be ameliorated as soon as practicable. In this vein, correlation statistics appeared as a mere temporary expedient eventually to be replaced by full observation of all relevant factors, with an absorption into a strictly nomothetic psychology as the final goal. In contrast to this, we must propose the statistical approach as an ultimate norm for psychology as a whole,[29] rather than as a somewhat embarrassing substitute in the relatively specific field of differential psychology. Such a reorientation is established in its more positive aspects by expanding the field of application of statistics to include the functional study of achievement through a correlation of stimulus variables with response variables as outlined in section 7. This shift defines what may be called a "probabilistic functionalism"[30] in the study of psychological laws (see sec. 20).

Situational generality in the representative design of experiments.—The call for unrestricted vicarious functioning in studying gross behavioral adjustment, voiced in section 8, injects a new element into the discussion of these issues. It is the requirement of normalcy, naturalness, "closeness to life" (*Lebensnähe*), or, with a more methodological slant, that of "situational representativeness."[30] According to the much-stressed requirement of "representative sampling" in differential psychology, individuals must be randomly drawn from a well-defined population; in the same manner, the study of functional organism-environment relationships would seem to require that not only

687

mediation but especially also focal events and other situational circumstances should be made to represent, by sampling or related devices, the general or specific conditions under which the organism studied has to function. This leads to what the writer has suggested to call the "representative design of experiments."[30] The correlation coefficient can then be employed as a measure of functional stimulus-response relationships as it has been accepted in the past as a measure of differential intertrait relationship. Such a generalized statistical conception of psychology would serve to establish a unity of outlook and research design hitherto lacking within this discipline.

Any generalized statement of relationship requires specification of a "reference class" or "universe" from which the material is drawn. In case of strict "laws" as aspired to in the natural sciences reference is supposedly unlimited; thus we are apt to forget that correlations have meaning only when specified as to applicability. Certain analogies existing between the interpretation of representatively designed psychological research and the problems of the generality of results connected with relativity physics have been pointed out by Hammond.[32]

By its extension of the principle of sampling from individuals to situations, representative design countermands a number of preconceived methodological notions of nomothetic experimentation, among them the "rule of one variable," i.e., the studying of one factor at a time. It further countermands the more general inclination to design experimental research in accordance with formalistic-"systematic" patterns which are too narrow to bring out the essentials of behavioral functioning although they may involve a shift to "multivariate" or to Fisher's "factorial" design.[31]

As a "methodological demonstration" of a deliberate switching from systematic experiment to a broad representativeness of stimulus conditions, the writer has undertaken a study of perceptual size-constancy in the manner of a representatively designed experiment in social perception.[30] Correlations of .95 and over—high for differential psychology but probably not uncommon for functional stabilization mechanisms—were obtained between measured distal object sizes and their perceptual estimates in a representative sample of daily-life situations involving a wide variety of sizes and distances; correlations of the estimates with proximal (retinal) image sizes were low by comparison.

The Molar Approach: Probability Laws, Representative Design

Macro-statistical approach in psychology vs. micro-statistical approach in physics.—Much has been written about the recent shift in physics from "strict" laws—also called "causal" or "deterministic" laws—to a statistical, probabilistic point of view. In a general way such a shift lends support to probabilistic claims or programs in other sciences. Yet there are important differences, and the analogy must not be pressed, especially in view of what has been said above in this section concerning the association between the statistical and the molar rather than the molecular approach in psychology.

In thermodynamics as well as in quantum physics—the two major points of inroad of the statistical approach in physics—it is the microscopic features that present themselves either as intrinsically not fully predictable or at least as in need of statistical lump treatment for practical reasons (see Philipp Frank, this *Encyclopedia*, Vol. I, No. 7). Univocality is for most practical purposes restored when we proceed from the microscopic to the macroscopic level, through the involvement of extremely large numbers of particles.

Psychological correlations, both of the differential and of the functional-stabilization kind, on the other hand, are erratic primarily when viewed macroscopically. It must also be noted that the main difficulty lies not in multiple dependency of reactions on a series of factors—a fact relatively easily coped with by the nomothetic approach—but in the impossibility or theoretical-methodological inadvisability (or both) of exerting (active or passive) control over the relevant, contributing factors. Bohr[33] concurs that the situation in physics is "essentially different from the recourse to statistical methods in the practical dealing with complicated systems" (to which latter alternative the situation in psychology essentially belongs).

We may illustrate the difference by an example. In studying the differential psychological problem of the inheritance of personal traits it is found that the "intelligence quotients" of fathers and sons correlate about .4. Two features must be emphasized here. First, in order to ascertain this result, one must test *individual* fathers and sons, and, what is more, the computation of the correlation coefficient utilizes the results individually. And, second, while the coefficient itself may be taken as an over-all evaluation of one among several contributing factors operating with a certain consistency if large numbers of individuals are investigated, the important psychological task of predicting of *individual* I.Q.'s retains some intrinsic hazards. In thermodynamics (in contradistinction to the kinetic gas theory designed to "explain" its laws) the individual molecule maintains its identity neither in observation nor in prediction, and the relation is not expressed by a correlation coefficient. To duplicate this situation in our example from psychology, one would have to devise measures of intelligence for the pooled resources of large populations of fathers and of sons which could not be separated into their component individuals, and hence the problem of individual prediction could not even be raised.

It may be added in this context that the physicist Bopp[33] has recently suggested an extension of correlation statistics which would be suitable for the description of quantum processes. Both classical correlation and quantum physical correlation constitute special cases of a more general correlation theory.

Breaking the gross functional units of behavior up into fragments is apt to increase precision. Thus we have relatively satisfactory laws of simple color vision or of the functioning of a single depth cue (especially when the cue is easily identifiable physiologically, as is, e.g., binocular disparity), while the more complex and vitally more relevant functional stabilization mechanisms of perceptual size- or color-constancy are, in spite of their impressive achievement, nonetheless found to be far from foolproof. The historical neglect of the complex in favor of the simpler problems constitutes but one of the cases of escape from the behaviorally macroscopic, or "molar," into the behaviorally microscopic or "molecular" level. In fact, some of the most outstanding formalizers of psychological theory of the present, Hull, Rashevsky, and to some extent even Lewin, have paid the price of becoming relatively molecular by a confinement of their scope to the organism as contrasted to its environment, or even to the central layer, sometimes with a quasi-physiological slant, in return for undeniable progress in the formalization of psychological laws (see sec. 22).

Briefly we may say that, while physics tends to be microstatistical and macro-nomothetic, psychology tends to be macrostatistical and micro-nomothetic. The fact that physics has better chances to find strict laws when becoming more macroscopic while psychology has better chances when becoming more microscopic is not a paradox, however; micro-psychology deals with phenomena of an approximately similar order of magnitude or coarseness as those of macro-physics, while both above and below this stratum there is less stringent orderliness. Yet, by comparison within its own size order, purposive behavior forms a region of outstanding texture. The existence of outstanding regions of focusing and of regions of comparative vicariousness in the organization of behavior itself as discussed above (sec. 6) is another type of the inhomogeneous distribution of orderliness in the world. In the light of historical perspective, it is psychology that seems to be called upon to deal with the relatively (although far from absolutely) orderly types of highly macroscopic life-occurrences as schematically represented by the lens model.

10. Descriptive and Reductive Theories. Law, Inference, Explanation

Synoptic description. Functional explanation.—The most elementary form of describing a set of facts is by simple enumeration. A more economical way is sometimes offered by such comprehensive synoptic devices as mathematical equations fitted to the data. These and related summary descriptions of relations define what is known as an "empirical law." Psychologists from Ebbinghaus to Hull and Woodrow (see secs. 15 and 22) have in this manner attempted to establish psychological laws of varying levels of generality.

Subsuming a new observation under a previously observed or hypothesized functional relationship (or arc) may be called "functional explanation." In particular, reference to a terminal focus in functional explanation may be called a "whither"-explanation; backward reference to an initial focus may be designated as "whence"-explanation. If the initial focus is assumed rather than observed, we have what Spence[1] calls a "response inferred construct." The difficulties in making central inferences from responses can be alleviated by proper consideration of the fact that under the definition of behavior such inferences cannot be made from one trial or along a single mediational track but must use the principle of vicarious functioning (sec. 6); thus varied and repeated accidents may establish a suicidal "purpose," although the occurrence of a single accident could not automatically be "explained" in this manner.[85]

Whither-explanations of behavior in terms of needs, wishes, or purposes are quite similar to those practiced and readily accepted in another macroscopic social-biological discipline, economics.[34] An example is the explanation of railroad communication in terms of "transportation needs."

Intervening variables and hypothetical constructs.—In their theories of behavior, Tolman and more recently also Hull have spoken of "intervening variables"; mostly these are assumed for the least accessible, central region in the network of organism-environment relationships. As was argued by MacCorquodale and Meehl,[1] an intervening variable in the original sense is "abstractive" in that it refers to a quantity obtained by a specified manipulation of the values of empirical variables; it will involve no hypothesis as to the existence of nonobserved entities or the occurrence of unobserved processes, physiological or other; "it will contain, in its complete statement for all purposes of theory and prediction, no words which are not definable either explicitly or by reduction sentences in terms of the empirical variables; and the validity of empirical laws involving only observables will constitute both the necessary and the sufficient conditions for the validity of the laws involving these intervening variables." By contrast, the authors propose that, wherever the theory contains a physiological or other "surplus meaning" in the sense of Reichenbach, we speak of "hypothetical constructs." Spence has been skeptical about the value of such surplus meaning in psychological theorizing, while others have become increasingly sympathetic toward it.

"Why" vs. "how." Technological-reductive explanation.—Functional explanations may also be called "why"-explanations in contradistinction to the "how"-explanations in terms of micro-mediational, technological detail (see sec. 8). The former deal with the grand strategy of the organism; the latter, with the more subordinate aspect of its tactics. Tracing, or hypo-

thetically filling in, intervening processes helps to reduce gross functional relationships to familiar mechanisms and may thus be designated as "reductive" (this reduction to detail is not to be confused with the reduction to observation in operational definitions; see sec. 4).

In our above example of railroad communication, how-explanation is exemplified by references to cylinders and wheels rather than to transportation needs; this corresponds to the emphasis on anatomical or physiological detail of sense processes or nerves in early experimental psychology (sec. 15). Since in historical perspective micro-mediationism seems to have tended to sidetrack psychology from the proper study of focal and vicarious patterns, we may speak of the underlying scientific urge or compulsion as *explanatio praecox*.

In his discussion of operationism, Feigl[1] distinguishes a series of "levels of explanation" from straight description through empirical laws—also called low-level, surface, or phenomenological laws—to theories of a higher order. The latter, being the most reductive, also require the highest abstraction; examples are the Maxwell electromagnetic wave theory or the Einstein theory. The term 'high' is somewhat misleading, since it also may remind one of the opposite, that is, the higher-level-of-complexity, functional laws and explanations of macro-behavior; these are at a relatively descriptive level and thus of the lower type in the terminology of Feigl.

In the continuous transition between description and reduction such theories as those of Hull (sec. 22) may reach the halfway mark between the two extremes. Use of a quasi-physiological or quasi-neurological terminology should not detract from the fact that the terms concerned are almost exclusively used in the sense of what Feigl has called "promissory notes" against observational reduction which in all likelihood will never be cashed. They are thus but an embellishment or a subjective expression of belief in the possibility of ultimate mechanistic reduction. Mechanical models of behaving organisms such as Tolman's "schematic sowbug" or Boring's "robot" are largely achievement-centered and do not go very far into micro-mediational detail.[35]

Ultimate observational reduction of the intra-organismic portions of behavior must always be physiological. Among the possible interests in this direction, the preponderantly central physiological identification of focal variables or dispositions such as intellectual ability or libido appears to be more legitimate from the standpoint of functional psychology than that of peripheral or micro-mediational detail.

"Dynamic" psychologists, with their concentration on the "why" aspects of behavior, such as Woodworth (see sec. 21), have urged that psychology—just like, say, geology—should not be considered a "fundamental" science in the sense of reductionism. Acceptance of the wisdom of this self-restriction should help the development of the molar approach as a research style in its own right. Such patterns of descriptive synopsis as the lens model would then come close to the level of psychological ultimates (see sec. 11).

III. Misconceptions of Exactitude in Psychology

For a while, psychology has tried to copy not only the basic methodological principles but also the specific thematic content of physics, thus nipping in the bud the establishment of the somewhat more specific methodological directives outlined in Chapter II. As Murchison has put it, psychology has tried to do *what* physics does rather than trying to do *as* physics does. The ensuing fallacies, or "biases," surrounding the concept of objectivity, operationism, or scientific exactitude in psychology may be grouped in three major syndromes. These range from thematic physicalism proper to some more derived preconceptions, hesitations, and inhibitions.

11. Thematic Physicalism

Analogizing and emulative physicalism. The molecular and the nomothetic bias.—To the outside observer, two policies stand out most prominently in the classical physics of Galileo and Newton (or in the high-school physics of our day). The first is given by the attempt to break up matter, at least conceptually, into small particles or mass-points. If cultivated uncritically or at the expense of other aspects of scientific procedure, this may lead to an unduly atomistic, elementaristic, microscopic, or, as the psychologist prefers to call it, a molecular attitude. The second policy is the nomothetic, i.e., law-stating, resynthesis of these particules which culminates in such allegedly singular principles of strict and universal validity as the law of gravitation.

In an effort to do for "mind" what physics has done for "matter," psychology began to drift into elementaristic sensationism (see secs. 1 and 2). It further attempted to establish the law of association of ideas by space-time contiguity as the sole determiner of the course of thought. (Contiguity-association is best exemplified by Plato's reference to the image of a friend elicited by the perception of a lyre this friend had been observed to play in the past.) The rather naïve transfer of elementarism and nomotheticism from matter to mind, that is, from one of the two "ontologically" contrasting realms postulated in philosophical dualism to the other, may be labeled "analogizing physicalism."

A second subvariety of thematic physicalism, to be called "emulative," is the preoccupation with elements and strict laws in objective behavior research. As does analogizing physicalism, emulative physicalism springs from an overawed "me too" attitude of a psychology struggling in the wake of an outdated image of the older natural sciences. But while the first has fallen short of the requirements of methodological physicalism, the second has by and large managed to become part of unified science, although it has remained unable to sever the umbilical cord so far as thematic identity is concerned. Superficial behaviorization is accomplished by shifting from sensation to the (discriminatory) reflex as the basic element and from association-of-ideas to conditioning (and other associationistic principles of learning) as the one basic law.

Statistics as the resolution of the nomothetic-idiographic dichotomy.—A particular confusion of issues seems to underlie the widespread appeal of the nomothetic bias. As the present writer has pointed out elsewhere,[1] the confusion is one "between univocality of observation and communication, on the one hand, and the univocality of prediction." From the standpoint of methological rather than thematic physicalism, however, a statistical correlation coefficient or any other probability law is just as exact, that is to say, just as public and palpable in its meaning as a strict law, although here exactitude has as its object the uncertainties of life. All the vagueness lies in the behavior, in the organism-environment relationships described rather than in the researcher or in his scientific activity by which the behavior in question is approached.[36]

The distinction between natural and cultural sciences (*Geisteswissenschaften*) is traditionally linked with that between the strictly nomothetic (and explanatory) approach which in principle claims absolute predictability and the idiographic (case-descriptive) approach which, strictly speaking, does not know of predictability. Yet "clinical" psychology with its emphasis on understanding the single case is possible only by the recognition of a certain self-consistency of the individual throughout shorter or longer stretches of his personal history on a probability basis, and with the tacit or explicit help of a more generalized study of individual differences and their interrelationships. Even the spatially unique topographic aggregates of geography show marked consistency in time and thus deal with "local" probability laws[77] of a statistical character rather than with bits of information strictly isolated from each other. Generally in modern science, the unrealistically absolutized nomothetic-idiographic dichotomy is resolved on the common ground of the probability approach.

The confusion of objectivity with strict nomotheticism seems widespread even among the most sophisticated of exact scientists. Thus the "cyberneticist," Wiener (see sec. 23), in a review of Philipp Frank's recent book on the philosophy of science,[28] critically characterizes the notion of probability as one "which is never a perfectly operational notion." In criticizing the present writer's probabilistic functionalism, Hull[1] has maintained that it entails a doubt in the "existence" of laws of behavior. However, espousal of the probabilistic point of view in psychology implies that the term 'strict law' ceases to be part of the grammar of the specialized language dealing with the functionalistic aspects of gross behavior. It is no longer appropriate to investigate strict laws in psychology; their repudiation is a methodological or representational, not an ontological, statement. It is related to real skepticism about lawfulness of behavior in the same manner as is the methodological statement, "consciousness is not an adequate subject-matter of scientific inquiry" (as guardedly espoused in this presentation) to its naïvely metaphysical alternate, "consciousness does not exist" (as sometimes found, alongside with the first-mentioned version, in the writings of classical behaviorists such at Watson). The compatibility of physical law and correlation-statistical approach in psychological ecology has been concretely demonstrated by the writer[30] (1947, Fig. 10).

The spell of the nomothetic ideal has in certain respects extended from the natural sciences to the *Geisteswissenschaften* and their attempted application in psychology. In the so-called "understanding" psychology of personality the coherent architecture and self-consistency of the individual are idealized to the point of suggesting logical deductibility of traits (*Wesensschau*, see sec. 17). Since at the same time the uniqueness of the individual or "existential type" is also idealized, one may be tempted to characterize this approach as "idionomothetic." Here too a statistical damper will have to be applied.

The sociological and ethical roots of the nomothetic idol of the natural sciences (and of its derivative, mechanism; see below), and the increasingly descriptive emphasis in modern physics, has been pointed out by Zilsel (this *Encyclopedia*, Vol. II, No. 8, pp. 61 ff., 90 ff.).

Quantification and topological language.—Since the time Descartes had alleged the nonspatiality of "mind" as the "thinking substance" in contrast with the "extended" character of "matter," the intrinsic nonquantifiability of sensation and of other conscious "qualities" has remained an almost unchallenged axiom. On this account, Kant expressed doubt in the possibility of a scientific psychology altogether, whereas Herbart, although devising an elaborate system of hypothetical quantitative laws, went along to the extent of doubting the possibility of experimentation in psychology. Criticism of the first quantitative psychological law, the Weber-Fechner equation stating the proportionality of "sensation" to the logarithm of the stimulus, was also largely based on this "quantity objection."

Ever since Hering, however, phenomenologists have insisted that there is a "visual space" in which experience is organized; they and the gestaltpsychologists have described the inhomogeneity and "anisotropy" of this perceptual, or "behavioral," space as compared with the physicist's space. Such recent investigators as Stevens[3, 23] have pointed to the surprising consistency and to what may be called the "transitivity" of such introspective operations as the bisecting or doubling of subjective auditory intervals. Facts of this kind tend to mellow the absoluteness of the quality-quantity dichotomy (see also sec. 4).

The major argument against the quantification bias in objective psychology is based on mathematics itself. As has been pointed out by Russell (this *Encyclopedia*, Vol. I, No. 1, p. 41), measurement and the use of numbers go beyond the essential requirements of the objective approach as defined by univocality of observation and of communication. Relational terms such as 'between,' 'above,' 'ancestor,' designating order in a series, is all that is required. The usefulness of such "comparative" or "topological" concepts has also been pointed out by Hempel and Oppenheim.[81] Psychological statistics makes extensive use of them in such techniques as "rating" and "ranking." The nature of psychological scales has been discussed by Stevens and by Coombs.[37]

Reductive micro-explanation and mechanism. Organic field theory vs. machine theory. Emergence vs. vitalism.—The ultimate ideal of micro-explanation is "mechanism," traditionally defined as reduction of gross functional relationships to the known basic principles of physics and chemistry. In dogmatically insisting on mechanism without sufficient means of empirical proof, the classical behaviorist confuses the unity of science in terms of the requirement of univocality of observation (methodological physicalism, sec. 4) with the unity of laws as we proceed from inanimate to biological phenomena (see Brunswik[1]). As we had to reject another confusion, that between the univocality of observation and the univocality of prediction given by strict vs. probabilistic laws (see above), physicobiological unity of laws must likewise not be taken for granted in an a priori manner in the framework of the objective sciences. As was pointed out by Carnap (this *Encyclopedia*, Vol. I, No. 1, pp. 60–61), the compatibility of biological or psychological laws with, or their derivability from, the system of physical laws cannot be inferred from the reducibility-of-terms but must be left to an empirical investigation. With reference to what has been pointed out concerning the statistical character of psychology, one may add, with Nagel (this *Encyclopedia*, Vol. I, No. 6, p. 24), that "the use of probability statements requires no commitment, even by implication, to any wholesale 'deterministic' or 'indeterministic' world-view."

Assertion of ultimate dualism of explanatory principles is called "vitalism." As an empirical problem, the mechanism-vitalism alternative is, like all problems of reductive explanation, a molecular rather than a molar problem. Its treatment requires minute control of energy exchange and of the technology of such biological phenomena as self-regulation, restitution, or regeneration. If an approach to it is at all feasible, it is a matter for the biologist proper, not the psychologist. Present-day behaviorists such as Tolman, Skinner, and others have therefore tended to ignore the controversy, although there can be little doubt that as scientists at large they expect ultimate mechanistic reducibility of behavior.

Any empirical verification of mechanistic or of vitalistic statements would require naming the concrete operations involved in testing them. Whenever such procedures are not given, or when the hypothetical entities or forces are explicitly declared to be unobservable (as in the case of Driesch; see Zilsel, this *Encyclopedia*, Vol. II, No. 8, p. 77) or even to be nonspatial—somewhat in the manner of Descartes's spaceless *res cogitans*—vitalism ceases to have scientific meaning and becomes a "metaphysical" pseudo-problem, that is, one dealing with assumed entities incapable of operational verification or falsification.

To be distinguished from vitalism is the emphasis, on the part of gestalt-psychology and related movements, upon interaction of physical forces in a molar, dynamic context. "Field" theories (see secs. 10 and 18) are contrasted with the more "machine-like" explanatory schemes supplied by the traditional reflex-arc or association theories. Both types of theory are physicalistic in an explanatory sense, i.e., naturalistic, and thus in the wider sense mechanistic, although the field theories do not subscribe to the particular mechanical models suggested by the machine theories. The terms 'organismic' or 'wholistic,' in the past sometimes applied to vitalistic conceptions, are increasingly being used as a synonym for 'molar.'

Vitalism must also be distinguished from what is becoming known as "emergentism." In proceeding from inanimate to biological and sociological occurrences there appears to be, on the descriptive level, an increasing "novelty" of structures and laws. Although in this sense the complex patterns do not seem to derive from the simple, no special explanatory principles may be needed. As has been pointed out by Feigl, even such a basic phenomenon as the law of gravitation requires a certain complexity of scope to make it discoverable, to wit, the taking into consideration of two rather than just one body. One body, when related to a frame of reference, would merely suffice to state, say, the law of inertia. Similarly, the kinetic theory of heat is based upon statistical hypotheses superimposed upon, and not anticipated in, Newton's mechanics, without actually transgressing the framework of mechanism in the broad sense of the word (see Frank, this *Encyclopedia*, Vol. I, No. 7, p. 26). In this sense such structures as the lens pattern (secs. 6 and 8) may turn out to be properties emergent on the molar level of approach; very likely they may be fitted into the physical conception of the universe with no more difficulty than the somewhat less complex physical principles just mentioned.[38]

The above example of gravitation as an extremely low-complexity case of "emergence" is attributed by Feigl to a casual conversational remark by the physicist, Bridgman; the obvious intent was to caricature the efforts of some emergentists posing as vitalists. In a more serious vein, Feigl stresses that there are no frames of reference without mass of their own that would lend themselves to the determination of the inertial movement of the body in question. His example was merely to bring out the fact that even in phenomena which are generally considered non-emergent, one could by sufficiently narrowing the basis of evidence make them emergent.

As has been pointed out by Bertalanffy,[39] the recent extension of thermodynamics from closed to open systems, advanced especially by Prigogine since 1946, has led to "fundamentally new principles" under which "self-regulation" and "equifinality" can be subsumed. Especially equifinality—given when the "final state may be reached from different initial conditions and in different ways" in the sense of our "terminal focusing" (see sec. 6)—has often been considered the main proof of vitalism. As the author points out, equifinality is indeed impossible in closed systems; it is for this reason that equifinality is, in general, not found in inanimate systems. Organisms have in the past erroneously been likened to closed systems with their tendency toward equilibrium; actually they are—as had already been pointed out by the biophysicist, Hill, in 1931—open systems in a relatively time-independent, quasi-stationary "steady state" maintained by a continuous flow of component materials. This "dynamic" equilibrium is usually very far from static equilibrium, although its net result may appear to be an absence of reactions in the normal state. Bertalanffy further explains that "according to definition, the second law of thermodynamics—the law of increasing chaos or 'entropy'—applies only to closed systems, it does not define the steady state." In this sense classical thermodynamics is not a consummate doctrine but rather "fragmentary"—as fragmentary, we may add, as a perception or learning theory that ignores distal reference. Open systems may thus "spontaneously develop toward states of greater heterogeneity and complexity" in accordance with a principle that is inherent in physical processes and that is "new" or "emergent" only in the sense that it requires an expansion of scope in order to become discoverable. The "supposed violation of physical laws does not exist, or, more strictly speaking . . . it disappears by the extension of physical theory." Basic developments in physical theory may in this manner be instigated by biological considerations.

In a similar vein, Schrödinger[40] points out how the organism is continually extracting orderliness, or "negative entropy," from outside to counterbalance the increasing disorder toward which the organism would degrade as a closed system. This happens in a manner perhaps surprising but "not alien" to physics; it constitutes just as integral a part of physical science as the classical laws.

Should the structural characteristics of behavior stressed so far in this presentation, especially that of equifinality, prove too unspecific for its complete description, the search for reduction models other than the lens analogy (Chap. II) would have to start afresh. On principle, the outcome of such a search is always uncertain. But it is encouraging to know that efforts at eventual resolution have thus far never remained unrewarded.

Misunderstandings of the principle of parsimony.—The striving for a unified theory of behavior in Watson, Hull, and others is sometimes done in the name of the principle of "economy of thought." In the search for simplicity it is often overlooked, however, that the principle of parsimony applies most properly to "tautological" systems of description, i.e., to systems which describe the same body of objective observations in different ways, and perhaps also to ultimate, reductive explanation, rather than to the relatively macroscopic phenomena of learning or of behavior in general. Furthermore, our choice of the simpler of two logically equivalent theories—such as of the heliocentric in preference to the geocentric view of the universe, or of the relativity theory in preference to classical physics augmented by "additional hypotheses"—merely follows convenience, and none of the rivals can have greater claims as to reality adequacy or "truth" than the other.[28] An analogue to this in psychology is the rotation from one system of axes to another in factor analysis (see sec. 20).

Most controversies in psychology, including that between the association and the gestalt-completion theory (sec. 18), contain differences of opinion about fact or are based on different selection of fact altogether and thus cannot be resolved by merely altering the representation of fact. The relation of association to completion theory has been compared to that between the theory of gravitation and electromagnetic field theory; as has been pointed out by Spence,[1] however, there is little intrinsic similarity between mechanics and associationism, on the one hand, and electromagnetic field theory and gestaltpsychology, on the other. In physics, the duality of gravitation and electro-magnetism has happily been kept in suspense up to recent attempts at a "unified field theory"; the latter combines the two principles, rather than eliminating one of them at the expense of the other. There is no reason why the plurality of gross dynamic principles in psychology should not likewise be left to remain unresolved indefinitely unless there is empirical disproof of presently claimed observation. (Concerning Lloyd Morgan's version of the "canon of parsimony" see the next section.)

12. Fear of Preliminaries

The iconoclastic bias. Operational redefinitions.—Revolutionary movements like behaviorism tend to develop in an emotionally overdramatized fashion, raising fears of the past in a variety of ways. A comparatively minor way of severing traditional ties is the elimination of palpable verbal symbols associated with older schools of thought. In early behaviorism there was fear that we may be slipping back into "medievalism" (Watson) when employing terms with an objectionable past.

All too often, however, replacement of terms merely leads to highly "visible" yet timid neologisms rather than to a true reconsideration or revision of previous notions. Furthermore, heuristic advantages to be gained from the "apperceptive mass" attached to traditional terms are unnecessarily renounced. Finally, there are usually numerous small and slow changes of meaning in developing a concept ("successive redefinition," see Lenzen, this *Encyclopedia*, Vol. I, No. 5, pp. 44 ff.; Bloomfield, *ibid.*, No. 4, p. 47; Ness[11]), witness the gradual physicalization of such multifacet vernacular terms as 'warm,' 'work,' or 'clock.' If we were to insist on relabeling the maturing concept at each turn of its meaning, considerable cumbersomeness and inflexibility would have to be expected. This would be especially true in the behaviorization of introspective or speculative terms now getting under way in psychology.

So far as the doctrinaire rechristening of psychology or of its basic concepts is concerned, Bekhterev's "reflexology" or Watson's "behaviorism" are some of the earlier examples. Their effect was, at least in the beginning, the establishment of a "school" set off against the rest of psychology rather than the revamping of psychology as a whole. The term 'behavioristics,' originally suggested as a substitute for 'psychology' by Neurath, has gained some foothold lately, although likewise with a limited field of application (see Boring[50]).

According to Carnap, both in psychology and in the methodology of science terms like 'observation,' 'thinking,' 'doubt,' 'understanding,' 'emotion,' etc., are practically indispensable. And they are admissible (1) as explicanda, i.e., for rough use, in great numbers, and (2) as explicata after redefinition, in a limited number of selected cases.

Psychological objectivists have in recent years increasingly made a point of ostentatiously re-using originally introspectionistic or even vitalistic terms such as those just mentioned. Examples are the operational redefinitions of 'purpose' and 'demand' (Tolman[74]; see also sec. 20), of 'hypothesis' (Krech[41]), of 'intentionality' (Brunswik[77]), of 'faculty' (Thurstone[81]; see also sec. 20), or of 'anxiety' (Estes and Skinner[41]). All these cases are characterized by an effort to preserve for the objectified meaning as much of the fruitful conceptual ramifications and connotations of the old concept as possible. This productive halo may have been hidden behind the mask of metaphysics or introspectionism and may have been temporarily lost in the insensitive hands of the classical behaviorist with his predilection for dogmatically peripheralistic translations (see sec. 13).

The anti-introspectionistic, anti-speculative, and anti-philosophical biases. Reinterpretation vs. flat rejection of metaphysics.—A special case of relinquishing imperfect modes of approach regardless of their potential fruitfulness is given by the early behaviorists' over-all rejection of introspectionism, and of speculation and philosophy in general. Aside from the fact that the "private" character of conscious data is a relative rather than an absolute matter (see sec. 4) and the line between science and not-science may therefore be difficult to draw, subjectivistic philosophy and introspective psychology is proving itself time and again an inexhaustible nursery for the supply of novel problem patterns and for reorientation regarding the basic aims and outlook of psychology. The best illustration of this

703

fact is the reoccurrence, in objective psychological research, of patterns first conceived of, decades or even centuries earlier, in the more flexible medium of philosophy or of introspective psychology and developed there up to the limit of the potentialities inherent in that medium (see Chaps. IV and V).

In many such cases of parallelism of the subjective and the objective approach, stimulation may have been implicit and unnoticed in slowly maturing "climates" of opinion or of interest. There also may be little overt terminological resemblance between the different approaches. No scientific problem may be confounded with its predecessor(s) in metaphysics, of course; but from a genetic-historical point of view many scientific problems appear as the more mature, adult stages of philosophical controversies, thus reaping the harvest of earlier frustrated arguments. It is in this sense that psychology may be said to be the legitimate heir of a good part of philosophy. Accordingly, there are increasing signs of rebellion against the "armchair taboo" in psychology.[42]

An ontogenetic parallel to certain aspects of the history of ideas is furnished by an experiment on the development of criticism in children.[43] Initial total rejection of objectionable statements—much in the vein of the negativism of early behaviorism—was found to be followed by specified rejection and/or positive rectification of the statements concerned, and still later by "higher forms of criticism." These latter include "productive reinterpretation," that is, the assignment of a new, figurative meaning which could be accepted as true, and "social reference," consisting in an effort to understand the statement in the light of the special external or internal conditions under which its author was assumed to have made it.

The general parallelism of the child's conception of the world with that of ancient philosophers has been pointed out in striking detail by Piaget.[44] Like Aristotle, ten-year-old children tend not to recognize the principle of inertia in their efforts to explain the flight of a projectile. Smaller children fail to consider still more elementary principles of "conservation," such as that of number when spatial arrangement is changed. The nature of scientific thought as a maturation process with an inherent sequence of necessary yet imperfect stages thus becomes the more evident.

In certain respects, such as in the undue lingering of introspectionism well into the twentieth century under the influence of dualistic philosophy, one may concur with Philipp Frank's statement that "the content of yesterday's metaphysics is today's common sense and tomorrow's nonsense" (this *Encyclopedia*, Vol. I, No. 7, p. 10). Yet, the balance sheet of metaphysics as a stimulant to psychology is probably definitely on the positive side. Metaphysical antitheses, such as especially that of empiricism versus rationalism, stand behind a host of scientifically meaningful controversies, among them those of sensationism versus phenomenology (secs. 1 and 2), of Helmholtz' empiricism versus Hering's nativism in the theory of space perception,[23] of association (conditioning) versus the internal field dynamics of closure (sec. 18), and of behaviorist environmentalism versus hereditarianism (secs. 13 and 18). To paraphrase Feigl,[1] it is only the "negativism" inherent in any false perfectionism that will cling to the letter of metaphysical formulations, readily dismissing them as pseudo-problems. A true "positivism" realizes and constructively develops their inherent potentialities.

Descriptive vs. vitalistic teleology. Anthropomorphism vs. animism.—Reference to terminal foci (see secs. 6 and 8) is sometimes being called "teleology," especially by its critics. However, terminal reference was found to be an essential requirement of the functional approach so long as it remains purely descriptive or is used only for the functional type of explanation. An ojectionable form of terminal reference is given only in the case of micro-explanation in terms of dogmatically assumed extra-natural, vitalistic principles of a metaphysical kind (see sec. 11).[45] Deliberate and extensive use of the term 'teleology' in the description of calculating machines and related mechanisms has recently been made in "cybernetics" (sec. 23).

Historical affiliation of objectivistic and anti-teleological tendencies in psychology is revealed by the fact that Lloyd Morgan, in his criticism of the "anecdotal" type of animal psychology prevalent in the nineteenth century, conceived of the objective approach primarily as one that keeps description at the lowest possible level (canon of parsimony) and avoids "anthropomorphistic" interpretation of animal behavior. Metaphysical-explanatory anthropomorphism, or "animism," is indeed objectionable; the operational anthropomorphism of the molar behaviorists, on the other hand, has been vindicated at the beginning of this section.

13. Hostility to Theory and to Central Inference

The dubious reputation of introspection and philosophical speculation with the objectively minded psychologists of the turn of the century also led to a tendency to throw overboard all theorizing. This was undoubtedly aided by antitheoretical movements within the physical sciences of the immediately preceding period as exemplified by Mach's aversion against such inferential constructs as the atom.

The observational bias. Emphasis of "data."—Early American functionalism and Watsonian behaviorism urged restriction of psychology to observed "facts." In this they followed in the footsteps of Comte, whose early "positivism" in many ways anticipates operationism and behaviorism, and who uses 'positive' as a synonym for 'observational,' 'preinferential,' 'undebatable.' Actually, the blind gathering and intercorrelation of data has led to selective emphasis upon regions relatively accessible to observation, thus sacrificing relevance to convenience, and has opened the door to tacit presuppositions of a more theoretical kind as well.

Köhler[46] has described this attitude as the outgrowth of a general "philosophy of caution." One may add that any safety-first approach is bound to be "inhibitory" rather than "excitatory," constrained rather than relaxed, intellectually puritanical rather than venturesome in a "Dionysian" way; it is genocidal in that it polices away the seeds of imagination and thought which Bridgman,[47] his operationism notwithstanding, came out to defend as the necessary "private" component in the establishment of science.

Peripheralism. "Glorification of the skin."—In consequence, Watson tended to "belittle the brain and the central nervous system" (Heidbreder[50]) and to shift the "definition" of even such cogently central activities as emotion and thinking to the sensory or motor periphery. Practical accessibility thus became confounded with importance, in a manner aptly criticized as "glorification of the skin."[4]

By dogmatically translating 'thinking' as 'laryngeal habits' or 'implicit verbalization' (for discussion see Murphy[50]) Watson's behaviorism became the objectivist counterpart of medieval nominalism, and of English empiricism and of structuralism with their assumption of sensory (pictorial or verbal) images as the constituents of thought. This is in sharp contrast with the "conceptualist" leanings of the Würzburg school with its emphasis on an independent representation of the abstract and general in thinking (sec. 2); over and beyond this, vicarious functioning of alternative verbal "formulations" of identical thoughts has been experimentally demonstrated for memory as well as for thinking proper by Karl and by Charlotte Bühler.[48] All this has helped to pave the way for the centralism of the molar behaviorists (sec. 20).

Physiologism as an escape from functionalism.—Next to the establishment of the peripheral anchors of behavior, the observational step-by-step tracing of mediating physiological processes has also attracted the attention of classical behaviorists as well as of such general theorists as Pratt,[49] mostly with an eye on the nomothetic micro-explanation of behavior.

The more remote ("distal") stimuli that precede, as well as the more remote effects, products, and permanent institutions that result from, behavior proper are likewise plainly measurable and are often accessible with great ease. Characteristically, however, classical behaviorism has made little use of these opportunities, although most of its representatives have paid lip service to them. Among them is especially Bekhterev,[70] who wishes to study "the complete correlation of the organism with the external world," may the latter be close or remote in space or time. The prime reason for this neglect must be sought in molecularism rather than in observationism. The functional-behavioral units studied by the inclusion of distal variables bridge over arcs too long for comfort, and they seem to provoke the dreaded teleological description.

Associationism and environmentalism.—It is likely that the relatively direct origin of contiguity association and conditioning in external chains of events had much to do with the appeal of this principle to empiricists of the more restricted type. Related to this is the emphasis on environment rather than heredity, on "nurture" rather than "nature," on learning rather than maturation,[69] in classical behaviorism. Again, there is reluctance to acknowledge factors which are not directly controllable, or so removed or otherwise unknown that dealing with them could leave loopholes for rationalistic "innate ideas" or vitalistic "instincts." The existence of a well-defined "anti-instinctivist" movement within classical behaviorism supports this interpretation.

IV. Traditional Approach and Constructive Crisis in Psychology

14. Historical Schema and Interpretative Hypotheses

Objectivity, molarity, regional reference.—The subsequent historical analysis[50, 51, 52] will be in terms of the three basic aspects developed up to section 8: (1) objectivity; (2) molarity, involving both the functional arc and the width of vicarious mediation; and (3) regional reference, involving the distinction of central, peripheral-proximal and distal areas, both on the reception and on the effection side and, as will be seen, both along the situational and along the historical axis (Figs. 2–10).[53]

Five working hypotheses.—The following interpretative hypotheses are offered and will be explicitly or implicitly considered during the remainder of our presentation:

1. There seems to be a continuous change-over from subjectivism to objectivism in psychology (from left to right in the main table containing Figs. 2–7, the first two of which were already discussed in secs. 1 and 2).

2. Within each of these two avenues of approach there seems to be a development from an emphasis on confined core events through peripheralism to functionalism, that is, broadly, from a static and molecular to a dynamic and molar type of approach (from top to bottom in the table).

3. Progress along one of these two perpendicular directions seems frequently accompanied by a standstill or even a regression along the other, but in the end this standstill turns out to be only temporary.

4. The successive subjective phases appear as encapsulated, relatively casual miniature anticipations of the reference patterns found in their physicalistic counterparts (connected with the former by slanted lines in the table), the latter culminating in a central-distal functionalism.

5. There seems to be a time lag between these formally corresponding introspectionistic and objectivistic stages which decreases as history proceeds, perhaps from as much as a few cen-

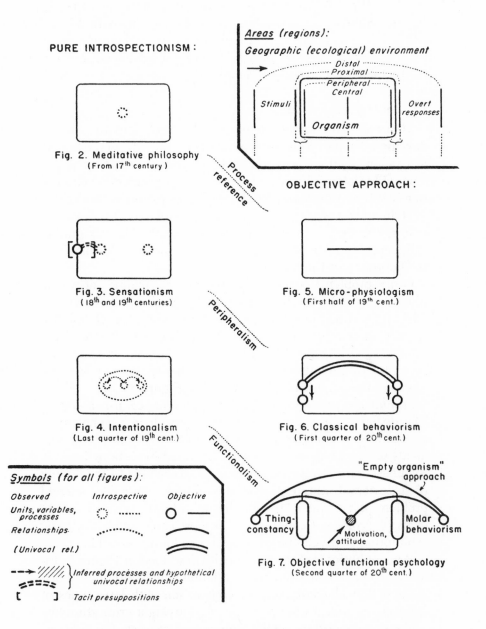

PURE INTROSPECTIONISM:

Fig. 2. Meditative philosophy
(From 17th century)

Areas (regions):

Geographic (ecological) environment

Distal
Proximal
Peripheral
Central

Stimuli

Organism

Overt
responses

OBJECTIVE APPROACH:

Process
reference

Fig. 3. Sensationism
(18th and 19th centuries)

Fig. 5. Micro-physiologism
(First half of 19th cent.)

Peripheralism

Fig. 4. Intentionalism
(Last quarter of 19th cent.)

Fig. 6. Classical behaviorism
(First quarter of 20th cent.)

Functionalism

Symbols (for all figures):

Observed	_Introspective_	_Objective_
Units, variables, processes		
Relationships		
(Univocal rel.)		

}_Inferred processes and hypothetical univocal relationships_

[   ]   _Tacit presuppositions_

"Empty organism"
approach

Thing-
constancy

Motivation,
attitude

Molar
behaviorism

Fig. 7. Objective functional psychology
(Second quarter of 20th cent.)

Figures 2 to 7. Major stages of introspective and objective psychology turies at the beginning to several decades in the more recent stages (compare chronological references given with Figs. 2 and 5, 3 and 6, 4 and 7), so that in many cases there is a spiral recurrence of analogous principles on more advanced levels of methodological perfection (see sec. 19).

15. Sensory Psychology

Micro-mediational physiologism.—A strong impetus for the development of experimental psychology comes from the biological and medical sciences, among them most directly from physiology. Physiological problems possessing psychological implications are best exemplified by the studies of Johannes Müller and of Helmholtz on the speed of nervous conduction. Intra-organismic, micro-mediational process interest of this kind is graphically represented in Figure 5 by a straight line placed well within the boundary of the organism.

Other problems conceived in the general manner of microphysiologism are the brain localization of peripheral—sensory and motor—functions, as well as most aspects of Helmholtz' sensory anatomy and physiology. The latter deal with the short-span relationships between proximal stimulation and peripheral physiological excitation and with the technology of neural transmission as schematically represented in Figure 8.

Stimulus-response psychophysics, physiological psychology.— The declared core of the classical approach in psychology is structurally to analyze, inventorize, and classify the basic elements of consciousness, as briefly described in sections 1 and 2. Auxiliary to the study of consciousness as such is (1) the finding of its external stimulus correlates—the psycho-environmental problem of Fechner's "psychophysics" (1860)—as well as (2) of its physiological counterpart, in most cases under the basic assumption of psycho-physiological parallelism (sec. 11).[54]

Psychophysics is diagrammatically described in Figure 8. Interest centers about a functional relation which, although short-arc and thus elementaristic by comparison with later types of research, establishes the principle of studying a gross stimulus-response relationship, while problems of sensory and nervous technology are removed to a secondary position. One of the terms of the relationship, here the response, is conceived of as a conscious content, sensation; verbal or other expressive behavior is interpreted as a tool of introspection. The programmatic outlook of classical psychophysics is thus physicalistic with respect to the stimulus, while it remains mentalistic concerning the response; in short, it proposes to be S-objective, R-subjective. Since stimulus-response correlation is under strictly nomothetic aspects without realization of the principle of

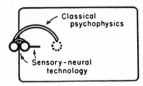 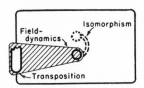

Fig. 8. Sensory psychology   Fig. 9. Gestalt psychology
(Second half of 19ᵗʰ cent.)   (First quarter of 20ᵗʰ cent.)

Figures 8 and 9. Part-objective stages of experimental psychology

vicarious functioning, a double arc is used in Figure 8 to represent the basic quantitative psychophysical "law" proposed by Weber and Fechner (see sec. 11; the double arc may at the same time be taken to represent the "constancy hypothesis" in a somewhat different manner than in Fig. 3).

Introspectionism, elementarism, sensationism, associationism. —A further nomothetic contribution is Ebbinghaus' study of memory for nonsense syllables (1885). By this study, contiguity association is brought within the orbit of psychological experimentation. This completes the picture of nineteenth-century psychology as an intersection of introspectionistic, elementaristic, sensationistic, and associationistic tendencies (Boring,[50] Bühler[55]).

16. Intentionalism, Early American Functionalism, Psychoanalysis

The second major period of psychology proper has been characterized as an abundance of divergent new growth, all in protest against one or the other major feature or combination of features of classical experimental psychology. Bühler[55] has spoken of this phase as the "constructive crisis" (*Aufbaukrise*) of psychology.

Intentionalism.—Brentano[56] has prepared the shift from sensationism to phenomenological introspection (sec. 2) by emphasizing the dynamic component of such psychological "acts" as perceiving or judging over their static sensory "content." The focus of introspection thus moves back to the central region where it had been with the meditative philosophers. However, there is a new stress on "directedness toward objects"; these objects are conceived as "inexistent" in mind (in the sense of momentary consciousness at large). Brentano has raised this "intentionality" to the major criterion of the subject matter of psychology; William James[57] has made it one of the chief characteristics of his "stream of thought."

Brentano's "objects" are not to be confused with the physicist's constructs, or "reality," which likewise are sometimes said to be immanent or inexistent in consciousness. In the related "*Gegenstandstheorie*" of Meinong anything our thinking may be concerned with as a problem, whether in reality possible or impossible, such as a round square, is called an object. Russell has criticized intentional objects as establishing an ontological "third Reich" that is neither mental nor physical. Uexküll and Bock[79] speak of the "monadological" introjection of the organism-environment relationship into the subject. Brentano himself hesitated to assign to the intentional relationship more than "quasi-relational" status (using the term '*Relativliches*' rather than '*Relation*'). This terminological trick constitutes recognition, on his part, of the fact that his intentional objects lack anchoring in what is now called an "independent approach," that is, the measurement and correlational scrutiny of the proximal or distal stimulus variables attained by a certain type of perceptual response (see sec. 20); indeed, intentional objects have their sole basis of observation, if not in analytic introspection, then in an encapsulated phenomenology or other revised form of introspection or retrospection.

In fact, the distinction between intentional objects and perceptual contents or acts can hardly be maintained save for some comparatively subtle differences in introspective tone. In Figure 4 the directly experiential character of the intentional objects is indicated by the use of dotted circles pointed at by dotted arrows (for intentionality as a relation), especially the one to the left; content, act, and object are inclosed within a dotted oval which is to represent the basic schema of dramatization in intentionalistically conceived central time-slices, conscious or dispositional, of which Lewin's "psychological life-space" (sec. 22) is another example.

Early American functionalism.—The encapsulated introspectionism characteristic of pure intentionalism was counterbalanced, toward the turn of the century, by the functionalism of Dewey, Mead, Angell, Carr, and others at the University of Chicago.[58] In criticizing the traditional molecular physiological emphasis on the reflex arc, Dewey[52] saw the key to the unity of psychological activity in its "function" or "successful issue," the "organization of means with reference to a comprehensive end." Reference to the region of distal results was thus established along with a utilitarian, adjustment-centered biological conception of psychology which may be traced to Darwin's views on the struggle for existence. (Concerning the "law of effect"[59] see sec. 18; on the recent expansion of functionalism in America see Chap. V.)

Psychoanalysis as depth-psychology.—Beginning in the 1890's and headed by Freud,[60] the psychoanalytic movement challenges the ordinary, controlled type of consciousness as the true representative of motivational dynamics. While in this sense psychoanalysis is definitely anti-introspectionistic, it shows partial adherence to conventional introspective methods by its relying on verbal reports of dreams or of the purposely disinhibited stream of ideas ("free association") as some of its "manifest" data. Peripheralism definitely is given up in favor of the more central emphasis on motivation. Associationism is challenged in that, in the eyes of psychoanalysis, dreams and the so-called free association is neither contiguity association nor free; rather, the course of ideas under disinhibition is considered to be largely governed by "repressed" instinctual tendencies, or "wishes," mostly in symbolic disguise. As is also true for gestalt-psychology (see sec. 18), the anti-associationism of psycho-analysis is not complete, however; contiguity association and such secondary principles of classical association theory as "recency" (experiences of the preceding day) are conceded to be modifiers in the specific choice of dream symbols. Although a few of the acknowledged symbols (e.g., water as a symbol for birth) are traced to space-time contiguity, most of the sex symbols are based on formal, geometrical similarity.

Reference to wishes and instincts as prime movers of the course of ideas and of behavior constitutes an element of intentionalism in psychoanalysis. Acquaintance of Freud with the intentionalism of Brentano has recently been definitely established.[61] Although attempts at reductive physiological identification of instincts or wishes are infrequent in psychoanalysis, the suspicion of vitalistic (animistic) leanings is unfounded. There is occasional explicit theorizing by Freud "in the spirit of 'physicalistic' physiology."[61]

In the earlier phases of psychoanalysis manifest introspection, as well as both proximal and distal behavior (Murray's[84] "actones" and "effects"; see sec. 21), were used primarily as a basis for inferences regarding the underlying "latent," largely repressed motivational depth stratum and its disturbances, as well as for inferences regarding the historic-genetic background of these disturbances, which was mostly assumed to be found in early childhood (Fig. 10). As was pointed out by Frenkel-Brunswik,[62] "concentration of scientific effort upon the central was not truly attempted before psychoanalysis. Correspondingly, psychoanalysis has, especially in its beginnings, comparatively neglected the distal results and achievements of behavior." In the same vein, Freud's daughter Anna[63] had bemoaned the fact that in early psychoanalysis the value of the scientific and therapeutic work done was seen to be in direct proportion to the "depth" of the psychic strata upon which attention was focused. She points out that whenever interest was transferred from the deeper to the more superficial psychic strata, from the id to the ego, it was felt that here was a beginning of apostasy from psychoanalysis as a whole; with problems such as that of the adjustment to the outside world, with health and disease, psychoanalysis was not properly concerned. The approach of early psychoanalysis is in this sense fragmentary. So far as the unit of behavior suggested in Figure 1 is concerned, depth-psychology is limited to the initial half; this is elaborated in the right-hand part of Figure 10.

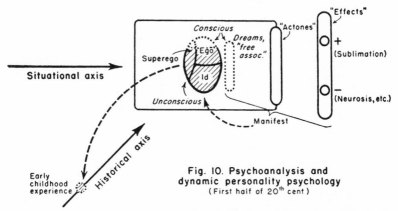

Fig. 10. Psychoanalysis and dynamic personality psychology
(First half of 20th cent.)

Vicarious functioning in psychoanalysis.—While dual focusing and the functional arc are thus neglected, a further basic feature of the lens model of behavior, vicarious functioning, has found such sweeping and unparalleled recognition in psychoanalysis —and in the crucial medium of the expression and satisfaction of drives at that—that this recognition may well be considered the most important contribution of this school to psychology.

As was first pointed out by Frenkel-Brunswik,[62] in an effort to integrate psychoanalysis with academic personality psychology under the auspices of logical empiricism, "no process of interpretation . . . takes surface phenomena at their face value. It rests, rather, on 'minimal cues.' It embraces a wide variety of circumstantial evidence—as furnished, for instance, by the techniques of psychoanalysis (free association and transference). A motive is considered central only after it has been recognized in a large number of behaviorally, or distally, diverse or even opposing features revealed in succession (such as an alternation between aggression and over-protectiveness)—if only these features have in common the same symbolic value empirically established by generalized evidence. . . . Psychoanalysis . . . has shown a great resolving power in conceptually bringing together an apparently chaotic variety of behavioral features relevant to life. It was in the service of this process of reduction that such important mechanisms as repression, denial, reaction-formation, sublimation, projection, and displacement were discovered." Most of the other concepts of psychoanalysis likewise refer to such mechanisms, stressing either the variety in the kinds of substitution (among them symbol, transference, condensation, rationalization, hysteric conversion, regression, cathexis, narcissism), or the width and flexibility of vicarious functioning (among them fixation, censorship—a more accurate translation of *Zensur* than the customary personalized form, 'censor'—choice of neurosis, frustration, resistance).

Further subsumable under the realization of the lens principle is the development, in Freud, from early efforts to treat hysterical symptoms directly by hypnosis, to a genuinely psychoanalytic approach aimed at the underlying dynamic core (which may also use hypnosis but in a different manner). Recognition of this core as a repressed wish implies that it is likely to find vicarious channels of expression and thus to create new symptoms unless itself removed. This development is paralleled by a general shift from "symptomatic" to "causal" treatment in medicine at large and by a general trend in psychology from peripheral-phenotypical to central-genotypical emphasis (see secs. 20–22).

As is to be expected from the wide variety of vicarious mechanisms, no single diagnostic cue is likely to have the character of foolproof evidence. Correspondingly, the depth-psychological establishment of motivational dynamics on the basis of concrete manifestations is far from being a process of univocal, atomistic "symbol-hunting"; it rather possesses all the earmarks of the probabilistic inferences so characteristic of recent functionalism (see secs. 7 and 9). It further follows that the range of circumstantial evidence for the theory as a whole must be extensive. Indeed, evidence in support of symbolism ranges from the slow direct uncovering of hidden connections in the process of psychoanalytic therapy, to folklore, myth, and jokes. Projective interpretation has occasionally been extended to metaphysical systems, e.g., in Freud's[64] discussion of Empedocles' love-hate philosophy as a "cosmic phantasy." (Note that 'projection' is applied to what such philosophers as Avenarius have called "introjection," and vice versa.) Analogies from biology at large have also been considered.[65]

On the other hand, the haziness of distal reference in psychoanalysis to which we have referred above may in part be the result of some remnants of a tacit "constancy hypothesis" (sec. 2)—as opposed to the full recognition of vicariousness—in the traditional conception of the drive-to-result correlation, "tacitly assuming that biologically relevant distal results are always due to the corresponding drive [and implying] that central causes directed toward such distal results will never lead to other effects. The sources of disturbance of the central-distal correlation are, however, so great that . . . there are too many central causes leading into side tracks (such as maladjustment) and blind alleys (such as compulsion symptoms), and [that there are] too many distal results not due to central causes explicitly directed toward them" (Frenkel-Brunswik[62]). In terms of the lens model, pathological reactions, or any "going off in vacuo" of behavior (to use a phrase by Tolman), are conceived as outgoing stray rays due to the imperfection of the "lens" (Fig. 1). The terminal foci are obscured in psychoanalysis unless it can be shown that they become organized into more or less autonomous substitute foci. Actually, a good part of psychoanalysis is devoted to the discovery of environmental conditions of quiescence which play the role of substitute goals or new terminal foci. Aside from neurosis and other channels harmful or neutral to the superordinate aspects of maintenance and propagation of life, there are positive, or at least socially acceptable, outlets, the outstanding example being the "sublimation" of the more primordial drives into cultural or scientific endeavors (Fig. 10). However, the concept of sublimation has remained relatively sterile and vague in the writings of psychoanalysis.[85]

Beginnings of functionalism in psychoanalysis.—In the early twenties, some years after being exposed to the rising functionalistic-behavioristic atmosphere in the United States, Freud shifted his emphasis from the introspectively oriented dimension conscious-unconscious to the more objectivistically conceived hierarchy of three dynamic functions within the person, the id, ego, and superego. As is shown in the central part of Figure 10, this part being borrowed from a textbook elaboration of Freud's own diagrams, consciousness comprises much of the ego and part of the superego. The three dynamic institutions represent, or take care of, respectively, the primary drives, reality relationships, and the culturally imposed conscience. Especially the concept of the ego thus prepares a turn to functionalism; as Murray[62] has put it, the ego is in the main defined by "listing what it *does* (repression, adaptation, etc.)." After pointing out the predominantly depth-psychological orientation of psychoanalytic theory (sec. 16), Anna Freud[63] goes on to explain that at least therapeutically psychoanalysis has always been interested in the "ego," the "reality principle," and its aberrations, and thus has proceeded from the roots to the out-

717

come of human activity. She lists some of her father's own contributions to ego psychology and develops them in a system of "defense mechanisms." Distal results of behavior are thus becoming a programmatically recognized region of reference in psychoanalysis.

Alternative hypotheses as to primary drives. Level of abstraction.—Freud's original system has often been set in opposition to the alternative systems of Adler and of Jung.[66] Yet Adler's major contribution, the attempt to shift earlier reduction schemes from the sex drive to the desire for mastery and prestige, is merely a change of content. The shift is comparable to Heraclitus' substitution of fire for Thales' water as the basic physical element. Little is changed in the level of abstraction or in the basic textural aspects of our diagrammatic scheme in Figure 10. The same holds for Jung's tracing of historical-genetic lines to cultural or "racial" rather than individual factors.

On the other hand, Adler's emphasis on social drives foreshadows the modern American trend toward situationism (sec. 21). And Jung's generalization of the concept of "libido" which continues Freud's own policy to broaden the concept of "sex" beyond its customary meaning helps to prepare the ground for an eventual treatment of motivation in terms of objectively established factors or related "nonpictorial" constructs removed from direct experience or common-sense behavioral observation (see sec. 20).

17. The Decline of Methodological and Nomological Dualism

Phenomenology, intuitionism, and "understanding" psychology.—The crisis of the metaphysical dualism of mind and matter brought about by behaviorism is paralleled by a crisis of the methodological dualism of introspection and physical measurement brought about, within introspectionism itself, by the frustrations of the phenomenological approach (see sec. 2). Related both to descriptive phenomenology and to Husserl's philosophically oriented "intuition of essences" (*Wesensschau*) is the "understanding" psychology of Dilthey and of Spranger.[67] In stressing its ties with the humanities (*Geisteswissenschaften*), it is said to be one of "two kinds of psychology"; the other, "explanatory" psychology, is seen as allied with the natural sciences by the use of experiment, statistics, and measurement and is dogmatically belittled as an offshoot of the "economic" rather than the "theoretical" value-orientation. Theories of personality are derived from metaphysical systems of values and related sources, leaving much to subjective empathy and arbitrary speculation.

Although allegedly patterned after mathematical deduction, intuition of essences is in fact camouflaged or submerged induction characterized by sloppy policies of inference and a tendency to premature systematization; it is for this reason that *Wesensschau* is in practice flagrantly ambiguous. Spranger's above-mentioned attempt intuitively to reduce the modes of thought of the exact sciences to the economic value system makes use of vague analogies between the equivalence of units in the process of measurement, and the universal exchangeability of goods in a system of values in which "everything has its price." Countermanding the derogatory intent of this derivation, one could point with at least equal cogency to the affinity of the concept of "natural law"—a style element of the natural sciences intrinsically interwoven with the concept of "measurement"—with the moral code and thus with what Spranger would have to subsume under his "religious" value system. Aside from the fact that for this latter affinity verification in historic-genetic terms seems feasible (for reference see sec. 11), one intuition would here be left to stand against another intuition, much in the way one introspection was found to stand against another introspection in the controversy concerning, say, the pictorial versus nonpictorial character of our thought-experiences (sec. 2).

Objective personality psychology shows increasing readiness to take up the problems raised—although not satisfactorily answerable—by the understanding approach. The success of such operational transformations has reassigned intuition of character structure, along with phenomenology, to its legitimate function as a fruitful and perhaps indispensable propaedeutic discipline to objective study in psychology.

Hormic psychology.—In contradistinction to intuitionism with its methodological dualism, vitalism (sec. 11) may be said to espouse a "nomological" dualism in that it postulates different reduction principles for the laws of psychology than may be furnished by physics. There is a vitalistic tradition in psychology from Aristotle to Driesch and McDougall.[68] The latter has seen the "instincts" as endowed with a unique driving power of their own, declared to be irreducible to physicochemical causality; hence the term 'hormic' (from the Greek equivalent for 'urge'). In the immediate present, vestiges of psychophysical interactionism are limited to occasional one-sided interpretations of such terms as 'psychosomatic disease,' although it is mostly recognized that no more need be implied than a promissory note for the tracing of behavioral disturbances to micro-physiological or hidden rather than gross organic or overtly palpable causes.

18. Antithetic Divergence: Gestaltpsychology and Classical Behaviorism

Gestalt theory of perception.—Evolving from the Graz school (sec. 3), gestaltpsychology reaches its full impact from the 1910's on in the so-called Berlin school of Wertheimer, Köhler, and Koffka.[69] Both schools recognize vicarious functioning of stimulus patterns in perception which possess intrinsic geometrical or melodic similarity. Since this principle of "transposition" ignores "families" of cues the members of which do not formally resemble each other but are held together merely by association (see secs. 6 and 20), recognition of vicariousness remains limited to one of its comparatively trivial aspects (fragmented oval in Fig. 9).

In the more modern versions of gestaltpsychology the perceptual response continuum is seen as broken up into a discrete series of relatively stable, preferred rallying points (*Prägnanzstufen*) characterized by clear-cut, salient organization or "good" form as defined primarily by geometric-ornamental simplicity and regularity, such as symmetry and closure. Although these crystallizations define positive contributions of the reacting organism to perception superseding and specifying the negative aspects of ambiguity and illusion stressed in the earlier phases of gestaltpsychology, stimulus-response correlation is further obscured (omission of the functional arc in Fig. 9).

In trying to explain the role of the subject in perceptual organization, the Berlin school has attempted to replace the older mentalistic "production" theory of gestalt as held by Meinong and others of the Graz school by a reductive theory of physiological "field" dynamics (indicated in Fig. 9 by the shaded area between stimulus and response). The general principles involved are developed in Köhler's theory of "physical *Gestalten.*" Combined with this is the extension of psycho-physiological parallelism to the perception of form ("isomorphism"). Mach's emphasis on the tendency toward geometrical regularity in closed physical systems approaching equilibrium ties in well with the particulars of the law of *Prägnanz.*

In warning us of projectively ascribing organizational properties of the perceptual response such as figural unity (sec. 3) to the stimulus aggregate ("experience error"), Köhler points out that there is no direct transmission, to the perceiver, of physical gestalt properties present in fellow-organisms or other objects. All unity must be newly constructed in the responder in accordance with his intrinsic, "autochthonous" brain dynamics. The subject is thus seen as basically out of contact with the dynamics of the environment, a fact also revealed by the rejection of the concept of learned "cues" hinted at above.[30] While psycho-physiological isomorphism receives all possible attention, the intricate problems of psycho-environmental (central-distal) stimulus-response coordination are, by both Köhler and Koffka, summarily dismissed by allusions to a vaguely conceived kind of pre-established harmony (or extended isomorphism) between the structural principles of the surroundings and the field dynamics within the organism.[69] The frame of reference of gestaltpsychology remains thus as encapsulated within the organism as was that of classical psychophysics (Fig. 8). The absence of genuine distal reference in gestaltpsychology has been explicitly pointed out by Heider.[73]

Closure (completion) as internal determination of the course of thought and action.—A particularly important development in gestaltpsychology is the extension of the law of *Prägnanz* from perception to thinking, thus evolving a principle set in opposition to contiguity association. In interpreting his experiments with chimpanzees, Köhler assumed that, by spontaneous dynamic reorganization of the perceptual field under the pressure of motivation, sticks or branches of a tree could be "seen" into the unfilled gap between animal and food. This completion or "closure" would in turn lead to the execution of appropriate action. The problem is thus solved by internal determination without benefit of association based on past experience. This view represents a physicalistic revival of the rationists' "mind," thought to be capable of producing a priori knowledge that had not passed through the senses.

Köhler further points out that the principle of completion tends to produce constructive "good errors" even if it goes astray in a practical sense. Some animals fill the gap between hand and goal visually though not mechanically by pushing out one stick by means of another, thus showing "insight" into the principle of the solution. By contrast, contiguity association is belittled and presented as a potential handicap that frequently leads to "poor errors," such as when the ape brings a box, previously found helpful in the gathering of food hung from the ceiling, in a futile effort to obtain a piece of food placed outside the cage.

Peripheral behaviorism.—Preceded by the "reflexology" of Pavlov and Bekhterev,[70] the early phase of American behaviorism is inaugurated by Watson[71] about coincident with Wertheimer's gestalt classic on apparent movement. Peripheral reference (previously discussed in sec. 13) is adopted from structuralism, the conception of a univocal functional arc from classical psychophysics; both are extended to the right and thus objectified, sensation merely shifting place to become "discrimination" by overt reflex, including external glandular secretion as well as what critics of classical behaviorism have designated as "muscle twitches." The functional arc runs straight from the sensory to the motor, by-passing the central region (Fig. 6).

A basic learning principle, Pavlov's "conditioning," deals primarily with the establishment of a new stimulus (e.g., the ringing of a bell) for a certain response (e.g., salivation) ordinarily elicited by another stimulus, after repeated presentation of the two stimuli in space-time contiguity (left arrow in Fig. 6). A symmetrical case (right arrow) is given when a previously unconnected response is linked to an existing stimulus,

such as in learning, by Thorndike's "trial-and-error and accidental success," say, to press a lever effecting escape in a situation of confinement.[3] Both cases imply a measure of vicarious
functioning and thus transcend a strict one-one relationship,
but the outlook on the problem of learning remains nonetheless
nomothetic (double arc in Fig. 6) in the tradition of thematic
physicalism (sec. 11). Both conditioning and trial-and-error
learning are but operational redefinitions of, or elaborations on,
the traditional association by contiguity in its broader sense,
with stimuli and responses taking the place of "ideas" (Thorndike's "connectionism").

Opposition to the law of effect in classical behaviorism.—Trial-
and-error and related forms of learning involving responses that
may be instrumental in changing the stimulus form the basis
of Thorndike's "law of effect." According to this law, contiguity
association of the response with satisfaction (or annoyance) is
a basic mechanism in the stamping-in (or stamping-out), the
"reinforcement" (or extinction), of the motor behavior of
which satisfaction is a result. This principle injects an element
of distal reference into learning theory in the spirit of the utilitarianism of American functionalism (sec. 16), to which movement Thorndike is related. The stigma of teleology and hedonis-
tic introspectionism attached to the law of effect prevented
Watson from accepting it, thus seriously curtailing the scope
and level of complexity of early behavioristic research. However, as was recently pointed out by Hilgard,[3] Thorndike was
actually ahead of rather than behind his behaviorist critics by
having given an irreproachable operational definition of a satisfying state of affairs, characterizing it as "one which the animal
does nothing to avoid, often doing things which maintain or
renew it."

V. Convergence toward an Objective Functional Approach

19. General Trends toward Realization of Norms

Temporary reciprocity of rigor and scope.—Of the two major
crisis schools arising against structuralism and defining the

second phase of modern psychology, classical behaviorism may be said to have overcome the introspectionism of nineteenth-century psychology but to have merely shifted the atomism, sensationism, and associationism to the overt behavioral level without altering its underlying peripheralism. Gestaltpsychology, on the other hand, has played the opposite role of overcoming atomism in its sensationistic and associationistic aspects without touching the basically introspectionistic conception of psychology. This divergent pattern of development may in some observers have reinforced the traditional intuitionist belief in the incompatibility of objectivity and molarity. The two have thus been seen as a kind of conjugate variables, so that any increase in objectivity would imply a "complementary" loss of scope. Yet, the inverse relationship observed soon turned out to be temporary and obviously due only to relatively insignificant psychological limitations inherent in formative stages of development. During a third phase of modern psychology, beginning in the 1930's, there is a "convergence" of the predominantly Anglo-American tradition of empiricist rigor (behaviorism, pragmatistic emphasis on overt action, statistical scrutiny) with the predominantly Continental stress on complexity and richness of scope. This convergence possesses all the earmarks of a genuine "synthesis" of the preceding divergent, mutually "antithetical" movements (to apply Hegel's notions of the dialectics of the creative process). It is not an eclectic intercombination of selected fragments of existing schools; rather, abstract methodological features are united for the first time in formations of striking novelty of style.

From metaphysics to methodological positivism.—Of most immediate interest from the logical empiricist point of view is the decline in the explicit entanglement of psychology with metaphysics, and a rise in what Allport[72] calls "methodological positivism."

His and Bruner's systematic content analysis of leading periodicals[72] has documented the "special lure" of operationism among psychologists. There also is an increasing emphasis on "mental process as a construct" that tends to "avoid the hypostatization of mental process." The employment of "postulational and geometric methods" likewise shows a marked rise after having become practically extinct during the period of intolerance toward theorizing at the beginning of this century. "Explicit treatment of the body-mind problem" shows a decline, said to reflect a "significant shift in viewpoint. In the earlier literature solutions to the problem were boldly offered in monistic or dualistic terms; today the fashion is to deny the existence of the body-mind problem, the denial being generally effected with the aid of Vienna logic."

Central-distal vs. peripheral reference.—More specifically, recent developments in psychology emerge as a combination of a de-emphasis of the peripheral region and the establishment of a central-distal or at least a central frame of reference which takes cognizance of the predominantly central-distal focusing of behavior itself (see sec. 6).[73]

20. Distal-Central Reference in Molar Behaviorism, Probabilistic Functionalism, Factor Analysis

In this section are presented empirical theories which are implicitly or explicitly probabilistic and which, with the exception of factor analysis, move on an intermediate level of formalization. All of them explicitly include objectively measured distal variables such as stimulus objects and/or behavioral results (e.g., the reaching of food, or the score on a test). Usually, but not always, there also is central reference, either of a direct, physiological or of an inferential nature.

Molar behaviorism.—The first large-scale and explicitly worked-out recognition of the compatibility of the objective and the molar approach is given by Tolman in his "purposive," "molar," or "operational" behaviorism.[74] Since its teleological emphasis on the reaching of "ends" in behavior remains at the empirical level, it is not to be confused with McDougall's vitalistic, "hormic" purposivism (secs. 11, 17, 21). Boring[23] has aptly characterized Tolman's system as the logical positivist-operationist-semanticist revolution within behaviorism.

The most prominent initial foci in Tolman's system are the "cognitive maps" (see sec. 8), generalized spatial orientation schemas represented by the shaded central unit-mark in Figure 7. They are "intervening variables" inferred primarily under the impact of the realization of the comparative nonspecificity (vicarious functioning) of the particular peripheral-proximal "means" as compared with the relative stability (focal character) of the ends reached. Together with the latter they establish a central-distal reference pattern which sharply contrasts with Watson's earlier peripheralist sensory-motor emphasis. The theoretical shift is the direct result of the unsuccessful attempts in classical behaviorism to find "the" peripheral focus of learning, and of the ensuing experiments, at California and elsewhere, demonstrating the intersubstitutability of kinesthetically different forms of locomotion (such as of running and swimming through a "maze") or of cues in different sense departments.[3, 74] In Figure 7 the new pattern emphasis is represented by the functional arc at right that spans over the oval indicating the motor boundary as a region of vicarious functioning and thus of comparative irrelevance so far as executional detail is concerned. Reference to ends achieved constitutes utilitarian emphasis and thus renders Tolman's system an integral part of objective functional psychology.

It must be stressed that Tolman's intervening variables are not conceived in an introspectionistic manner; that is, they are not taken as criteria indicating the presence of consciousness. They merely take their cue from introspectionism so far as the formative stages of inquiry are concerned; in contrast to classical behaviorism, however, an attempt is made to behavioristically recast the structural elements of our intellectual heritage as congenially as possible (for examples see sec. 12).

In further contrast with classical behaviorism, motivational factors such as "hunger" are actually varied rather than held constant or otherwise taken for granted. They are being operationally anchored in such historical variables as "time since last feeding"; in Figure 7 this is indicated by the oblique arrow (compare also with Fig. 10). Knowledge of the stimulus configuration, of the ends reached, and of the antecedents of motivation constitutes a legitimate three-pronged basis of inference for the intervening variables defining the internal, cognitive-emotional situations. Vicarious functioning enters, at least in principle, into the necessary scrutiny of the inferences, especially those based on criteria on the response side, by virtue of its incorporation in Tolman's operational redefinition of "purpose" as outlined above (sec. 6; see also sec. 10).

Behaviorism, supported, refined, and broadened by a properly interpreted operationism and methodological positivism, has thus become capable of absorbing the basic molar conceptions of an earlier, vague functionalism and purposivism. To a lesser extent, the resultant system of Tolman has also drawn on gestalt-psychology proper. In one way this has been the case by de-emphasizing, in the theory of learning, the earlier contiguity-associationistic principles of frequent repetition—law of exercise—and of reinforcement by reward—law of effect—in favor of a recourse to figure-like standing out within the perceptual field which (as was first shown by Muenzinger) can be brought about by punishment as well as by reward—law of emphasis. In another way, gestalt influences are apparent in the renunciation, hinted at above, of chain-reflex notions of learning in favor of the "insight"-like transferability of organized segments of spatial patterns. However, the all-important gestalt principle of closure does not loom large in the system of Tolman; molar behaviorism thus remains basically associationistic even though it is generalized rather than specific, selective ("noncontinuous") rather than indiscriminately stimulus-bound association that is being emphasized. On the other hand, molar behaviorism goes beyond gestaltpsychology in extending vicarious functioning from simple, transposition-like instances of vicariousness to a more general exchangeability of "means" and of "signs" which would be inconceivable without association; the concept of "sign-Gestalt" used by Tolman in this context has added the semantic depth of distal reference to the mere field orientation of the radical Gestaltists.

Organization in depth is also evident in the concept of "latent learning." It is concerned with the ability to acquire new re-sponse-systems in the absence of the specific motivating agent, say, hunger, and with the demonstration of such acquisition by subsequent introduction of an eliciting motivation. Conceptually, latent learning involves what Carnap[6] has called "dispositional predicates" as exemplified in physics by such concepts as that of 'magnetizeability' (in contradistinction to actual magnetization). Momentarily observable behavioral performance is de-emphasized in a manner similar to Lewin's de-emphasis of the "phenotype" or Freud's de-emphasis of the "manifest."

The multifacet trend in present-day psychology toward a revival of emphasis on motivation is also apparent in the introduction of the concept of "operant" behavior, as contrasted with the commonly emphasized "respondent" behavior, by Skinner.[80] It represents an effort to find recognition for the "spontaneity" of the individual in objective psychology.

Brain-and-achievement studies.—A physiological rather than inferential form of molar behaviorism evolves from the experiments by Lashley and by others on the influence of brain lesions upon higher learning.[75] The central unit-mark in Figure 7 would have to be replaced by an unshaded one to describe the observational character of the central emphasis. The central physiological data are being correlated with the reaching of high-order distal goals such as arrival at the end of a complicated maze.

Along with peripheral variables, some central variables also turn out to be vicarious in that different brain areas are found to be "equipotential." However, the amount of brain area damaged —regardless of its location within wide limits—proves a highly focal central variable by virtue of its being highly correlated with the degree of complexity of a maze an operated animal is capable of mastering.

From intentionalism to thing-constancy research.—The lens model extending left of center in Figure 7 represents the perceptual constancies discussed in sections 3 and 6. The problem-complex involved is known since Helmholtz; the studies in color constancy of David Katz, Bühler, and others[3, 76] represent one of the dominant interests of Continental psychology since the 1910's. Thing-constancy research has largely been independent of gestaltpsychology both in personnel and in outlook. It has established probabilistic central-distal focusing and proximal vicariousness in the cognitive domain[77] and thus has gone as much beyond nomothetic proximal-central and part-vicarious gestalt-psychology as the latter had gone beyond nomothetic proximal-central but nonvicarious classical psychophysics (cf. Figs. 8 and 9 and the left part of Fig. 7).

Realization of the vicissitudes of probabilistic reconstitution of the object world leads to an emphasis on the degree of achievement, or the utility value, rather than on the technological detail of the thing-constancies. This is in the spirit of American functionalism. Boring[23] has characterized the present writer's emphasis on the "functional use" of perception and the attendant comparative "ignoring of the brain" as an approach which is "less physiological and more biological," in a Darwinian sense, than that of gestaltpsychology. By virtue of the central-distal pattern of reference of constancy research, this approach appears as a symmetrical counterpart to molar behaviorism (Fig. 7), although there is no direct historical relationship between the two.

This symmetry has been elaborated in a joint paper by Tolman and the present writer[78] in which the probabilistic character of molar behavior has also been acknowledged and set in parallelism to that of the perceptual constancies.

Perhaps nowhere is the formal analogy of objective research patterns with, and at the same time their historical descendence from, subjectivistic predecessors clearer than in the field of thing-constancy. Figure 4 contains left of center the encapsulated introspective antecedent to constancy-research, given by the intentionalisms of Brentano, Meinong, and James (sec. 16), to which Figure 7 is like the unfolding of a seed. It is perhaps for this pictorial analogy that Hilgard[3] has recently labeled the present writer's outlook upon thing-constancy as a subvariety of "act psychology." However, while Brentano's "intentional objects" had to be classified above as derived from introspection and not checked by measurement, the "attained objects"[77] of thing-constancy research are independently constituted stimulus variables in the sense of physicists' constructs which have been found statistically to correlate with the responses. It is in this manner that the old introspective-metaphysical question as to whether we "see" the retina or the outside world can be given an operational meaning which is testable, in terms of "achievement," by functional validity coefficients.

"Psychology in terms of objects," "psychology of the empty organism," mental testing.—Determination of stimulus-response correspondences easily leads to a shift of emphasis from the re-

sponses to the stimuli correlated with them. Suggestions in this direction, although with an accent on proximal more than on distal stimuli, have been made by Bekhterev[70] and, somewhat less unequivocally, also by Uexküll.[79] The present writer[77] has spoken of a "psychology in terms of objects" (*Psychologie vom Gegenstand her*) in which organisms are described, and differentiated from one another, by reference to the—predominantly distal—stimulus or result variables with which they have "attained" stabilized relationships. By applying this approach to distal-to-distal functional arcs bridging over the entire organism without descending into it (Fig. 7), one may further gain in scope and at the same time get around the hazardous construction of intervening variables. Such an approach may be characterized as "without, yet about, the organism"[1] in that it is concerned with relationships established by the organism although not with their organismic anchorage. In basic conception, this approach is closely related to what Boring[35] has called "psychology of the empty organism," in referring to Skinner's[80] proposal that "psychologists had better give up the nervous system and confine their attention to the end-terms." In specifying this proposal by urging positive ascertainment of focusing, and of the width of vicarious functioning in the proximal versus the distal region(s), we can avoid focal-arc atomism (sec. 8) in spite of the ignoring of intra-organismic mediation. The full-fledged pattern of functionalistic research can be realized in this manner, thus removing what must seem the most cogent basis for criticism of the empty-organism approach (see also sec. 10 on economics vs. technology).

Whenever psychological research dispenses with mediational pillars within the organism, the subject must be defined indirectly or in terms of a more or less broadly conceived class. In animal experimentation, one may refer to species. In mental testing—perhaps the most prominent illustration of the empty-organism approach—one may specify individuals or populations in terms of such generic concepts as nationality, chronological age, normalcy, or socioeconomic status. The crucial point is that this kind of research is conceived of without reference to, or scrutiny of, organismic detail or the particulars of the actual execution of the test. On the other hand, the test problems themselves are precisely specified, and so are the solutions that have issued from the individuals—and that define the test-"scores." All this is in terms of objects attained, much in the way one might characterize the over-all behavior of cats by pointing out that mice (distal stimulus) disappear (distal result) in their presence.

Neo-facultative factor analysis.—In the mathematical analysis of test results thus obtained, "factors" may be extracted through statistical intercorrelation of the responses of large samples of individuals. Some psychological statisticians prefer to remain strictly within the distal region by considering the factors merely as self-contained abstractive "condensations" or mathematical artifacts without inferential implication. Others, notably Thurstone, tend to see the factors as intervening variables or, more specifically, as hypothetical constructs. In this sense, factors would represent "functional unities" within the organism and serve as operationally defined successors to the "faculties" which had become disreputable in the wake of speculative psychology.[81]

One of the objections to be made to faculty psychology from the point of view of the philosophy of science, to the effect that it constitutes an unnecessary (metaphysical) duplication of observed behavior, can be met by pointing to the many-many rather than one-one character of the relations involved; in the general case, each factor emerges as the result of a multitude of underlying test performances, and vice versa. The suggested regional separation of the dispositional or covert from the manifest or overt traits then appears justified in the light of the principle of vicarious functioning; two individuals with identical scores on a certain test may have produced them by quite different methods or patterns of factor-endowment, and it is only through combination with further tests that some degree of univocality is restored to the appraisal of underlying factors in the individuals.

One of the chief difficulties in the consideration of factors as dispositional "realities" lies in the fact that from a purely mathematical point of view a body of data can be factored in many different ways. Thus Spearman's original assumption of a "general factor," *g*, which is common to (nearly) all intellectual activities, is in a formal sense compatible with Thurstone's multiple-factor theory which postulates a minimum overlap of common factors responsible for the performances in the various tests. A resolution of this dilemma cannot be achieved without invoking the principle of parsimony (see sec. 11); this is actually done by Thurstone in his method of "simple structure" which curtails the possibility of arbitrary "rotation" of the axes representing the factors. Thurstone considers the fact that the "primary" factors obtained by this method have reappeared in successive studies with different test batteries as an empirical confirmation of their "psychological uniqueness." He expresses the hope that "mentality is, in this sense, not formless, but that it is structured somehow into constellations of processes which will eventually be identified."

Thurstone is conversant with the fact that, by proceeding from the computation to the "interpretation" of factors, he may transcend "objectivity" of construction. There are two major problems of interpretation. One is the physiological, genetic, endocrinological, sociological, or other independent identification (see sec. 10) of the factors in terms of what Tryon has called "radical components." The second problem of interpretation—more delicate from the standpoint of objectivistic language than the first—is that of the "naming," or "psychological meaning," of the factors. In this respect factors seem to be somewhat like the abstract, "nonpictorial" constructs in modern physical theory (see Philipp Frank, this *Encyclopedia*, Vol. I, No. 7). However, while in physics the difficulty usually lies with "visualization" of the construct in a more or less distinctly spatial sense, in the case of factor analysis it is more the introspective empathy into the factors or the possibility of their identification with common mental or behavioral operations which is deficient.

As in physics, this is no objection to the usefulness of such constructs. Thurstone prefers naming his factors by "well-known concepts"—such as facility with numbers, or self-restraint in making rational inferences—so as to render them "psychologically challenging," rather than just labeling them by a series of letters. And whether or not, say, the number factor may have to be characterized more broadly as "facility with a certain type of chain association" (as was suggested by Landahl) will, according to

Thurstone, eventually resolve into an empirical check concerning its nature by means of further tests.

21. Dynamic Personality Theory

In this section we will deal with convergence products of centralism, functionalism, and psychoanalysis, mostly characterized by a relatively low level of formalization.

Dynamic psychologies, cognitive and motivational.—Various psychologies with centralist leanings have, at one time or another, resorted to designating themselves as "dynamic." In the case of Köhler[69] the term refers to central-physiological "field"-interaction as it plays a part in cognitive organization (sec. 18). Developments toward a "dynamic psychology of cognition" have been surveyed by Heidbreder,[82] who compares her own studies on the "attainment of concepts" with Goldstein's emphasis on abstraction and the approaches by Woodworth (see below) and by the present writer (Chap. II and sec. 20). More recently the term 'dynamic' has acquired a more motivational rather than cognitive connotation. This is the case with Lewin's "vectors" and "valences" (sec. 22), which deal with relatively transient motivational states; the theory of "personality dynamics," to be discussed in a moment, deals with more enduring emotional and motivational patterns conceived under the direct or indirect influence of psychoanalysis (sec. 16) with its elaboration of the core of the "person" which is left relatively undifferentiated in the system of Lewin.

An earlier variant of a motivationally oriented "dynamic psychology" is that of Woodworth,[83] one in Heidbreder's list of seven major psychological systems.[50] It stresses the need to expand the peripheral type of stimulus-response or S-R psychology into an S-O-R and eventually into a W-S-Ow-R-W psychology, whereby 'O' stands for the organism including the drives, 'W' for the world, and the 'w' attached to O for the organism's "situation-and-goal set." Woodworth's complete scheme thus anticipates the fivefold sequence from distal stimuli to distal results as presented at the upper right of our main table of figures and is thus close to modern functionalism. Like Tolman and most other modern functionalists, Woodworth believes that the science of psychology has little to do with the question of mechanistic reducibility of motivational factors. By the same token, however, macro-mediation is not given sufficient attention. In consequence, there is no proper recognition of vicarious functioning and of the comparative irrelevance of the peripheral stimulus and response regions, so that in reality the system shrinks to what Woodworth discusses as a W-O-W pattern; it remains too sketchy for the actual execution of a full-fledged W-(S)-O(R)-W scheme, as the molar behaviorism and probabilistic functionalism referred to in section 20 and in Figure 7 might be described in Woodworth's notation.

Need-effect psychology. Experimental and statistical approaches to personality dynamics.—The instinct theory and some aspects of the procedure of psychoanalysis including the principle of "projective" interpretation of manifest data have exerted considerable influence upon the theoretical and clinical psychology of personality, either directly or through Murray and his Harvard collaborators.[84] Murray has properly exposed the comparative irrelevancy of the mediating peripheral-proximal "actones" (see Fig. 10); yet his approach is not free of unscrutinized inferences regarding the hypothetical central "needs" on the basis of distal behavioral "effects." In spite of occasional warnings to the contrary on the part of Murray himself, his system is dominated by a kind of tacit constancy hypothesis regarding the coordination of the need and the effect region albeit with a direction of inference opposite to that of psychoanalysis (see sec. 16). As was argued by Frenkel-Brunswik,[62] the "casting of a duplicate copy . . . of a certain overt and essentially distal pattern . . . into the central region, by just adding . . . the phrase 'need for' . . . is in fact not central, but distal psychology." She has therefore attempted to establish an "independent" approach to the central and has statistically studied the "and/or" families of phenotypically sometimes quite unrelated "alternative manifestations" of the same need in overt behavior; operational justification can thus be lent to the concept of motivation in accordance with patterns of vicariousness first developed in psychoanalysis.[85]

Numerous efforts are being made in dynamic psychology[86] to test a variety of further psychoanalytic hypotheses by objective methods.[87] Evidence, when collected with a proper understanding for the underlying psychoanalytic assumptions, is on the whole encouraging. For such topics as repression the technique of unfinished tasks developed in the school of Lewin has proved

734

useful; the importance of early infancy in the establishment of the adult personality has been verified in experiments on animals. There also was direct experimental confirmation of the crucial claims concerning symbolism.

Situational-cultural approach.—The original historic-genetic emphasis of psychoanalysis on "infantile phantasies carried on into adult life (for) imaginary gratifications"[63] has recently given way not only to a more functionalistic ego-psychology (sec. 16) but also to a more "situational" emphasis in the tracing of causes that affect the central picture (see Fig. 10). This development, although to a certain extent represented by Murray's emphasis on external "press" on the individual, finds its major expression outside of psychology proper, in Fromm and in Horney, as well as in a group of cultural anthropologists, notably Kardiner.[88]

Recognition of the importance, in establishing basic personality patterns, of the impact of maternal care, sibling interactions, paternal dominance, and of conduct more indirectly related to social institutions perpetuates psychoanalytic emphasis on infancy and early childhood. But specific tracing of how the earliest components of action systems were learned and integrated injects an element of Pavlovian thinking. The emphasis on "reality systems" which take care of dealings with the outer world and with other people forms a counterpart to the "projective systems" in religion, folklore, and the arts and establishes contact with the ego-psychology of the orthodox modern psychoanalysts.

22. Nomothetic Encapsulation in Topological Psychology, Postulational Behavioristics, Mathematical Biophysics

Discussion in this section concentrates on what is commonly known as psychological "theories" in the narrower, formal sense of the term. Nomothetic rigor is, so far as the actual body of the theory is concerned, in each case attained at the price of withdrawal within boundaries too narrow for adequate scope. Confinement is within the central region in the case of Lewin, and within a peripheral-central system in the case of Hull.

Encapsulated centralism in topological-dynamic psychology.— Gestaltpsychological, molar behavioristic, and formal-nomothetic features are combined in the theories of Lewin.[89] In his "topological" psychology he attempts graphic representation of the cognitive and motivational central time-slice which he calls the "life space" or "psychological environment." This is in preparation for a "dynamic" analysis in terms of hypothetical directional forces or "vectors" which leads to the development of laws allowing prediction of the future states of the system. Here we are to study "systematic" (or nomothetic) causation in terms of generalized essential conditions rather than "historical" (or idiographic) causation which traces individual occurrences through their actual, more incidental past.

The theory contains reference neither to the objective stimulus conditions nor to the motor execution and the further results of a response. The system and its laws thus remain confined to a postperceptual, yet prebehavioral, "contemporaneous field" in the central area (see Lewin[1] and Brunswik[1]). This field possesses a solipsistic or monad-like if intentionalistic character (see sec. 16). For this reason, Lewin's topological diagrams correspond to the dotted oval representing the theater of reference of intentionalism in Figure 4, especially to its right-hand side describing directedness toward a "goal" not yet attained. The organism itself appears in the life-space by means of its relatively undifferentiated central reflection, the "psychological person." The goals have different "valences," and the path to them is more or less "segmented" by "barriers" into gestalt-like unitary "regions" (the term used differently than in this monograph, where it refers to objective organism-environment stratification) which are anticipated to lie between person and goal. Lewin's shying-away from the peripheral and "phenotypical" in favor of the underlying "genotypical" central events has induced Hull[1] to concur with the present writer in his criticism of Lewin's system as an "encapsulated" one.

Programmatically Lewin's drawings are meant to refer to an objective constellation of central forces within the organism, rather than to a primarily conscious reality as is the case with intentionalism proper (act-psychology). However, Lewin's actual source of verification of the life-space is one of the following three token ones: (1) introspective phenomenology, as critically claimed by Spence;[1] (2) causal one-to-one response-inference without scrutiny in terms of vicarious functioning, in that occurrence of the first or the most prominent member of the respective habit-family-hierarchy. with its potential or incidental reaching of the distal end is taken as a criterion (this procedure, also hinted at by Spence, was more explicitly criticized in sections 6 and 11); (3) surreptitious one-to-one inference from the stimulus. The last-named fallacy is exposed by one of the most frequently heard objections against Lewin, accusing him of *de facto* confusion of geographical with behavioral (life-space) environment in spite of lip service to their conceptual separation.

In fairness to Lewin as an experimentalist, it must be added that he and his collaborators have extensively studied "substitution" of tasks and other aspects of vicarious functioning.[89] Perhaps the clearest expression of Lewin's *theoretical* detachment from the distal aspects of behavior is given by his explicit repudiation of the study of achievement. He has declared the question as to whether or not an animal actually reaches its goal to be psychologically irrelevant (see Brunswik[1]). In stating that "one will have to give up, in principle, to organize processes defined merely in terms of 'achievement' (*Leistung*) under unitary, psychological laws,"[89] Lewin in effect subscribes to our considerations of sections 7 and 9; the difference is that we do not consider the impossibility of a strictly nomothetic treatment of achievement sufficient reason for the excommunication of this aspect from psychology.

While Lewin in effect states that after having *conceived* of a certain situation the individual *will want* to do such and such, the molar behaviorist or probabilistic functionalist in effect states that when *placed* in a certain situation the individual *is likely to arrive* at such and such end (which will be part of the criterion of his "wanting" to do so). The former type of intrasystemic prediction, considerably narrower in scope than the latter, may possibly be made with nomothetic certainty; the latter, encompassing as it does the intricate problem of mastery of the causal texture of the environment on the part of the organism, is bound to remain probabilistic.

It is one of the assets of the system of Lewin that quantification of the life-space is rejected, and Euclidean, metric space replaced by the non-quantitative yet mathematical device of topological space based on relations of order rather than on measurement proper (see sec. 11). This is well in keeping with the phenomenological interpretation of Lewin's life-space suggested by Spence. In fact, regardless of the somewhat disputed mathematical soundness of topological psychology, topological representation may be the long-sought answer to the introspectionist dream of finding a mathematical, and thus univocal, "objective" means of communication for data of consciousness in spite of their intrinsic—absolute or relative—non-quantifiability. In granting this, one should not forget that objective language does not in itself guarantee objectivity of the observations which are to be communicated and that no result can be more objective than the most subjective of its single steps of observation, inference, or communication (see sec. 3).

Convergence toward an Objective Functional Approach

Peripheral-central encapsulation in formalized learning theory.—While Watson may be said to have succumbed to all the biases of objectivity listed in Chapter III, and Tolman to have avoided them all, the founder of a third major movement within the framework of behaviorism, Hull, has successfully overcome hostility to theory and to central constructs but in turn displays all the characteristics of emulative physicalism.

In the first of two books,[90] published in 1940 in collaboration with members of his group at Yale, an effort is made to develop a "hypothetico-deductive" theory of so-called mechanical or "rote" learning in attempted analogy to the axiomatic method in geometry. Under the stimulus of Woodger (see this *Encyclopedia,* Vol. II, No. 5) symbolic logic is used in addition to more conventional mathematical derivation with the intent of assuring utmost "clarity" in the sense of objective rigor (sec. 4).

An effort is made to embrace both the dynamic factors assumed by Ebbinghaus for rote memory and those of Pavlov for conditioned reflexes within a unified system based on "postulates"; these are to be expanded and recast in accordance with new evidence collected with the express purpose of experimentally checking or rechecking the theory. The principle of habit-formation through reinforcement by contiguous need-reduction (effect) is stated in the fourth postulate. Definition of such "unobservables" as "stimulus trace," "excitatory potential," and "inhibitory potential" is seen as a parallel to that of such physical constructs as gravity, energy, or atom. The list of eighteen stated postulates is set in analogy to that of the thirty-nine estimated by Malisoff to underlie mathematical physics.

In a second book, more encompassing and more directly concerned with "principles of behavior,"[90] Hull further proceeds to reconcile the learning principles of Pavlov with those of Thorndike, especially the law of effect (sec. 18). This as well as extensive use of the utilitarian concept of the 'survival' of self-maintaining organisms characterizes the system as functionalistic at least in intent. But very few of the experiments cited are concerned with higher, genuinely adjustive types of learning problems. In any event, the system stays within the framework of associationistic tradition; there is token recognition of stimulus patterning and other gestalt principles, primarily through such relatively casually superimposed principles as that of "afferent neural interaction" and "behavior oscillation." As Skinner has pointed out in his critical review,[90] these additions lessen the rigor of the system by blurring "the concrete manifestation of empirical laws."

Hull shares with Tolman the inclusion of motivational factors, such as needs and drives, into the behaviorist system. He also has adopted Tolman's terms 'molar' and 'intervening variable.'[1] In general, the conceptions and inferential procedures of the two authors resemble each other in many respects more closely than appears on the surface; this has also been recognized by Bergmann and Spence[1] and by Spence[1] along with the fact that Hull's method is not axiomatic in the strict sense of the term but merely a gradual adjustment of tentative hypotheses. However, while Tolman's definition of purposive behavior lays stress on vicariously mediated distal results in the constitution and inferential ascertainment of intervening variables (see sec. 6), Hull's major concern is with the interrelationships of these intervening variables among one another and with peripheral-proximal boundary conditions as they may be formulated in a system of laws. Hull's theory is entirely contained within the limits of the organism, as can best be seen from the concluding diagram of his *Principles*, depicting his entire system. The adjustiveness of learning, as reflected in such concepts as "consummatory response," finds representation only in its intra-organismic reflections, that is, with the central end of the feedback loop rather than with the loop in its distal, probabilistic portions. He is thus paying with molecular encapsulation for progress along nomothetic lines (concerning some basic confusions in Hull's nomotheticism see sec. 11). The entire technique of stressing, in the establishment of central variables, the customary type of univocal equations is an indication of the limitations of the system. The importance of adaptive probabilities is only vaguely acknowledged in the last few pages preceding the summary of his *Principles*.

The neglect of the principle of vicarious functioning becomes especially obvious in Hull's diagrams used in discussing intervening variables.[1] He seems to take for granted the legitimacy of inferring these variables from proximal stimuli or responses in a simple one-track or parallel-track manner in the way of a constancy hypothesis (see sec. 2). This methodological naïveté, and the neglect of well-focused distal reference inherent in this naïveté, not only holds for Hull's own system but is also present in his attempted representation of Tolman's and of the present writer's procedures along with those of Carnap.

In consequence of all this, Hull has, in his formalized writings, lost sight of what in the opinion of the present writer is his most significant contribution to psychological theory, namely, his concept of "habit-family-hierarchy" (sec. 6). As was pointed out by Ritchie,[90] Hull's present analysis is concerned with segments of behavior too small to require the introduction of such truly molar concepts. As temporarily during the first wave of behaviorist enthusiasm, there is a loss in essential level of complexity in the wake of a gain in rigor (see secs. 18 and 19).

Many of Hull's theoretical constructs are labeled in terms remindful of physiology. Examples are 'neural interaction,' 'efferent impulse,' 'reaction potential.' As has been pointed out by Spence,[90] however, the surplus meaning accruing by implicit or explicit allusion to a hypothetical neural identification may add little to the theory.

Somewhat akin to one of Hull's own efforts is an attempt by Woodrow[91] to find a single "generalized quantitative law" covering a wide variety of topics. In this case the suggested principle extends, among others, to learning, psychophysics (presenting an alternative to the Weber-Fechner law), and the development of intelligence. It combines two hypothetical factors, one for the physiological ceiling inherent in all performance, the other representing what might crudely be labeled "sensitivity." Woodrow thus follows the example of Hull and of Rashevsky (see below) in proceeding from an abstractive synopsis of the curve-fitting type to assumptions concerning underlying dynamics. The formula itself is an exponential function with a large number of parameters. Since, contrary to the "universal" physical constants, such as the speed of light, Woodrow's parameters vary from one context to another, the formula can be fitted to an extremely wide range of curve shapes, thus reducing the value of synopsis which would seem to lie in comparative specificity.

An alternative procedure, more in keeping with the tenets of this monograph, is that of Estes,[92] who attempts, for a variety of problems in the domain of learning, to "systematize well established relationships at a peripheral, statistical level of analysis. . . . Likelihood of responding [is] taken as the primary dependent variable; . . . independent variables are statistical distributions of environmental events. Laws of the theory state probability relations between momentary changes in behavioral and environmental variables."

An example of approaches in which physiological orientation becomes dominant is given by some recent attempts to synthetize psychology about the concept of "homeostasis" as developed by a general manner by Cannon and as increasingly applied to explanatory problems in psychology, psychopathology, and the psychoanalytic theory of instincts.[93]

Rashevsky's quasi-neurological mathematical biophysics.—Extensive use of quasi-neurological models including hypothetical hierarchies of brain centers is made in the type of mathematical biophysics developed by Rashevsky with his collaborators at the University of Chicago and by others.[94] It is broader in psychological scope than that of Hull yet less well known among psychologists. Starting from a simplified schema of cellular occurrences, it has succeeded in mathematically deducing experimental results in the field of higher mental functions with considerable specificity and accuracy. This holds for problems of thinking, gestaltpsychology (especially ambiguity and the transposition of the "universals" of form), perceptual thing-constancies, aesthetics, equipotentiality of brain areas, along with elementary sensory discrimination and conditioning. More recently the theory has also been extended to social psychology, social ecology, and sociology in general.

The variety of possible and not necessarily incompatible theoretical approaches to one and the same psychological problem is well illustrated by the efforts to explain the aesthetic appreciation of geometric forms. Birkhoff[95] had postulated, in a somewhat intuitive way, that certain geometric characteristics possess a certain arbitrary degree of pleasantness or unpleasantness. A different approach is made by the mathematical biologist. He begins with a set of assumptions about the mechanism of the central nervous system and tries to specify the processes that correspond to pleasantness or unpleasantness. Rashevsky assumes that four hypothetical brain centers take care, respectively, of the stimulus factors of (1) verticality, (2) horizontality, (3) angularity, and (4) length, and he arrives at predictions well in agreement with experimental results. However, an independent factor analysis of the stimulus bases for judging Birkhoff's figures[96] leads to the isolation of four quite differently conceived factors.

741

These are (1) smoothness (as contrasted with "blockishness" which is disliked)—possibly also to be interpreted as "complexity of design within a simple outline"—(2) simplicity in a geometric-conventional sense, (3) symmetry (especially rotational and diagonal and only secondarily horizontal and vertical symmetry), and (4) odd points or upward-reaching tallness. The discrepancy between the two approaches raises once more the problem of the "reality" of factorial constructs (see sec. 20).

McCulloch and Pitts[94] have elaborated on structural similarities between logic and nervous activity, thus bridging over to the mathematical theory of communication. In this latter theory (to be discussed in sec. 23), each "freedom of choice" defines a unit or "bit" of information, the term condensed from the binary digits, 0 and 1, which are used to replace the ten-digit series of customary arithmetic. Since firing in nerve elements follows an all-or-none principle and may thus be idealized as dichotomous, McCulloch[97] has pointed out that "neural events and relations among them can be treated by means of (two-valued) propositional logic" in a strictly nomothetic manner and that all psychic events have an inherently "intentional," "semiotic," or "purposive" character.

In their earlier paper the authors further point to the presence of feedback circles (see sec. 23) and stress that the inclusion of disjunctive relations leads to uncertainties in the disentanglement of causal relations which prevents complete determination of preceding states of the world or of ourselves. "Even if the net be known, though we may predict future from present activities, we can deduce neither afferent from central, nor central from efferent . . . activities." The asserted "ignorance implicit in all our brains" would seem to have significant implications not only as to the necessarily probabilistic character of perceptual and behavioral extrapolations on the part of the organism but also as to the risks involved in the use of "response-inferred" constructs in psychological theory (see sec. 10).

Micro-probabilistic approach in the analysis of neural nets and of social interaction.—The analysis of McCulloch and Pitts just referred to has been taken up by Rashevsky as a microscopic underpinning to the fundamental equations of mathematical biophysics much in the manner in which the microscopic kinetic theory of gases serves as a reductive support to the relatively macroscopic thermodynamics with its more direct access to empirical verification (see sec. 9). Together with Landahl, the first-named authors have undertaken to develop a formal method for converting logical relations among the actions of neurones in a net into statistical relations among the frequencies of their impulses.[94] The correspondence is likened to that of Boole between the algebra of logic and that of probability.

In this manner the earlier biophysical study of the average effect of large numbers of hypothetical neurones was paralleled by a study of "random nets" whose structure is "not a clearly defined topological manifold such as could be used to describe a circuit with explicitly given connections, since concern is not with 'this' neurone synapsing on 'that' one, but rather with 'gross' distribution of tendencies and probabilities associated with points or regions in the net." Starting from net-tendencies, probabilistic dynamics based on fundamental assumptions about the behavior of individual neurones (such as threshold, synaptic delay) are applied to derive the probable characteristics of behavior (or order out of chaos in the sense of Wiener; see sec. 23). This statistical "rather than deterministic" biophysics was developed by Anatol Rapoport and by Shimbel.[94] It has in the main retained its basically microstatistical character even in its subsequent extension to the dynamics of social interaction in the sense that the underlying individual occurrences do not attain observational identity (see sec. 9). The approach must thus not be confused with the macro-probabilistic concern with individual probability predictions in cybernetic communication theory (sec. 23).

23. Brain Models and Statistical Extrapolations in Cybernetics and Communication Theory

Negative feedback circles in servomechanisms.—Of great importance to psychological theory are the considerations developed in the course of the construction of computing machines and related mechanical and electrical "brains" during the second World War. The approach has been labeled "cybernetics" by

Wiener,[97] after the centrifugal "governor" which establishes automatic self-control of the intake of steam into an engine. One of the basic principles involved in such "servomechanisms" is known in engineering as "feedback." Feedback occurs when part of the effects of a given process are channeled into a "circular" loop so as to revert upon the source of the process in question. In the case of the steam engine and of all healthy biological self-compensation the feedback is inverse or "negative," that is, it neutralizes excessive conditions so as to establish a periodic-stationary stabilization or "steady state." Writing on what Charles Bell had called "nervous circle," the mid-nineteenth-century French neurophysiologists defined the "reflex" as an activity which, originated by a change in some part of the body, proceeded to the central nervous system, whence it was reflected to that part of the body where it had been initiated, and there diminished, stopped, or reversed the change that had given rise to it.

The psychological reader will easily recognize the similarity between the idea of feedback and the pivotal concept of American functionalism (sec. 16), that of adjustment through the setting of causes which in their aftereffects reflect upon the organism as correctives of its initial conditions. But while in psychological functionalism the circular loop is apt to include smaller or larger portions of the surroundings before it boomerangs—mostly in a beneficial though sometimes in a harmful manner—the feedback loops of computing machines are contained within the system itself. In classifying the "teleological mechanisms" described in cybernetics at large, McCulloch[97] distinguishes "appetitive" negative feedback devices, in which the "circuit passes through regions external to the system," from such simpler cases of feedback as the "merely homeostatic" which keep some internal parameters of the system constant, or the "servo," in which the value of the parameter or end sought by the system can be altered from without the circuit (whereby they may become instruments of superordinate systems). The self-contained case corresponds to the peripheral-central model with encapsulation within the organism discussed in section 22. It easily falls into the simplified pattern of a nomothetic single-path ideology; this is, to be sure, entirely adequate in the case of computational tasks involving merely the relations between "input" (peripheral stimulus) and "output" (peripheral response) without venturing beyond the boundary of the system proper.[98] Significantly, Wiener's book *Cybernetics* bears the subtitle "control and communication *in* the animal and the machine" (italics ours).

Both mathematical biophysics and cybernetics have claimed to be able to cover, with their machine-like conceptions, the internal dynamics of gestalt processes. On the other hand, Bertalanffy[39] has argued that "feedbacks, in man-made machines as well as in organisms, are based upon structural arrangements. Such mechanisms are present in the adult organism, and are responsible for homeostasis. However, the primary regulability, as manifested, for example, in embryonic regulations, and also in the nervous system after injuries, etc., is based upon direct dynamic interactions."

Macro-statistical extrapolation of extrasystemic sequences in appetitive mechanisms.—It is of utmost significance that in a number of specific instances cyberneticists have transgressed the limits of the machine boundary and have demonstrated the suitability of their conceptions in dealing with problems in which the spatial or temporal environment is made part of a larger system. In "appetitive" closed paths (see above) the machine or the organism is no more than a link, and predictive problems as to the events to be expected in the surroundings come to the forefront. It is primarily in such extended contexts that cybernetics becomes involved in a "recasting and unifying of the theories of control and communication . . . on a statistical basis" (Wiener[97]).

While in principle everything may be under strict control within the machine, the remote space-time surroundings are in the general case known to the system by extrapolation only, that is, predicted with some uncertainty. As psychological functionalism, when actually carried out, has thus been found to be forced into probabilism, a cybernetics with ecological involvement must contain probabilistic elements. A guided missile can at best compute the *probable* future positions of the airplane it is chasing, but at the same time it must do so in order to be efficient.

For guns to shoot at the place where the target would (probably) be when the shell arrived, analysis of past observational data becomes necessary. This means introducing the macro-statistical approach we have tried to point up as essential for the objective study of organism-environment relationships in behavior (sec. 9). And it also means incorporation of "memory" or of "learning" from past success or failure into the devices under discussion, which may go so far as continuously to modify behavior as results appear.

The extrapolations into the future are based on the "stochastic" character of the sequences involved, that is, the presence of statistical regularities which, when the length of the sample is reasonably large, will allow probability-prediction provided that the over-all character of the process remains fairly stable. Wiener has demonstrated that there is an "optimal" prediction based on such past experience of the system. He goes on to explain that the autocorrelations and intercorrelations of data in time series used in such analyses may detect the existence of causal connections, and their lags in one or both directions, provided the correlation is less than perfect. In this manner we may also detect feedback, including inverse feedback, in ecology, anthropology, and sociology. The data need be no more than decisions, actions, or opinions in time, provided we have runs of sufficient length. Man should learn to recognize and to decrease the gain in those reverberating circuits that build up to open aggression. By means of long-time runs of data we can determine whether the mechanism of negative feedback accounts for the "stability and purposive aspects" of the behavior of groups.

A first step in the direction of such bivariate correlation is made by Wiener in his analysis of multiple time series and, in particular, double time series:

"Such series occasionally make their appearance in engineering applications of the theory, but they are most conspicuous in the statistical applications, both economico-sociological and meteorologico-geophysical, since in both instances the relative lead of one time series with respect to another may well give much more information concerning the past of the second than of its own. For example, on account of the general eastward movement of the weather, Chicago weather may well be more important in the forecasting of Boston weather than Boston weather itself."

It may be noted that even here the two variables (or groups of variables) used in the example are of the same kind or physical dimension(s) or denomination(s) ("weather"); in such cases comparisons or correlations are not yet as genuinely bivariate (or multiple) as are those between psychological cues and objects or means and goal-attainments.

Ancillary to the emphasis on unidimensional sequential distribution predominant in cybernetics is a pioneer study on "statistical behavioristics" by Miller and Frick.[99] Their actual exposition encapsulates within the single region of overt responses. Shannon's sequential statistics as exemplified by the study of characteristic stochastic word-sequences in common English are applied to an analysis of the "courses of action"—related to the "strategies" in the playing of games[100]—and the "dependent probabilities" resulting from the fact that the preceding occurrence of a response does not always return the system to the original state. Following the suggestion by the early behaviorist, Hunter, that a response influenced by the penultimate of the preceding responses indicates a symbolic process, it is made quantitatively palpable by means of these data that seven-year-old children have a greater ability for symbolic processes than do rats, while the latter also possess that ability to a certain extent. An "index of stereotypy" of behavior used in this process may become useful in the application of Tolman's or Krech's objective criteria of purposiveness or of the presence of "hypotheses" (see sec. 12).

An example of the fact that not only bivariate correlation but even the basic schema of vicarious functioning has entered the scope of cybernetically oriented work outside psychology proper (albeit on a relatively nontechnical level) is given by a paper comparing various alternative sets of flight instruments in insects and in modern aircraft.[101] The issue concerned in this example is, to be sure, primarily that of the varying availability of cue systems rather than the more intrinsic and complex statistical one of the dependability of a certain type of cue if and when it is available.

As to this dependability of cues, the fact that air movement past the insect can be used to simulate its normal passage when flying, and will in turn lead to the characteristic stereotyped response pattern, may be taken as a drastic and somewhat atypical analogue to the possibilities of deception wherever input (or cue)—say, air pressure—is uncritically taken as a univocal representant of its presumed distal referent—say, air speed. The danger of being misled in prediction could be taken care of only in a probabilistic manner by statistical correlation so long as consideration remains confined to the limited set of variables available to the particular mechanism or subsystem concerned.

Thinking, perception, telecommunication.—To take for granted univocal interrelations among different types of events is to a certain extent justified so long as one has in mind only the psychological texture typical of "thinking" in the sense of explicit logical reasoning. It ceases to be adequate when we include the less ideally executed patterns of thought or the compromise type of probabilistic quasi-reasoning on the basis of insufficient evidence which is implicit in the more primitive form of cognition usually called "perception" (see sec. 7). Ideal rational thinking manages to isolate cues of highest dependability and thus is enabled to switch from vicarious to single-track functioning with not only no loss but even a gain in univocality. This lends a narrowly machine-like quality to discoursive thinking; it is precisely this quality which is represented in the "machines that think" of cybernetics and in some of the concepts of mathematical biophysics (see sec. 22).

In airplanes or guided missiles which continue to "think" while flying or seeking out their target the mediation processes may well in first approximation be considered as patterned in a similarly idealized manner. The "posits" of perception and of overt behavior, on the other hand, are in some cases more like ordinary passive missiles intrusted to a highly erratic medium at a stage far removed from the goal and imbued with considerable uncertainty as to reaching it. In communication engineering, the closest analogue to this is the distortion of telephone messages and related problems of distant communication. A more generalized exposition of those of the mathematical principles of communication which are of particular relevance to the understanding of vicarious functioning in psychological mechanisms has been given by Shannon and by Weaver.[102]

Using the vocabulary of the special brand of telecommunication engineering from which the theory has taken its start, perceptual cues and behavioral means are like "signals" in "coded messages." The perceived objects and behavioral results which correspond to the message are mediated through "noisy channels." These latter are contaminated with interferences or constraints of their own which reduce the sender's freedom of choice (also called lack of bias, degree of randomness, or "entropy") in the medium that must be molded to carry the message. The result is "equivocation" (or "entropy of the message relative to the signal"). It is then "not in general possible to reconstruct the message with certainty by any operation on the signal."

The inherent tangledness of the causal texture of the environment of a behaving organism may be seen as a specific type of "noise." We are reminded of Heider's[20] juxtaposition of the ideal "medium," such as the electromagnetic field, which is pliable to one kind of influence with a minimum of interference, with "things" which have firm properties of their own. These latter can be imposed upon the medium, thus becoming messages in the sense of communication theory. In all this, the undesirable uncertainty arising from structural or statistical properties of the medium is in inverse relationship to the desirable uncertainty which arises by virtue of freedom of choice of the message to be transmitted.

Shannon's diagram showing the fanning-out of "reasonable causes" (messages, inputs) for a given "high probability received signal" or effect, and of "reasonable effects" (signals, outputs) from a given "high probability message" or cause "in a channel," bears formal resemblance to the present writer's diagram showing the univocal and equivocal types of "coupling between intra- and extraorganismic regions"[77] which can also be read into our diagram of the lens model (Fig. 1).

If the richness of variability, or the "capacity," of a channel is less than the richness of variability, or the "entropy," of the source from which it accepts messages, then it is "impossible to devise codes which reduce the error frequency as low as one may please. . . . However clever one is with the coding process, it will always be true that after the signal is received there remains some undesirable (noise) uncertainty about what the message was." The channel then is "overloaded."

We may add that the crux of organismic adjustment obviously lies in the fact that distal perceptual and behavioral mediation must by the nature of things in the general case rely on overloaded channels, with the ensuing limited dependability.

Redundancy as an antidote to equivocation.—However, the authors[102] point to one means by which the chances of error can be decreased. This is "redundancy," as exemplified by repetition. The English language is about 50 per cent redundant, by virtue of grammatical and other structural properties which make certain sequences of letters or words more likely than others. It would be possible to save about one-half the time of telegraphy by a proper encoding process, provided one was going to transmit over a noiseless channel. When there is noise on a channel, however, there is some real advantage in not using a coding process that eliminates all the redundancy. For the remaining redundancy helps combat the uncertainty of transmission.

We may add that vicariousness of psychological cues and means may be viewed as a special case of receiving or sending messages through redundant, repetitive channels, thus reducing the probability of errors, that is, the set of possible causes, or effects, that could result in, or be produced by, the type of event in question. Vicarious functioning is thus indeed of the essence of behavior.

Relevant to our above discussion of an "objective language" in science and its close relationship to statistical reliability and validity (secs. 4 and 7) is the following quotation from Weaver:[102] "Language must be designed (or developed) with a view to the totality of things that man may wish to say; but not being able to accomplish everything, it too should do as well as possible as often as possible. That is to say, it too should deal with its task statistically."

In the manner described, communication theory may well contribute to the efforts, stressed in the present paper, to determine the structural and functional properties of the unit of behavior in abstract terms. Such determination will in turn contribute toward an explicit recognition not only of the rules and restrictions but also of the licenses and liberties of the objective as well as of the molar approach. It will further contribute to the much-needed establishment of psychology as a discipline of distinctive, well-circumscribed internal coherence and formal unity of purpose within the more broadly unitary framework of science at large.

Bibliographical Notes

Chapter I

1. Of the problems subsumed under "general methodology" and presented in the first three chapters, those of the status of intervening variables, of strict versus probability laws, and of the field and achievement concepts in psychology have been discussed in a Symposium on Psychology and Scientific Method, held at the Sixth International Congress for the Unity of Science, University of Chicago, 1941, with Brunswik, Hull, and Lewin as the participants (*Psychological Review*, Vol. L, No. 3 [1943]); for further discussion of intervening variables and the nomothetic point of view see K. W. Spence, "The Nature of Theory Construction in Contemporary Psychology," *Psychological Review*, Vol. LI (1944), and "The Postulates and Methods of 'Behaviorism,' " *ibid.*, Vol. LV (1948). "On a Distinction between Hypothetical Constructs and Intervening Variables" see K. MacCorquodale and P. E. Meehl (*Psychological Review*, Vol. LV [1948]). (All papers listed in this paragraph are republished in Marx.[52])

The interest in "operationism" follows the appearance of P. W. Bridgman's *Logic of Modern Physics* (New York, 1927). For an early survey of the repercussions of this and of related movements upon psychology see S. S. Stevens, "Psychology and the Science of Science," *Psychological Bulletin*, Vol. XXXVI (1939); also of importance is G. Bergmann and K. W. Spence, "Operationism and Theory in Psychology," *Psychological Review*, Vol. XLVIII (1941) (both papers also in Marx[52]), and a Symposium on Operationism in which psychology is represented by Boring, Pratt, Skinner, and Israel, and the philosophy of science in the broader sense by Bridgman and Feigl (*Psychological Review*, Vol. LII, No. 5 [1945]) (for a discussion of "levels of explanation" see Feigl's "Rejoinder"). Relations to the logical positivism of the "Vienna circle" are discussed by H. Feigl, "Logical Empiricism," in *Twentieth Century Philosophy*, ed. D. Runes (New York: Philosophical Library, 1943); this and the above-mentioned papers by Feigl are also found in *Readings in Philosophical Analysis*, ed. H. Feigl and W. Sellars (New York: Appleton-Century-Crofts, 1949). See further Bloomfield (this *Encyclopedia*, Vol. I, No. 4, p. 13). For the general aspects of the "quest for certainty" see J. Dewey's classic bearing the same title (London: Allen & Unwin, 1930).

2. See H. Reichenbach, "Rationalism and Empiricism: An Inquiry into the Roots of Philosophical Error," *Philosophical Review*, Vol. LVII (1948). For general orientation see B. Russell, *History of Western Philosophy* (New York: Simon & Schuster, 1945), pp. 564 ff. See also K. J. W. Craik, *The Nature of Explanation* (Cambridge: Cambridge University Press, 1943), p. 10.

3. The standard text in *Experimental Psychology* by R. S. Woodworth (New York: Henry Holt, 1938) contains reference to the majority of experimental findings referred to in this monograph. Thinking, including the completion theory (in the Würzburg version) and the problems of phenomenological introspection, is treated in chapter 30 (quotation from p. 788). The first half of chapter 6 gives a vivid description of the shift from peripheral to the central emphasis in the behavioristic study of maze learning. On learning see further E. R. Hilgard, *Theories of Learning* (New York: Appleton-Century-Crofts, 1948).

4. See A. F. Bentley, "The Human Skin: Philosophy's Last Line of Defense,"

Bibliographical Notes

Philosophy of Science, Vol. VIII (1941). In another paper, published by the same author jointly with John Dewey, under the title "Transactions as Known and Named," *Journal of Philosophy*, Vol. XLIII (1946), esp. pp. 535–36, 541 ff., 546, a "transdermally transactional" point of view is taken which has much in common with the central-distal functionalism described later in the present monograph. For an application to linguistics see the same authors' paper on "Specification," in the same volume. See also their *Knowing and the Known* (Boston: Beacon Press, 1949).

5. The view of "things" as a system of sensory "aspects" as presented in B. Russell, *Our Knowledge of the External World* (Chicago: Open Court, 1935), pp. 87–90, is still within the orbit of structuralist sensationism.

6. R. Carnap, "Testability and Meaning," *Philosophy of Science*, Vol. III (1936) and Vol. IV (1937), esp. pp. 420, 440, 449, and 466.

7. F. S. C. Northrop, *The Meeting of East and West* (New York: Macmillan Co., 1946).

8. E. Frenkel-Brunswik, "Intolerance of Ambiguity as an Emotional and Perceptual Personality Variable," *Journal of Personality*, Vol. XVIII (1949).

9. W. S. Hulin, *A Short History of Psychology* (New York: Henry Holt, 1934).

10. M. Schlick, *Allgemeine Erkenntnislehre* (Berlin: Springer, 1918; rev. ed., 1925); A. S. Eddington, *The Nature of the Physical World* (New York: Macmillan Co., 1928), chap. xii.

11. A. Ness, *Erkenntnis und wissenschaftliches Verhalten* (Oslo: Academy of Science, 1936).

12. C. Morris, *Signs, Language, and Behavior* (New York: Prentice-Hall, 1946).

13. W. Dubislav, *Naturphilosophie* (Berlin, 1933), p. 49.

14. M. F. Washburn, *The Animal Mind* (New York: Macmillan Co., 1908), chap. ii. See also H. Carr, "The Interpretation of the Animal Mind," *Psychological Review*, Vol. XXXIV (1927).

15. E. K. Strong, *Vocational Interests of Men and Women* (rev. ed., Stanford: Stanford University Press, 1946).

16. A. Ness, *"Truth" as Conceived by Those Who Are Not Professional Philosophers* (Oslo: Academy of Science, 1938).

Chapter II

17. A. P. Weiss, *A Theoretical Basis of Human Behavior* (Columbus, Ohio: Adams & Co., 1925), pp. 346–47.

18. W. S. Hunter, "The Psychological Study of Behavior," *Psychological Review*, Vol. XXXIX (1932).

19. E. B. Holt, *The Freudian Wish* (New York: Henry Holt, 1915); L. T. Hobhouse, *Mind in Evolution* (New York: Macmillan Co., 1926); M. F. Meyer, *Psychology of the Other-One* (Columbia: Missouri Book Co., 1921).

20. Aside from the considerations of the classical behaviorists on vicarious functioning, the ground for the development of the lens analogy was laid by two papers of F. Heider, "Ding und Medium," *Symposion*, Vol. I (1927), in which a regional stratification of the environment in general physical terms, emphasizing the pliable "messenger" character of such "media" as light-rays, is attempted, and "Die Leistung des Wahrnehmungssystems," *Zeitschrift für Psychologie*, CXIV (1930), 381. For a lens model similar to the present one see Brunswik.[77]

21. B. Russell, *Introduction to Mathematical Philosophy* (New York: Macmillan Co., 1919).

22. H. Reichenbach, *Experience and Prediction* (Chicago: University of Chicago Press, 1938). Concerning partial causes and effects see also "Die Kausalstruktur der Welt und der Unterschied zwischen Vergangenheit und Zukunft," *Bayrische Akad. der Wissenschaften* (1925). For Reichenbach's general views on probability see further his *Theory of Probability* (Berkeley: University of California Press, 1949).

For an introduction into philosophy of science see H. Reichenbach, *The Rise of Scientific Philosophy* (Berkeley and Los Angeles: University of California Press, 1951); see also Frank.[28]

23. E. G. Boring, *Sensation and Perception in the History of Experimental Psychology* (New York: Appleton-Century Co., 1942), esp. pp. 11–13, 83–90, and the concluding sections of chaps. 7 and 8.

24. H. Werner, *Comparative Psychology of Mental Development* (New York: Harper & Bros., 1940; rev. ed., Chicago: Follett, 1948).

25. C. L. Stevenson, *Ethics and Language* (New Haven: Yale University Press, 1944).

26. I. Krechevsky, "Brain-Mechanisms and Variability," *Journal of Comparative Psychology,* Vol. XXIII (1937). (Three papers.)

27. R. von Mises, *Probability, Statistics and Truth* (New York: Macmillan Co., 1939).

28. P. Frank, *Das Kausalgesetz und seine Grenzen* (Vienna: Springer, 1932; French trans., Paris: Flammarion, 1936). For general orientation see the same author's *Modern Science and Its Philosophy* (Cambridge: Harvard University Press, 1949), pp. 117 ff., 176. For an earlier discussion of simplicity and conventionality in the formation of hypotheses see H. Poincaré, *Science and Hypothesis* (French original, 1902; trans., Lancaster, Pa.: Science Press, 1905).

29. An early hint at the intrinsically statistical nature of functional psychology as the "exact science of the probable in the domain of self-preservative behavior" may be found in E. A. Singer, *Mind as Behavior* (Columbus, Ohio: Adams & Co., 1924), pp. 67 ff. For general orientation concerning the methodological status of the statistical approach in science see C. W. Churchman, *Theory of Experimental Inference* (New York: Macmillan Co., 1948), and F. S. C. Northrop, *The Logic of the Sciences and the Humanities* (New York: Macmillan Co., 1948). F. Kaufmann's *Methodology of the Social Sciences* (London: Oxford University Press, 1944) does not quite come to grips with the problem of the specific character of the probability approach in the social sciences.

30. E. Brunswik, *Systematic and Representative Design of Psychological Experiments* (Berkeley: University of California Press, 1947), esp. pp. 23, 32 ff., 41 ff., and Figs. 7 and 10. A second edition, now in preparation, will, among other additions, contain a report on experiments attempting to make palpable some of the differences between perception and explicit reasoning hitherto published only in abstracted form (*American Psychologist,* Vol. III [1948], and "Remarks on Functionalism in Perception," *Journal of Personality,* Vol. XVIII [1949]; see the latter paper also on the rejection of the "cue" concept by gestaltpsychology).

31. R. A. Fisher, *Design of Experiments* (Edinburgh: Oliver & Boyd, 1935).

32. K. R. Hammond, "Relativity and Representativeness," *Philosophy of Science,* Vol. XVIII (1951); see also the present writer's "Note" on Hammond's paper in the same number of the journal.

33. N. Bohr, "On the Notions of Causality and Complementarity," *Dialectica,*

Bibliographical Notes

Vol. II (1948). For a generalization of correlation statistics which makes possible the inclusion of quantum physics see F. Bopp, "Quantenmechanische Statistik und Korrelationsrechnung," *Zeitschrift für Naturforschung*, Vol. 2a (1947).

34. See T. Haavelmo, *The Probability Approach in Econometrics* (*Econometrica Supplement*, 1944), and R. Frisch, "Statistical vs. Theoretical Relations in Economic Macro-dynamics" (mimeographed for the Business Cycle Conference, Cambridge, England, 1938, as referred to by Haavelmo).

35. A tropistic model for certain facts of higher learning has been presented by E. C. Tolman in "Prediction of Vicarious Trial and Error by Means of the Schematic Sowbug," *Psychological Review*, Vol. XLVI (1939). For a human "robot" see E. G. Boring, "Mind and Mechanism," *American Journal of Psychology*, Vol. XLIX (1946).

Chapter III

36. Even in a discipline relatively as "fundamental" as chemistry, H. Polanyi ("The Value of the Inexact," *Philosophy of Science*, Vol. IV [1937]) considers it "ill-advised" to let one's self be "frightened by physicists into abandoning all vague methods and to restrict [one's self] to the field where exact laws pertain." For a general discussion of problems of vagueness see C. G. Hempel, "Vagueness and Logic," *Philosophy of Science*, Vol. VI (1939), and M. Black, *Language and Philosophy: Studies in Method* (Ithaca: Cornell University Press, 1949).

37. S. S. Stevens, "On the Theory of Scales of Measurement," *Science*, Vol. CIII (1946); C. H. Coombs, "Psychological Scaling without Unit of Measurement," *Psychological Review*, Vol. LVII (1950).

38. A detailed discussion of emergentism is given in G. Bergmann, "Holism, Historicism, and Emergence," *Philosophy of Science*, Vol. XI (1944).

39. L. von Bertalanffy, "The Theory of Open Systems in Physics and Biology," *Science*, Vol. CXI (1950). For a more detailed presentation of the discussion related to vitalism see the same author's *Theoretische Biologie*, Vol. II (Berlin: Borntraeger, 1942).

40. E. Schrödinger, *What Is Life?* (Cambridge: Cambridge University Press, 1944).

41. I. Krechevsky, "Hypotheses in Rats," *Psychological Review*, Vol. XXXIX (1932). 'Anxiety,' 'inhibition of the hunger motive,' and 'decrease in the frequency of pressing the food-lever' are operationally equated to one another by W. K. Estes and B. F. Skinner, "Some Quantitative Properties of Anxiety," *Journal of Experimental Psychology*, Vol. XXIX (1941). See further S. Koch, "The Logical Character of the Motivation Concept," *Psychological Review*, Vol. XLVIII (1941).

42. D. B. Klein, "Psychology's Progress and the Armchair Taboo," *Psychological Review*, Vol. XLIX (1942); republished in *Twentieth Century Psychology*, ed. P. L. Harriman (New York: Philosophical Library, 1946).

43. From a thesis by Anna B. Brind, reported in E. Brunswik, "Experimente über Kritik," *Bericht über den 12-ten Deutschen Kongress für Psychol.*, ed. G. Kafka (Jena: Fischer, 1932).

44. J. Piaget, *The Child's Conception of the World* (New York: Harcourt, Brace & Co., 1929).

45. A distinction between descriptive and explanatory teleology essentially similar to the one suggested here has been given by A. D. Weber and D. Rapaport, "Teleology and the Emotions," *Philosophy of Science*, Vol. VIII (1941).

46. W. Köhler, "A Perspective on American Psychology," *Psychological Review*, Vol. L (1943) (William James Jubilee Number).

47. P. W. Bridgman, "Science: Public or Private," *Philosophy of Science*, Vol. VII (1940).

48. For a brief presentation of the Bühler studies on memory and the dynamics of thought see E. Brunswik, *Experimentelle Psychologie in Demonstrationen* (Vienna: Springer, 1935), pp. 120–23, 138–39; there also is discussion of instances of misleading feelings of clarity, based primarily on experiments by Lindworsky and by Poppelreuter (pp. 134–35 and 137–38). On association versus completion experiments see also Rapaport,[87] especially pp. 53–54.

49. C. C. Pratt, *The Logic of Modern Psychology* (New York: Macmillan Co., 1939).

Chapter IV

50. An excellent survey of the development of the various standpoints in psychology is given by G. Murphy, *Historical Introduction to Modern Psychology* (rev. ed.; New York: Harcourt, Brace & Co., 1949). The major classical schools of psychology are reviewed in E. Heidbreder, *Seven Psychologies* (New York: Century Co., 1933). The standard work on the subject is E. G. Boring, *History of Experimental Psychology* (rev. ed.; New York: Appleton-Century-Crofts, 1950).

51. Firsthand summaries of a considerable number of schools by their originators or by close collaborators were assembled by C. Murchison in *Psychologies of 1925* and *Psychologies of 1930* (Worcester: Clark University Press, 1928, 1930).

52. Anthologies on contemporary movements and on the history of psychology are presented in *Psychological Theory: Contemporary Readings*, ed. M. H. Marx (New York: Macmillan Co., 1951), and in *Readings in the History of Psychology*, ed. W. Dennis (New York: Appleton-Century-Crofts, 1948).

53. A similar schematic representation of schools was used in E. Brunswik, "The Conceptual Focus of Some Psychological Systems," *Journal of Unified Science* (*Erkenntnis*), Vol. VIII (1939) (republished in Harriman[42] and in Marx[52]). A second paper anticipating certain aspects of the present monograph is E. Brunswik, "Points of View," in *Encyclopedia of Psychology*, ed. P. L. Harriman (New York: Philosophical Library, 1946).

54. W. Wundt, *Physiologische Psychologie* (1874, 1 vol.; 6th ed., 3 vols., Leipzig: W. Engelmann, 1908–11). A belated defense of structuralism is given in E. B. Titchener's posthumously published *Systematic Psychology: Prolegomena* (New York: Macmillan Co., 1929).

55. K. Bühler, *Die Krise der Psychologie* (2d ed.; Jena: Fischer, 1929).

56. F. Brentano, *Psychologie vom empirischen Standpunkt* (1874); re-edited by O. Kraus (2 vols.; Leipzig: Meiner, 1924–25).

For a summary of Brentano's and Meinong's views on intentional objects, as well as for a general discussion of the various concepts of "mind," see C. Morris, *Six Theories of Mind* (Chicago: University of Chicago Press, 1932).

57. W. James, *Principles of Psychology* (2 vols.; New York: Henry Holt, 1890).

58. J. R. Angell, "The Province of Functional Psychology," *Psychological Review*, Vol. XIV (1907) (republished in Dennis[52]).

59. L. Postman, "The History and Present Status of the Law of Effect," *Psychological Bulletin*, Vol. XLIV (1947); see also the ensuing controversy, with Meehl and with others (in subsequent volumes of the same *Bulletin*).

Bibliographical Notes

60. For a general introduction see Sigmund Freud, *Introductory Lectures on Psychoanalysis* (London: Allen & Unwin, 1922); this may be augmented by such indexed and glossaried selections from his works as the one edited by J. Rickman (London: Hogarth Press, 1937). The latter also contains a republication of *The Ego and the Id* (1923).

61. Concerning the beginnings of Freud as a physiologist and his general leanings toward the natural sciences see S. Bernfeld, "Freud's Earliest Theories and the School of Helmholtz," *Psychoanalytic Quarterly*, Vol. XIII (1944), and "Freud's Scientific Beginnings," *American Imago*, Vol. VI (1949), the latter paper also concerning Freud's acquaintance with Brentano.

62. In a Symposium on Psychoanalysis as Seen by Analyzed Psychologists, inaugurated by G. W. Allport (*Journal of Abnormal and Social Psychology*, Vol. XXXV [1940]), Boring, J. F. Brown, Frenkel-Brunswik (quotations are from pp. 179–83), Landis, and Murray (quotation from p. 165) represent major psychological movements or emphasize the theoretical aspects of the controversy, while psychoanalysis is represented by Alexander and by Sachs.

63. A. Freud, *The Ego and the Mechanisms of Defense* (London: Hogarth Press, 1937).

64. S. Freud, "Analysis Terminable and Interminable," *International Journal of Psychoanalysis*, Vol. XVIII (1937).

65. The rising methodological awareness of psychoanalysts concerning their procedures finds an expression in S. Bernfeld, "The Facts of Observation in Psychoanalysis," *Journal of Psychology*, Vol. XII (1941). The same author's "Zur Revision der Bioanalyse," *Imago*, Vol. XXIII (1937), deals with analogies to psychoanalysis from biology, first stressed by Ferenczi.

66. A. Adler, *The Neurotic Constitution* (New York: Moffat, Yard & Co., 1917); see also Murchison.[51] Cf. C. G. Jung, *Psychology of the Unconscious* (New York: Moffat, Yard & Co., 1916).

A survey of psychoanalysis from its beginning to the present, including the systems or contributions of Adler, Jung, Rank, Ferenczi, Reich, Horney, Fromm, and Sullivan is given in C. Thompson, *Psychoanalysis: Evolution and Development* (New York: Hermitage House, 1950).

67. E. Spranger, *Types of Men* (rev. German ed., 1921; trans. from 5th ed., 1928). See also H. Klüver's contribution to the earlier editions of Murphy.[50] For the general history of personality psychology see G. W. Allport, *Personality* (New York: Henry Holt, 1937), chap. iii.

68. W. McDougall, *Introduction to Social Psychology* (London: Methuen, 1908).

69. A comprehensive presentation of Berlin gestaltpsychology is given in K. Koffka, *Principles of Gestalt Psychology* (New York: Harcourt, Brace & Co., 1935); the passage on correspondence with the environment referred to in sec. 18 is on p. 305. Among the books of W. Köhler, *The Mentality of Apes* (London, 1927) is perhaps the best known; his *Dynamics in Psychology* (New York: Liveright, 1940) gives in chap. ii a short presentation in English of his theory of "physical Gestalten." The "maturation" as contrasted with the "learning" aspect of development is stressed in K. Koffka, *The Growth of the Mind* (New York: Harcourt, Brace & Co., 1924). See also W. D. Ellis, *A Source Book of Gestalt Psychology* (New York: Harcourt, Brace & Co., 1938). The relationship of gestaltpsychology to Kant is pointed out in W. Metzger, *Gesetze des Sehens* (Frankfurt: Kramer, 1936). For the views of the rival Leipzig group see Sander.[51]

An analysis of the concept of gestalt from the point of view of logical positivism is given by K. Grelling and P. Oppenheim, "Der Gestaltbegriff im Lichte der neuen Logik," *Erkenntnis*, Vol. VII (1938).

70. The most conspicuous use of the term 'objective' in psychology is found in the title of V. M. Bekhterev's pioneer work, *Objective Psychology* (Russian original, 1907; French and German trans., 1913). See also Murchison.[51]
A history of "Psychological Objectivism" is given by C. M. Diserens, *Psychological Review*, Vol. XXXII (1925). Concerning the behavioristic character of ancient Chinese psychology see S. Fernberger, "Fundamental Categories as Determiners of Psychological Systems: An Excursion into Ancient Chinese Psychologies," *Psychological Review*, Vol. XLII (1935).

71. The major treatises by J. B. Watson are *Psychology from the Standpoint of a Behaviorist* (Philadelphia: Lippincott, 1919), in which a scrupulous attempt is made to avoid terms that imply consciousness and the conception of the human individual as a stimulus-response machine is developed, and *Behaviorism* (New York: Norton, 1924; rev. ed., 1930), which contains not only the most radical exposition of Watson's environmentalism but also the most accurate statement of his theoretical position.

To be distinguished from Watson's system is J. R. Kantor's "interbehaviorism"; see his "Current Trends in Psychological Theory," *Psychological Bulletin*, Vol. XXXVIII (1941).

72. G. W. Allport, "The Psychologist's Frame of Reference," and J. S. Bruner and G. W. Allport, "Fifty Years of Change in American Psychology," both in *Psychological Bulletin*, Vol. XXXVII (1940).

Chapter V

73. The distinction between "central" and "peripheral," an old standby in general psychology, has been emphasized as a watchword in modern system-making especially by Murray.[84] "Distal" versus "proximal" stimulus reference is the leading motive in a discussion and classification of contemporary schools by F. Heider, "Environmental Determinants in Psychological Theories," *Psychological Review*, Vol. XLVI (1939).

74. E. C. Tolman, *Purposive Behavior in Animals and Men* (New York: Century Co., 1932; republished by University of California Press, 1949); concerning the definition of purposive behavior and the preceding efforts by R. B. Perry ("Docility and Purposiveness," *Psychological Review*, Vol. XXV [1918]), and by McDougall, see pp. 14 ff.

For Tolman's philosophy-of-science orientation see especially his "Psychology vs. Immediate Experience," *Philosophy of Science*, Vol. II (1935); "An Operational Analysis of 'Demands,' " *Journal of Unified Science* (*Erkenntnis*), Vol. VI (1936); and "Operational Behaviorism and Current Trends in Psychology," *Proc. 25th Anniversary Inaug. Grad. Stud., Univ. Southern California* (1936), pp. 89–103, reprinted, among others of Tolman's writings, in *Collected Papers* (Berkeley: University of California Press, 1951).

Concern with vicarious functioning is also implicit in one of Tolman's most recent comprehensive papers, "Cognitive Maps in Rats and Men," *Psychological Review*, Vol. LV (1948). See further "There Is More than One Kind of Learning," *Psychological Review*, Vol. LVI (1949).

75. K. S. Lashley, *Brain Mechanisms and Intelligence* (Chicago: University of Chicago Press, 1929).

76. D. Katz, *Die Erscheinungsweisen der Farben* (*Zeitschrift f. Psychol. Supplement*, 1911; Engl. trans. from 2d ed., *The World of Colour* [London: Kegan Paul, 1935]); K. Bühler, *Die Erscheinungsweisen der Farben* (Jena: Fischer, 1922).

Bibliographical Notes

77. E. Brunswik, *Wahrnehmung und Gegenstandswelt: Grundlegung einer Psychol. vom Gegenstand her* (Vienna: Deuticke, 1934). Some of the major points are summarized in the same author's "Psychology as a Science of Objective Relations," *Philosophy of Science*, Vol. IV (1937); *Errata*, Vol. V (1938). See also Brunswik.[30]

78. E. C. Tolman and E. Brunswik, "The Organism and the Causal Texture of the Environment," *Psychological Review*, Vol. XLII (1935).

79. J. v. Uexküll, *Umwelt und Innenwelt der Tiere* (Berlin: Springer, 1909). See further J. v. Uexküll and F. Bock, "Vorschläge zu einer subjektbezogenen Nomenklatur in der Biologie," *Zeitschrift für die ges. Naturwissenschaft* (1935).

80. B. F. Skinner, *The Behavior of Organisms* (New York: Appleton-Century, 1938). Short-cutting of the organism by the psychologist is first suggested in the same author's "The Concept of the Reflex in the Description of Behavior," *Journal of Genetic Psychology*, Vol. V (1931).

81. A brief, nontechnical introduction to factor analysis is L.L. Thurstone's address, "The Vectors of Mind," *Psychological Review*, Vol. XLI (1934). Further general discussion, including the defense (chiefly against Godfrey Thomson, R. C. Tryon, and A. Anastasi) of the factors as real "functional unities" of a dispositional kind, is found in his "Shifty and Mathematical Components," *Psychological Bulletin*, Vol. XXXV (1938), and "Current Issues in Factor Analysis," *ibid.*, Vol. XXXVII (1940). For a presentation of methods see his *Multiple Factor Analysis* (rev. ed.; Chicago: University of Chicago Press, 1947), quotations from pp. 70, 145, and, for major results, his *Primary Mental Abilities* (Chicago: University of Chicago Press, 1938).

From the point of view of logical positivism, a discussion of problems of individual differences has been given by C. G. Hempel and P. Oppenheim, *Der Typusbegriff im Lichte der neuen Logik* (Leiden, 1936).

82. E. Heidbreder, "Toward a Dynamic Psychology of Cognition," *Psychological Review*, Vol. LII (1945).

83. R. S. Woodworth, *Dynamic Psychology* (New York: Columbia University Press, 1918). The diagrammatic letter sequences are cited from Woodworth's *Psychology* (4th ed.; New York: Henry Holt, 1940).

84. See the first two chapters of H. A. Murray and collaborators at the Harvard Psychological Clinic, *Explorations in Personality* (New York: Oxford University Press, 1938).

85. E. Frenkel-Brunswik, *Motivation and Behavior* (Genetic Psychology Monographs [1942]); see also the same author's paper in *Perception: An Approach to Personality*, ed. R. R. Blake and G. V. Ramsey (New York: Ronald Press, 1951).

86. For a comparison of systems within personality psychology proper see S. Rosenzweig, "Converging Approaches to Personality: Murray, Allport, Lewin," *Psychological Review*, Vol. LI (1944).

87. For an early summary see R. R. Sears, *Survey of Objective Studies of Psychoanalytic Concepts* (Social Science Research Council Monographs, No. 51 [1943]). Concerning symbolism see D. Rapaport, *Organization and Pathology of Thought* (New York: Columbia University Press, 1951), Part III.

88. For the situational-sociological approach in psychoanalysis see K. Horney, *New Ways in Psychoanalysis* (New York: Farrar & Rinehart, 1941), and A. Kardiner, with R. Linton, C. DuBois, and J. West, *Psychological Frontiers of Society* (New York: Columbia University Press, 1945).

89. The two major aspects of K. Lewin's theoretical system receive mention in the titles of his *Dynamic Theory of Personality* (New York: McGraw-Hill Book

Co., 1935) and *Principles of Topological Psychology* (New York: McGraw-Hill Book Co., 1936). The first of these contains translations or republications of previous papers, among them of "The Conflict of Aristotelian and Galileian Modes of Thought in Contemporary Psychology," as well as reference to experimental studies directed by Lewin. A discussion of law and experiment in psychology and a criticism of the use of achievement concepts are presented in "Vorbemerkungen über die seelischen Kräfte," *Psychologische Forschung*, VII (1926), 306–7. See further *Conceptual Representation and Measurement of Psychological Forces* (Contributions to Psychological Theory, No. 4 [1938]). A theoretical defense of the system is given in "Formalization and Progress in Psychology," *University of Iowa Studies in Child Welfare*, Vol. XVI, No. 3 (1940).

In reply to criticism by Garrett, Lewin (*Psychological Review*, Vol. XLVI [1939]) acknowledges the similarity of his "vector" concept with "excitation tendency" and "drive." For further criticism see I. D. London, "Psychologists' Misuse of Auxiliary Concepts of Physics and Mathematics," *Psychological Review*, Vol. LI (1944). For Lewin's partial recognition of the "historical" study of origins in psychoanalysis see his "Psychoanalysis and Topological Psychology," *Bulletin of the Menninger Clinic*, Vol. I (1937).

90. C. L. Hull, C. I. Hovland, *et al.*, *Mathematico-deductive Theory of Rote Learning* (New Haven: Yale University Press, 1940); and C. L. Hull, *Principles of Behavior* (New York: Appleton-Century, 1943), comprehensive presentation on pp. 381 ff.

Reviews and critical discussion are found in *Psychological Bulletin* (Koch, Ritchie, 1944; on mathematico-deductive theory: Hilgard, Marhenke, Fitch, 1940); the *Journal of Genetic Psychology* (Leeper, 1944; Welch, 1945); and in the *American Journal of Psychology* (Skinner, 1944). For general discussion, emphasizing "postulationism" as an important correlate to an "operationism [which] has not emphasized deductive rigor," yet warning of its "psychological" dangers, see F. W. Hall, "Some Dangers in the Use of Symbolic Logic in Psychology," *Psychological Review*, Vol. XLIX (1942). The superfluousness of Hull's neurophysiological model of receptor-effector connections is suggested by K. W. Spence, "Cognitive vs. Stimulus Response Theories of Learning," *Psychological Review*, Vol. LVII (1950).

Hull's concept of "habit-family-hierarchy" is developed in two articles published shortly before his all-out effort at formalization (*Psychological Review*, Vol. XLI [1934]).

Hull's approach to the concept of value is illustrated by his paper on "Value, Valuation, and Natural-Science Methodology," *Philosophy of Science*, Vol. XI (1944).

For Hull's sympathetic approach to psychoanalysis see the "Memorandum to Psychology" for May 2, 1940 (mimeographed by the Department of Psychology, Yale University).

91. H. Woodrow, "The Problem of General Quantitative Laws in Psychology," *Psychological Bulletin*, Vol. XXXIX (1942).

92. W. K. Estes, "Toward a Statistical Theory of Learning," *Psychological Review*, Vol. LVII (1950).

93. W. B. Cannon, *The Wisdom of the Body* (New York: Norton, 1932). For a survey of developments in psychology see C. O. Weber, "Homeostasis and Servo-mechanisms for What?" *Psychological Review*, Vol. LVI (1949).

94. N. Rashevsky, *Mathematical Biophysics* (Chicago: University of Chicago Press, 1938; rev. ed., 1948), and *Mathematical Biology of Social Behavior* (Chicago: University of Chicago Press, 1951).

Bibliographical Notes

See the *Bulletin of Mathematical Biophysics* (since 1939) for such further relatively recent developments as W. S. McCulloch and W. Pitts's "A Logical Calculus of the Ideas Immanent in Nervous Activity," their joint paper with Landahl (Vol. V [1943]), and the related statistical biophysics and probabilistic sociology of A. Rapoport, Shimbel, and others (from Vol. X [1948] on).

95. G. Birkhoff, *Aesthetic Measure* (Cambridge: Harvard University Press, 1933).

96. C. M. Harsh and J. G. and R. Beebe-Center, "Further Evidence Regarding Preferential Judgment of Polygonal Forms," *Journal of Psychology*, Vol. VII (1939).

97. N. Wiener, *Cybernetics: Control and Communication in the Animal and the Machine* (New York: Wiley & Sons, 1948). For the statistical turn in theorizing see the same author's *Extrapolation, Interpolation, and Smoothing of Stationary Time Series* (New York: Wiley & Sons, 1949), quotations from pp. v and 23.

A Symposium on Teleological Mechanisms held in 1946 (*Annals of the New York Academy of Science*, Vol. L, No. 4 [1948]) includes the psychologist L. K. Frank, the psychiatrist and mathematical biophysicist W. S. McCulloch, and the ecologist G. E. Hutchinson. For a discussion which is alert to patterns of vicarious functioning without explicit use of the concept and which stresses the limitations as well as possible future expansions of computing machines see E. C. Berkeley, *Giant Brains: Machines That Think* (New York: Wiley & Sons, 1949).

For a nontechnical survey of psychological problems of communication see G. A. Miller, *Language and Communication* (New York: McGraw-Hill, 1951).

An attempt to apply communication theory to psychiatry is given in J. Ruesch and G. Bateson, *Communication: The Social Matrix of Psychiatry* (New York: Norton, 1951).

98. Further linked with cybernetics is J. D. Trimmer's *Response of Physical Systems* (New York: Wiley & Sons, 1950). The confinement of nomothetic endeavor to the internal aspects of the system as seen under the impact of external influences is strikingly similar to the encapsulated theorizing and model-making in most of modern theoretical psychology.

99. G. A. Miller and F. C. Frick, "Statistical Behavioristics and Sequences of Responses," *Psychological Review*, Vol. LVI (1949).

100. J. von Neumann and O. Morgenstern, *Theory of Games and Economic Behavior* (2d ed.; Princeton: Princeton University Press, 1947).

101. T. H. Waterman, "Flight Instruments in Insects," *American Scientist*, Vol. XXXVIII (1950). The article includes reference to von Frisch's work on the "language" of bees, which is also relevant in this context.

102. C. E. Shannon and W. Weaver, *The Mathematical Theory of Communication* (Urbana: University of Illinois Press, 1949); quotations are from pp. 41, 111–12, 116–17.